TENERIFE
SEIA'19

Sensors and Electronic Instrumentation Advances:

Proceedings of the 5th International Conference on Sensors and Electronic Instrumentation Advances

25-27 September 2019,
Adaje, Canary Islands (Tenerife), Spain

Edited by Sergey Y. Yurish

IFSA Publishing

Sergey Y. Yurish, *Editor*
Sensors and Electronic Instrumentation Advances
SEIA' 2019 Conference Proceedings

E-mail (for orders and customer service enquires): ifsa.books@sensorsportal.com

Visit our Home Page on http://www.sensorsportal.com

ISBN: 978-84-09-17492-8
BN-20191230-XX
BIC: TJFC

Contents

Foreword

On behalf of the SEIA' 2019 Organizing Committee, we introduce with pleasure these proceedings devoted to contributions from the 5th International Conference on Sensors and Electronic Instrumentation Advances (SEIA'2019) held in Adeje, Tenerife (Canary Islands), Spain. The conference is organized by the *International Frequency Sensor Association (IFSA)* – a professional association serving for the sensor industry and academy during 20 years in technical cooperation with the IFSA Group companies: *IFSA Publishing, S.L.* (Spain) and *F2D, Ltd.* (Ireland), and media partners: Open Access MDPI journals *Sensors* (ISSN 1424-8220) and *Biosensors* (ISSN 2079-6374), Switzerland. The conference program provides an opportunity for researchers interested in various applications of sensing and measurement to discuss their latest results and exchange ideas on the new trends. The main objective of the conference is to encourage discussion on a broad range of sensor related topics and to stimulate new collaborations among the participants.

Since its inception, the Conference has enjoyed an atmosphere of high quality technical presentations on a broad range of multidisciplinary topics related to sensors, transducers, measurements and their applications.

Extending the tradition that began in 2015 in Dubai, UAE, this annual conference on sensors attracts over 100 researchers and practitioners in the sensor, field, sensor technology and application from around the world including 5 keynote speakers from a distinguished researchers of industry from the USA, the Netherlands, Spain, Romania and Portugal who were invited to overview the progress in selected research trends. This year, we have included 95 papers from 31 countries into the conference proceedings, covering theory, design, device technology, and applications of sensors and sensing systems. To accommodate this range of interests, the 5th SEIA' 2019 Conference was organized in seven dedicated regular sessions, four special sessions on '*Gas Sensors*', '*IoT: Sensors, Networks and Applications*', '*New Trends in Measurement Science: Intelligent Measurements*', '*Frequency & Time*' and one poster session.

The proceedings contains all accepted papers of both: oral and poster presentations. We hope that these proceedings will give readers an excellent overview of important and diversity topics discussed at the conference.

We thank all authors for submitting their latest work, thus contributing to the excellent technical contents of the Conference. Especially, we would like to thank the individuals and organizations that worked together diligently to make this Conference a success, and to the members of the International Program Committee for the thorough and careful review of the papers. It is important to point out that the great majority of the efforts in organizing the technical program of the Conference came from volunteers.

Prof., Dr. Sergey Y. Yurish *Dr. Amin Daneshmand Malayeri*

(002)

Optical Sensor Systems with Micro-structured Grating in PMMA for POF-Applications

U. H. P. Fischer-Hirchert [1], **S. Höll** [2], **M. Haupt** [1], **M. Joncic** [3]
[1] Harz University of Applied Sciences, Friedrichstr. 57, 38855 Wernigerode, Germany
[2] Sicoya GmbH, Carl-Scheele-Straße 16, 12489 Berlin, Germany
[3] IAV GmbH, Carnotstraße 1, 10587 Berlin
Tel.: + 49 3943659 351, fax: + 49 3943659 5351
E-mail: ufischerhirchert@hs-harz.de

Summary: Data communication over Polymer Optical Fibers (POF) is limited to only one channel for data transmission. Therefore the bandwidth is strongly restricted. By using more than one channel, it is possible to break through the limit. This technique is called Wavelength Division Multiplexing (WDM). It uses different wavelengths in the visible spectrum to transmit data parallel over one fiber. Two components are essential for this technology: A multiplexer (MUX) and a demultiplexer (DEMUX). The multiplexer collects the light of the different sources to one fiber and the demultiplexer separates the light at the end of the fiber into the different fiber output ports. In this paper, we show the realization of a grating spectrometer realised by injection moulding in PMMA, working in the visible spectrum, respectively. We present the results of a demonstrator of an integrated polymeric spectrometer device produced with injection moulding in combination with hot embossing. The paper discusses the results of the different development steps, the measurements done with the first demonstrator and the challenges related to the injection moulding process.

Keywords: Optical demultiplexer, PMMA injection moulded demultiplexer, Hot embossing, Polymeric fiber WDM systems, POF over WDM.

1. Introduction

Polymer Optical Fibers (POF) are used in various fields of applications. The core material consists of PMMA (Polymethylmethacrylate), while the cover is made of fluorinated PMMA. The whole fiber has a diameter of 1 mm. POFs are used for optical data transmission based on the same principle as glass fiber. As a communication medium they offer a couple of advantages related to other data communication systems such as copper cables, glass fibers and wireless systems, and have great potential to replace them in different applications.

Namely, in comparison with glass fibres (GOF), POFs have the advantage of easy and economical processing and are more flexible for optical connections [1]. However, one advantage of using glass fibres is their low attenuation, which is below 0.2 dB/km in the infrared range. The larger core diameter of POFs leads to higher mode dispersion and thus to higher attenuation across the electromagnetic spectrum. This increased attenuation leaves only one remaining transmission window, namely the visible spectrum of light (400 – 700 nm). Hence, POFs are best suited for the use in short distance data communication.

Here, POFs can outperform the current standard of copper cable as communication medium. On the one hand, they feature lower weight and space. On the other hand, POFs are not susceptible to electromagnetic interference. [2-3]. For these reasons, POFs are already used in various application domains, for example in the automotive sector and for in-house communication [4-7]. At present, the great potential of the POF is not available as the alternative techniques offer transmission rates up to 10 GB over copper and up to 40 GB over glass fibers in the network area. The WDM technique offers an approach to achieve these high data rates also in the POF range

For WDM two essential components are needed: a multiplexer and a demultiplexer. To create a functional demultiplexer for POF, several preconditions must be fulfilled. Firstly, a mirror must focus the divergent light beam coming from the POF. The shape of this mirror cannot be spherical because of the appearing spherical aberration. Instead, a toric shape of the mirror prevents spherical aberration [8-10].

The second function is the separation of the different transmitted wavelengths, which can be achieved by a diffraction grating. This principle is illustrated in Fig. 1. The light is split into different orders of diffraction. The first order is the important one to regain all information. There, the outgoing POFs can be placed.

To separate the channels at the output ports, one interesting option for high multimode transmission systems is to use an optical grating. Here, the optical grating is placed on an aspheric mirror, which focuses the monochromatic parts of light into the outgoing fibers. In order to keep the advantage of cost-effective POFs it is necessary to mass-produce the MUX and DEMUX component at reasonable prices. For polymers, injection molding is the only technology, which offers high potential to achieve this goal.

Before starting the production of the mold insert, a demonstrator of the DEMUX is fabricated by directly machining it in the PMMA material by means of diamond turning technique (see Fig. 2). Thus, the same diamond-turning technology is used for the manufacture of the mold insert. This step is done due to validate the simulation results with the produced component.

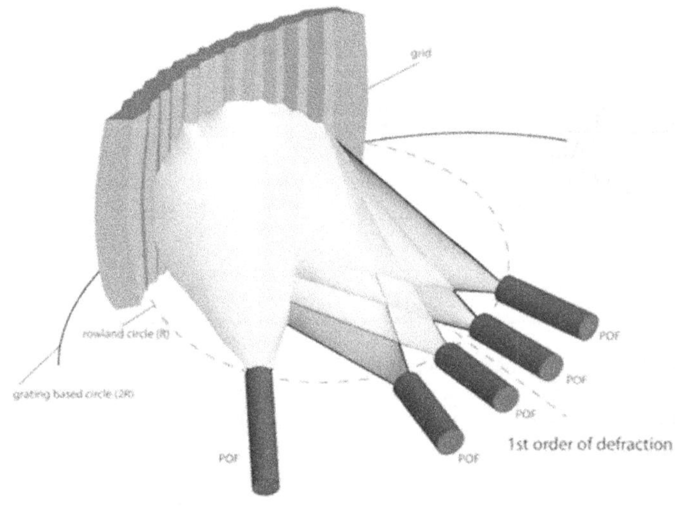

Fig. 1. Rowland set-up of integrated PMMA spectrometer.

Fig. 2. Prototype of integrated PMMA spectrometer.

2. Manufacturing of the Demonstrator

By using the injection molding process, the manufacturing of the mold insert is the most important factor. Due to the three-dimensional toric structure of the grating planar manufacturing methods like lithography, especially LIGA (German acronym for Lithographie, Galvanoformung, Abformung – Lithography, Electroplating, and Molding) cannot be used. LIGA is used to manufacture planar spectrometers based on the glass fiber technology [11-14]. But in our case, the three-dimensional grating needs another machining method. Especially the microstructure of the grating and the exact curve shape of the toric surface require high precision.

Fig. 3. 3D Sawtooth like simulation of the grating of the prototype.

The microstructure has the shape of a saw tooth with a pitch between the teeth of 2.5 µm. Fig. 3 shows an enlarged 3D-Model of the grating. After investigate several machining methods only the diamond turning meets the high demands of the micro structured grating.

3. Optical Measurements

Prior to the production of the DEMUX some preliminary investigations have taken place to find the best suitable material for the demultiplexer. Therefore, both the processability of the material and the optical parameters had to be considered in detail. The injection molding process was tested with a thick-walled mold.

This test tool had the same shape as the final DEMUX, except for the grid. The test runs were carried out with an injection-molding machine from Babyplast 6-10P. This device was able to inject precisely small parts. Table 1 lists all the materials used for the study. Further parameters such as the respective melt volume rate (MVR) and light transmittance (according to the manufacturer's specification) are depicted. The test was additionally used to find the optimized injection molding process parameters for the material.

In addition, the optical quality of the polymer materials must be investigated. Therefore, a mold for injection molding test plates was designed. The test plates had a thickness of 2 mm. The mold is used to make samples from each material listed in Table 1. The DIN EN ISO 13468-2 standard describes the measurement of the optical transmission of polymer materials. Therefore, the test plates are designed to meet this standard.

Transmission measurements were carried out with all test plates. The results are shown for 405 nm in Fig. 4. It can be seen that both ZEONEX types and PMMA POQ62 show the highest value for the light transmission. PMMA POQ62 is a polymer grade with high purity of polymer granulates. The measurement is made at a wavelength of 405 nm because it is one of the wavelengths used for the WDM system.

Table 1. Injection molding materials for the integrated spectrometer element.

Name	Type	MVR [cm³/10 min]	Transmission [%]
Plexiglas 6N	PMMA	12	92
Plexiglas POQ62	PMMA	21	92
Topas 5013L-10	COC	48	91.4
Topas 6013M-07	COC	14	91
ZEONEX F52R	COP	22	92
ZEONEX 350R	COP	26	92
Makrolon LED2245	PC	35	90

In order to measure the position of the focal points of the different separated wavelength a special measurement setup was build. It uses a parallel-kinematics precision alignment system to align a fiber at the surface of the hemisphere. A fixed input-fiber is used to couple white light into the DEMUX as shown in Fig. 6. In this figure can be seen that the separated wavelength are focused at a ring on the hemisphere.

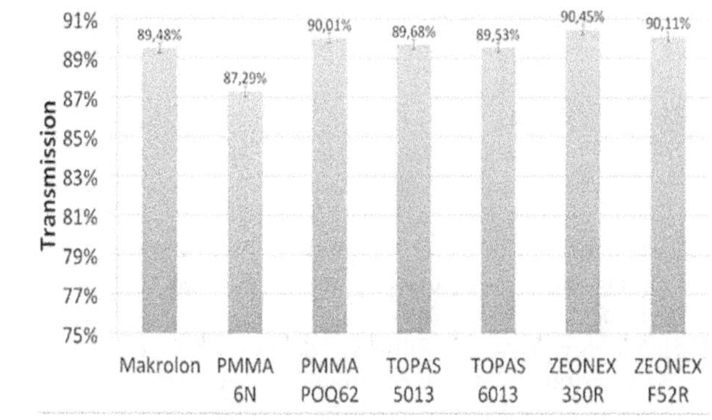

Fig. 4. Transmission of different material at 405 nm

This ring is scanned by the fiber on the alignment system. The light of the scanning fiber is analyzed by a spectrometer. The positions of the wavelengths measured by the setup are depicted in Fig. 6. In comparison to the simulation a shift of the positions can be recognized. Nevertheless, the separation of the wavelengths was measured and confirms the functionality of the demultiplexer.

4. Manufacturing of the Demonstrator

By using the injection molding process, the manufacturing of the mold insert is the most important factor. Due to the three-dimensional toric structure of the grating planar manufacturing methods like lithography, especially LIGA cannot be used [14].

In our case, however, the three-dimensional grating requires a different processing method. The microstructure of the grating and the exact shape of the toric surface require a particularly high precision of manufacture.

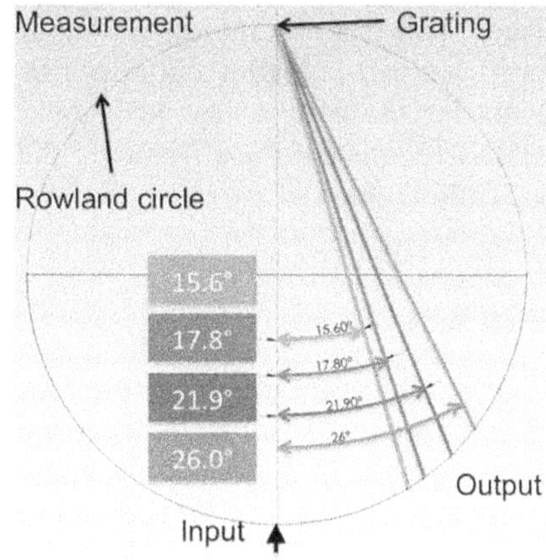

Fig. 5. Measurement results of the focal points for different wavelengths (405 nm violet ray; 450 nm blue ray; 520 nm green ray; 650 nm red ray.

The microstructure has the shape of a sawtooth with a distance between the teeth of 2.5 μm. Fig. 3 shows an enlarged 3D model of the grating. An in-depth investigation of various processing methods has shown that only the diamond turning fulfills the high requirements of the production of the microstructural lattice. The diamond twisting technique is a special machining method using a single crystal diamond-cutting tool. It is also possible to produce a surface with an optical quality at the edge of the optical component. It offers several advantages:

- True three-dimensional contour generation;
- Accuracy of one part in 106 with absolute accuracy of 1 part in 108 on a single axis for ideal conditions;
- Surface finish of 5 nm Ra for a range of materials and as good as 1 nm Rz;
- Ability to generate surfaces with variable aspect ratios;
- Feature sizes that exceed the limits of optical microscopy. [15, 16].

Fig. 6. High quality structures of the grating.

A metallization process was used to analyze the surface of the lattice. The surface was sputtered with a thin aluminum layer depicted in Fig. 6. It is now possible to measure the shape of the surface with a white light interferometer and to examine the lattice structure under the scanning electron microscope (SEM).

The metallized surface of the grating is shown in Fig. 5. It can be seen that the structure on the left side has a dull and mat surface instead of the glossy residue of the surface. This is a first indication that the surface roughness in this part is higher and does not meet the requirements for the component precision. The first visual impression was then confirmed by the analysis under the SEM.

4. Conclusions

In summary, injection moulding the spectrometer element for POFs poses several challenges, especially the microstructure of the grating on a three-dimensional surface. It is shown that the current manufacturing process is able to produce the structure size and the exact radius needed for the spectrometer used as a DEMUX for POF over WDM. The optical measurements to prove the principle were done. These confirm the separation of the wavelength in the visible spectrum.

The device is also a very good candidate for optical sensor applications. The component can be used both in the registration of paint colors of the automotive industry or paint industry, as well as in the recognition of the textile inks in the textile industry. Furthermore, there is the possibility of measuring residual proportions in the waste water of dishwashers for dosing of rinse water and detergent.

Acknowledgements

We gratefully acknowledge the funding by the German Ministry of Education and Research (BMBF) under grant number 16V0009 (HS Harz) /16V0010 (TU BS). All injection moulded parts are done with the support of the Institute of Micro and Sensor Systems at the Otto-von-Guericke University Magdeburg and Prof. Bertram Schmidt

References

[1]. W. Daum, J. Krauser, P. E. Zamzow, O. Ziemann, POF Handbook: Optical Short Range Transmission Systems, *Springer-Verlag*, 2008.

[2]. H. S. Nalwa (Ed.), Polymer Optical Fibres, *American Scientific Publishers*, California, 2004.

[3]. J. Marcou, Club des Fibres Optiques Plastiques (CFOP) France, Plastic Optical Fibres – Practical Applications, *John Wiley & Sons*, Masson, 1997.

[4]. J. Brandrup, E. H. Immergut, E. A. Grulke, Polymer Handbook, 4th Ed., *Wiley-Interscience*, 1999.

[5]. R. T. Chen, G. F. Lipscomb, WDM and photonic switching devices for network applications, *Proceedings of SPIE*, Vol. 3949, 2000.

[6]. J. Colachino, Mux/DeMux Optical Specifications and Measurements, Lightchip Inc. White Paper, *Lightreading*, 2001.

[7]. A. H. Gnauck, A. R. Chraplyvy, R. W. Tkach, J. L. Zyskind, J. W. Sulhoff, A. J. Lucero, et al., One terabit/s transmission experiment, in *Proceedings of the Conference on Optical Fiber Communication (OFC'96)*, 1996.

[8]. U. H. P. Fischer-Hirchert, Photonic Packaging Sourcebook: Fiber-chip Coupling for Optical Components, Basic Calculations, Modules, *Springer*, Berlin, Heidelberg, 2015.

[9]. U. H. P. Fischer, M. Haupt, WDM over POF: The inexpensive way to breakthrough the limitation of bandwidth of standard POF communication, *Proceedings of SPIE*, Vol. 6478, 2007, 64780I.

[10]. U. H. P. Fischer, M. Haupt, Integrated WDM system for POF communication with low cost injection moulded key components, *OSA Technical Digest*, 2010, p. AWB2

[11]. M. Stricker, G. Pillwein, J. Giessauf, Focus on precision – Injection molding optical components, *Kunststoffe International*, Vol. 4, , 2009, pp. 15-19.

[12]. J. P. Ferguson, S. Schoenfelder, Micromoulded spectrometers produced by the LIGA process, in *Proceedings of the IEE Colloquium on Microengineering in Optics and Optoelectronics*, 1999, pp. 11/1-11/4.

[13]. M. A. Davies, C. J. Evans, R. R. Vohra, B. C. Bergner, S. R. Patterson, Application of precision diamond machining to the manufacture of microphotonics components, *Proceedings of SPIE*, Vol. 5183, 2003.

[14]. D. Dornfeld, S. Min, Y. Takeuchi, Recent advances in mechanical micromachining, *CIRP Annals*, Vol. 55, Issue 2, 2006, pp. 745-768.

[15]. M. Stricker, G. Pillwein, J. Giessauf, Focus on precision – Injection molding optical components, *Kunststoffe International*, Vol. 4, , 2009, pp. 15-19.

[16]. M. A. Davies, C. J. Evans, R. R. Vohra, B. C. Bergner, S. R. Patterson, Application of precision diamond machining to the manufacture of microphotonics components, *Proceedings of SPIE*, Vol. 5183, 2003.

(003)

High Performance of Cellulose Nanofibers/ Poly(3,4-ethylenedioxythiophene):Poly(4-styrenesulfonate)/ Metal Oxide/Ionic Liquid Actuators

N. Terasawa
Inorganic Functional Material Research Institute, National Institute of Advanced Industrial Science and Technology (AIST), 1-8-31 Midorigaoka, Ikeda, Osaka 563-8577 Japan
Tel.: + 81727517914, fax: + 81727518370
E-mail: terasawa-naohiro@aist.go.jp

Summary: This study proposes novel actuators based on cellulose nanofiber/poly(3,4-ethylenedioxythiophene):poly (4-styrenesulfonate)/hydrous ruthenium oxide/ionic liquid (CNF/PEDOT:PSS/RuO_2/IL) electrodes and aims to investigate the effect of hydrous ruthenium oxide on the electrochemical and electromechanical properties of the proposed actuators. The proposed system contains an electrochemical capacitor electrode, which acts as both a faradaic capacitor (FC) and small electrostatic double-layer capacitor (EDLC). This hybrid capacitor is based on the FC (RuO_2) mechanism, and a base polymer (PEDOT:PSS) and a CNF skeleton is used instead of carbon nanotubes (CNTs). Therefore, this device functions differently from traditional CNT/poly(vinylidene fluoride-co-hexafluoropropylene) (PVdF(HFP))/IL actuators, which are only used as EDLC units, and CNF/PEDOT:PSS/IL (without RuO_2) actuators, where PEDOT:PSS plays the role of both FC and base polymer. Devices built using the proposed actuators exhibit higher strain values than those exhibited by devices based on CNF/PEDOT:PSS/IL, i.e. without RuO_2.

Keywords: Actuator, Cellulose nanofiber, PEDOT:PSS, Ruthenium oxide, Ionic liquid, High performance.

1. Introduction

Recently, electrochemical capacitors (ECs) have attracted considerable research attention because of their long lifetimes, high power densities, and the ability to bridge the power and energy gaps that exist between fuel cells or batteries and standard dielectric capacitors [1, 2]. ECs can be divided into two main types: faradaic capacitors (FCs) and electrostatic double-layer capacitors (EDLCs). Furthermore, it is also possible to fabricate a hybrid capacitor – a unit that simultaneously exhibits the characteristics of both FCs and EDLCs, with one of these two capacitor types responsible for the primary mechanism [3].

To develop high-power electrochemical capacitors, previous research has investigated various metal oxides as possible electrode materials [4]. Among them, ruthenium oxide (RuO_2) has been widely used for various applications because of its good catalytic properties in electrochemical and photochemical processes as well as in high-charge storage capacity devices [5, 6].

Furthermore, it has been demonstrated that the use of amorphous hydrous ruthenium oxide ($RuO_2.xH_2O$) as an electrode material leads to a considerably higher specific capacitance than that recorded for RuO_2 [4]. The charge-storage mechanism of RuO_2H_2O is believed to be different from that of RuO_2 because the charge is stored beyond, at, or near a solid electrode surface. For example, the cellulose nanofiber/ poly(3,4-ethylenedioxythiophene):poly(4-styrenesulfonate)/hydrous ruthenium oxide/ionic liquid (CNF/PEDOT:PSS/$RuO_2(xH_2O)$/IL) electrodes would be expected to have a higher specific capacitance than that of the CNF/PEDOT:PSS/IL (without $RuO_2(xH_2O)$) electrodes, and the CNF/PEDOT:PSS/$RuO_2(xH_2O)$/IL actuators having this structure would exhibit higher strain and generated stress values than those obtained using only CNF/PEDOT:PSS/IL (without $RuO_2(xH_2O)$).

Increasing environmental concerns have resulted in the fundamental research on bionanofibres and their applications drawing increased attention [7]. In addition to natural cellulose fibres, chemically and mechanically modified cellulose fibres have been used for decades in various fields, ranging from commodities, such as paper and textiles, to high-tech materials. Therefore, CNFs are expected to replace carbon nanofibers, as they can form highly entangled CNF open mesopore networks in electrodes. However, CNFs are not electroconductive, which is a limitation of the material.

The doping of PEDOT with poly (4-styrenesulfonate) (PSS) results in the formation of PEDOT:PSS, which is an important type of conductive polymer (CP); it has the ability to form colloidal particulate dispersions in water. PEDOT:PSS has been employed in numerous organic and polymeric electronic/optical devices [8-10] and PEDOT: PSS-based electrodes have been studied [11, 12] and confirmed to be capable of converting electrical energy to mechanical energy [13]. Hence, it is expected that insulator CNFs when modified with PEDOT:PSS will become electroconductive.

Recently, researchers have extensively investigated soft materials that are capable of generating mechanical energy from electrical energy. These materials have been used for several applications, such as in robotics for prosthetics, tactile and optical displays, microelectromechanical systems, and medical devices [14]. In this context, electroactive polymers (EAPs) have many attractive characteristics that render them suitable for these applications, and in particular for use in actuators and sensors. Previously, the production of a dry actuator [15, 16] was reported via a facile process in which a so-called 'bucky gel' underwent layer-by-layer casting [17]. This material was composed of IL-containing SWCNTs and possessed a gel-like consistency at room temperature. More specifically, the bimorph-structured dry device was composed of an IL gel electrolyte layer supported on a polymer and positioned between polymer-supported bucky-gel electrode layers. This configuration permitted rapid, long-lasting movement in response to minimal applied voltages in air under ambient conditions. Furthermore, in a previous study, a CNF/PEDOT:PSS/IL actuator that exhibited high strain and maximum generated stress was developed [18].

In view of all the aforementioned information, novel CNT-free, hybrid, self-standing actuators with the CNF/PEDOT:PSS/$RuO_2(xH_2O)$/IL structure were developed herein by exploiting the advantages of the component materials. In addition, we investigated the effects of $RuO_2(xH_2O)$ on the electrochemical and electromechanical properties of the capacitors and actuators.

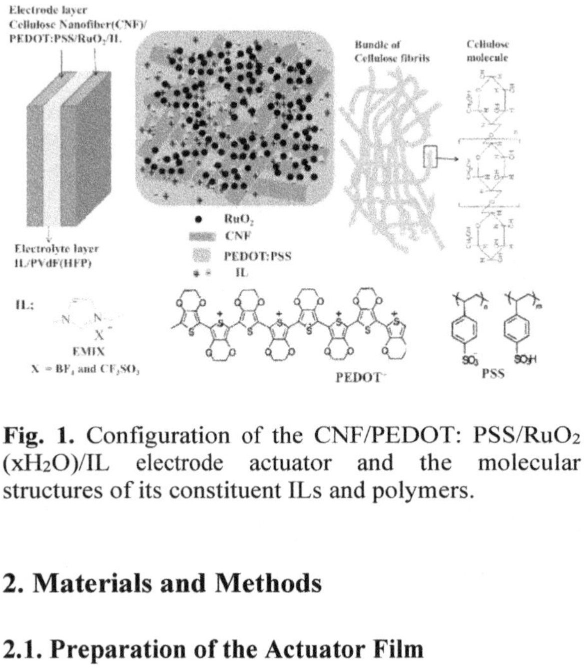

Fig. 1. Configuration of the CNF/PEDOT: PSS/RuO_2 (xH_2O)/IL electrode actuator and the molecular structures of its constituent ILs and polymers.

2. Materials and Methods

2.1. Preparation of the Actuator Film

The configuration of the CNF/PEDOT:PSS/RuO_2/IL actuators developed in this study is depicted in Fig. 1. Typically, the CNF/PEDOT:PSS/RuO_2/IL electrode layer (50/100/100/300) comprises 9 wt.% CNF, 18 wt.% PEDOT:PSS, 18 wt.% $RuO_2.xH_2O$, and 55 wt.% IL. The individual layers were prepared by mixing BB (500 mg of a 10 wt.%), PEDOT:PSS (100 mg), $RuO_2.xH_2O$ (100 mg), and EMI[BF_4] or EMI[CF_3SO_3] (300 mg) in H_2O (9 mL), and this mixture was further ultrasonically dispersed for more than 5 h. Subsequently, to obtain an electrode film with a thickness of 150-175 μm, the resulting gelatinous mixture (2.4 mL) was cast in a Teflon mould in the order outlined above. Electrode layers containing a CNF/PEDOT:PSS/RuO_2/IL layer with different ratios of the individual layers were prepared using the same procedure. The gel electrolyte layers were fabricated by casting a solution (0.3 mL) containing either of the two ILs and PVdF(HFP) (100 mg each) mixed with MP (1 mL) and PC (250 mg) in a Teflon mould (2.5×2.5 cm). To yield gel electrolyte films with thicknesses ranging from 20 to 30 μm, the solution was subjected to evaporation and complete removal of the solvent was achieved *in vacuo* at 80 °C. Finally, to obtain an actuator film with a thickness of 250-300 μm, the electrode and electrolyte layers comprising the same IL were hot-pressed.

2.2. Displacement Measurements

The displacement measurements were conducted according to a previous study.[34] The resulting displacement parameter (δ) was further used to compute the strain difference between the bucky-gel electrode layers (ε) by assuming planar (i.e. undistorted) cross sections along the actuator based on the following expression:

$$\varepsilon = 2d\delta/(L^2 + \delta^2), \tag{1}$$

where d denotes the actuator thickness and L denotes the free length [20].

3. Results and Discussion

CNF/PEDOT:PSS/RuO_2/EMI[BF_4] and CNF/PEDOT:PSS/RuO_2/EMI[CF_3SO_3] electrodes with CNF/PEDOT/RuO_2 ratios of 50/100/50, 50/100/100, 50/100/200, and 50/100/0 were synthesised according to the procedure detailed above. The specific capacitance C (C_I scaled to the weight of PEDOT:PSS) of each electrode was measured; the values are summarised in Table 1. The C values recorded for the CNF/PEDOT:PSS/RuO_2/IL electrodes were in the range of 300-520 F g^{-1} at a slow sweep rate of 1 mV s^{-1} and were observed to be dependent on the IL species. The increase in the values exhibited by the CNF/PEDOT:PSS/RuO_2/IL electrodes was observed to be directly proportional to the increase in the ratio of the RuO_2 component (Tables 1 and 2).

The C values for the CNF/PEDOT:PSS/RuO_2/IL electrodes were considerably higher than those for the CNF/PEDOT:PSS/IL electrodes (*ca.* 24 F g^{-1}) [19]; furthermore, the C values for the CNF/PEDOT:PSS/RuO_2/EMI[CF_3SO_3] electrodes were higher than those for the CNF/PEDOT:PSS/RuO_2/EMI[BF_4] electrodes. The values for the (50/100/50), (100/100/50),

(50/100/100), and (100/100/100) CNF/PEDOT: PSS/RuO_2/EMI[BF_4] electrodes that were similar to each other were independent of the amount of CNF component in the system (Tables 1 and 2). In terms of the mechanism, the specific capacitance values of the CNF/PEDOT:PSS/RuO_2/IL electrodes are marginally related to the EDLC mechanism of PEDOT:PSS but mainly related to the FC mechanism involving the RuO_2. The electrode in the proposed actuator system represents a hybrid EDLC and FC electrochemical capacitor because RuO_2 acts as an FC and PEDOT:PSS acts as a base polymer EDLC layer. In addition, the CNFs act as a base polymer and skeleton. Therefore, the functional mechanism of this device differs from that of the CNF/PEDOT:PSS/IL electrode (without RuO_2) actuators, which are mainly driven by the FC mechanism of PEDOT:PSS and only marginally related to the EDLC mechanism. The reversible oxidation and reduction of RuO_2 [4] and EDLC mechanism can be expressed as follows:

Table 1. Specific capacitance, $C = C_l$, divided by the weights of PEDOT:PSS ($F\,g^{-1}$), values of the CNF/PEDOT:PSS/RuO_2/IL and CNF/PEDOT:PSS/IL electrodes (applied triangular voltage: ± 0.5 V and sweep rate: 1 $mV\,s^{-1}$).

CNF/PEDOT/Ru	50/100/50	50/100/100	50/100/200	50/100/0
EMI[BF_4]	302.6	425.4	476.9	23.8*
EMI[CF_3SO_3]	-	459.0	517.2	24.2*

*[19]

Table 2. Specific capacitance values, $C = C_l$, divided by the weights of the PEDOT:PSS ($F\,g^{-1}$) components in the CNF/PEDOT:PSS/Ru_2O/EMI[BF_4] (50-100/100/50-100) electrodes (applied triangular voltage: ±0.5 V and sweep rate: 1 $mV\,s^{-1}$).

	50/100/50	50/100/100	100/100/50	100/100/100
EMI[BF_4]	302.6	425.4	318.0	429.9

$$RuO_2 + \delta EMI^+ + \delta e = RuO_{2-\delta}(EMI^+)_\delta \quad (0 < \delta < 2), \qquad (2)$$

$$EMI\ [IL\ anion] = EMI^+ + [IL\ anion]^- \qquad (3)$$

Table 3 presents the electrical conductivities of the electrode layers comprising RuO_2, PEDOT:PSS, CNFs, and ILs. The CNF/PEDOT:PSS/RuO_2/IL (50-100/100/50-200) and CNF/PEDOT:PSS/IL (50/100/0) electrodes exhibited similar values independent of the IL species used in the system. The electrical conductivities of the electrode layers comprising RuO_2, PEDOT:PSS, CNFs, and ILs. The CNF/PEDOT:PSS/RuO_2/EMI[BF_4] (50-100/100/50-100) electrodes exhibited similar values independent of the amount of CNFs used in the system. The values for the CNF/PEDOT:PSS/RuO_2/IL electrodes were observed to be lower than those for the previously reported PEDOT:PSS/VGCF/IL electrodes, which were determined to have values of 10–12 $S\,cm^{-1}$ [19], but were similar to the values determined for the PEDOT:PSS/IL electrodes [21]. Therefore, the CNF-coated PEDOT:PSS electrodes were observed to have similar values to those of the PEDOT:PSS/IL electrodes, whereas the CNF-coated PEDOT:PSS electrodes have lower values in comparison to those of the carbon nanotube (CNT)-coated PEDOT:PSS electrodes.

Table 3. Electrical conductivity ($S\,cm^{-1}$) measurements of the CNF/PEDOT:PSS/Ru_2O/IL (50/100/50-200) electrodes.

CNF/PEDOT/Ru	50/100/50	50/100/100	50/100/200	50/100/0
EMI[BF_4]	6.9	5.7	3.4	4.5*
EMI[CF_3SO_3]	-	7.3	3.9	6.6*

* [19]

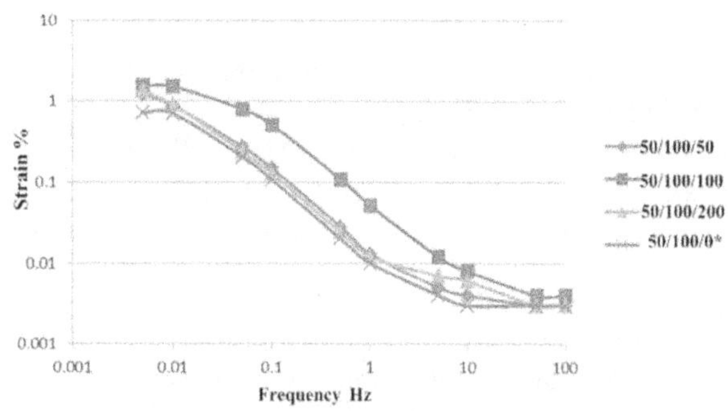

* [19]

Fig. 2. Strain, as calculated from the peak-to-peak displacements generated in the CNF/PEDOT: PSS/RuO_2/EMI[BF_4] (50/100/50-200) electrode actuators as functions of the applied triangular voltage (±2 V) frequency.

Fig. 2 shows a plot of the measured strains of the CNF/PEDOT:PSS/RuO_2/EMI[BF_4] (50/100/50-200) electrodes versus the frequency of the applied triangular voltage (±2 V). As depicted, the strain exhibits a clear dependence on the voltage frequency. At low frequencies (i.e. 0.01-0.005 Hz), the CNF-coated PEDOT:PSS in the electrode layer was considered to be fully charged. In contrast, at high frequencies, the time available was not enough to complete the charging. The CNF/PEDOT:PSS/RuO_2/EMI[CF_3SO_3] (50/100/100-200) and CNF/PEDOT:PSS/RuO_2/IL (100/100/100-200) electrodes exhibited identical trends. At all frequencies, the strain of the CNF/PEDOT:PSS/RuO_2/EMI[BF_4] (50/100/100) electrode actuator exhibited a higher performance than those exhibited by the CNF/PEDOT:

PSS/RuO_2/EMI[BF_4] (50/100/50 and 200) and CNF/PEDOT:PSS/EMI[BF_4] electrode actuators.

Table 4 summarises the maximum strain values for the five actuators with electrodes comprising RuO_2 and two ILs. These values depended on the amount of RuO_2 in case of EMI[BF_4]. The maximum strains observed for the CNF/PEDOT:PSS/RuO_2/EMI[BF_4] electrode actuators were greater than those observed for the CNF/PEDOT:PSS/EMI[BF_4] (without RuO_2) electrode actuators, whereas the maximum strains observed for the CNF/PEDOT:PSS/ RuO_2/EMI[CF_3SO_3] electrode actuators were slightly greater than those observed for the CNF/PEDOT:PSS/EMI[CF_3SO_3] (without RuO_2) electrode actuators. The values were observed to be dependent on the IL species.

Table 4. Maximum strains (%) of the CNF/ PEDOT: PSS/RuO_2/IL and CNF/PEDOT:PSS/IL electrodes.

CNF/PEDOT/Ru	50/100/50	50/100/100	50/100/200	50/100/0
EMI[BF_4]	1.19	1.56	1.35	0.72*
EMI[CF_3SO_3]	-	0.58	0.52	0.46*

* [19]

The maximum strain values measured for the CNF/PEDOT:PSS/RuO_2/EMI[BF_4] electrode actuators were observed to exceed those of the CNF/PEDOT:PSS/RuO_2/EMI[CF_3SO_3] electrode actuators. In addition, the maximum strain values measured for the CNF/PEDOT:PSS/RuO_2/EMI[BF_4] electrode actuators (50/100/50-100) exceeded those observed for their (100/100/50-100) counterparts (Table 5), and these values were observed to be dependent on the amount of CNFs used in the systems. Furthermore, the maximum strain values for the CNF/PEDOT:PSS/RuO_2/EMI[BF_4] electrode actuators (50/100/50-200) were found to be greater than those observed for the previously reported PEDOT:PSS/VGCF/IL [22] and PEDOT:PSS/IL [21] electrode actuators.

As mentioned previously, in the CNF/PEDOT:PSS/RuO_2/IL actuators, the RuO_2 functions as an FC electrode, the PEDOT:PSS polymer functions as an electroconductive base polymer, and the CNF polymer acts as a skeleton, whereas in the CNF/PEDOT:PSS/IL actuators, the PEDOT:PSS polymer functions as the base polymer and FC electrode and the CNF polymer acts as the skeleton.

Furthermore, a previous study [19] considered the motion of the CNF/PEDOT:PSS/IL actuator to occur mainly because of the swelling and the expansion of the anode layer according to the FC mechanism with a minor contribution from the EDLC mechanism, resulting in the actuator being deflected toward the anode. In this actuator system, the electrode acts as an electrochemical capacitor, operating partially via the EDLC mechanism of PEDOT:PSS but mainly via an FC mechanism.

Table 5. Maximum strain (%) values of the CNF/PEDOT: PSS/RuO_2/EMI[BF_4] (50-100/100/50-100) electrodes.

	50/100/50	50/100/100	100/100/50	100/100/100
EMI[BF_4]	1.19	1.56	0.29	0.67

In this study, the CNF/PEDOT:PSS/RuO_2/IL actuator motion was considered to occur mainly due to the swelling and expansion of the anode layer according to the FC mechanism with a minor contribution from the EDLC mechanism. Thus, the actuator is deflected toward the anode. In the proposed actuator system, the electrode represents an electrochemical capacitor that partially undergoes an EDLC mechanism but mainly undergoes an FC mechanism, i.e. RuO_2 behaves as an FC and PEDOT:PSS behaves mainly as an electroconductive base polymer. Additionally, the CNF framework serves as a replacement for SWCNTs.

4. Conclusions

Here, we successfully developed a novel CNT-free, self-standing, and CNF-coated PEDOT:PSS electrode actuator containing RuO_2 (CNF/PEDOT:PSS/RuO_2/IL). This mechanism is in direct contrast to that proposed for PEDOT/VGCF/RuO_2/IL gel electrodes, where the FC mechanism involving RuO_2 was mainly observed; for CNF/PEDOT/IL gel electrodes, where the FC mechanism involving RuO_2 was mainly observed; for CNF/PEDOT/IL gel electrodes, where the FC mechanism involving PEDOT:PSS was observed; and for conventional actuators featuring copolymer-supported PVdF(HFP)/SWCNT/IL gel electrodes, which only exhibited the EDLC effect. Furthermore, the CNF/PEDOT:PSS/RuO_2/IL actuators exhibited maximum strain values that exceeded those exhibited by the CNF/PEDOT:PSS/IL devices. The flexible and robust films described herein are CNT-free and self-standing. Further, they exhibit excellent performance and have the potential for application as electrode materials in wearable energy-conversion devices.

Acknowledgements

This work was partly supported by the KAKENHI Grant-in Aid for Scientific Research C (17K05983) from JSPS.

References

[1]. C. Largeot, C. Portet, J. Chmiola, P. Taberna, Y. Gogotsi, P. Simon, Relation between the ion size

and pore size for an electric double-layer capacitor, *J. Am. Chem. Soc.*, Vol. 130, 2008, pp. 2730-2731.

[2]. S. Kandalkar, D. Dhawale, C. Kim, C. Lokhande, Chemical synthesis of cobalt oxide thin film electrode for supercapacitor application, *Synth. Met.*, Vol. 160, 2010, pp. 1299-1302.

[3]. G. Wang, L. Zhang, J. Zhang, A review of electrode materials for electrochemical supercapacitors, *Chem. Soc. Rev.*, Vol. 41, 2012, pp. 797-828.

[4]. J. P. Zheng, T. R. Jow, A new charge storage mechanism for electrochemical capacitors, *J. Electrochem. Soc.*, Vol. 142, 1995, pp. L6-L8.

[5]. B. Cornell, D. Simonsson, Ruthenium Dioxide as Cathode Material for Hydrogen Evolution in Hydroxide and Chlorate Solutions, *Electrochem. Soc.*, Vol. 140, 1993, pp. 3123-3129.

[6]. S. Trasatti, G. Buzzanca, Ruthenium dioxide, A new interesting electrode material. Solid state structure and electrochemical behaviour, *J. Electroanal. Chem.*, Vol. 29, 1971, pp. A1-A5.

[8]. A. Isogai, T. Saito, H. Fukuzumi, TEMPO-oxidized cellulose nanofibers, *Nanoscale,* Vol. 3, 2011, pp. 71-85.

[9]. A. G. MacDiarmid, "Synthetic metals": A novel role for organic polymers (Nobel lecture), *Angew. Chem. Int. Ed.,* Vol. 40, 2001, pp. 2581-2590.

[10]. D. Hohnholz, H. Okuzaki, A. G. MacDiarmid, Plastic electronic devices through line patterning of conducting polymers, *Adv. Funct. Mater.*, Vol. 15, 2005, pp. 51-56.

[11]. H. Sirringhaus, T. Kawase, R. H. Friend, T. Shimoda, M. Inbasekaran, W. Wu, E. P. Woo, High-resolution inkjet printing of all-polymer transistor circuits, *Science*, Vol. 290, 2000, pp. 2123-2126.

[12]. R. K. Cheedarala, J.-H. Jeon, C.-D. Kee, I.-K. Oh, Bio-inspired all-organic soft actuator based on a π-π stacked 3D ionic network membrane and ultra-fast solution processing, *Adv. Funct. Mater.*, Vol. 24, 2014, pp. 6005-6015.

[13]. S.-S. Kim, J.-H. Jeon, H. I. Kim, C.-D. Kee, I.-K. Oh, High-fidelity bioelectronic muscular actuator based on graphene - mediated and TEMPO-oxidized bacterial cellulose, *Adv. Funct. Mater.*, Vol. 25, 2015, pp. 3560-3570.

[14]. J. Zhou, M. Mulle, Y. Zhang, X. Xu, E. Q. Li, F. Han, S. T. Thoroddsen, G. Lubineau, High-ampacity conductive polymer microfibers as fast response wearable heaters and electromechanical actuators, *J. Mater. Chem. C*, Vol. 4, 2016, pp. 1238-1249.

[15]. Y. Bar-Cohen (Ed.), Directions for Development of the Field of Electroactive Polymer (EAP), *SPIE Press*, Washington, DC, 2011.

[16]. T. Fukushima, K. Asaka, A. Kosaka, T. Aida, Fully plastic actuator through layer-by-layer casting with ionic-liquid-based Bucky gel, *Angew. Chem. Int. Ed.*, Vol. 44, 2005, pp. 2410-2413.

[17]. I. Takeuchi, K. Asaka, K. Kiyohara, T. Sugino, K. Mukai, T. Fukushima, T. Aida, Electromechanical behavior of fully plastic actuators based on bucky gel containing various internal ionic liquids, *Electrochim. Acta*, Vol. 53, 2009, pp. 1762-1768.

[18]. T. Fukushima, A. Kosaka, Y. Ishimura, T. Yamamoto, T. Takigawa, N. Ishii, T. Aida, Molecular ordering of organic molten salts triggered by single-walled carbon nanotubes, *Science,* Vol. 300, 2003, pp. 2072-2074.

[19]. N. Terasawa, K. Asaka, Self-standing cellulose nanofibers/poly(3,4-ethylenedioxythiophene):poly(4-styrenesulfonate)/ionic liquid actuators with superior performance, *RSC Adv.*, Vol. 8, 2018, pp. 33149-33155.

[20]. Q. Pei, O. Inganas, Electrochemical applications of the bending beam method. 1. Mass transport and volume changes in polypyrrole during redox, *J. Phys. Chem.*, Vol. 96, 1992, pp. 10507-10514.

[21]. N. Terasawa, K. Asaka, High-performance PEDOT:PSS/Single-Walled Carbon Nanotube/Ionic Liquid actuators combining electrostatic double-layer and faradaic capacitors, *Langmuir*, Vol. 32, 2016, pp. 7210-7218.

[22]. N. Terasawa, K. Asaka, Superior performance PEDOT:poly (4-styrenesulfonate)/vapor-grown carbon nanotube/ionic liquid actuators exhibiting synergistic effects, *Sen. Actuators B*, Vol. 248, 2017, pp. 273-279.

(004)

A Flexible Acoustic Sensing System and Its Application to IIoT – Manufacturing Field Site

Y. Serizawa [1] **and Y. Shomura** [1]

[1] Hitachi America Ltd. R&D, IoT Edge Laboratory, 2535 Augustine Drive, 3rd Floor, Santa Clara, CA, USA

Tel.: + 1-408-986-6300

E-mail: yasutaka.serizawa@hal.hitachi.com

Summary: Acoustic data is a valuable source of information that can be used to develop various IIoT (Industrial IoT) applications such as machine state classification, anomaly detection, mobile tracking, and so on. Traditionally, especially in manufacturing field sites, sensing data acquired by sensors have been stored on stages with kinds of file formats, which has also been consumed by batch-type analysises as data sources. This system architecture has been one of the causes of lowering production efficiency such as long unplanned downtime of production machines. Accordingly, we have developed an acoustic streaming data processing system that is both scalable and reusable. The developed system has pluggable system architecture where data acquisition part, signal processing part, and analytics part can be mutually communicated with common APIs, which provides flexible combination of streaming data processing in multiple parts. The scalability of developed system was demonstrated on two example applications of machine state classification and anomaly detection. The successful operation of both applications on the same acoustic sensing system indicates that pluggable APIs work as expected, which suggests the developed system can reduce time and cost for establishment of applications significantly.

Keywords: Acoustic sensing system, Acoustic analytics, Streaming data processing, Data acquisition, Flexibility, Scalability.

1. Introduction

The market of IIoT especially in manufacturing field has been growing increasingly. Given this trend, recently many manufacturers have become interested in whole process optimization using IT technologies. Therefore, introduction of IoT system has been widely considered especially in manufacturing factory shop floor in order to improve productivity and increase added value of products. However, it has been found that fully organized information (data) doesn't exist in existing manufacturing systems to monitor total condition/status of overall shop floor in real-time manner.

According to this situation around manufacturing IoT, we have focused on extraction of information from factory field site with microphones from the view point of creating sensor-position agnostic applications. Sensing system should be scalable and reusable in order to be applied to multiple environments seamlessly. Therefore, we considered non-contact type sensors such as microphones which have promising features for reusable applications.

In this background, our research has been focused on streaming data processing system with pluggable APIs. Many works on sophisticated analytics have been disclosed and provided kinds of applications (eg. [1, 2]). Almost all of them provide sophisticated analytics to solve each problem. However, they provides only analytics part from IoT acoustic sensing system view point. In order to make acoustic sensing system flexible, or scalable, it should have pluggable APIs between function blocks as well as function modules, and configuration functionality from analytics part. In this paper, the basic analytics-centric architecture of an acoustic streaming data processing system which provides pluggable APIs and packaged functions, and verification results of its scalability with applying to example applications, are reported.

2. Acoustic Streaming Data Processing System

2.1. Data Processing in Manufacturing Site

In many cases of IIoT, especially in manufacturing field site, there are still many static system for data analysis utilized. Acquired data tend to be stored on databases, or any formats of files (eg. xml, xls, ...) and are processed when problems occur or are checked by manufacturing operators periodically (eg. daily, or weekly), which leads to long unplanned downtime, inefficient ways of manufacturing operation, and unoptimized assignment of resources.

Taking this situation as a general problem in manufacturing field site, we have proposed streaming data processing system providing real-time data processing and having analytics-centric APIs, which enables rapid establishment of applications especially on manufacturing use cases (scalability for application establishment). As previously described, acoustic sensors have been selected in order to support scalability as the first sensing device for this system. Fig. 1 shows conventional system for acoustic data processing, where data are stored and utilized by processing with file-based media and configuration interfaces are located on each part of data processing. This would be basically be the same system architecture as one in manufacturing field site (batch type processing scheme). In Section 2.2, our implementation of streaming data processing system compared with this conventional scheme is described.

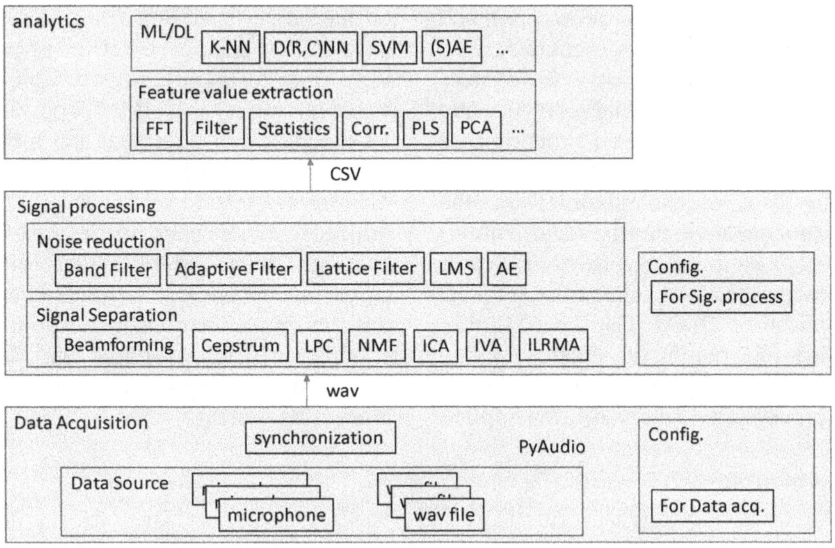

Fig. 1. Conventional acoustic data processing system.

2.2. Proposed Streaming Data Processing System

Fig. 2 shows detail description of an implementation of our acoustic streaming data processing system. In this system, gRPC [3] has been utilized to reduce message size and support scalability in terms of programming languages. One of the feature of gRPC is characterized by serialization of data using protocol buffers instead of JSON and its high-speed communication capability (in other words, reduction of message size). Utilizing protocol buffers also provides fixed data structure designated by system, which can specify the data structure to other system as an API. In our implementation of acoustic streaming data processing system, protocol buffer provides an API of data acquisition part and data processing part to analytics part, which means that this system is analytics-centric design. gRPC server and client are deployed on each part (analytics, signal processing, and data acquisition), which enables flexible communication between individual parts. Within gRPC, request from client and response from server are exchanged based on protocol buffer as shown in Fig. 3. This structure enables flexible combination of signal processing and data acquisition from analytics point of view. Additionally, owing to broker functionality on signal processing part and data acquisition part, analytics can choose any kind of data from any kind of data source (eg. data from device (raw streaming data), processed data with designated signal processing, or files).

3. Verification

In order to verify scalability of the developed acoustic streaming data processing system for multiple applications, in this chapter verification results of enabling multiple applications (analytics) are shown with two example applications of machine state classification and anomaly detection of a motor. A MEMS-type 8-channels circular synchronized microphone array with 16 kHz sampling rate and int16 sound encoding was utilized for both use cases.

3.1. Machine State Classification

To verify the capability of enabling this application, we used an industrial motor for manufacturing field site. We used 'stop' state, 'low' speed state, and 'high speed' state as 3 different states of the motor. Fig. 4 shows spectrogram and averaged FFT results of the acoustic signal. A little difference can be seen between states.

With taking the spectrogram and averaged FFT into account, we developed a 4 layers neural network which consists of an input layer, two hidden layers, and an output layer, whose number of nodes are 129, 128, 64, and 3 respectively, where sequential model was utilized. The second and third models were based on sigmoid function, and final one on softmax function. This neural network was simple all connected neural network developed with OSS Keras / TensolFlow [4, 5]. In this ML process, only the result of STFT was utilized as input of neural network, and out output nodes correspond to each state of the motor (stop, low, high). The classification results of streaming signal corresponding to each motor state are shown in Fig. 5, which shows 100 % classification accuracy. Classification was executed, then the result was output every 1 second. Streaming chunk size was 1024, which is actually the same size as STFT.

3.2. Anomaly Detection

The difference between "normal" and "anomaly" was demonstrated by the extent of tightening a screw which makes the motor mounted on a metal base plate, where anomaly state was defined as the motor with loosed screw. Fig. 6 shows raw waveform data of normal and anomaly status, where no significant difference can be found. Fig. 7 shows spectrogram and averaged FFT of normal and anomaly signals derived from a microphone respectively. Some differences between normal and anomaly around 200 Hz and 2,000 Hz can be found. Here, acquired synchronized raw signals were processed on a MVDR beamformer to make the beam focus on the motor. Then, the supervised NMF (nonnegative matrix factorization) with multiple bases, was applied to extract the clearer motor sound signal separated from other noise sources. Finally, in this application, GMM (Gaussian Mixture Model) was applied for training in order to judge anomaly degree (score) of normal and anomaly signals. In GMM process, mel filterbank was applied on the power spectrum from FFT results in order to have clearer difference between normal and anomaly. Regarding to feature values, which are independent variable input to GMM, power spectrum with 15-channels mel filterband and time-series difference (delta feature value) of power spectrum were applied. Therefore, a feature value vector with 30 elements was utilized to extract anomaly degree of each signal. Here, although many sound signal separation and data clustering technologies have been reported [4-6], relative simple methods were applied in order to make system verification simple and include basic data processing components into streaming data processing system.

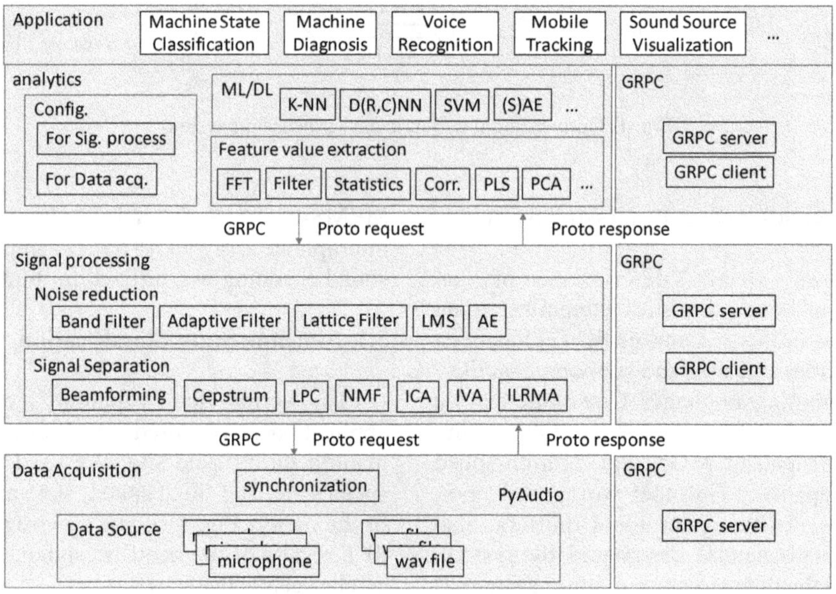

Fig. 2. System implementation.

Property	Contents	Description	Direction
Command	STOP_STREAMING	STOP_STREAMING	Req.
	START_STREAMING	START_STREAMING	Req.
	SUSPEND_STREAMING	SUSPEND_STREAMING	Req.
	RESUME_STREAMING	RESUME_STREAMING	Req.
Config.	FilterType	Processes in signal processing part	Req./Resp.
	SamplingFreq	Sampling frequency	Req./Resp.
	ChunkSize	Streaming data size	Req./Resp.
	BitWidth	Number of bit per sample	Req./Resp.
	Encoding	Data type	Req./Resp.
Stream	AudioStream	Streaming data with requested property	Resp.

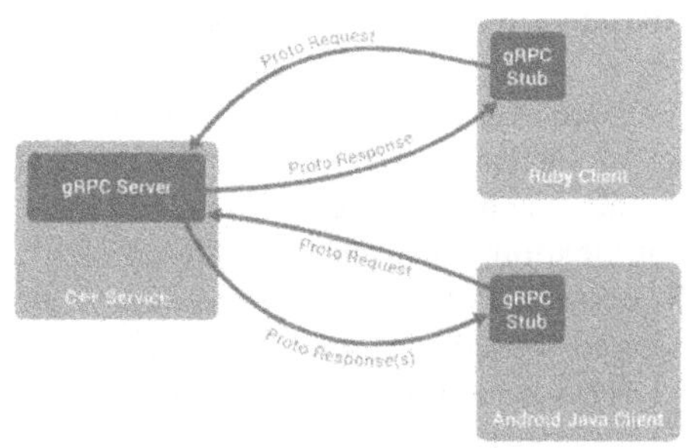

Fig. 3. Request/Response property on protocol buffer, and gRPC procedure [3].

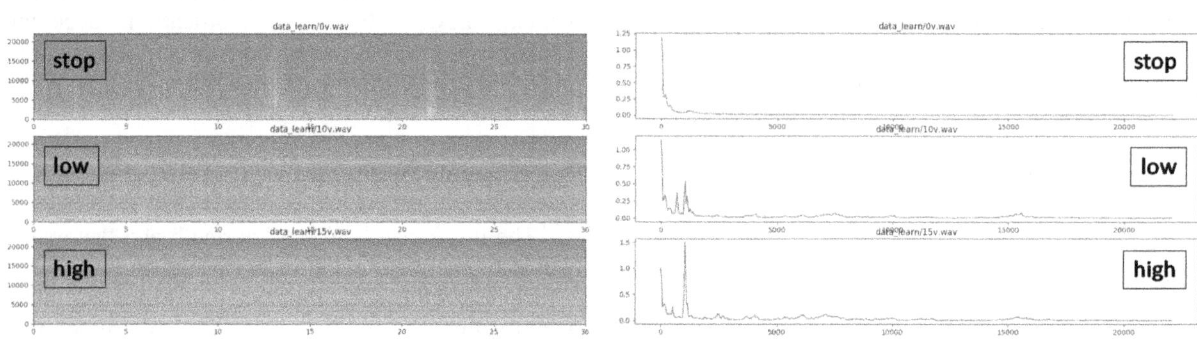

Fig. 4. Spectrogram (left) and averaged FFT (right).

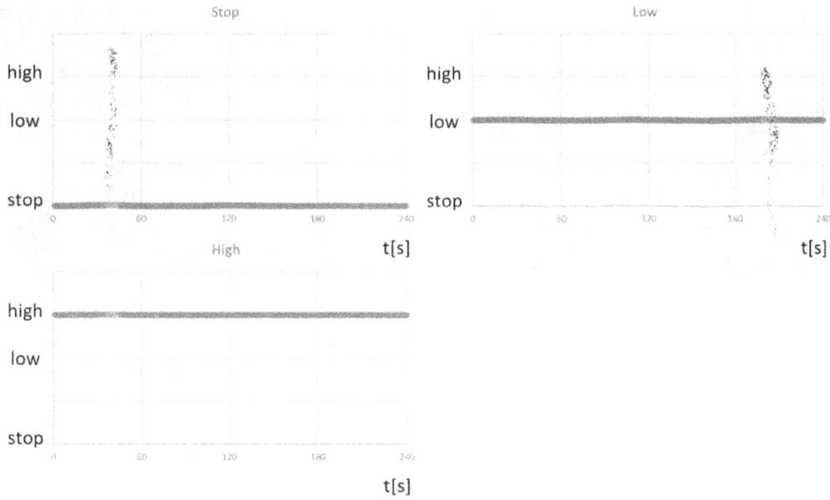

Fig. 5. Classification results.

In Fig. 7, there seems to be some difference between normal and anomaly state on the 4th channel of mel filterbank. However, as shown in Fig. 8 and Fig. 9, distribution of power spectrum and delta value of 4th mel channel actually shows no clear separation of normal and anomaly signals. Fig. 8 shows distribution of normal signal with fitting result of one Gaussian distribution, and in Fig. 9 distribution of normal and anomaly signals with fitting result of two Gaussian distribution is shown.

Fig. 10 shows anomaly degree of original overall waveform (600 sec) of normal and anomaly as a result of this sequence of signal processing. This result shows obvious higher anomaly degree in case of anomaly state than normal state, which indicates that anomaly defined in this verification test can be successfully detected with this method. Additionally, this results can be obtained in time series manner (not static manner), which means developed system can provide real-time feedback of anomaly detection to manufacturing field site.

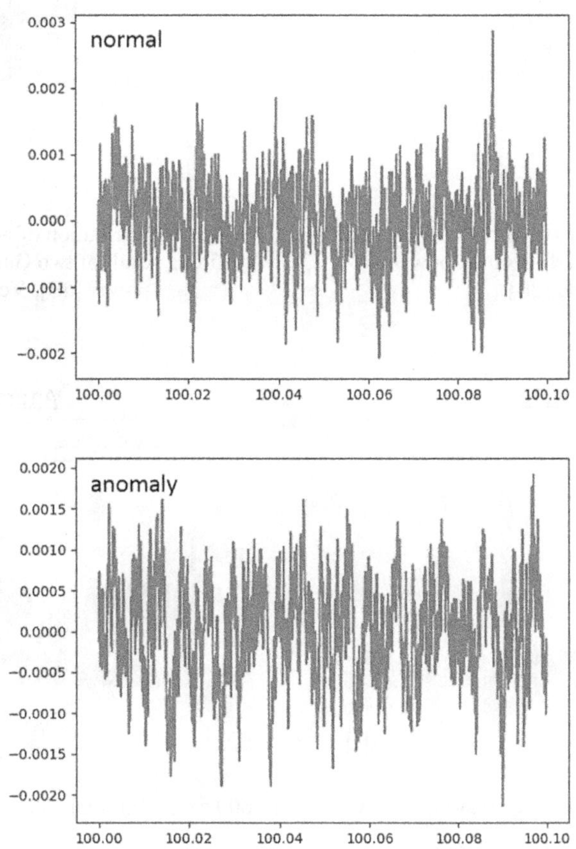

Fig. 6. Raw data of normal/anomaly signal.

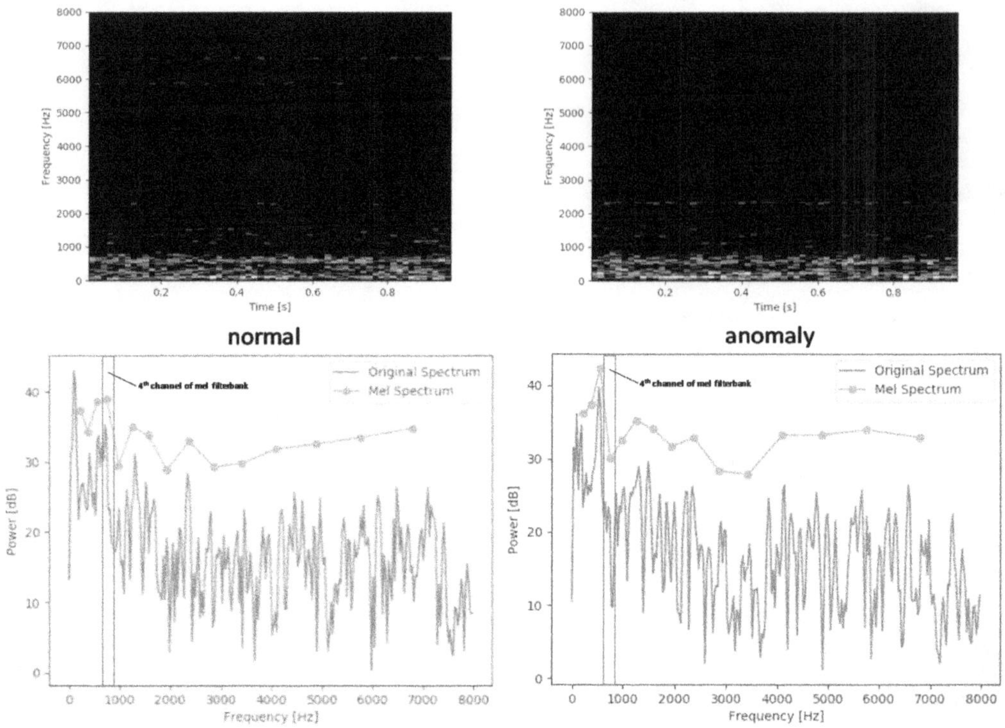

Fig. 7. Spectrogram and averaged FFT of normal and anomaly signals.

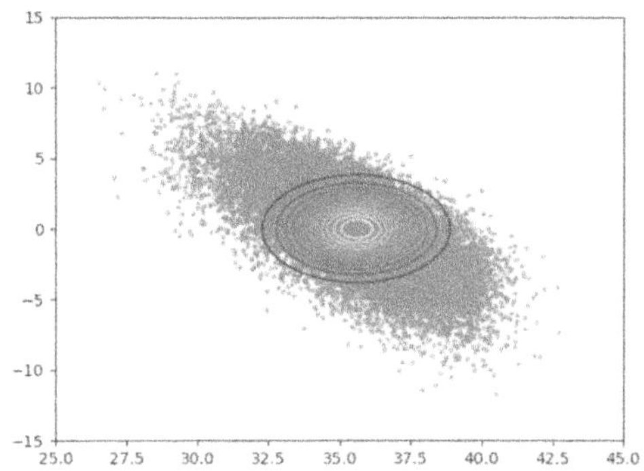

Fig. 8. Distribution of normal signal with fitting result of one Gaussian distribution (Horizontal: power [dB], Vertical: delta [dB]).

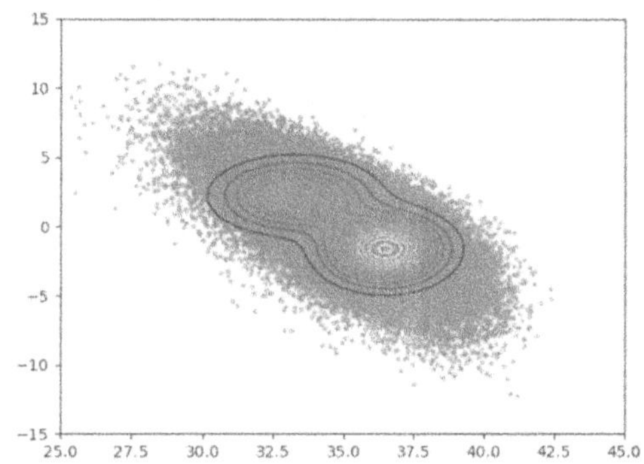

Fig. 9. Distribution of normal and anomaly signals with fitting result of two Gaussian distribution (Horizontal: power [dB], Vertical: delta [dB]).

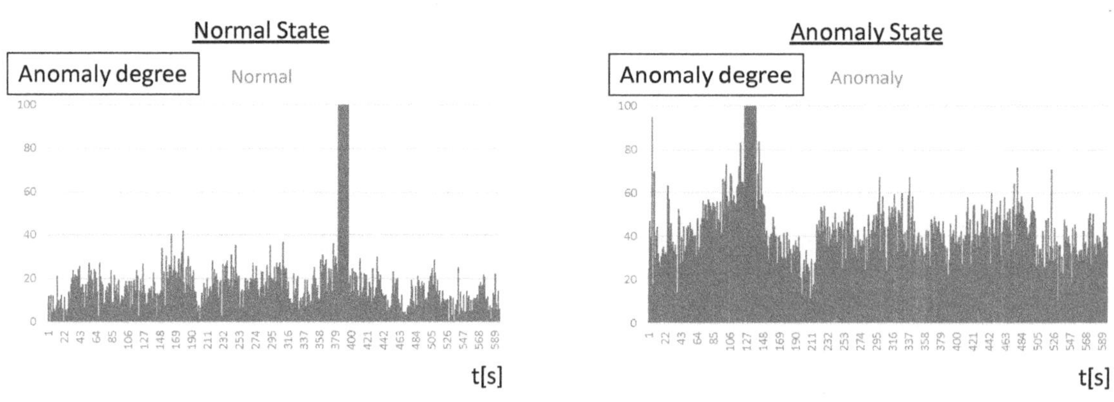

Fig. 10. Anomaly degree of overall waveform (600 sec) of normal and anomaly signal.

4. Conclusions

In this research, we have developed acoustic streaming data processing system to make IIoT system scalable and reusable. Two example implementations for IIoT applications are described to demonstrate streaming signal / data processing and scalability in terms of multiple applications operated on the same streaming processing system.

References

[1]. S. Pasha, et al., A deep learning approach to the acoustic condition monitoring of a sintering plant, in *Proceedings of Asia-Pacific Signal and Information Processing Association Annual Summit and Conference (APSIPA ASC'18)*, Honolulu, USA, 12-15 Nov. 2018, pp. 1803-1809.

[2]. E. Wangkanklang, et al., The progress investigation of Ishikawa mixing and grinding machine using sound signal processing, in *Proceedings of IEEE 7th Global Conference on Consumer Electronics (GCCE'18)*, Nara, Japan, 9-12 Oct. 2018, pp. 513-514.

[3]. gRPC/Proto buffer, Google, https://grpc.io/

[4]. A. Saeed, Urban Sound Classification, Part 1, http://aqibsaeed.github.io/2016-09-03-urban-sound-classification-part-1/

[5]. F. Shaikh, Getting Started with Audio Data Analysis (Voice) using Deep Learning, https://www.analyticsvidhya.com/blog/2017/08/audio-voice-processing-deep-learning/

[6]. M. Togami, et al., Noise robust speech dereverberation with Kalman smoother, in *Proceedings of the IEEE International Conference on Acoustics, Speech and Signal Processing (ICASSP'13)*, Vancouver, BC, Canada, 26-31 May 2013, pp. 7447-7451.

[7]. R. Takashima, et al., Separation of vibration-derived sound signals based on fusion processing of vibration sensors and microphones, in *Proceedings of the* 25th European Signal Processing Conference (EUSIPCO'17), Kos, Greece, 28 Aug.-2 Sept. 2017, pp. 2428-2432.

[8]. Y. Kawaguchi, et al., Non-negative novelty extraction: a new non-negativity constraint for NMF, in *Proceedings of the 16th International Workshop on Acoustic Signal Enhancement (IWAENC),* Tokyo, Japan, 17-20 Sept. 2018, pp. 256-260.

(005)

An Ultra-wideband Band-pass Filter for Cosmic Microwave Background Observations

Javier De Miguel-Hernández [1,2], **Roger J. Hoyland** [1,2]
[1] Instituto de Astrofísica de Canarias, E-38200 La Laguna, Tenerife, Spain
[2] Departamento de Astrofísica, Universidad de La Laguna, E-38206 La Laguna, Tenerife, Spain
E-mail: jmiguel@iac.es

Summary: The QUIJOTE Experiment has been built to study the polarization of the Cosmic Microwave Background (CMB). The study of the CMB provides one of the most powerful tools in modern cosmology. In particular, CMB polarization offers a new window onto the inflationary phase of the Universe, some 10-35 seconds after the Big Bang. Measurement of the CMB spectrum to an accuracy that will reveal interesting science (micro-kelvin variations) is particularly difficult because there are many systematics that lead to effects higher than this level. There are several instruments that have been designed for CMB spectral measurement (FIRAS-COBE, ARCADE, PIXIE). The present plan is to extend the goals of QUIJOTE by building a new and unique microwave spectrometer in the frequency range 10-20 [GHz], with the aim of characterizing the spectral signals associated with the reionization epoch and dark ages in the Universe. This document describes the first component for the Tenerife Microwave Spectrometer (TMS), an ultra-wideband bandpass filter (BPF). A filter prototype covering the 10-20 [GHz] band has been successfully designed, manufactured, and measured in the laboratory to compare results and simulations.

Keywords: Cosmic microwave background, Radio-astronomy, Microwaves, Cosmology, Ultra-wide band, Filters.

1. Introduction

In this section, the TMS is briefly described. Fig. 1 shows the preliminary schematic diagram of a microwave spectrometer for measuring CMB spectra between 10 and 20 [GHz]. In order to achieve sufficient sensitivity the front end of a radiometer is cooled in a cryostat to 4-10 [K]. The cryogenic reference load, opto-mechanics, and ultra-LNAs are cooled by a 4 [K] closed cycle helium gas cooler and maintained at a constant temperature slightly higher than 4 [K] by actively heating the cold stage.

The radiometer consists of a pair of feedhorns, one pointed at a cold black body and the other looking out of the cryostat through a transparent window. These feedhorns are both followed by broad-band 10-20 [GHz] waveguide OMTs to couple the two linear polarizations. The polar outputs of both sky and load horns are fed to either arm of a radiometer correlator that employs both cold and warm gain and 180° broad-band phase switch to switch the outputs. The correlating elements are 90° hybrid couplers or ridged waveguide hybrid tees. Each correlator contains two similar LNAs with a noise temperature lower than 10 [K] followed by further amplification and an 180° phase switch in each branch. The phase switches are housed in the BEM. Finally, the load and sky signals are decorrelated in a similar 90° hybrid coupler or ridged waveguide hybrid tee to that of the input. The outputs of these two hybrids are sent to the novel full-band FT-spectrometers bank. The DC output is amplified and then digitized for post processing in an ADC unit.

So, the first element to design is a microwave filter covering the entire band, of the experiment; that is, with a 2:1 bandwidth factor, which is challenging since most microwave filters are designed to cover a few tens or hundreds of megahertz.

In the following sections the given solution is described, leading to an excellent device which satisfies all the system requirements.

2. System Requirements

2.1. Requirements

In order to satisfy the general requirements of the project, the design specifications are:

- Pass band attenuation: 3.01 [dB];
- Stop band attenuation: 40 [dB];
- Impedance: 50 [Ω];
- Pass band frequency: 15 [GHz];
- Pass band width: 10 [GHz];
- Stop band frequency: dependent on the filter order (N) ($N < 9$ is desirable because of size requirements).

2.2. Design of the BPF

The design of microwave filters has been adequately treated by multiple authors (e.g. [1]). An ultra-wideband bandpass filter (10 to 20 [GHz]) will be developed in this project through the use resonant LC circuits, microwave inverters, and both Richard's Transformations [2] in order to convert lumped elements to transmission line sections and Kuroda's Identities [3] to physically separate filter elements by using transmission line sections. A complete treatment of these theoretical transformations can be found by consulting the references indicated at the end of this document.

It is important to note that since there are no commercial devices optimized for the ultra-wide 10-20 [GHz] band, the filter must be designed from scratch for this experiment.

The design will be computer-assisted by CST Microwave Studio Suite®.

Some typical terminated periodic structures (i.e. topologies) for microwave filters have been tested (*Minimum Segments, Minimum Stubs, Stepped Impedance, Open Stub Resonators, Parallel Edge, Hairpin Resonators,* etc.). Because of the ultra-wide band that is desired for this filter, it has been found that the topology offering the best response is formed by *Shunt Stub Resonators*. This filter topology, with origins commonly attributed to crystals and/or composite filters, uses the resonance properties of the LC circuits in order to make a frequency selection over a received signal, i.e. there exists a resonant frequency at which the inductive and capacitive reactances are equal in magnitude. Therefore, the impedance will be minimal and equal to the ohmic resistance.

A Butterworth type filter (also referred to as a maximally flat magnitude or binomial filter) was designed (using the Instertion-loss IL method) for the 8th order, which optimized the stop band frequency in the range considered. For this, a theoretical circuit was first designed using lumped elements (resistors, capacitors, inductors, etc.) following the theoretical framework published by many authors [1].

This circuit is shown in Fig. 2. It is important to note that the capacitance, impedance, and resistance values shown in Fig. 2 are the consequence of a theoretical design and have not been already optimized.

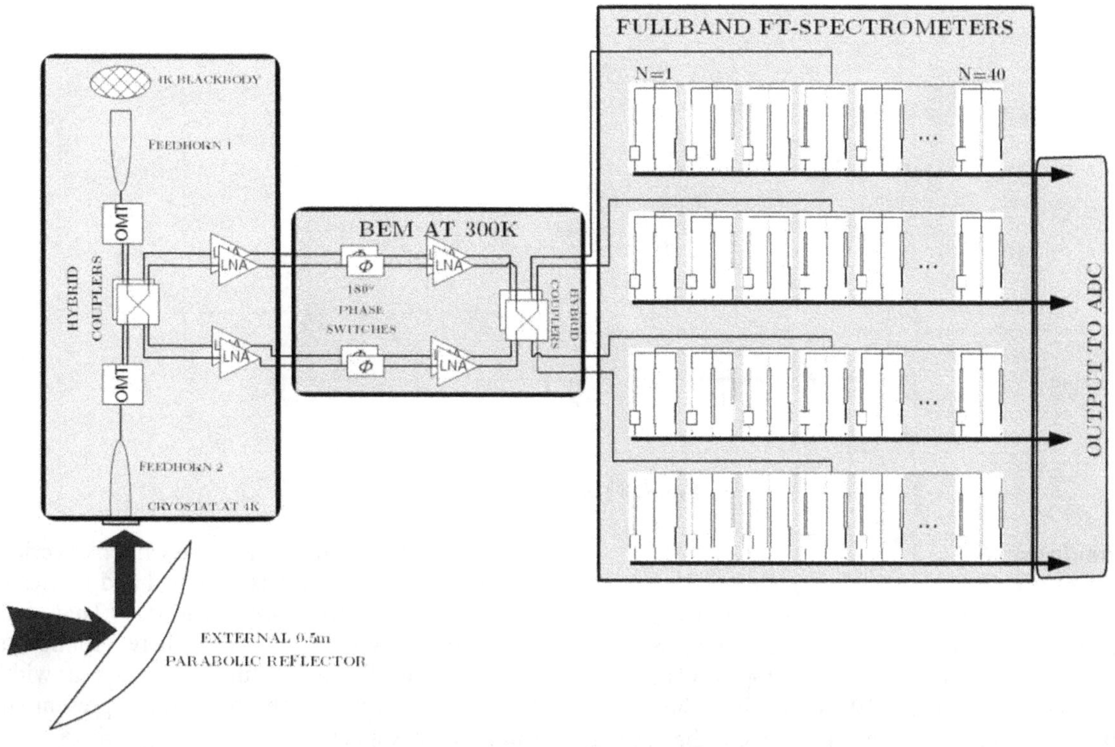

Fig. 1. TMS general schematic.

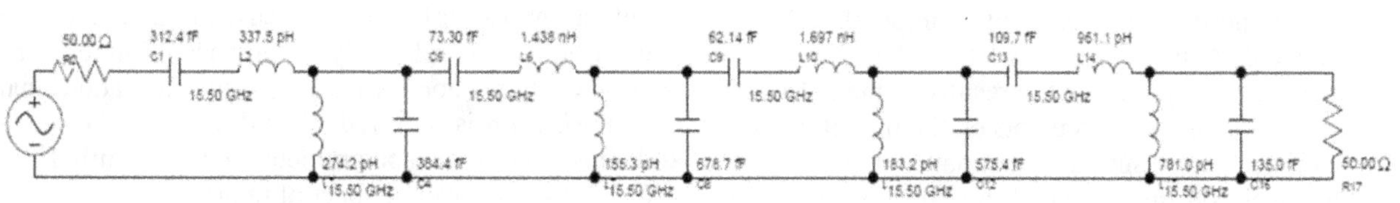

Fig. 2. Theoretical design of 8th order BPF based on lumped elements.

However, the response of the concentrated elements (resistors, capacitors, etc.) is typically not good over wide bandwidths. The elements making up a lumped element system are theoretically concentrated at singular points in space. Therefore, the physical quantities, current and voltage, are functions which depends only on time (and not space). The real filter implementation could now be obtained using Richard's Transformations in order to replace lumped elements with short circuit (S.C.) or open circuit (O.C.) stub transmission lines. The Kuroda Identities then allow us to obtain a more practical filter implementation. A progressive process of design optimization based on simulations and successive improvements is then carried out.

Note that shielding to avoid electromagnetic interactions between stubs and resonances in cavities has been added in the design shown in Fig. 2.

3. Results

The filter was fabricated with an LPKF machine model ProtoMat-S100 on a PCB (Rogers 4003-C) circuit. The objective of the final stage is to compare the simulated model with the manufactured model and the requirements, in case of discrepancies between them, to return to the design stage with the objective of investigating and solving them. Several iterations were necessary. An image of the manufactured filter and the results of the test measurements made on it are shown in Figs. 3 and 4. For the measurements a KeySight model PNA-X N5245A machine was used.

Fig. 3. 8th order BPF prototype while measuring in the laboratory using SMA-port adaptors.

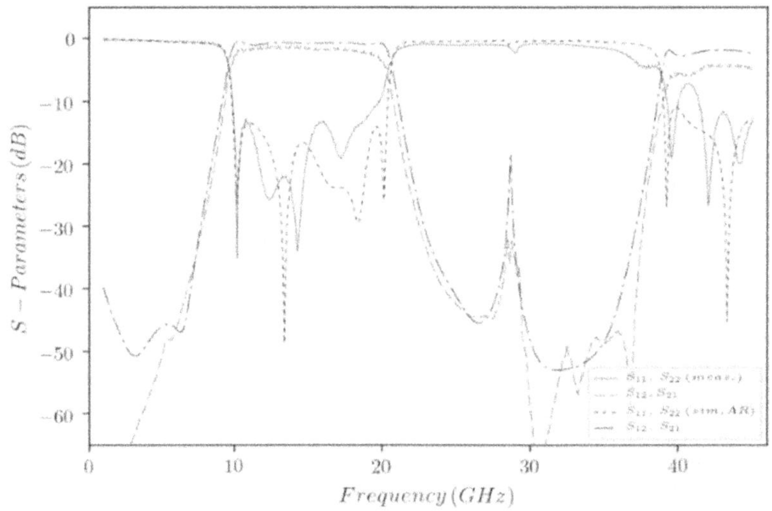

Fig. 4. 8th order BPF results.

4. Conclusions

The preliminary design requirements are successfully satisfied, with an excellent adaptation, correct attenuation in the pass and stop bands and an attenuation better than -10 [dB] over the entire 10-20 [GHz] range. Note that, given the design characteristics of the stubs (where the length of the transmission lines is $\lambda/8$ at the cut-off frequency), the response is periodically repeated at intervals of four times the cut-off frequency (Fig. 4).

It has been verified that the results of the software simulation and the measurements on the manufactured filter are reasonably similar. This makes it possible to double-check, on the one hand, by checking that the software is reasonably reliable and, on the other hand, that significant errors have not been made during the manufacturing and assembly phase, with the manufacturing tolerances being within a reliable limit.

With a view to future improvement of the design, a trial with computer-assisted optimization has been carried out, with the result that the filter design would be difficult to improve. This enables us to conclude that the design is excellent, and that the results are difficult to surpass.

A high quality ultra band-pass filter covering the range 10 to 20 [GHz] was designed and fabricated to meet the technical specifications. This was a technological challenge, since there are not many references to bandpass filters of such a width in frequency, and those that exist do not present such a satisfactory response.

In addition, it is a robust design of astonishing simplicity and very cheap to manufacture, which would allow its replacement in case of damage due to the filter is relatively compact (dimensions 25.6×3.6 mm approximately), so no additional miniaturization is required. This device will be ready to be installed in the spectrometer and efficiently filter the CMB in the desired spectral range.

Acknowledgements

The author wishes to acknowledge the supervision and support of and all the technicians, engineers, scientists, and administrative staff of the Instituto de Astrofísica de Canarias who have participated in the Quijote-CMB Experiment.

References

[1]. D. M. Pozar, Microwave Engineering, 4th Ed., *University of Massachusetts at Amherst,* 2011.

[2]. P. I. Richards, Resistor-transmission-line circuits, *Proceedings of the IRE,* Vol. 36, Issue 2, 1948, pp. 217-220.

[3]. K. Kuroda, Derivation methods of distributed constant filters from lumped constant filters, *Experimental Wireless and the Wireless Engineer*, Vol. 7, Oct. 1930, pp. 536-541.

[4]. S. Butterwworth, Theory of filters amplifiers, in *Proceedings of the Joint Meeting of Kansai Branch of IECE,* Japan, Oct. 1952, p. 32.

[5]. G. L. Matthaei, L. Young, E. M. T. Jones, Microwave Filters, Impedance-Matching Networks, and Coupling Structures, *Artech House,* Dedham, Mass., 1980.

[6]. J. A. Rubiño-Martín, R. Rebolo, et al., The Quijote CMB experiment, in Highlights of Spanish Astrophysics V, *Springer*, Jan. 2010, pp. 127-135.

(006)

An Ion Beam Guidance Control Tool Proposal

L. Bengtsson [1]
[1] University of Gothenburg, Dept. of Physics, Kemivägen 9, SE-412 96 Göteborg, Sweden
Tel.: +46317869128
E-mail: lars.bengtsson@physics.gu.se

Summary: A versatile tool for ion beam line up is proposed. This tool will be used at the CERN-ISOLDE facility in negative ion experiments. Apertures in the ion optics are used to guide the beam and traditionally the current is measured on the apertures in order to detect the presence of an ion beam. By slitting the aperture into four separate pieces, the aperture current can also be used to determine the direction of the misalignment and the size of the ion beam diameter. A 4-channel pico ammeter with optical LED indication of the ion beam position is suggested in an analog/digital hybrid design. The design utilizes two charge/current-to-digital chips in a microcontroller design.

Keywords: Ion optics, Aperture, Pico ammeter, Guarding technique, Microcontroller.

1. Introduction

In ion-beam experiments, the process of lining up the ion beam is typically a long-suffering and tedious job since it requires guiding the ions through a long range of ion optics [1]. Fig. 1 illustrates an example of the kind of ion optics that is frequently used.

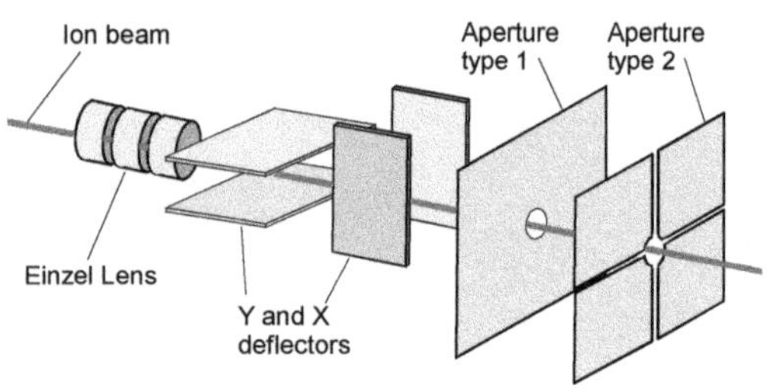

Fig. 1. Ion beam optics example.

Enzel lenses are used to focus the beam and deflector plates are used to control the beam trajectory in the X and Y directions. Apertures are used to narrow the beam diameter but are also useful indicators of the beam quality and direction. Ideally, the current registered on any aperture plate should be zero. The apertures' current is a crucial parameter to monitor during ion beam line-up.

Ion beam currents are of the order of nano amps at best and sub-pico amps at worst and hence this requires sensitive ammeters with careful design layouts.

Commercial sub-pico ammeters are very expensive; a desktop instrument could cost as much as $5,000 [2] and in order to line up an ion beam, multiple ammeters are required. For these reasons, a versatile, non-expensive multi-channel ammeter with pico ampere capability has been requested.

There are in principle two basic ammeter circuits; the shunt ammeter and the feedback ammeter [3]. This work is based on the DDC112 chip from Texas Instruments [4] which contains two capacitive-feedback ammeters [5]. The instrument should primarily be used to line up the ion beam through the ion optics to the target. Hence the accuracy, resolution and absolute values are not crucial but the tool should be able to indicate ion currents of the order of 1 pA or less.

1.1. Problem Description

A typical aperture plate has a circular hole in the center (left side of Fig. 2). This aperture design will narrow the ion beam and by tapping off its current an indication of misalignment is achieved but it will not reveal the direction of the misalignment. For that reason, the aperture is split into four separate pieces, see Fig. 2, right side.

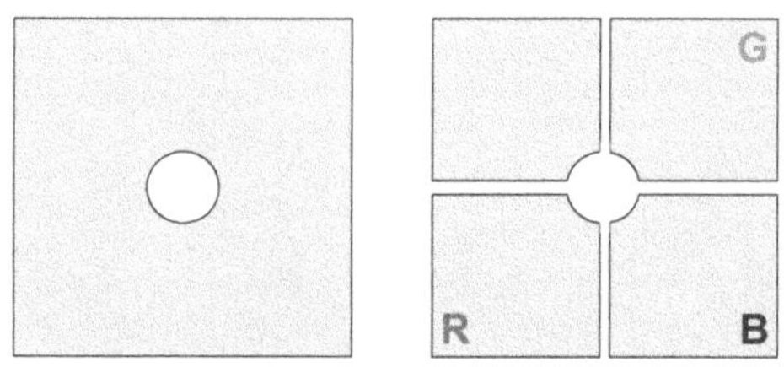

Fig. 2. Ion optics apertures.

By measuring the current separately on all four aperture plates, it is possible to determine the direction of the misalignment and also draw some conclusions about the beam quality; a large current on one plate indicates a misalignment in that direction and a current on all plates indicates a wide beam. The disadvantage is that in order for the quad-aperture to be really useful, it requires that the four currents are measured simultaneously; an inexpensive four-channel pico ammeter is required.

1.2. Proposed Solution

An analog/digital hybrid 4-channel picoamp tool will be designed. Each one of the four aperture plates will be color-coded (Yellow, Green, Red and Blue). Four pico ammeters will be implemented on the same PCB based on two DDC112 chips. The digitized current is transferred to a microcontroller via a synchronous serial interface (SPI compatible) and can optionally be relayed to a host computer, but this is primarily astand-alone desktop instrument intended as an ion beam guiding tool. The instrument panel will have four sets of four colored LEDs as illustrated in Fig. 3.

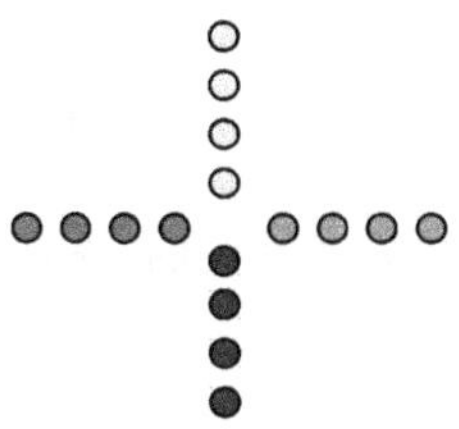

Fig. 3. The instrument panel has four arrays of LEDs.

Each colored LED array corresponds to the ion current on the corresponding aperture plate. This will offer a convenient optical indication of the ion beam (mis)alignment. Some examples are illustrated in Figs. 4 through 6. Notice that a high degree of activity on the LEDs indicates a misaligned or broad ion beam and a low activity (or no activity) indicates a clean passaged through the aperture.

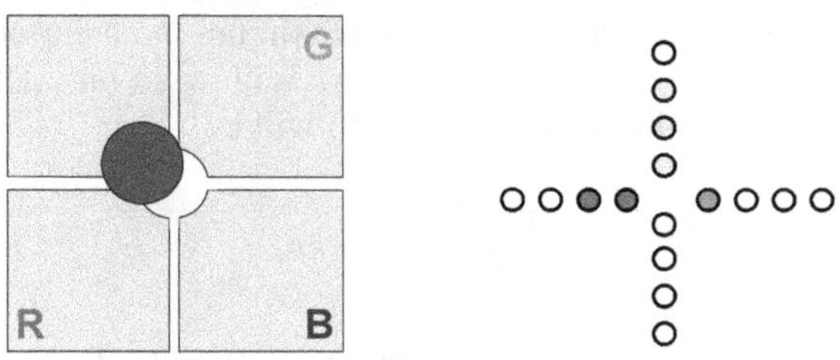

Fig. 4. Ion beam too broad and too far to the upper left.

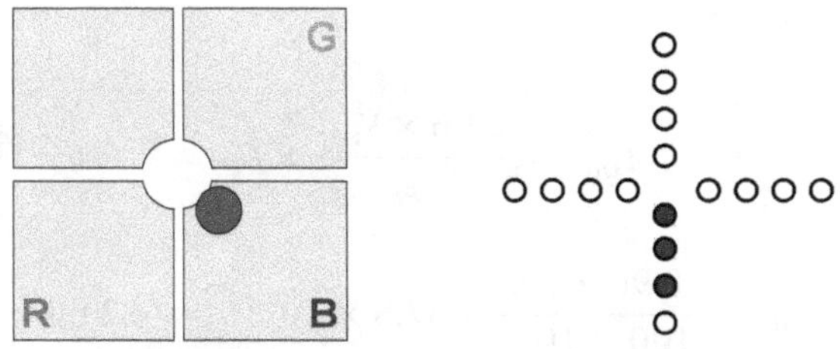

Fig. 5. Ion beam too far to the lower right (but size is probably ok).

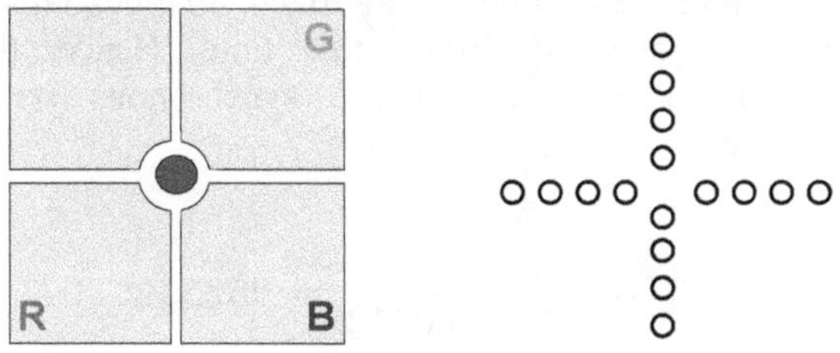

Fig. 6. No LEDs turned on indicates that the ion beam passes the aperture in the center of the Z direction (unobstructed).

2. Design

2.1. Hardware

At the core of the design are two DDC112 chips from Texas Instruments [4]. This 28-pin SO-chip has two charge input channels (sharing the same 20-bit ΣΔ ADC) and each channel comprises two classical analog charge integrators; while one is being digitized, the other one is being charged. Hence, the input charge/current is seamlessly monitored. The range depends on the feedback capacitor and the integration time. Internal feedback capacitors are available but external capacitors can also be used. The integration time is provided by the user by toggling the CONV pin. Fig. 7 illustrates the pin layout and Fig. 8 is a (simplified) model of input channel 1.

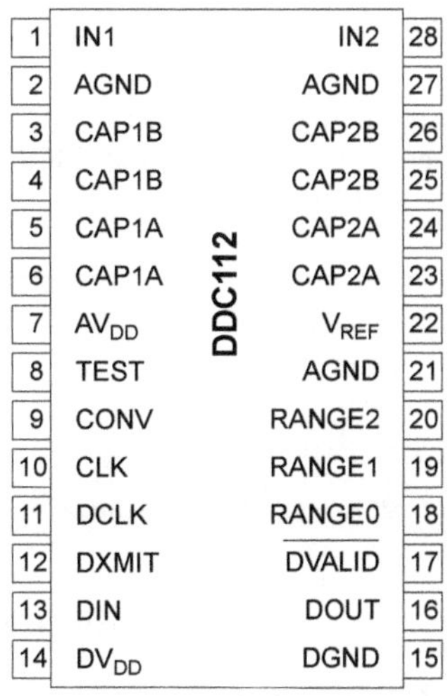

Fig. 7. Current-to-digital converter (DDC112).

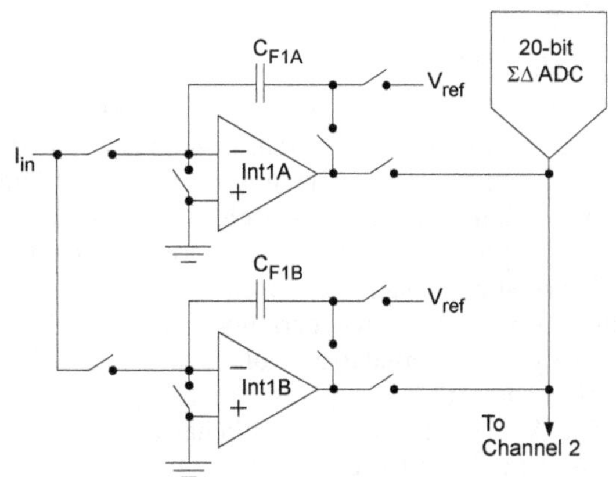

Fig. 8. Charge-input integrator.

The switches in Fig. 8 alternately charges/ discharges the feedback capacitors and forward the integrators' outputs to the ADC for digitalization. The switches are controlled by an internal synchronous state machine in order to charge/discharge the capacitors and to relay the two integrator outputs to the ADC alternately. Fig. 9 illustrates the timing diagram; input integrators A (on both channels) are integrated simultaneously while integrators B are being digitized by the AD converter. The $\overline{\text{DVALID}}$ signal indicates when both channels have been digitized and data is stored in a 40-bit shift register.

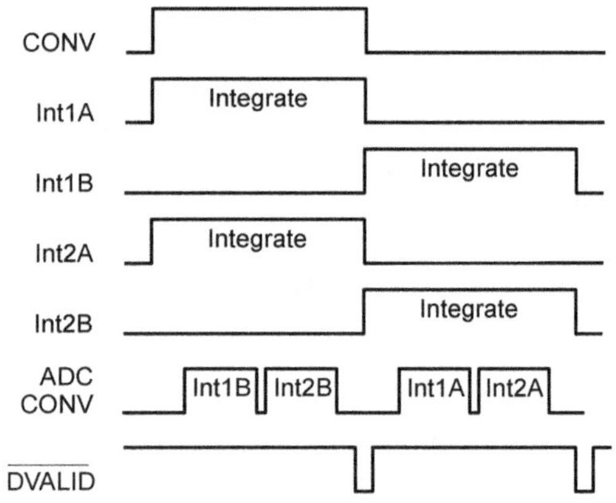

Fig. 9. Integration timing diagram.

The user retrieves data through the synchronous serial interface (pins DCLK and DOUT). This is illustrated in Fig. 10 [6].

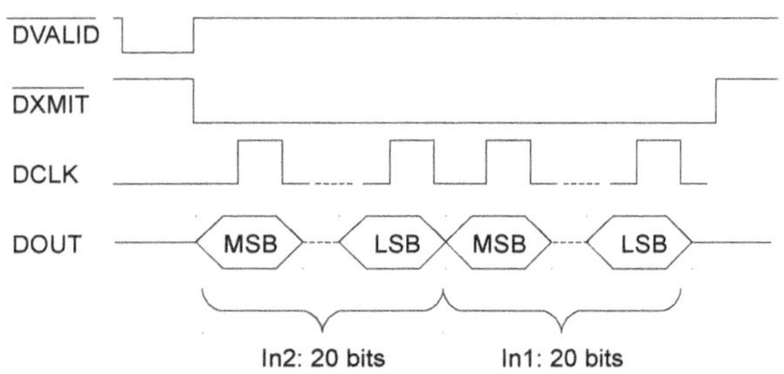

Fig. 10. Timing diagram of data retrieval.

Transmission is initiated by the user setting the $\overline{\text{DXMIT}}$ signal low and data is shifted out on the DOUT pin synchronous to the rising edges on the DCLK pin. Notice in Fig. 10 that the $\overline{\text{DVALID}}$ signal is automatically reset when transmission is initiated.

The primary concern in this design is to detect negative ion currents or electrons. However, sometimes positive ion experiments are conducted and for that reason the instrument needs to be able to detect both positive and negative currents. The DDC112 chip can easily be converted to a bipolar ammeter by adding an offset current to the input, see Fig. 11 [7].

The ADC in the DDC112 chip has a resolution of 20 bits; if the input current equals the full scale current (I_{FS}), the ADC output will be 0xFFFFF (2^{20} -1). V and R in Fig. 11 will produce an offset current V/R and this current is added to the sensor current. In order to make the instrument bipolar, V and R are dimensioned so that $V/R = I_{FS}/2$.

The two currents from the sensor and the offset branch are added in the junction and the sum of the two currents are fed to the DDC112 chip. The consequence of that is that when the sensor input current $I_{SENS} = -I_{FS}/2$, the input current to the DDC112 chip is zero. Table 1 illustrates the consequence of the offset current. From Table 1 it is obvious that the instrument is now bipolar.

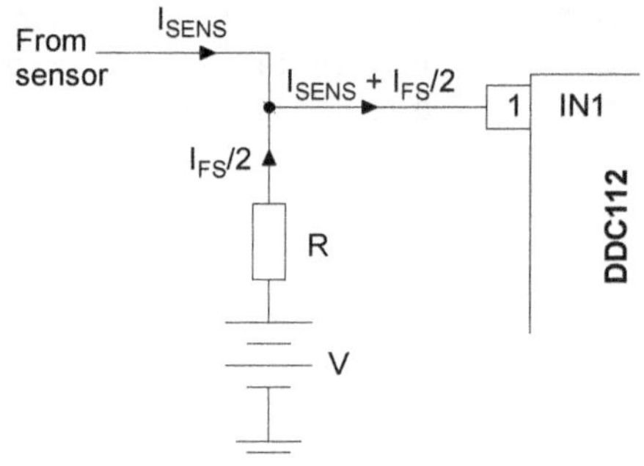

Fig. 11. Bipolar input.

Table 1. ADC output vs sensor current.

I_{SENS}	Input current	ADC output
$I_{FS}/2$	I_{FS}	0xFFFFF
0	$I_{FS}/2$	0x7FFFF
$-I_{FS}/2$	0	0x00000

2.2. Current Range Calculations

The aim is to create a bipolar instrument with a full-scale range of ±50 nA. Due to the bipolar circuitry in Fig. 11, this means that the maximum input current to the DDC112 chip is 100 nA. With a 20-bit ADC resolution, the current resolution will be

$$\frac{100 \text{ nA}}{2^{20}} = 95 \text{ fA}, \tag{1}$$

which is less than the expected noise level (≈ 2 pA).

If we choose internal feedback capacitors 87.5 pF, the required integration time for a full-scale range of 100 nA is [4]

$$T_{int} = \frac{0.96 \times V_{ref}}{I_{FS}} \times C_F, \tag{2}$$

$$T_{int} = \frac{0.96 \times 5.0}{100 \times 10^{-9}} \times 87.5 \times 10^{-12} = 4.2 \text{ ms}$$

This time is important not only for the integration; from Fig. 10 we can see that in order to retrieve the old data before the next data is digitized, 40 clock pulses must be generated during this time. Hence, the minimum clock rate on the synchronous serial interface is

$$f_{DCLK,min} = \frac{40}{0.0042} = 9523 \text{ Hz}$$

In order to allow for some overhead, the DCLK clock rate will be at least 20 kHz.

2.3. The Microcontroller Design

The two DDC112 charge integrating chips are connected to an 8-bit PC16F1779 microcontroller [8] as described in Fig. 12.

Notice in Fig. 12 how the two DDC112 chips have been cascaded; the data of the first one will be shifted through the second chip. The data is transferred to the microcontroller via a synchronous serial interface and from Fig. 10 it is clear that signals DXMIT, DCLK and DOUT complies with a standard SPI interface (or a clocked USART interface).

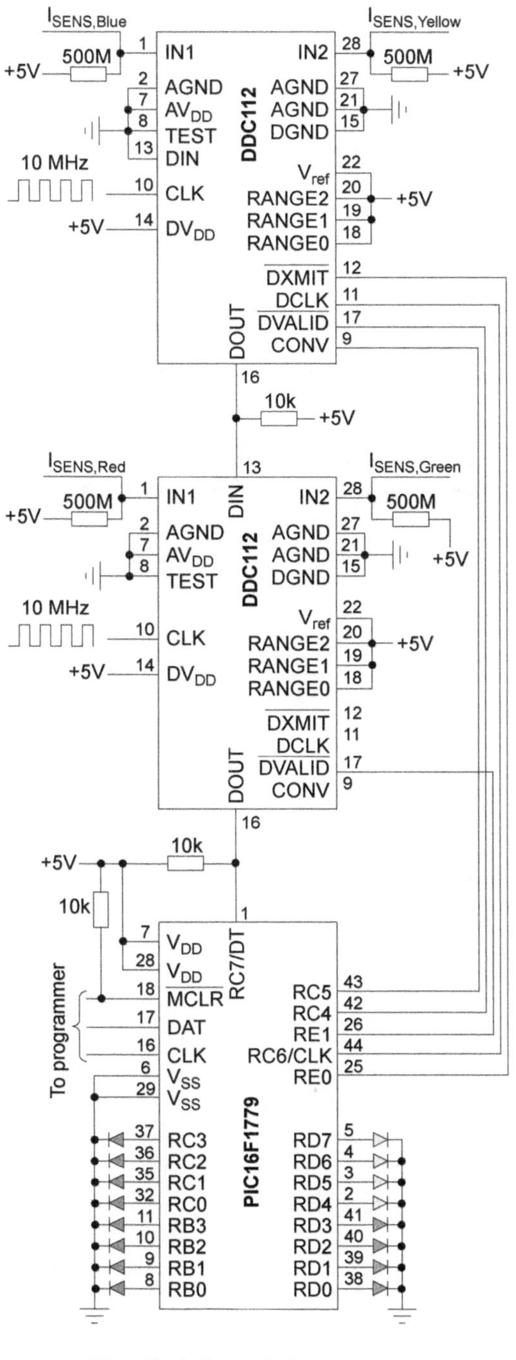

Fig. 12. 4-channel pico ammeter.

In this first prototype, data will be translated to different diode levels, four on each collimator plate, and the high-resolution DDC112 chip may appear to be an overkill. However, the use of the DDC112 chips makes the design easier (compared to using op amps) and since they make the absolute current values available (high-resolution), it would be easy to modify the design to either display them locally on an LCD or transfer them to a host computer. Primarily, knob will be added to allow for at least three different full-scale ranges (50 nA, 5 nA and 1 nA).

In all sub-nano current designs, it is important to protect the inputs from stray wire capacitances. In order to facilitate an expedient PCB layout for this purpose, the two analog inputs are located at the top of the DDC112 chip and the next two pins are analog ground [3]. This makes it easy to surround the analog inputs by a shielding copper layer on the PCB [4] as illustrated in Fig. 13.

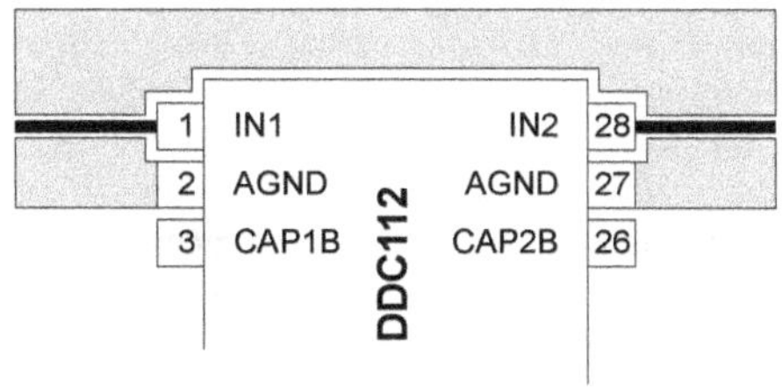

Fig. 13. Input signals are protected from stray capacitances.

3. Related/Previous Work

The DDC112 chip is a common solution to low-current/low-charge problems. It has been used in precise current integrators in scanning tunneling microscopes and atomic force microscopy [9], measuring sub-picoamp currents from Faraday cup detector [10], ion beam monitoring in deep proton lithography [11] and as photometric analysers in optical telescopes [12]. The idea explored in this work, where an ion optics collimator plate is split into four symmetric pieces, was first suggested by Snowden and Barber [13] but their application was for microamp currents and used shunting resistors in a bridge to detect the current. The main difference of this work is that it utilizes the versatile DDC112 chip (based on charged integration), the option to transfer absolute current values to a host computer and the convenient visual indication of the beam position fascillitated by the front panel LEDs.

4. Conclusions

A microcontroller based, 4-channel pico ammeter is being developed for the purpose of facilitating convenient line up of ion beams in particle physics experiments at the GANDALPH setup at the CERN-ISOLDE facility. Instead of devolving to discrete feedback opamp solutions, two, 2-channel charge integrating chips are employed. This solution is less expensive, more robust, simplifies PCB layout and facilitates a digital interface for retrieving the absolute current value.

A prototype will be presented in October 2019 and beta testing will be performed in the Atomic Physics Laboratories in Gothenburg during the fall before the instrument is shipped to the CERN ISOLDE facility later this year.

References

[1]. D. A. Skoog, J. J. Leary, Principle of Instrumental Analysis, 4th Ed., *Saunders College Publishing*, Orlando, Florida, 1992, pp. 429-432.

[2]. Tektronix Inc., Keithley Series 6400 Picoammeters, https://www.tek.com/keithley-low-level-sensitive-and-specialty-instruments/keithley-series-6400-picoammeters

[3]. Low Level Measurements Handbook, 6th Ed., *Keithley Instruments Inc.*, Cleveland, Ohio, 2004.

[4]. Burr-Brown, DDC112: Dual Current Input 20-bit Analog-to-Digital Converter, *Texas Instruments*, 2004.

[5]. A. A. Mortuza, F. Pervez, K. Hossain, S. K. Sen, N. H. Mia, M. K. Basher, S. Alam, Pico-current measurement challenges and remedies: A Review, *Universal Journal of Engineering Science*, Vol. 5, Issue 4, 2017, pp. 57-63.

[6]. J. Todsen, Retrieving Data from the DDC112, *Burr-Brown Corporation*, Tucson, Arizona, 1998.

[7]. J. Todsen, Creating a Bipolar Input Range for the DDC112, *Burr-Brown Corporation*, Tucson, Arizona, 1999.

[8]. PIC16(L)F1777/8/9, *Microchip Technology Inc.*, Tucson, Arizona, 2015-2016.

[9]. K. Raczkowski, T. Piasecki, M. Rudek, T. Gotszalk, Design and evaluation of precise current integrator for scanning probe microscopy, *Measurement Science and Technology*, Vol. 28, Issue 3, 2017, pp. 1-9.

[10]. S. Agosteo, E. Borsato, F. Corso, A. Fazzi, F. Gonella, M. Intrioni, I. Lippi, M. Lorenzoli, L. Modense, F. Montecassiano, M. Pegoraro, A. Pola, V. Varoli, P. Zotto, Performance of a proton irradiation chamber, *Nucl. Instr. and Methods in Phys. Res. Sec. A: Accelerator, Spectrometers, Detectors and Associated Equipment*, Vol. 664, Issue 1, 2012, pp. 193-202.

[11]. P. Vynck, B. Volckaerts, M. Vervaeke, H. Ottevare, P. Tuteleers, L. Cosentino, P. Finchiaro, A. Pappalardo, A. Hermanne, H. Thienpont, Beam monitoring enhances Deep Proton Lithography: Towards high-quality micro-optical components, in *Proceedings of the Symposium IEEE/LEOS Benelux Chapter*, Amsterdam, 2002.

[12]. R. S. Romaniuk, Space and high energy experiments advanced electronic systems, *Int. J. Electron. Telecomm.*, Vol. 58, Issue 4, 2012, pp. 441-462.

[13]. K. Snowdon, C. Barber, Simple accelerator beam position monitor, *J. Phys. E: Sci. Instr.*, Vol. 12, Issue 10, 1979, 923.

[14]. Ultralow Input Bias Current Operational Amplifier, Rev. H, *Analog Devices Inc.*, Norwood, Maine, 2008.

[15]. Monolithic Precision, Low Power FET-Input Electrometer Op Amp, Rev. A., *Analog Devices Inc.*, Norwood, Maine, 1997.

[16]. Microchip Technology Inc., PIC16(L)F1516/7/8/9, 2016, http://ww1.microchip.com/downloads/en/DeviceDoc/40001452F.pdf

5th International Conference on Sensors Engineering and Electronics Instrumentation Advances (SEIA' 2019), 25-27 September 2019, Canary Islands (Tenerife), Spain

(007)

Femtosecond Laser Inscribed Tilted In-fiber Beam Splitter Used for Reflective Mach-Zehnder Interferometer

Hua Zhang [1], D. N. Wang [1] and Chunliu Zhao

[1] China Jiliang University, College of Optical and Electronic Technology, Hangzhou, China

Tel.: + 8657186835782, fax: + 8657186835752

E-mail: dnwang@cjlu.edu.cn

Summary: We propose a reflective in-fiber Mach–Zehnder interferometer, which is based on femtosecond laser inscribed tilted beam splitter in single-mode fiber (SMF). The incident light propagating in the fiber core is split into two parts by the beam splitter; one passes through the beam splitter and travels along the fiber core. The other part is partially reflected by the in-fiber beam splitter to the cladding-air interface, where it experiences total internal reflection and is directed back into the fiber core. The two parts of light beams meet and interfere before being reflected back into fiber core at the fiber end face. The device demonstrated is compact, robust, easy-fabricated and can be used in environmental refractive index (RI) sensing.

Keywords: Fiber beam splitter, Mach-Zehnder, Sensor.

1. Introduction

Optical fiber Mach-Zehnder interferometer (MZI) is of particular interest for many biochemical and environmental sensing applications due to its high sensitivity, simplicity and corrosion resistance [1].

The fiber in-line MZI interferometers commonly utilize a transmission type structure [2-5], where the sample needs to be placed between the light source and the detection end during the system test, which imposes a limit of application in a narrow and enclosed space. A "reflective" MZI structure is then needed for the detection of samples in a small space.

Here we propose and demonstrate an in-fiber reflective MZI based on femtosecond laser inscribed tilted beam splitter in SMF. The incident light propagating in the fiber core is split into two parts by the beam splitter; one passes through the beam splitter and travels along the fiber core. The other part is partially reflected by the in-fiber beam splitter to the cladding-air interface, where it experiences total internal reflection and is directed back into the fiber core. The two parts of light beams meet and interfere before being reflected back into fiber core at the fiber end face. The device demonstrated is compact, robust, easy-fabricated and can be used in environmental refractive index (RI) sensing.

2. Principle of Operation

The incident light beam propagating in the fiber core is reflected is split into two parts by the beam splitter, as shown in Fig. 1, where L, d and z [$z = 2d\tan(90° - 2\theta)$] are the length of the tilted beam splitter, the distance between the center of fiber core and the outer edge of the cladding and the distance between the center of the beam splitter and the fiber end face, θ and ϕ [$\phi \geq arcsin(n_{Air}/n_{Clad})$] are the angle between the beam splitter and the central fiber axis, and the incident angle at the interface of fiber cladding and air, respectively. The light intensity can be written as:

$$\mathrm{I} = I_1 + I_2 + +2\sqrt{I_1 I_2}\cos\left[\frac{2n_{Clad}{}^{d}/_{\cos\phi} - n_{Co}z}{\lambda}\right], \quad (1)$$

where I_1 and I_2 are the intensities of the two reflected light beams, n_{Co}, n_{Clad} are the RIs of SMF core and cladding respectively.

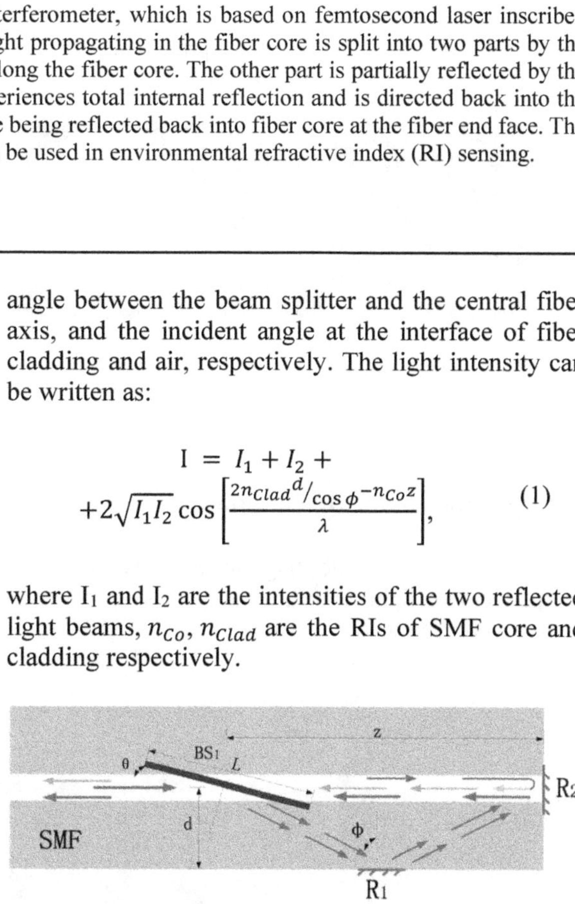

Fig. 1. Operation principle of the reflective MZI.

3. Device Fabrication

The fabrication process for the proposed device is shown in Fig. 2(a). The SMF is placed on a laser micromachining platform. Firstly, the beam splitter having a specific tilted angle θ = 16° with the central axis of the fiber was engraved by a femtosecond laser operating at the central wavelength of 800 nm, with pulse energy of 500 μJ and pulse duration of 35 fs, at 5 kHz repetition rate and set in a 10 μm/s writing speed, focused through an objective lens of 20× with a numerical aperture (NA) of 0.4, and the length of the beam splitter was ~127 μm. Three layers of oblique lines with a depth of 4 μm were engraved. Each layer consists of five inclined lines with a spacing of 1 μm.

Finally, the fiber is precisely cut by a femtosecond laser to ensure the exact distance (~200 μm) between the center of the beam splitter and the end face of the fiber, as shown in Fig. 2(b). The microscope images of the beam splitter and fiber end are displayed in Fig. 3(a) and 3(b), respectively.

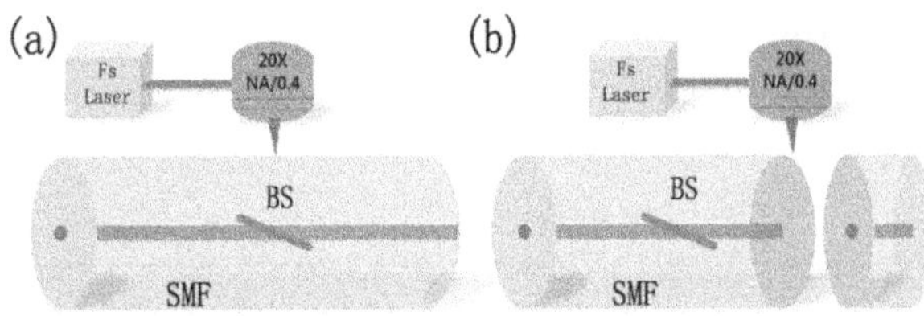

Fig. 2. Femtosecond laser fabrication process of the tilted beam splitter in SMF with 20X lens.

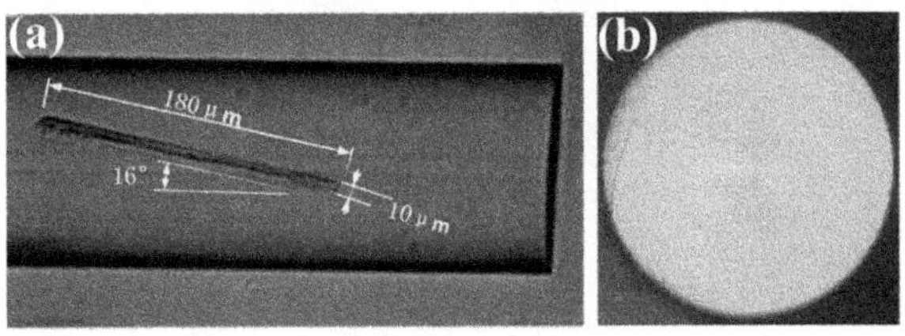

Fig. 3. Device samples used in sensing experiments (a) Microscope image of tilted beam splitter. (b) Fiber end view.

4. Experiment and Discussion

Fig. 4 is a schematic diagram of an experimental setup. Incident light from a broadband source (BBS) with a wavelength range from 1452 to 1652 nm is introduced to the fiber with the MZI structure via a circulator, and the reflected interferometer output is directed to an optical spectrum analyzer (OSA) with a resolution of 0.02 nm to record its spectrum.

In the temperature test, the fiber device sample is placed into an electric oven and increase the temperature from 25 °C to 70 °C gradually in steps of 5 °C to measure its temperature response. The temperature sensitivity obtained from the linear fit within the range between 20 °C and 75 °C is 9.91 pm/°C.

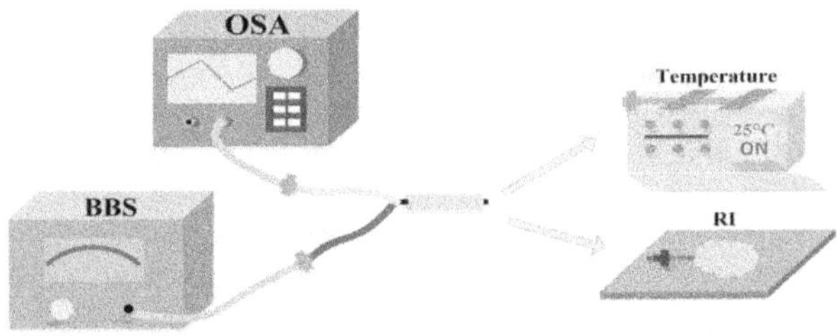

Fig. 4. Schematic diagram of the experimental system for temperature sensing test and RI sensing test.

In the testing of the device response to RI change. The device sample was immersed into a series of index matching oils (Cargille) at a constant temperature of 20 °C, ranging from 1.30 to 1.38 in steps of 0.01. The relationship between spectral intensity and RI is depicted in Fig. 8(b), where a good linearity can be found and the sensitivity obtained is -50.24 dB / RIU (RI units).

Thus, the simple structured device can be used for RI sensing in a convenient manner.

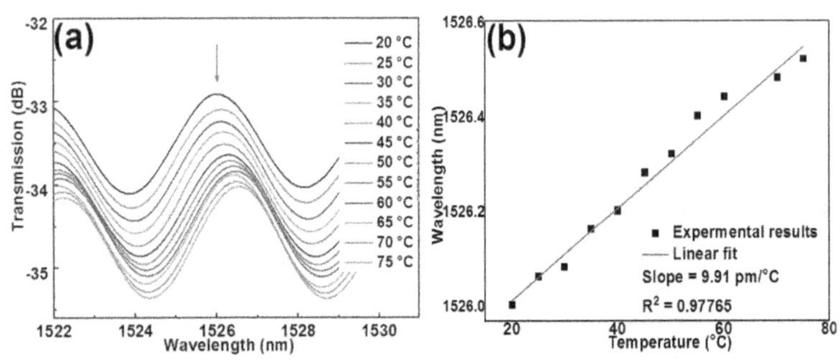

Fig. 5. The reflection spectrum and a linear relationship between shift wavelength dip and temperature.

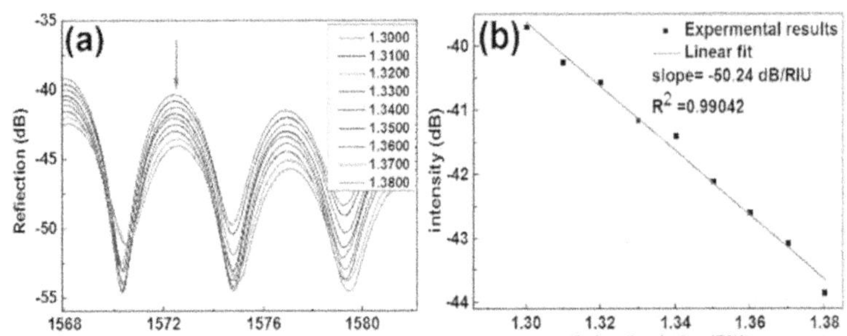

Fig. 6. The reflection spectrum and a linear relationship between shift wavelength peak and RI.

4. Conclusions

In conclusion, a reflective MZI interference structure is proposed and demonstrated, which is formed by a tilted beam splitter fabricated by femtosecond laser. Such a device can be used for environmental RI measurement, with sensitivities – 50.24 dB/RIU. The device is compact in size, robust in mechanical strength, easy to fabrication, and convenient in operation, which makes it attractive for many sensing applications.

Acknowledgements

We acknowledge the support from the National Natural Science Foundation of China (Grant No. 61661166009).

References

[1]. B. H. Lee, Y. H. Kim, K. S. Park, J. B. Eom, M. J. Kim, B. S. Rho, H. Y. Choi, Interferometric fiber optic sensors, *Sensors (Basel)*, Vol. 12, Issue 3, 2012, pp. 2467-2486.

[2]. J. H. Lim, H. S. Jang, K. S. Lee, J. C. Kim, B. H. Lee, Mach-Zehnder interferometer formed in a photonic crystal fiber based on a pair of long-period fiber gratings, Opt. Lett., Vol. 29, 2004, pp. 346-348.

[3]. Z. B. Tian, S. S. H. Yam, H-P Loock, Single-mode fiber refractive index sensor based on core-offset attenuators, *IEEE Photon. Technol. Lett.*, Vol. 16, 2008, pp. 1387-1389.

[4]. T. Y. Hu, Y. Wang, C. R. Liao, D. N. Wang, Miniaturized fiber in-line Mach-Zehnder interferometer based on inner air-cavity for high-temperature sensing, *Opt. Lett.*, Vol. 37, 2012, pp. 5082-5084.

[5]. W. W. Li, W. P. Chen, D. N. Wang, Z. K. Wang, B. Xu, Fiber inline Mach-Zehnder interferometer based on femtosecond laser inscribed waveguides, *Opt. Lett.*, Vol. 42, 2017, pp. 4438-4441.

(008)

On-fiber Polymer Whispering Gallery Mode Resonator Microprinted by Femtosecond Laser

Changrui Liao [1], Zhengyong Li [1], Lei Xu [2], Ying Wang [1], Jun He [1], Zhiyong Bai [1], Shen Liu [1], Yiping Wang [1]

[1] Key Laboratory of Optoelectronic Devices and Systems of Ministry of Education and Guangdong Province, College of Physics and Optoelectronic Engineering, Shenzhen University, Shenzhen 518060, China

[2] School of Electronic and Communication Engineering, Shenzhen Polytechnic, Shenzhen 518060, China

E-mail: cliao@szu.edu.cn

Summary: We demonstrate a novel method for producing whispering gallery mode resonators on a microfiber using femtosecond laser two-photon polymerization technology. The typical resonance was observed by coupling the microresonator with a tapered fiber in a region where they were physically separated. Unlike conventional WGM devices, this photosensitive resin-based resonator was deposited on a microfiber, whereby the structural integration was improved.

Keywords: Whispering gallery mode, Optical fiber devices, Femtosecond lase microprint.

1. Introduction

The physical description of whispering gallery modes (WGMs), first introduced by Rayleigh, relied on acoustic ray interpretation. This included extensive investigations of multiple internal reflections along the inner gallery boundary [1]. Mie [2] and Debye [3] provided the first in-depth theoretical analysis of optical WGM resonators, based on Maxwell's equations, at the turn of the century. After several decades of development, WGM theory has become more systematic and accurate. The unique spectral properties of WGMs, including narrow line widths, tunability, high Q-factors, and high stability under environmental conditions [4, 5] have attracted significant interest in recent years. WGM resonators have the ability to confine and couple light using total internal reflection. As such, a robust and efficient coupling methodology is critical for optimizing resonator performance.

In this study, we experimentally demonstrate a photosensitive resin-based WGM resonator with a Q-factor of 1.9×10^3. The ring resonator was printed on a microfiber using Fs laser two-photon polymerization (TPP) technology. The resulting microfiber exhibited stronger evanescent fields, which coupled light into the ring resonator. The light was confined by continuous total internal reflection (TIR), resulting from the high refractive index (RI) of the photosensitive resin compared with the surrounding media. This TIR process excited WGMs in the resonator cavity.

2. Operation Principle

When the SMF is tapered to several micrometers, the diameter of the guiding mode is larger than the tapered region, which can let out a part of light propagating along the surface of the microfiber. As such, microfibers are more suitable for coupling light into resonators. As shown in Fig. 1, a circular ring was deposited on the microfiber. Input light propagates along the fiber core and leaked light enters the circular ring, inducing resonance. Finally, a fraction of this light recouples into the microfiber. The resulting WGM resonance spectrum was measured using an optical spectrum analyzer. In order to guarantee the light resonating in the ring cavity the phase change for the light should be an integer multiple of 2π:

$$2\pi nr = m\lambda, \tag{1}$$

where n and r are the RI and radius of the ring resonator, respectively. Rearranging this equation gives the free spectral range (FSR), which can be written as

$$FSR = \frac{\lambda^2}{2\pi nr} \tag{2}$$

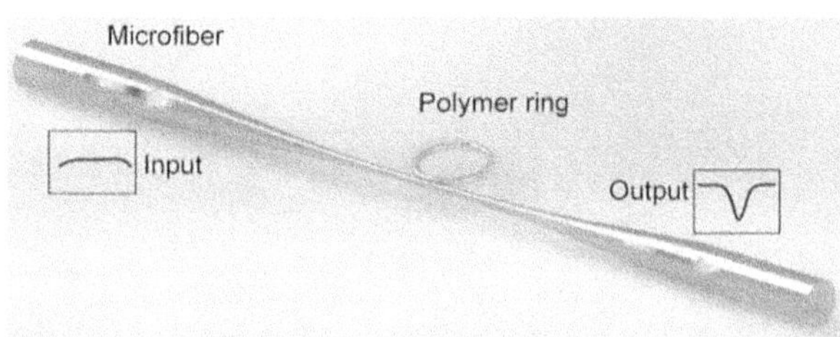

Fig. 1. Schematic diagram of the on-fiber polymer whispering gallery mode resonator.

3. Device Fabrication

The whole fabrication process includes two primary steps: drawing the microfiber and fabricating the ring resonators using Fs laser TPP technology. The SMF (Corning, SMF-28) used in these experiments has the core and cladding diameters of 8 and 125 μm, respectively. We designed and constructed a fiber-tapped system capable of producing submicron-diameter microfibers. In this experiment, we draw the microfiber with diameters of 2-4 μm and a homogeneous area length of ~1 cm. The microfibers were then immobilized on magnesium fluoride glass instead of silica glass. Because its RI is lower than the fiber, thereby limiting light propagating along the surface of the microfiber.

Fs laser TTP is a process in which a monomer is converted into a polymer network and cross-linking of polymer chains is introduced by free radical polymerization. The exposed region is then insoluble in the solvent and the microstructures can be printed on the substrate. The Fs laser beam is firstly focused on the microfiber and then shifted to the edge of microfiber. The power density used in the experiments is 20 mW/μm^2. Then the microfiber is translated according to designed track and velocity. When the laser fabrication is completed, the cover slide above the microfiber is carefully taken away and redundant resin can then be rinsed using isopropanol. A scanning electron microscope (SEM) was used to image the polymer structures, as shown in Fig. 2. The line width of the ring resonator was approximately 2 μm.

Fig. 2. SEM images of the polymer ring resonator.

4. Results and Discussion

Fig. 3 shows the transmission spectra of a ring resonator with a diameter of 20 μm. A series of resonance peaks are clearly present at long wavelengths. The corresponding FSR was measured to be 25.4 nm. According to the Eq. (2), the theoretical FSR was calculated to be 25.2 nm, which is in excellent agreement with experimental results. The Q-factor of the ring resonators were calculated to be 77 and at 1540 nm.

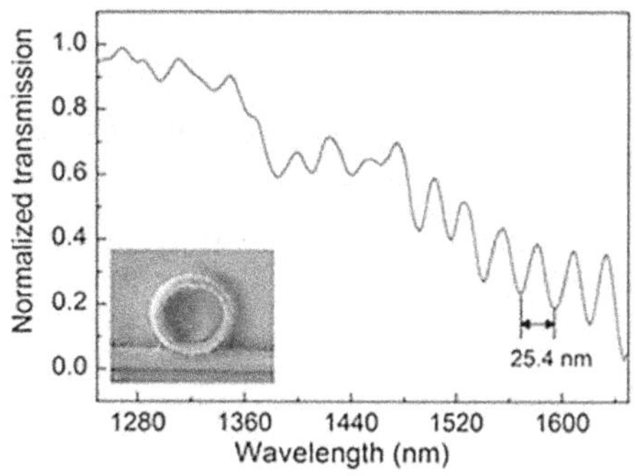

Fig. 3. The normalized transmission spectrum of the ring resonator as a function of wavelength.

5. Conclusion

We demonstrated an effective method for microprinting WGM resonators using Fs laser TPP technology. WGM resonance was observed as a polymerized ring resonator deposited on a microfiber. The ring resonator was experimentally investigated. The size and the material the resonators could influence the Q-factor.

Acknowledgements

National Natural Science Foundation of China (NSFC) (61575128); Natural Science Foundation of Guangdong Province (2018B030306003); Science and Technology Innovation Commission of Shenzhen (KQJSCX20170727101953680).

References

[1]. A. N. Oraevsky, Whispering-gallery waves, *Quantum Electronics*, Vol. 32, Issue 5, 2002, pp. 377-400.

[2]. G. Mie, Beiträge zur Optik trüber Medien, speziell kolloidaler Metallösungen, *Ann. Phys.*, Vol. 330, 1908, pp. 377-445.

[3]. P. Debye, Der lichtdruck auf kugeln von beliebigem material, *Ann. Phys.*, Vol. 335, 1909, pp. 57-136.

[4]. M. R. Foreman, J. D. Swaim, F. Vollmer, Whispering gallery mode sensors, *Advances in Optics and Photonics*, Vol. 7, 2015, pp. 168-240.

[5]. G. P. Lin, A. Coillet, and Y. K. Chembo, Nonlinear photonics with high-Q whispering-gallery-mode resonators, *Advancs in Optics and Photonics*, Vol. 9, Issue 4, 2017, pp. 828-890.

(009)

Microfiber/SiC-nanowire Coupler for All-optical UV Photodetection

Ying Wang, Changrui Liao, Longfei Zhang, Jun He, Zhiyong Bai, Shen Liu, Yiping Wang
Key Laboratory of Optoelectronic Devices and Systems of Ministry of Education and Guangdong Province, College of Physics and Optoelectronic Engineering, Shenzhen University, Shenzhen 518060, China
E-mail: yingwang@szu.edu.cn

Summary: Semiconductor nanowires are traditionally used for photodetection based on electrical test methods, which are inevitably disturbed by dark current noise. We propose an all-optical photodetection scheme based on a microfiber/SiC-nanowire directional coupling structure, which directly utilizes the refractive index change of SiC nanowire caused by photogenerated carriers instead of the change of photocurrent. The device exhibits a sensitivity of 103 pm/(W/cm^2) when irradiated by a 266 nm UV laser.

Keywords: SiC nanowire, All-optical photodection, Microfiber, Photogenerated carriers, Ultraviolet (UV).

1. Introduction

In recent years, various photodetectors based on semiconductor nanomaterials have shown broad application prospects, such as organic-inorganic hybrid perovskite material [1], ZnO nanowires [2], and black phosphorus [3]. Among these emerging nanomaterials, SiC nanostructured materials have received more and more attention due to their excellent physical and chemical properties [4, 5]. Compared to traditional narrow-bandgap material based detectors, such as Si detectors and other commercial semiconductor detectors, SiC-based photodetectors can operate stably under high temperature and harsh conditions without the need of extra protection [6]. At the same time, SiC can work normally under the command operation since it has a large electron saturation speed [7]. However, the intrinsic carrier concentration of SiC is on the order of about 10^{16}~10^{18} cm^{-3}, which is higher than that of Si (~10^{10} cm^{-3}) by about 6 orders of magnitude. This causes very high dark current of the photodetector, or very poor signal-to-noise ratio, and limits the practical application of SiC based photodetectors. In order to overcome the problem mentioned above, we propose a microfiber/SiC-nanowire directional coupler (MSNDC) for all optical photodetection, of which the response is not a photocurrent but an optical wavelength shift, naturally avoiding the influence of dark current. When illuminated by a 266-nm deep ultraviolet (UV) laser, results show that a sensitivity of about 103 pm/(W/cm^2) can be obtained.

2. Device Fabrication

The SiC nanowires used in the experiment are commercial available. A single SiC nanowire is directly attached to the uniform waist region of a microfiber by van der Waals force or electrostatic force. As subwavelength-diameter microfiber is needed for single-mode operation, an oxyhydrogen flame-heated cyclic taper system is used to gradually reduce the diameter of a single-mode fiber (SMF, SMF28e, Corning Inc.) to ~1 μm. The length of the uniform waist region is ~15 mm. A single SiC nanowire is picked up and transferred to the microfiber by a tungsten probe. The structure of the MSNDC is shown in Fig. 1(a). Figs. 2(b) and 2(c) show the optical microscopic image and SEM image of the MSNDC, respectively.

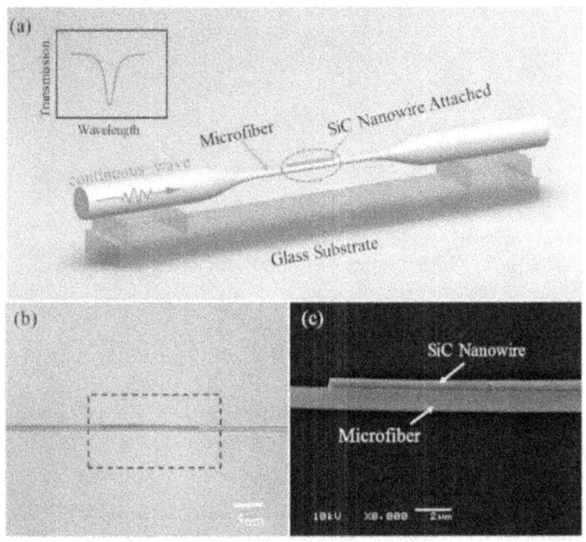

Fig. 1. Schematic illustration of the MSNDC all-optical photodetector. (b) Microscopic and (c) SEM image of the device.

3. Results and Discussion

The principle of its operation lies in the fact of guiding modes coupling between microfiber and SiC nanowire. In detail, as excellent optical waveguides, microfiber and SiC nanowire support the propagation of optical guiding modes in the near-infrared. The number of guiding modes is related to the diameter of the waveguide and the refractive index (RI) of the waveguide material. Once the SiC nanowire and microfiber are placed nearby in parallel, resonant dips originated from the mode coupling between the microfiber and SiC nanowire can be observed under phase-matching conditions in the transmission spectrum of the microfiber in MSNDC. Fig. 2 shows the transmission spectrum of two MSNDCs with SiC nanowire diameter of 652 nm and 703 nm, respectively.

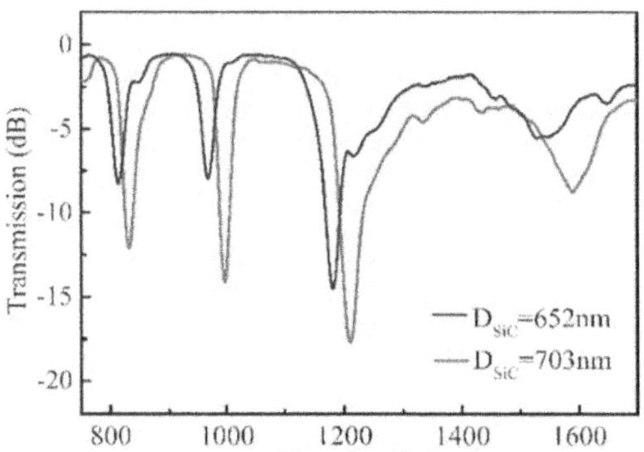

Fig. 2. Transmission spectra of two MSNDC devices, where the diameters of the SiC nanowires are 652 nm and 703 nm, respectively, and the microfiber diameter is ~1 μm.

When 266 nm UV light is irradiated onto the SiC nanowire of device, a large number of photogenerated carriers are induced, which will change the RI of SiC nanowire and lead the resonance dip shift of MSNDC. A schematic diagram of the experimental setup for all optical photodetection is shown in Fig. 3. The input of the MSNDC is connected to a supercontinuum white-light source, the output is connected to an optical spectrum analyzer, and a 266 nm single frequency deep UV laser is employed to irradiate the MSNDC through a focusing lens, whose focal length is 10 cm.

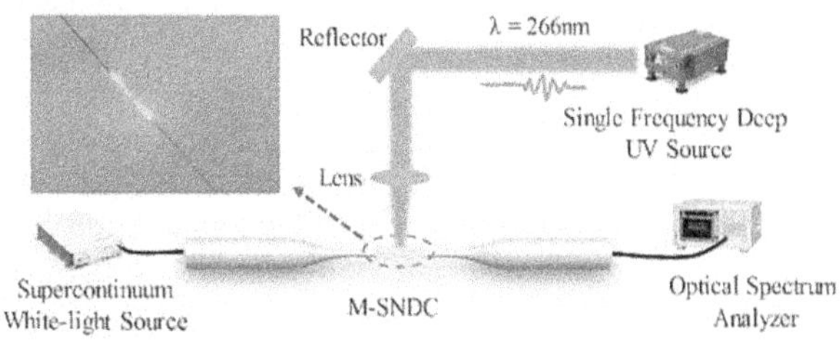

Fig. 3. The experimental setup for all-optical photodetector.

When MSDNC is irradiated by UV light, the resonance dip exhibits redshift with the laser power density increases, as shown in Figs. 4(a) and 4(c). Figs. 4(b) and 4(d) show the relationship between the resonant wavelength and the UV light intensity, and a sensitivity of approximately 103 pm(W/cm^2) and 73 pm(W/cm^2) can be obtained for lower order mode resonance and higher order mode resonance, respectively, through linear fitting.

The wavelength shift of the device is attributed to the RI change caused by photogenerated carriers in SiC nanowire. The existence of carriers in semiconductors causes large changes in absorption coefficients and can easily produce a considerable change in RI according to the Kramers-Kronig relationship. With the increase of carrier concentration under UV illumination, the band gap of SiC nanowire decreases gradually, resulting changes in absorption of SiC materials. This causes an RI increase of SiC nanowire and changes the phase-matching condition of MSNDC device, which produces a redshift of the resonance dipof the device.

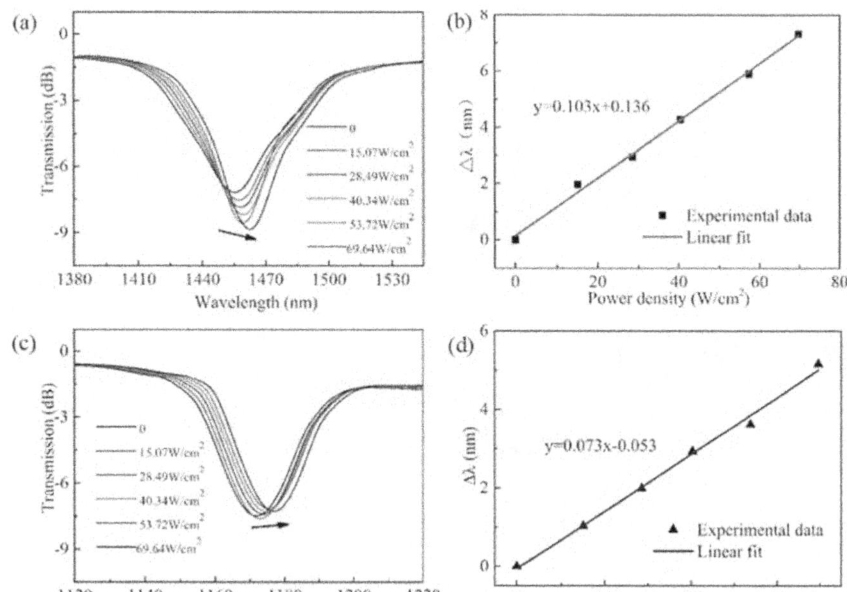

Fig. 4. (a) Transmission spectra and (b) resonant wavelength shift for lower order mode resonance at different light intensities. (c) Transmission spectra and (d) resonant wavelength shift for higher order mode resonance at different light intensities.

5. Conclusion

This paper proposes an all-optical photodetection scheme based on a MSNDC structure. The change in the RI of the SiC nanowire caused by the photogenerated carriers is utilized to achieve phtotodetection, instead of the change in the photocurrent, thereby effectively avoiding dark current noise. When the MSNDC is irradiated by a deep UV laser with a wavelength of 266 nm, its sensitivity reaches up to 103 pm/(W/cm^2). The proposed device does not require electrode integration and extra surface functionalization or modification of nanowires, which is very simple to fabrication and exhibits good performance. It shows good potential in all-optical photodetection, all-optical modulation and fiber lasers.

Acknowledgements

National Natural Science Foundation of China (NSFC) (61675137); Science and Technology Innovation Commission of Shenzhen (JCYJ20170818093743767).

References

[1]. L. Dou, Y. Yang, J. You, Z. Hong, W.-H. Chang, G. Li, Y. Yang, Solution-processed hybrid perovskite photodetectors with high detectivity, *Nature Communications*, Vol. 5, 2014, 5404.

[2]. X. Zhang, B. Liu, W. Yang, W. Jia, J. Li, C. Jiang, X. Jiang, 3D-branched hierarchical 3C-SiC/ZnO heterostructures for high-performance photodetectors, *Nanoscale*, Vol. 8, 2016, pp. 17573-17580.

[3]. M. Huang, M. Wang, C. Chen, Z. Ma, X. Li, J. Han, Y. Wu, Broadband Black-Phosphorus Photodetectors with High Responsivity, *Advanced Materials*, Vol. 28, 2016, pp. 3481-3485.

[4]. K. Zekentes, K. Rogdakis, SiC nanowires: Material and devices, *Journal of Physics D: Applied Physics*, Vol. 44, 2011, 133001.

[5]. T. Yang, L. Zhang, X. Hou, J. Chen, K.-C. Chou, Bare and boron-doped cubic silicon carbide nanowires for electrochemical detection of nitrite sensitively, *Scientific Reports*, Vol. 6, 2016, 24872.

[6]. W. Lien, D. Tsai, S. Chiu, D. G. Senesky, R. Maboudian, A. P. Pisano, J. He, Low-temperature, ion beam-assisted SiC thin films with antireflective ZnO nanorod arrays for high-temperature photodetection, *IEEE Electron Device Letters*, Vol. 32, 2011, pp. 1564-1566.

[7]. H. Morkoç, S. Strite, G. B. Gao, M. E. Lin, B. Sverdlov, M. Burns, Large-band-gap SiC, III-V nitride, and II-VI ZnSe-based semiconductor device technologies, *Journal of Applied Physics*, Vol. 76, 1994, pp. 1363-1398.

(010)

SAW Sensors Based on Nanoparticles Embedded in Polymer for VOCs Detection

C. Viespe, I. Constantinoiu, G. Popescu-Pelin
National Institute of Laser Plasma and Radiation Physics, Laser Department, Magurele, Romania
Tel.: + 40214574027, fax: + 40214574027
E-mail: cristian.viespe@inflpr.ro

Summary: Surface acoustic wave sensors (SAW) with two types of polymer sensing films, polyethyleneimine (PEI) and polydimethylsiloxane (PDMS), containing embedded ZnO, TiO_2 and WO_3 nanoparticles for detecting volatile organic compounds (VOCs) were fabricated and studied. The nanoparticles were obtained by pulsed laser ablation, using the same conditions for all types of materials. After incorporation of nanoparticles into the polymer, the suspension obtained was deposited by an airbrush method on the quartz substrate of the sensor. Sensors were tested for four types of VOCs: ethanol, toluene, acetone and dichloroethane. Those with PEI sensitive films have shown a superior sensitivity than those with PDMS. It has also been found that the sensors with WO_3 nanoparticles had the best results for almost all types of gas.

Keywords: Gas sensor, Nanoparticles, SAW, Sensitivity.

1. Introduction

Human health is a subject of great interest to our society today. It depends to a large extent on the quality of the environment in which it carries out its activities. An environment with a polluted atmosphere leads to the occurrence of lung diseases, skin diseases or other more dangerous diseases such as cancer and diabetes [1-2]. By monitoring air quality in the atmosphere, a number of such diseases can be prevented. In this context, the development of sensors capable of detecting substances harmful to the body at ambient temperature becomes a topical and a high interest field.

VOCs are part of a class of carbon-based chemical substances that volatilize very easily at room temperature and pressure. They come into the atmosphere through industrial processes, building materials, household or personal cleaning products, cosmetics, paints, varnishes, waxes, etc.

The SAW sensor is characterized by very good sensitivity, selectivity, reversibility, short response time, low cost, ease of manufacturing and wireless operation [3-4].

Use of composites with polymers and nanoparticles offers better properties to sensors, due to the synergistic effect of polymer and nanoparticles sensitivity. The sensitivity of polymeric films is explained by the large number of functional groups linked in the polymeric chain that can interact with the gas molecules.

The nanoparticles obtained by pulsed laser deposition have high purity and allow control of their morphology. Airbrush deposition method of polymer and nanoparticles suspension contributes to the uniform distribution on the surface of the sensor and also favors the more rapid evaporation of the solvent [5].

2. Materials and Methods

The SAW sensor is a "delay-line" type, designed as a two port-resonator and fabricated on a ST-X cut quartz substrate with an oscillating frequency of $\approx$ 69 MHz. The gold interdigital transducers (IDT) (150 nm) were deposited by photolithographic techniques onto chromium layers (10 nm), to ensure the adhesion of gold to the quartz. The configuration of IDT was "double-double", which consist in 50 electrodes pairs, 2500 µm acoustic aperture and 45.2 µm wavelengths.

The nanoparticles were produced by laser ablation using an Nd-YAG laser (EKSPLA NL301HT) having pulse durations of 5 ns and a repetition rate of 10 Hz. A laser wavelenght of 532 nm, 160 mJ energy, 14.5 J/cm2 fluence, was used to obtain the nanoparticles in an irradiation chamber with a controlled O_2 atmosphere.

Table 1. SAW sensors.

Sensor	Polymer	Solvent	Nanoparticles
S1	PEI	Ethanol	-
S2	PEI		ZnO
S3	PEI		TiO_2
S4	PEI		WO_3
S5	PDMS	Toluene	-
S6	PDMS		ZnO
S7	PDMS		TiO_2
S8	PDMS		WO_3

The sensitive layers, S1 and S5, were obtained mixing PEI with ethanol and PDMS with toluene, respectively in a concentration of 10 mg/ml (Table 1). Sensitive films for the other sensors (S2, S3, S4, S6, S7, S8) were obtained by mixing an amount of nanoparticles with the polymer solution in the above concentration, so as to obtain a concentration of 0.8 mg/ml. The suspension obtained was sonicated, then deposition was performed. The thickness of the layers, about 200-300 nm, was determined by WLI (White Light Interferometry).

3. Results and Discussions

Nanoparticle synthesis was performed for all types of oxides under the same conditions. In SEM images (Fig. 1) can be observed agglomerates of spherical nanoparticles of different sizes, depending on the nature of the material from which they are made. Thus, ZnO nanoparticles (Fig. 1, A) had dimensions between 10-20 nm, and those of WO_3 between 10-30 nm (Fig. 1, C). In terms of TiO_2, the obtained particles had not only nanometric dimensions but also micrometric dimensions. Instead, on grains with micrometric dimensions can be observed the presence of 10-20 nm nanoparticles. (Fig. 1, B).

Fig. 1. SEM image for ZnO (A), TiO2 (B), WO_3 (C) nanoparticles synthesized by PLD.

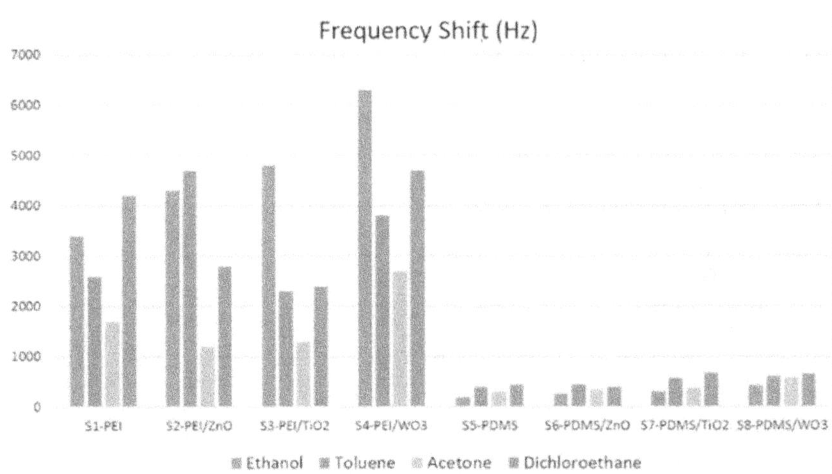

Fig. 2. Frequency shift of sensors at 800 ppm concentration of ethanol, toluene, acetone and dichloroethane.

Sensor responses were also influenced by the type of polymer used. From Fig. 2, which is the response of the sensors to the four gases in a concentration of 800 ppm, it can be observed a major difference in the frequency shift, for each sensor. For PEI sensors, the frequency shift is about 10 times higher than the frequency shift offered by the PDMS sensors.

4. Conclusions

SAW sensors with composite sensitive films of polymer and nanoparticles were fabricated for VOCs (ethanol, toluene, acetone and dichloroethane) detection. Oxide nanoparticles were obtained by laser ablation and their morphology was investigated by scanning electron microscopy. ZnO and WO_3 nanoparticles were obtained, with dimensions of approximately 10-30 nm, and for TiO_2 both nanometric and micrometric particles. The concentration of nanoparticles and polymer suspension was 0.8 mg/ml and it was deposited on the quartz substrate of the sensor by airbrush method. The results indicated that PEI has a sensitivity about 10 times higher than PDMS and the sensors with WO_3 nanoparticles (S4 and S8) had the best sensitivity for almost all types of gas tested. LOD of S4 sensor was 6 ppm for ethanol and 8 ppm for dichloroethane.

*

Acknowledgements

This work was supported by a grant of the Romanian Ministry of Research and Innovation, CCCDI-UEFISCDI, project number PN-III-P1-1.2-PCCDI-2017-0172/15PCCDI and project NUCLEU 16N / 08.02.2019.

References

[1]. Q. Zhao, Q. Wang, Y. Li, P. Ning, Senlin, Influence of volatile organic compounds (VOCs) on pulmonary surfactant monolayers at air-water interface: Implication for the pulmonary health, *Colloids and Surfaces A*, Vol. 562, 2019, pp. 402-408.

[2]. E. Duffy, A. Morrin, Endogenous and microbial volatile organic compounds in cutaneous health and disease, *Trends in Analytical Chemistry*, Vol. 111, 2019, 163e172.

[3]. M. Cole, I. Spulber, J. W. Gardner, Surface acoustic wave electronic tongue for robust analysis of sensory components, *Sensors and Actuators B*, Vol. 207, 2015, pp. 1147-1153.

[4]. J. Devkota, P. R. Ohodnicki, D. W. Greve, SAW sensors for chemical vapors and gases, *Sensors*, Vol. 17, 2017, 801.

[5]. C. Viespe, C. Grigoriu, Surface acoustic wave sensors with carbon nanotubes and SiO_2/Si nanoparticles based nanocomposites for VOC detection, *Sensors and Actuators B*, Vol. 147, 2010, pp. 43-47.

(011)

High-temperature Sensors Based on Femtosecond Laser-inscribed Fiber Bragg Gratings

Jun He, Xizhen Xu, Changrui Liao, Ying Wang, Yiping Wang
Key Laboratory of Optoelectronic Devices and Systems of Ministry of Education and Guangdong Province, College of Physics and Optoelectronic Engineering, Shenzhen University, Shenzheen 518060, China
E-mail: hejun07@szu.edu.cn

Summary: We inscribed various fiber Bragg gratings (FBGs) for high-temperature sensing. Three different femtosecond laser inscription methods, including phase-mask method, point-by-point inscription method, and line-by-line scanning method, were proposed for inscribing FBGs. A negative-index FBG was obtained with a reflectivity of 99.22 %, an insertion loss of 0.08 dB, and an operating temperature of up to 1000 °C. A type-II FBG was fabricated inn a pure-silica suspended core photonic crystal fiber. We also fabricated single-crystal sapphire FBGs (SFBGs) using a femtosecond laser line-by-line scanning technique. An SFBG with a reflectivity of 6.3 % was obtained and can withstand a high temperature of 1612 °C.

Keywords: Fiber Bragg gratings, Ultrafast laser, Fiber-optic sensors, High temperature sensors.

1. Introduction

Sensing in harsh environments, such as oil and gas fields, power stations, aero engines, and furnaces, requires high operating temperatures ranging from 400 °C to over 1600 °C. Various fiber Bragg gratings (FBGs)-based high temperature sensors have been reported in the past decades. The maximum operating temperature of all silica-based gratings is limited by the glass transition temperature of silica ($Tg \approx 1330$ °C) [1-5]. Sapphire fibers have a high melting temperature of 2045 °C, and could be developed for fabricating various high temperature sensors. Femtosecond laser has been used to inscribe FBGs in sapphire fibers.

In this letter, we fabricated various FBGs for high temperature sensing. At first, we fabricated negative index FBGs using femtosecond laser overexposure and thermal regeneration. A negative-index FBG was obtained with a reflectivity of 99.22 %, an insertion loss of 0.08 dB, and an operating temperature of up to 1000 °C. We also fabricated a type-II FBG in a pure silica suspended-core photonic crystal fiber. We also fabricated SFBGs using a femtosecond laser line-byline scanning technique. An SFBG with a reflectivity of 6.3 % was obtained and a serial array consisting of five different SFBGs was constructed. These SFBGs can withstand a high temperature of 1612 °C. Such FBGs could be further developed for promising high temperature sensors in aero engines.

2. Negative-index Fiber Bragg Gratings with Enhanced Thermal Stability

Negative-index fiber Bragg gratings (FBGs) were fabricated using 800 nm femtosecond laser overexposure and thermal regeneration. A positive index type I-IR FBG was first inscribed in H_2-free fiber with a uniform phase mask, and then a highly polarization dependent phase-shifted FBG (PSFBG) was created from the type I-IR FBG by overexposure. Subsequently, the PSFBG was annealed at 800 °C for 12 hours. A negative-index FBBG was obtained with a reflectivity of 99.22 %, an insertion loss of 0.08 dB, a blue-shift of 0.83 nm, and an operating temperature of up to 1000 °C [1, 2].

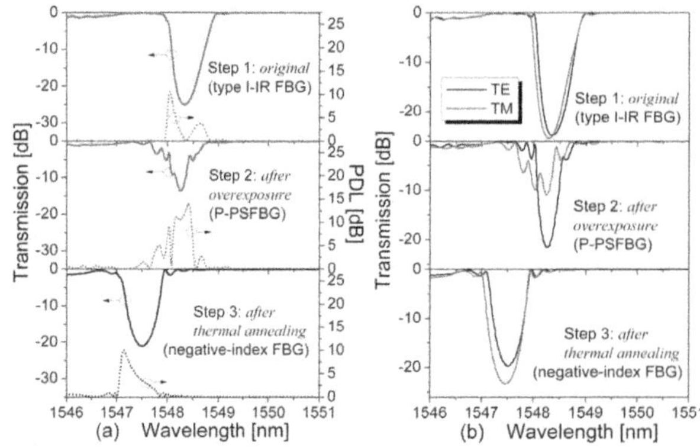

Fig. 1. Transmission spectra of the original type I-IR FBG (upper panel), highly polarization dependent phase-shifted FBG (middle panel), and negative--index FBG (lower panel).

3. Bragg Gratings in Purre-silica Suspended-core Photonic Crystal Fiber

We report a novel type-II photonic crystal fiber Bragg grating for high-temperature applications. The Bragg grating is inscribed in a low-loss in-fiber structure named suspended-core photonic microcell, which is post processed from a commercial pure silica photonic crystal fiber. Grating samples with core diameters of about 4 µm were made by using a focused near-infrared femtosecond laser and a phase mask, and then tested in a tube furnace from room temperature to about 1200 °C. The thermal response of the Bragg resonant wavelength was measure to be about 12 and 16 pm/°C, respectively, at the temperature around 100 °C and 1000 °C. The grating spectrum remained stable in a 10-h isothermal annealing at 1000 °C and started decaying at about 1120 °C with the rate of about 0.02 dB/min. This type of grating possesses flexibilities in both waveguide and grating structure design, exhibits good

high-temperature performance, hence would be promising platform for building wavelength-division-multiplexed fiber sensors and tunable devices with a wide temperature range [3, 4].

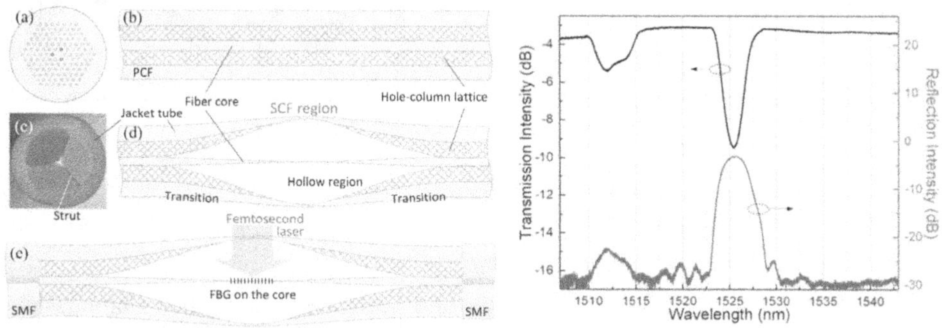

Fig. 2. Type-II FBG in a suspended-core photonic microcell created from a pure-silica photonic crystal fiber.

4. Sapphire Fiber Bragg Grating Sensors

We propose and demonstrate the fabrication of single-crystal sapphire fiber Bragg gratings (SFBGs) using a femtosecond laser line-by-line scanning technique. This approach provides a robust method for producing SFBGs at various Bragg wavelengths with an acceptable reflectivity. The spectrum characteristics of the SFBGs with various fiber diameters, track lengths, and grating pitch quantities were investigated. An SFBG with a reflectivity of 6.3 % was obtained via optimization of fabrication parameters. Additionally, a serial array consisting of five SFBGs at different wavelengths was successfully constructed. The high temperature response of these SFBGs was tested and the experimental results showed the SFBGs could withstand a high temperature of 1612 °C. Moreover, a temperature sensitivity of 36.5 pm/°C was achieved at the high temperature region [5].

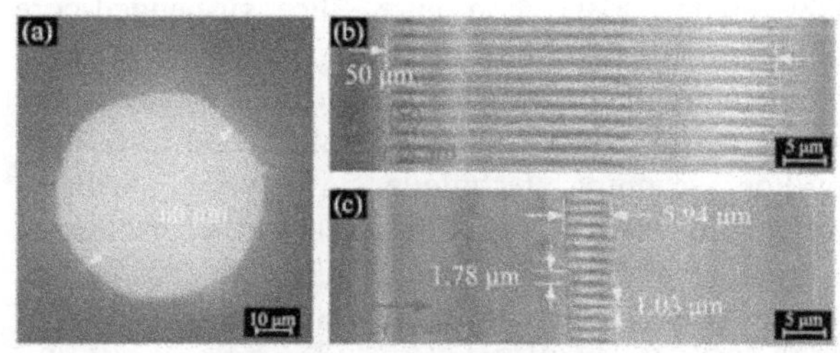

Fig. 3. Microscope images of a sapphire fiber Bragg grating.

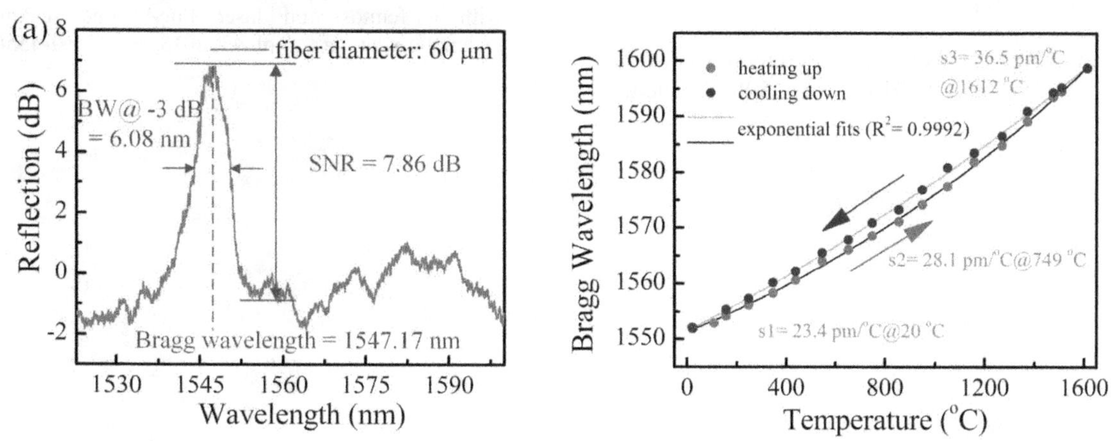

Fig. 4. Reflection spectra of SFBGs inscribed in sapphire fibers at the temperature response.

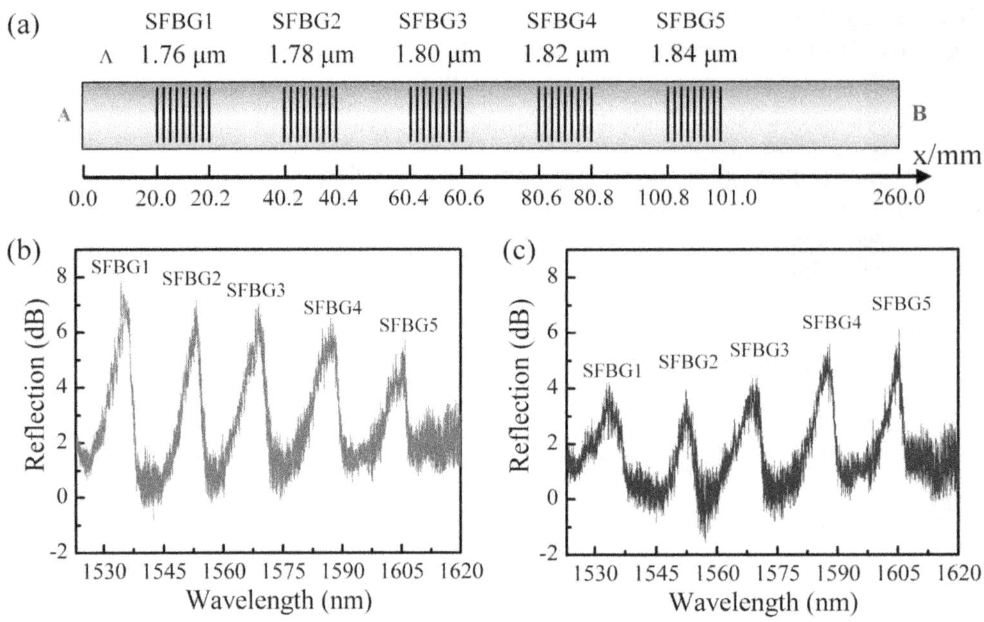

Fig. 5. A serial SFBGs array consisting of five SFBGs at different Bragg wavelengths along a single sapphire fiber.

5. Conclusion

We inscribed various FBGGs for high-temperature sensing with three different methods, including femtosecond laser phase-mask method, point-by point inscription method, and line-by-line scanning method. A negative-index FBGG was obtained with a reflectivity of 99.22 %, an insertion loss of 0.08 dB, a blue-shift of 0.83 nm, and an n operating temperature of up to 1000 °C. We also fabricated a FBG in a pure silica suspended-core photonic crystal fiber and can withstand a high temperature of ~1100 °C. Subsequently, we fabricated single-crystal SFBGs using a femtosecond laser line-by-line scanning technique. An SFBG with a reflectivity of 6.3 % was obtained and a serial array consisting of five different SFBGs was constructed. These SFBGs can withstand a high temperature of 1612 °C. Such SFBGs could be developed for promising high temperature sensors in aero engines.

Acknowledgements

National Natural Science Foundation of China (NSFC) (61875128, 918601388, 61635007); Science and Technology Innovation Commission of Shenzhen (JCYJ20180507182058432, JCYJ20170302143105991, JCYJ20160427104925452 2).

References

[1]. J. He, et al.,"Highly birefringent phase-shifted fiber Bragg gratings inscribed with femtosecond laser, *Opt. Lett.*, Vol. 40, 2015, pp. 2008-2011.

[2]. J. He, et al., Negative-index gratings formed by femtosecond laser overexposure and thermal regeneration, *Sci. Rep.*, Vol. 6, 2016, 233379.

[3]. C. Wang, et al., Bragg gratings inscribed in selectively inflated photonic crystal fibres, *Opt. Express*, Vol. 25, Issue 23, 2017, pp. 28442-284550.

[4]. C. Wang, et al., Bragg gratings in suspended-core photonic microcells for high temperature applications, *J. Lightwave Technol.*, Vol. 36, Issue 14, 2018, pp. 2920-2924.

[5]. X. Xu, et al., Sapphire fiber Bragg gratings inscribed with a femtosecond laser line-by-line scanning technique, *Opt. Lett.*, Vol. 43, 2018, pp. 45562-45565.

(012)

Portable Bioimpedance Device and Monitoring of Hydration in a Healthy Person Before and After Exercise

V. Leonov [1], M. Konijnenburg [2], H. Ha [2], B. Grundlehner [2] and N. Van Helleputte [1]
[1] Imec, Connected Health Solutions Belgium Department, Kapeldreef 75, 3001 Leuven, Belgium
[2] Imec The Netherlands, Connected Health Solutions NL Department, High Tech Campus 31, 5656 AE Eindhoven, The Netherlands
Tel.: +32 16 288 367
E-mail: vladimir.leonov@imec.com

Summary: The bioimpedance sensor recently developed at Imec was calibrated on the equivalent *RC* model of human body and tested on the volunteer undergoing repeated mild dehydration and rehydration. Some sources of variability in hydration assessment are observed that were not spotted and reported before. The specific protocols developed in this study enabled minimization of errors in altered hydration state. The bioimpedance method is shown to measure the hydration variation in a single person, with no need for averaging over large population. The predictable body fluid loss and gain was supported by dehydration and rehydration diets designed for accurate control of ion and water intake. The accuracy of fluid shift measurement down to a standard deviation of 300 ml is demonstrated, which exceeds capabilities of known methods and devices, including 'gold standards' like isotope dilution for hydration assessment.

Keywords: Hydration, Human being, Bioimpedance, Dehydration, Rehydration, Multifrequency.

1. Introduction

Bioimpedance methods have been widely studied in application to estimation of body composition, including hydration, in normally hydrated population. There was however a lack of proof presented that bioimpedance methods can assess hydration change with no need for averaging over a large population. In addition to demonstrated statistical studies, there is a strong demand for any measurement device and method to assess total body water (TBW) shift in a single healthy person, or in ambulatory application, at accuracy of about 1 liter. The task of hydration assessment is currently being solved through a set of hydration indices and symptoms (such as weight variation, edema, urine osmolarity, color and volume, mouth dryness, confusion, dizziness, tachycardia, thirst, fainting when standing, headache, tiredness, etc.), blood test (plasma osmolarity and sodium), or by using isotope dilution. The methods still lack accuracy. The conclusion about reliability of the methods can be found in the literature, e.g., [1] says: "As of now, there are no reliable tools to determine hydration status."

2. Bioimpedance Sensor and Setup

The recent hydration monitoring studies at Imec [2], which were conducted using commercial bioimpedance analyzers, lock-in amplifiers and earlier versions of wearable bioimpedance sensors developed at Holst Centre [3, 4] have indicated necessity of adjustment of the measurement protocol and deeper studying of accuracy of bioimpedance measurements of hydration status. In this work, an evaluation test board was built using the multiparameter signal acquisition SoC (System-on-Chip) MUSEIC v2 for personal health applications [5]. The chip with a size of 8 mm × 7 mm contains ten readout channels with the state-of-the-art accuracy, Fig. 1. In this work, the only multifrequency bioimpedance channel and the related current generator were used. The bioimpedance sensor in this work ran at frequencies of 16, 32, 64, and 128 kHz at a current of 10 μA. A low-pass digital (FIR) filter selectable from 10 Hz to 150 Hz was used at 10 Hz. Variable current and gain enable selection of the most suitable combination of sensitivity and dynamic range, depending on electrode location on the human body. A 5-electrode setup consists of two current electrodes, two potential electrodes, and a body bias electrode (to enhance system stability).

For characterization of the bioimpedance sensor, the volunteer's electrical properties between the forearms and those of electrode-skin couple have been measured using EG&G 5210 lock-in amplifier within the 1 Hz – 100 kHz range in a course of several days, averaged, and the electrical *RC*-model shown in Fig. 2 was obtained. It was used for calibration of the sensor and measurement of the system accuracy.

The impedance of electrode-skin couple is highly variable depending on sweating rate and skin blood flow. On average, at indoor temperatures and typical sedentary activity in the office, i.e., with no sweating, the impedance of Meditrace 200 wet-gel electrode and skin of the subject was modeled using the circuit shown in Fig. 2, bottom. For calibration of the bioimpedance sensor and system characterization, all five electrodes shown in Fig. 2 were replaced with electrode-skin circuits. At integration time of 16 s, a standard deviation of resistance measurements obtained on the *RC*-model at 64 kHz was 45 mΩ.

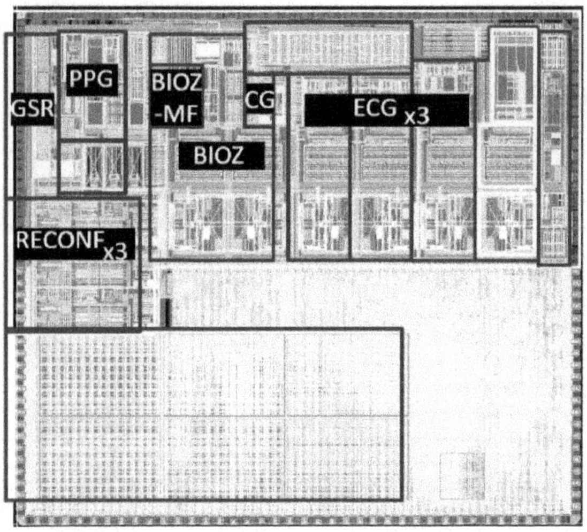

Fig. 1. MUSEIC v2 SoC containing sensor readouts, filters, controller, and data analysis modules. Abbreviations denote Current Generator (CG), and readout channels for: Galvanic Skin Response (GSR), Photoplethysmography (PPG), a fixed-frequency Bioimpedance (BIOZ), a multifrequency Bioimpedance (BIOZ-MF) within the 1 kHz – 1 MHz range. There are also three Electrocardiography (ECG) and three reconfigurable (RECONF) channels.

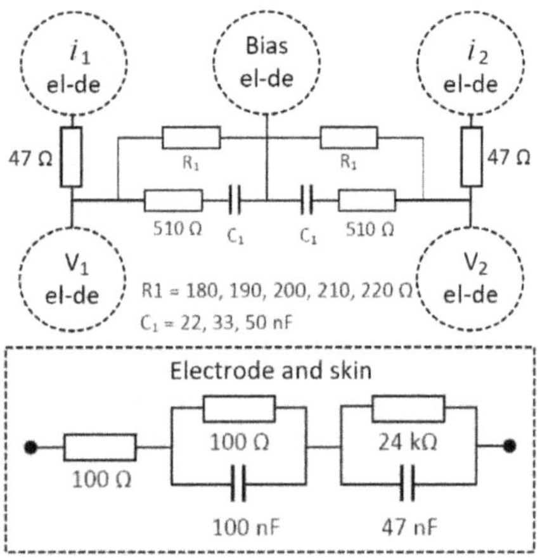

Fig. 2. Experimental *RC*-model of the studied subject and of electrode-skin couple suitable for 1 Hz – 100 kHz range. Typical R_1 was 200 Ω, and C_1 was 22 nF. The other listed *RC*-values were used for the system calibration.

3. Measurements, Protocols and Results

The impedance of a healthy and normally hydrated person fluctuates during a day near its stable average value [2]. The human body loses water and ions through insensible perspiration, sweating and urine. However, they are renewed on a regular basis through food and water intake. But before the ions from food reach the cardiovascular system, the body must secrete of the order of one liter of digestion fluid. Despite presence of many other factors affecting measurement results, the variation of bioimpedance due to food intake was detected through averaging of multiple measurements before and after food intake in a course of several days using commercially available Maltron BioScan 920 bioimpedance analyzer. Fig. 3 shows impedance measured at 5, 50 and 100 kHz. The trend line for left ankle-wrist plot correlates well with the trend line for wrist-wrist results. The maximum of impedance corresponds to about 30-60 min after meals. This is the time when digestion fluid is secreted, thereby increasing body impedance, but electrolyte from food is not yet fully absorbed.

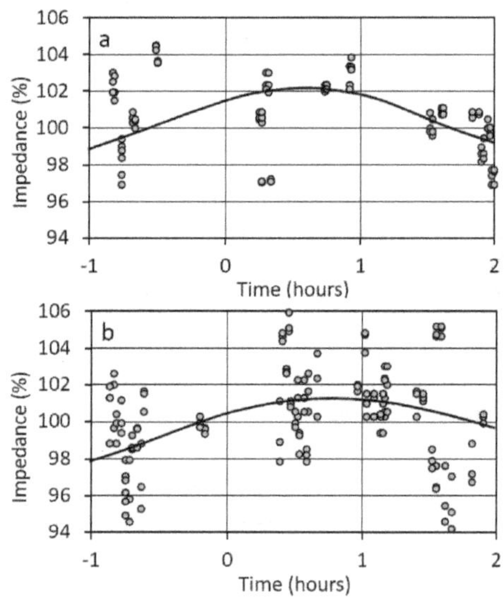

Fig. 3. Experimental results showing that food intake causes variation of body impedance between left ankle and wrist (a) measured in supine position, and between the forearms; (b) measured on a sitting person. Experimental points correspond to three frequencies, 5, 50 and 100 kHz plotted together. Lines is a polynomial fit. Zero time corresponds to the end of meals.

Fig. 3 represents the way of measurements typically used to illustrate the bioimpedance method, namely, a single measurement is chaotic, but large number of measurements on a population (or, repeated multiple measurement on the same person) enables observation of the trend. In this work, however, the task was to demonstrate ability of new sensor to detect bioimpedance variations on a single person, with no averaging over population, i.e., to demonstrate the device suitable for ambulatory applications.

Fig. 3 shows that there are certain daily variations of impedance. Furthermore, the study in [2] and in this work has shown that electrode placement accuracy and reproducibility of posture in different measurements are not perfect and introduce some additional variability of measurement results. Therefore, to easier spot dehydration on the background of usual impedance variations, the baseline of impedance was established prior to plan hydration variation in a test person. Fluid shift was obtained using mild exercise-related dehydration and limited fluid intake. The results are compared with the most accurate reference method for hydration assessment based on weight monitoring and intake of fluid and food. This reference method works only in a course of several days (or several weeks in case of precisely counted caloric intake balanced against energy expenditure in a healthy person). It consumes plenty of time for counting calories equal to energy expenditure, choosing appropriate food combinations providing required weight of sodium and potassium ions, and finally accounting for the rate of their absorption. However, this reference method enabled hydration calculation with about 200 ml precision, better than could be obtained using other methods.

The study preceding this project [2] has shown that uncontrolled consumption of sodium (too much table salt in food consumed in chaotic manner) resulted in dramatic daily variation of ions and water retention in the body. Accounting for unbalanced food intake in respect to energy expenditure, these factors create large variations of both reference subject weight and the impedance. Therefore, the protocol included matched diets with precise quantity of potassium and sodium per day. Furthermore, while the subject was undergoing dehydration, the sodium intake was minimized, but it was doubled during rehydration. The latter enabled faster rehydration compared to using balanced sodium intake.

As an example of the developed diets, the menu on exercise day included rice, juices (apple, grape, tomato), chocolate bars, sugar, butter, coffee, sunflower seeds, potato and pasta, all with no added salt (only 50 mg sodium, but 3.6-6 g potassium intake per day). Rehydration diet contained up to 6.7 g sodium/day thereby enabling better water retention and rapid rehydration.

Dehydration was provided through joint effect of limited water intake during 2-3 days, including 1-2 days of cycling in warm/hot weather for 40-50 km distance at about 5 ± 1 MET light effort. Low-sodium diet enabled some thirst decrease and promoted water rejection. The impedance measurements were conducted on a sitting or standing subject between forearms to avoid measurements on exercising muscles (legs) of cyclist. Fig. 4 shows variation of both resistance at 64 kHz and weight of the subject during a 7-day dehydration and rehydration experiment.

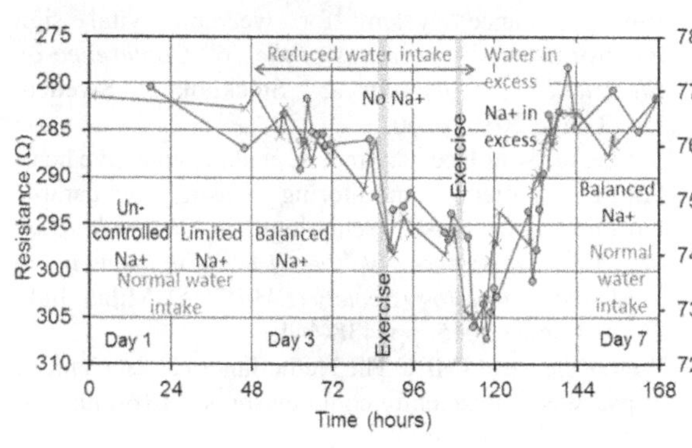

Fig. 4. Resistance between forearms at 64 kHz (circles) and weight (crosses) of a subject undergoing dehydration on days 3 – 5, cycling exercises on days 4 – 5, and rehydration on day 5 (after the second exercise) and on day 6.

Having dehydration baseline measured, it becomes possible to calculate hydration status using the subject's weight together with food and water intake. Fig. 5 illustrates such calculated variation of reference hydration plotted together with resistance variation from Fig. 4. The results are still noisy, i.e., there are some impedance variations in respect to the reference hydration. Accounting for imperfect calculation of hydration from weight and intake, a digital low-pass FIR filter is applied to both sets of data, resistance and hydration. The result of such noise reduction is shown in Fig. 6. The result of similar data processing of reactance is shown in Fig. 7 for the same experiment.

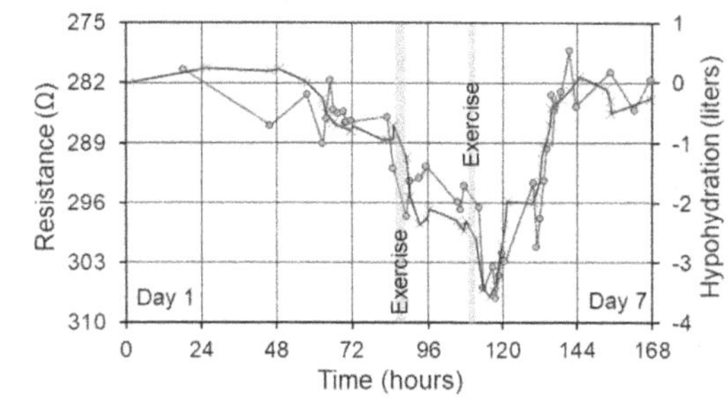

Fig. 5. Resistance between forearms at 64 kHz (circles) from Fig. 4, and the reference hydration (crosses) of the subject calculated using his weight, weight of consumed food and water.

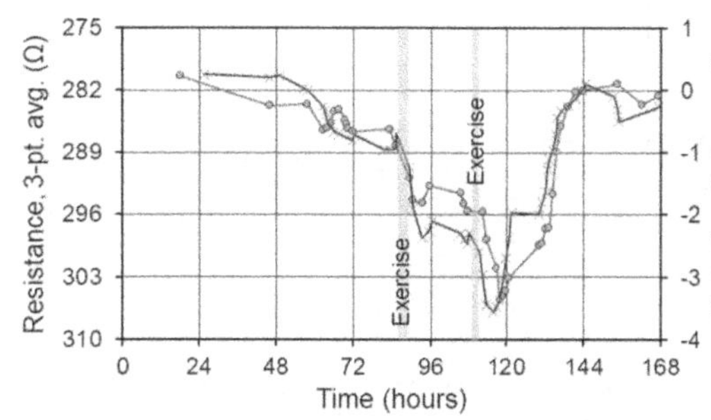

Fig. 6. Resistance between forearms at 64 kHz (circles) and the reference hydration (crosses). A low-pass filter (3-point moving average) is applied for both resistance and the reference hydration.

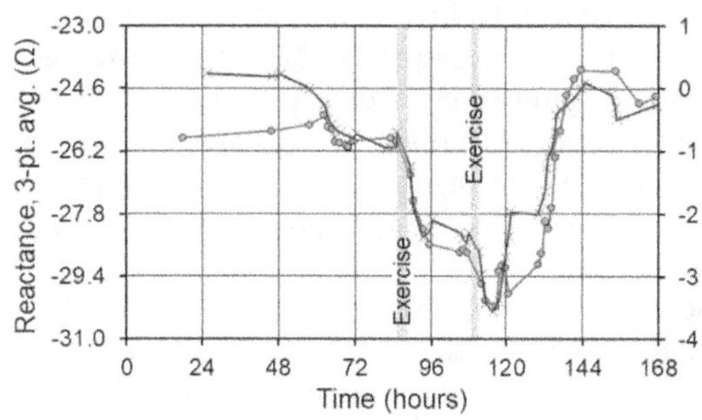

Fig. 7. Reactance between forearms at 64 kHz (circles) corresponding to resistance in Fig. 6, and the reference hydration (crosses). A low-pass filter (3-point moving average) is applied for both reactance and hydration.

The residual difference between hydration and resistance shown in Figs. 6, 7 can be explained by joint effect of different factors such as electrode placement accuracy, tiny variation of subject's posture, inaccuracy of the reference hydration calculation, and delay of hydration measured electrically in respect to the weight-and-intake reference during rehydration.

The results presented in Figs. 6, 7 can be re-plotted to obtain dependence of hydration on resistance or reactance. The example is shown in Fig. 8, where regression equation is obtained for calculation of hydration based on reactance. Using such equation, the results of hydration measurement with the sensor can be plotted against reference hydration, Fig. 9.

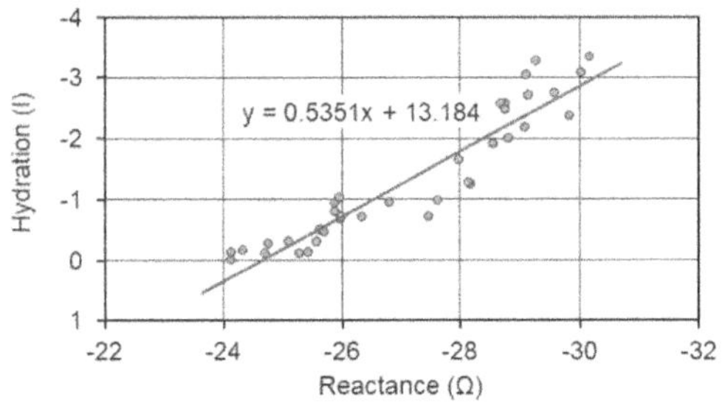

Fig. 8. Dependence of hydration on reactance between forearms at 64 kHz. A low-pass filter (3-point moving average) is applied for both reactance and hydration.

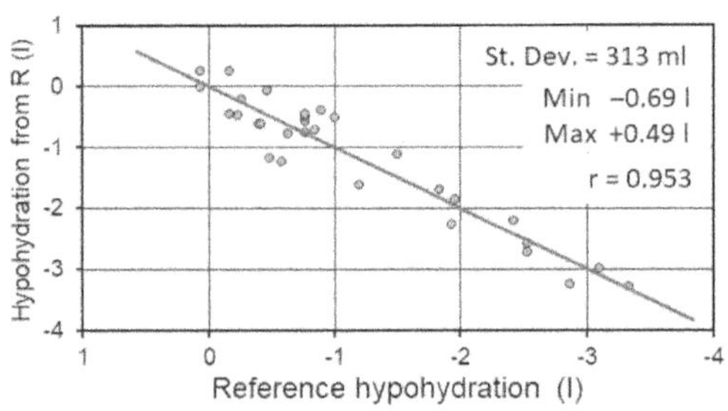

Fig. 9. Hydration measured using resistance at 128 kHz in one of the experiments and plotted versus reference hydration based on subject weight and intake with applied digital filter (3-point moving average) for both.

Application of near-zero sodium diets on exercise days allowed larger slope of dehydration, 2.9 %/l resistance rise at 16 kHz than with sodium intake as reported in [2], where 1.56 %/l was observed at 8-16 kHz. Application of data filtering, e.g., using a FIR filter, enabled further improvement of signal-to-noise ratio, where signal is hydration and noise is device-, electrode- and posture-related errors in the subject.

Several factors usually cause instability of impedance. Among them, it is necessary to mention variable sodium intake, which is related to heavily salted food products available in grocery stores. As a result, sodium intake becomes chaotic, sodium ions in the body largely vary after meals thereby affecting water retention, and the impedance fluctuates making hydration baseline noisy. The posture of the subject is not exactly reproduced from one measurement to another, and this is one of important factors introducing errors. Commercial impedance meters are frequently sensitive a little to electrode and skin impedances. In this relation, the body bias electrode, high CMRR and high input impedance implemented in the developed device helped to eliminate some of errors observed in bioimpedance meters on the market.

4. Conclusion

It is shown that the new bioimpedance device together with a diet-based protocol enables accurate hydration monitoring in a healthy subject, which varies during exercise. The achieved accuracy exceeds accuracy of the methods usually used for hydration shift assessment. It still needs both deeper studying and further validation on population. Of course, the measurement protocol and diets, which were necessary here for calibration of the device, could be simplified in the future for approval in hospitals and sport. Recently, the highly miniaturized wearable version of the device used in this work was presented [6].

Acknowledgements

The authors are thankful to L. Hons (Premed, Leuven) for providing medical surveillance and to C. Van Hoof (Imec) for enthusiasm, fruitful discussions, suggestions and effective support of the study.

References

[1]. A. Bak, A. Tsiami, C. Greene, Methods of assessment of hydration status and their usefulness in detecting dehydration in the elderly, *Current Research in Nutrition and Food Science*, Vol. 5, 2017, pp. 43-54.

[2]. V. Leonov, S. Lee, A. Londergan, et al., Bioimpedance method for human body hydration assessment, in *Proceedings of the 41st Conference of the IEEE Engineering in Medicine and Biology Society (EMBC'19)*, Berlin, Germany, 23-27 July 2019, pp. 6036-6039.

[3]. G. Squillace, S. Lee, V. van Acht, M. Vandecasteele, Bio impedance system for wearable vital sign monitoring, in *Proceedings of the 16th Conference on Electrical Bio-Impedance*, Stockholm, Sweden, 19-23 June 2016, p. 60.

[4]. S. Lee, G. Squillace, C. Smeets, et al., Congestive heart failure patient monitoring using wearable bio-impedance sensor technology, in *Proceedings of the 37th Conference of the IEEE Engineering in Medicine and Biology Society (EMBC'15)*, Milan, Italy, 25-29 August 2015, pp. 438-441.

[5]. Press Release at IMEC The Netherlands (Holst Centre), https://www.holstcentre.com/newspress/2016/museic-2/?highlight=museic

[6]. S. Lee, B. Grundlehner, R. Garcia van der Westen, et al., Nightingale V2: Low-power compact-sized multi-sensor platform for wearable health monitoring, in *Proceedings of the 41st Conference of the IEEE Engineering in Medicine and Biology Society (EMBC'19)*, Berlin, Germany, 23-27 July 2019, pp. 1290-1293.

(013)

Biofluorometric Gas-imaging System (Sniff-cam) for Body Volatiles

K. Iitani, K. Toma, T. Arakawa and K. Mitsubayashi
Institute of Biomaterials and Bioengineering, Tokyo Medical and Dental University, 2-3-10 Kanda-Surugadai, Chiyoda-ku, Tokyo 101-0062, Japan
Tel.: + 81-3-5280-8091, fax: + 81-3-5280-8094
E-mail: m.bdi@tmd.ac.jp

Summary: Gas-phase biosensors have been investigated not only for human volatiles (acetone, methyl mercaptan, trimethylamine, ethanol, isopropanol, etc.) but also for residential harmful VOCs (formaldehyde, toluene, nicotine) causing sick-house syndrome, etc. The biochemical gas-imaging system (sniff-cam) with enzyme immobilized mesh demonstrates a spatiotemporal gas-imaging for human volatiles (i.e. ethanol, acetaldehyde, etc. after drinking). A biofluorometric techniques was applied in order to improve the sensitivity of the gas-imaging system. The novel biofluorometric sniff-cam for ethanol was fabricated with ADH (alcohol dehydrogenase) immobilized mesh and the NADH visualization unit (UV-LED sheet array & high sensitive CCD), thus imaging human ethanol vapor not only exhaled air but also skin gas after drinking. The sniffer-device would be useful for conventional detecting the volatile biomarkers.

Keywords: Gas-phase biosensor, Biofluorometry, VOCs, Exhaled air, Body gas, Enzyme, UV-LED.

1. Introduction

Ethanol is volatile and exhaled ethanol concentration in proportion to the concentration in blood, transfer from the blood into the alveolar air sacs in the lung after drinking [1]. Then, it has possibility to determine the exhaled ethanol concentration and predict with a high degree of accuracy, the blood ethanol concentration at the same point of time. So, the determination of blood or exhaled ethanol concentrations is one of the most frequently requested tests in forensic medicine, and understanding of the pharmacokinetics of ethanol may give a clear appreciation of its biological effects on the nervous system [2].

In our previous study, we reported an imaging system of gaseous ethanol based on CL measurement employing an EM-CCD camera, and the system is applied in breath analysis [3-4]. In this paper, we developed a novel biofluorometric imaging system for high sensitive imaging of gaseous ethanol emissions from human breath and palm skin.

2. Construction of Biofluorometric Sniff-cam

This imaging system measures ethanol vapor concentrations as intensities of biofluorescence through an enzymatic reaction induced by alcohol dehydrogenase (ADH). The imaging system consisted of multi UV-LED excitation sheet and a high-sensitive CCD camera. The system applied to visualize and analyze to metabolism function at real time.

Cotton mesh was used for enzyme immobilization. Alcohol dehydrogenase (ADH) (E.C. 1.1.1.1) with activity of 100 units/mg of protein onto a cotton mesh substrate. 2-methacryloyloxyethyl phosphoryl choline copolymerized with 2-ethylhexyl methacrylate (PMEH). PMEH was synthesized by free radical polymerization method as previously reported [1]. In order to operate the imaging system (Fig. 1), PB (pH 8.0, 80 mmol/l) containing 20 mmol/l NAD^+ was prepared. A cotton mesh substrate was immobilized with ADH with glutaraldehyde cross-linkage for enzyme stabilization.

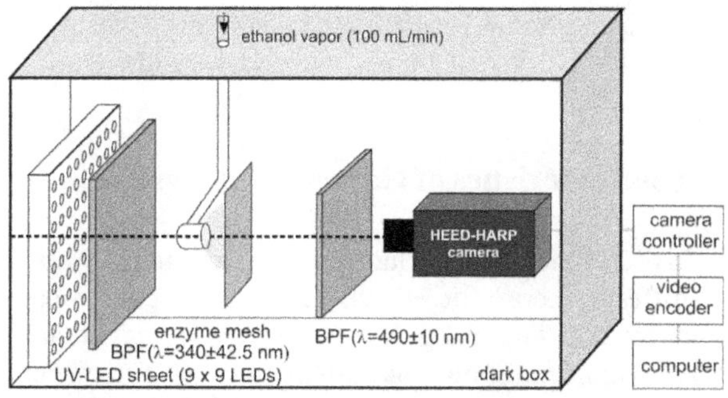

Fig. 1. Fluorometric imaging system for ethanol.

This system for imaging the spatial movement of gas was developed. We proposed a NADH dependent biosensor as one of the promising approaches to develop such an advanced sensor. Equation 1 is the ADH reaction for recognition of ethanol vapor.

$$\text{ethanol} + NAD^+ \xrightarrow{ADH} \text{acetaldehyde} + \boxed{NADH} + H^+ \quad \text{ex:340 nm} \quad \text{fl:491 nm} \tag{1}$$

It uses oxidized NAD as an acceptor and produces NADH for ethanol vapor. So ethanol can be measured by counting NADH, which is excited by 340 nm UV-LED and produces fluorescence of 490 nm [5]. Standard ethanol gases were prepared using a gas generator for characterization of the imaging system. The solution was stable when stored in the dark box. The fluorescent imaging system was constructed with a 9×9 multi UV-LED excitation sheet and a high-gain avalanche rushing amorphous photoconductor (HEED-HARP) CCD camera (Pioneer Co. Ltd. Japan), a HDD recorder, and a video encoder.

An enzyme of acetaldehyde type 2 deficiency involved in the metabolism of acetaldehyde, a toxic metabolite of ethanol. Skin ethanol patch test was performed following the method reported. A patch plaster of lint pads on an adhesive tape was prepared. 100 μL of 70 % ethanol was placed on one of the lint pads, and as a control, the same volume of distilled water was put on the second pad. The patches were taped to the inner surface of the arm for 7 minutes and then removed slowly. Patch areas that showed erythema after removal. In this experiment, an ALDH 2 negative volunteer was selected for further experiments after oral administration of alcohol.

We improved the gaseous ethanol imaging system to the transdermal ethanol imaging system for the human palm. We manufactured the pattern plate, which formed the cell of the same shape into dots form covering the enzyme immobilization mesh. The mesh with square patterns was soaked in NAD^+ solution. And palm part of the volunteer adhere to a cell plate after the drinking and recorded the emission of the skin gas by CCD camera. For measurement of transdermal ethanol from the palm, gaseous ethanol samples were collected after oral administration of alcohol and measured authorizing by the Human Investigations Committee of Institute of Biomaterials and Bioengineering, Tokyo Medical and Dental University. After fasting for at least 4 hours, a volunteer was given an alcoholic drink (0.4 g per kg body weight).

3. Characteristics of Gas-imaging System

A 2-dimensional fluorometric imaging system "Sniffer camera" for ethanol vapor was constructed and tested. The imaging system measures ethanol vapor concentrations as intensities of fluorescent change by enzymatic reaction induced by alcohol dehydrogenase (ADH). This imaging system uses ADH for recognition of ethanol vapor, and it measures ethanol vapor by measuring fluorescence of nicotinamide adenine dinucleotide (NADH), which is oxidized by enzymatic reaction on the mesh.

This fluorometric imaging system achieved the two-dimensional real-time imaging of ethanol vapor distribution (0.5-200 ppm). The system showed a rapidly and accurately responses and a visible measurement. The intensity of fluorescence was linearly related to the concentration of ethanol vapor. Fluorescence intensity of NADH is saturated because reacted NADH solution remained on the cotton mesh. In order to analyze change of fluorescent intensity, we applied a differential analysis between change of fluorescent images intensity and time. We selected differential analysis for changes of intensity by time (5 seconds). dI/dt has a peak before fluorescence was saturated. The intensity of the imaging system was included in the drunk driving (approximately 78 ppm). Also it was applied to the detection of exhaled breath after drinking (data not shown).

A high sensitivity fluorescent imaging of ethanol vapor allows us to successfully visualize the emissions dynamics of transdermal ethanol vapor. An ALDH2(-) volunteer's hand was placed stably on the acrylic cell plate after alcohol administration. Fluorescent image depended on ethanol vapor emissions from palm skin and evaluated using a high-sensitive imaging system. A fluorescent image of transdermal gaseous ethanol on the palm at 30 minutes after drinking is shown in Fig. 2. As a result of the experiments, light intensity of fluorescent image following emissions of transdermal gaseous ethanol rapidly increased from the palm part, which was confirmed from drinking later. Released ethanol gas was not constant in parts of the palm, and it was suggested that distribution of ethanol emissions ethanol from the human hand were based on the alcohol metabolism process.

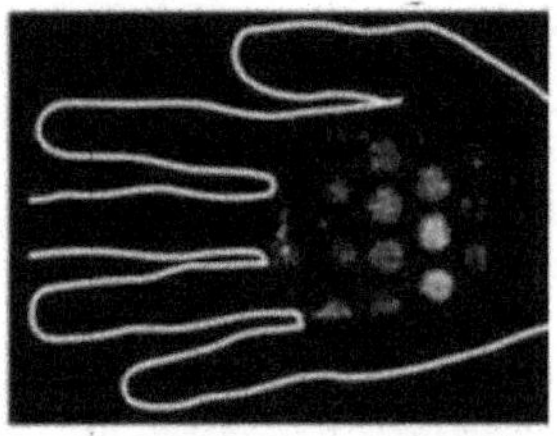

Fig. 2. Fluorometric image of transdermal gaseous ethanol.

4. Conclusions

We developed a biofluorometric gas imaging system (sniff-cam) for ethanol using an enzymatic reaction induced by alcohol dehydrogenase (ADH). This system measures ethanol concentrations as intensities of biofluorescent image by enzyme reaction. The imaging of gaseous ethanol was achieved at 0.5 ppm to detection limit. We applied a high sensitivity gaseous ethanol imaging system for measurement of skin ethanol vapor from a human palm.

Acknowledgements

These works were partly supported by Japan Society for the Promotion of Science (JSPS), by Japan Science and Technology Agency (JST) and by New Energy and Industrial Technology Development Organization (NEDO).

References

[1]. R. G. Hahn, A. Norberg, A. W. Jones, Rate of distribution of ethanol into the total body water, *Am. J. Therapeut*, Vol. 2, 1995, 50.

[2]. H. Kalant, Direct effects of ethanol on the nervous system, *Fed. Proc.*, Vol. 34, 1975, pp. 1930-1941.

[3]. X. Wang, E. Ando, D. Takahashi, T. Arakawa, H. Kudo, H. Saito, K. Mitsubayashi, 2D spatiotemporal visualization system of expired gaseous ethanol after oral administration for real-time illustrated analysis of alcohol metabolism, *Talanta*, Vol. 82, 2010, pp. 892-898.

[4]. T. Arakawa, X. Wang, K. Kajiro, K. Miyajima, S. Takeuchi, H. Kudo, K. Yano, K. Mitsubayashi, A direct gaseous ethanol imaging system for analysis of alcohol metabolism from exhaled breath, *Sens. Actuators B*, Vol. 186, 2013, pp. 27-33.

[5]. H. Kudo, M. Sawai, X. Wang, T. Gessei, T. Koshida, K. Miyajima, H. Saito, K. Mitsubayashi, A NADH-dependent fiber-optic biosensor for ethanol determination with a UV-LED excitation system, *Sens. Actuators B*, Vol. 141, 2009, pp. 20-25.

(014)

BaY_2F_8:Yb^{3+}, Ho^{3+}/Tm^{3+} Upconversion Phosphor for Optical Thermometer

H. C. Swart, Ashwini Kumar and Govind B. Nair
Department of Physics, University of the Free State, P. O. Box 339, Bloemfontein 9300, South Africa
Tel.: + 27 514012926, fax: +27 514013507
E-mail: swarthc@ufs.ac.za

Summary: BaY_2F_8:Yb^{3+},Ho^{3+} / Yb^{3+},Tm^{3+} phosphors were synthesized by the co-precipitation method. The upconversion luminescence of the Yb^{3+},Ho^{3+} ion pair was used to investigate the temperature-sensing properties by using the Fluorescence Intensity Ratio (FIR) method. The emission peaks centred at 538 nm and 549 nm, and originating from the $^5F_4 \rightarrow ^5I_8$ and $^5S_2 \rightarrow ^5I_8$ transitions, were considered for calculating the FIR. The $^5F_4$ and $^5S_2$ levels of Ho^{3+} are thermally coupled. The temperature dependent upconversion luminescence of these thermally coupled multiplets were carried out from 300 to 623 K with a 980 nm laser excitation source. For the BaY_2F_8:Yb^{3+},Tm^{3+} system the $^3F_2$, $^3F_3$ and $^3H_4$ excited states of Tm^{3+} are also thermally coupled. The emission bands corresponding to Tm^{3+}: $^3F_2$, $^3F_3$ / $^3H_6$ transition and $^1G_4$/$^3F_4$ showed different temperature dependent characteristics, which led to monotonous improvement of the related FIR with an increase of temperature. The temperature sensing behaviour and the mechanism of a nano-micro heater of the developed phosphor have been presented by exploiting the stark sublevels of the $^1G_4$ level of the Tm^{3+} ion, which are generally responsible for the blue upconversion emission. A maximum and a minimum sensor sensitivity at different temperatures for the developed multifunctional phosphors have been determined.

Keywords: Fluoride, Optical thermometry, Upconversion, Yb^{3+} ions, Ho^{3+} ions, Tm^{3+}, Co-precipitation.

1. Introduction

Recently, optical thermometry techniques gained plenty of attention for their application in non-invasive temperature sensing [1, 2]. Optical thermometry is generally performed by recording the temperature-dependent fluorescence intensity ratio (FIR) or fluorescence lifetime [3-5]. The FIR method involves the calculation of the intensity-ratio of two thermally coupled energy levels.

Upconversion (UC) luminescence from lanthanide ions is found to be an effective process, which can be utilized for optical thermometry. In this work, the temperature-dependent UC emissions of BaY_2F_8:Yb^{3+},Ho^{3+}/Yb^{3+},Tm^{3+} phosphors were utilized to determine the FIR. The results suggest its potential application for high-temperature sensing.

2. Experimental

BaY_2F_8:Yb^{3+},Ho^{3+} and BaY_2F_8:Yb^{3+},Tm^{3+} phosphors were prepared by a co-precipitation method. Highly pure nitrate precursors of $Ba(NO_3)_2$,$Y(NO_3)_3$, $Yb(NO_3)_3$, and $Ho(NO_3)_3$, along with NH_4F were used for the synthesis. The solutions of $Ba(NO_3)_2$,$Y(NO_3)_3$, $Yb(NO_3)_3$, and $Ho(NO_3)_3$ in double-distilled water were mixed thoroughly using magnetic stirring at 60 °C. Thereafter, a NH_4F solution in distilled water was added dropwise to the metal ion solution. The addition of NH_4F solution resulted in the formation of white precipitates. The resultant solutions with precipitates were kept for stirring for 2 h at 100 °C. The precipitates were washed with distilled water and then by ethanol using centrifugation. The precipitates were dried in an oven at 80 °C and then annealed in inert argon atmosphere at 600 °C for 6 h.

3. Results and Discussion

In the case of Ho^{3+}, the $^5F_4$ and $^5S_2$ are thermally coupled levels. In fact, these levels are separated just by a margin of 372 cm^{-1}. The fluorescence intensity arising due to the transitions from these two levels towards the ground state $^5I_8$ has been utilized to evaluate the temperature sensing property of BaY_2F_8:Yb^{3+},Ho^{3+} phosphor. The FIR for these two levels can be written as:

$$FIR = \frac{I_{538}}{I_{549}} = A\, exp\left(\frac{-\Delta E}{kT}\right), \tag{1}$$

where the symbols have their usual meanings [2, 6].

Eq. (1) can be written as:

$$\log(FIR) = \left(\frac{-\Delta E}{kT}\right) + \log A,$$

$$\log(FIR) = B\left(\frac{1}{T}\right) + \log A, \tag{2}$$

where, B = $\left(\frac{-\Delta E}{k}\right)$.

The variation of FIR as a function of temperature is plotted in Fig. 1. (Photo of the UC emission is given as an inset). All the experimental data were fitted in Eq. (2). From, the plot, the values of ΔE and A were found to be 387.16 cm^{-1} and 2.26, respectively.

The ability of $BaY_2F_8:Yb^{3+},Ho^{3+}$ phosphor for temperature-sensing can be identified by determining the sensitivity through the following equations:

$$S_R = \frac{1}{R}\frac{dR}{dT} = \left(\frac{-\Delta E}{kT^2}\right), \quad (3)$$

$$S_R = \frac{dR}{dT} = R\left(\frac{-\Delta E}{kT^2}\right), \quad (4)$$

where S_R is the relative sensitivity and S_A is the absolute sensitivity of the phosphor for temperature-sensing. Fig. 2 shows the variation of relative sensitivity (S_R) and absolute sensitivity (S_A) for $BaY_2F_8:Yb^{3+},Ho^{3+}$ phosphor as a function of temperature from 300 to 623 K. The values of S_R and S_A obtained for different temperatures are tabulated in Table 1.

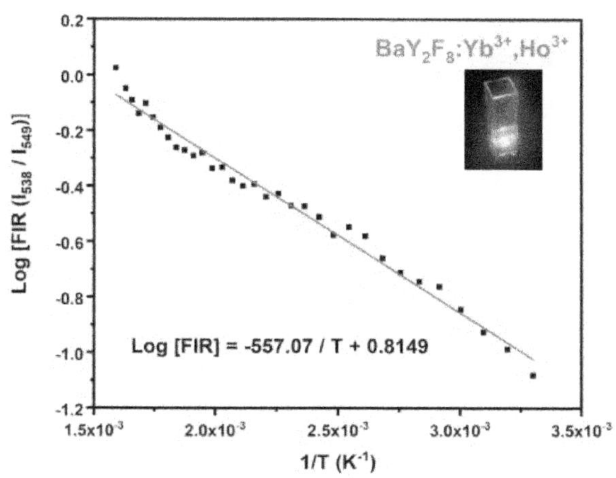

Fig. 1. Variation of FIR for $BaY_2F_8:Yb^{3+},Ho^{3+}$ phosphor as a function of temperature. UC emission photo as inset.

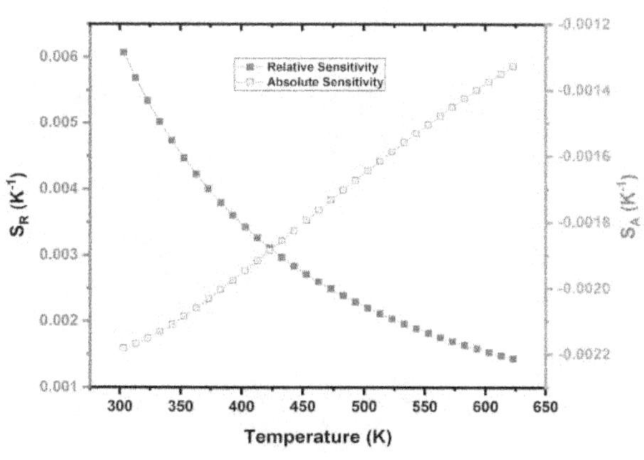

Fig. 2. Variation of relative sensitivity (S_R) and absolute sensitivity (S_A) for $BaY_2F_8:Yb^{3+},Ho^{3+}$ phosphor as a function of temperature.

Similar result for $BaY_2F_8:Yb^{3+},Tm^{3+}$ were obtained. The emission bands corresponding to Tm^{3+}: $^3F_2$, $^3F_3$ / $^3H_6$ transitions and $^1G_4$/$^3F_4$ showed different temperature dependent characteristics, which led to monotonous improvement of the related FIR with an increase of temperature. The temperature sensing behaviour and the mechanism of a nano-micro heater of the developed phosphor may be presented by exploiting the stark sublevels of the $^1G_4$ level of the Tm^{3+} ion, which are generally responsible for the blue upconversion emission.

Table 1. Relative sensitivity (S_R) and absolute sensitivity (S_A) for $BaY_2F_8:Yb^{3+},Ho^{3+}$ phosphor at different temperatures.

No.	Temperature (K)	S_R (10^{-4} K^{-1})	S_A (10^{-4} K^{-1})
1	303	60.6	15
2	403	34.3	27.6
3	503	22.3	42.7
4	623	14.4	58.8

4. Conclusions

$BaY_2F_8:Yb^{3+},Ho^{3+}$ / Yb^{3+},Tm^{3+} phosphors were successfully prepared by co-precipitation method. The temperature-dependent UC emission of the as-prepared sample was utilized for evaluating its potential for optical thermometry. The temperature-dependent UC emission from the thermally-coupled excited levels $^5F_4$ and $^5S_2$ were investigated in the temperature range 300 to 623 K. the results suggest that $BaY_2F_8:Yb^{3+},Ho^{3+}$ phosphor has the potential to be used for optical thermometry. Similar results were obtained for $BaY_2F_8:Yb^{3+},Tm^{3+}$.

Acknowledgements

This research is supported by the South African Research Chairs Initiative of the Department of Science and Technology and National Research Foundation of South Africa (Grant 84415). The Edinburgh Instruments FLS980 system used in this study was funded by the National Research Foundation of South Africa (Grant EQP14080486021, 93214). The PL system used for temperature-dependent studies is supported by the rental pool programme of the National Laser Centre (NLC). The financial assistance from the University of the Free State is highly recognized.

References

[1]. S. Zhou, S. Jiang, X. Wei, Y. Chen, C. Duan, M. Yin, Optical thermometry based on upconversion luminescence in Yb^{3+}/Ho^{3+} co-doped $NaLuF_4$, *Journal of Alloys and Compounds*, Vol. 588, 2014, pp. 654-657.

[2]. P. Singh, P. K. Shahi, A. Rai, A. Bahadur, S. B. Rai, Effect of Li^+ ion sensitization and optical temperature sensing in Gd_2O_3: Ho^{3+}/Yb^{3+}, *Optical Materials*, Vol. 58, 2016, pp. 432-438.

[3]. P. Pal, T. Penhouët, V. D'Anna, H. Hagemann, Effect of temperature and pressure on emission lifetime of Sm^{2+} ion doped in MFX (M = Sr, Ba; X = Br, I) crystals, *Journal of Luminescence*, Vol. 142, 2013, pp. 66-74.

[4]. H. Peng, M. I. J. Stich, J. Yu, L. Sun, L. H. Fischer, O. S. Wolfbeis, Luminescent europium(III) nanoparticles for sensing and imaging of temperature in the physiological range, *Advanced Materials*, Vol. 22, 2010, pp. 716-719.

[5]. Y. Shen, X. Wang, H. He, Y. Lin, C.-W. Nan, Effects of Sm3+ doping on the temperature-dependent fluorescence intensity ratio of Er^{3+}, Sm^{3+}-co doped-yttria stabilized zirconia, *Journal of Alloys and Compounds*, Vol. 536, 2012, pp. 161-165.

[6]. B. P. Kore, A. Kumar, L. Erasmus, R. E. Kroon, J. J. Terblans, S. J. Dhoble, H. C. Swart, Energy transfer mechanisms and optical thermometry of $BaMgF_4:Yb^{3+},Er^{3+}$ phosphor, *Inorganic Chemistry*, Vol. 57, 2018, pp. 288-299.

(015)

Novel Integrated Magnetic Sensor Based on Hall Element

Janez Trontelj [1], **Damjan Berčan** [1] **and Miha Gradišek** [1]
[1]University of Ljubljana, Faculty of Electrical Engineering, Laboratory for Microelectronics, Trzaska 25, 1000 Ljubljana, Slovenia
E-mail: janez.trontelj1@guest.arnes.si

Summary: A novel magnetic sensor based on the cluster of Hall elements is described. The improvements of the proposed sensor over the single Hall elements is in reduction of power consumption, in the increase sensitivity and consequently better signal to noise and offset ratio, and in simplified design of the magnetic microsystem.

Keywords: Hall element array, Magnetic sensor power consumption reduction, Magnetic sensor signal to noise optimisation.

1. Introduction

Recent studies (Fig. 1), confirm a steady growth of the market for magnetic sensors based on Hall element [1] primarily integrated into ASICs used for position and motion control. They are mostly used for automotive, and robots markets. The key advantage of Hall element is its robustness and ease of integration in the integrated circuits without any technology modification. On the other hand the disadvantage of Hall element is its relatively low sensitivity, high offset voltage and noise and relatively high power consumption. Any improvement in this area will further boost the application of Hall elements.

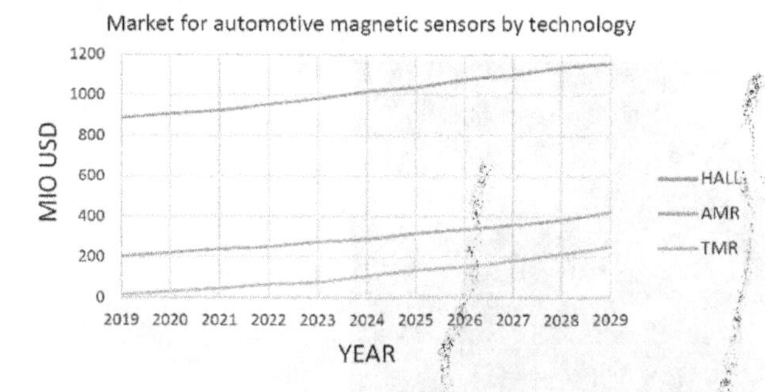

Fig. 1. Market production for various types of magnetic sensors for automotive magnetic sensors ASICs.

In this paper we are presenting the new magnetic sensor, based on Hall elements which will potentially dramatically change the share of Hall element based magnetic sensors over the competing techniques.

The proposed structure of the magnetic sensor addresses all mentioned deficiencies of a single Hall element.

The low sensitivity of single Hall element is improved by parallel operation of an array of Hall elements containing n Hall elements. This approach directly multiplies the sensitivity by n times.

However a parallel operation of Hall elements implies a non-desirable increase of power consumption. Power consumption management is therefore critical. In the paper the proposed power consumption management is discussed in detail showing the dramatic reduction of power consumption which can reduce the consumption of the sensor to a fraction of the single Hall element.

2. Novel Magnetic Sensor

In Fig. 2 the block diagram of the proposed sensor is shown. It consists of an improved single Hall element cell where the electronic circuit is added which allows the array to be simply concatenated in an expandable array. This resulting subcell is described in the paper and is patent pending [2].

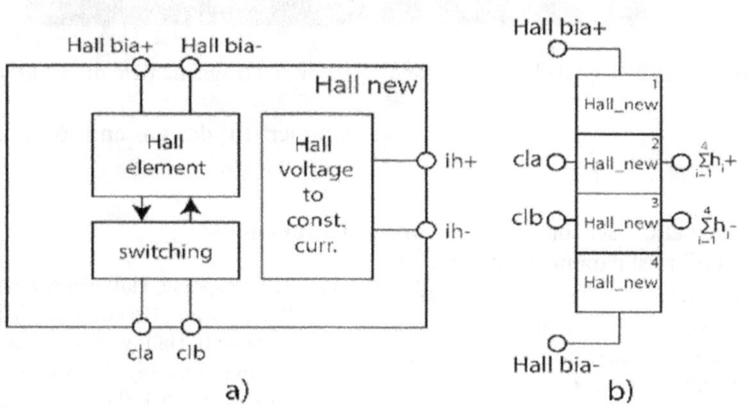

Fig. 2. Set-up scheme for soil impedance measurement.

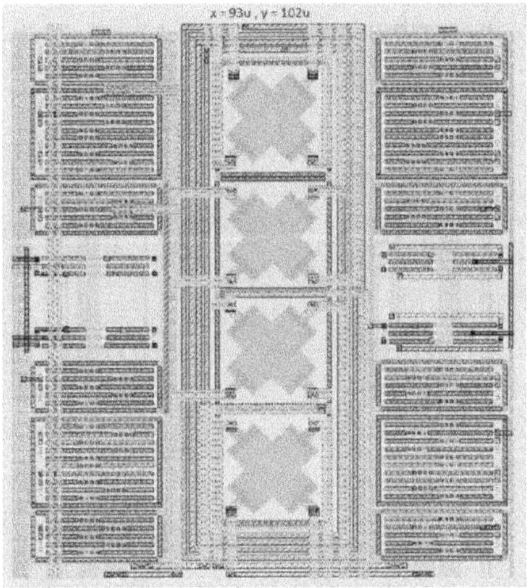

Fig. 3. Layout of the proposed magnetic sensor call in 0.35 μm CMOS technology.

The idea of the new sensor is to create a sensor cell with an array of Hall elements, but with simplified external electronic compared to the external electronic of a single Hall element. The structure of the proposed sensor is such that it allows simpler layout with less connections and that the total sensor size allows a single pixel dimensions of less than 100 μm × 100 μm, which is small enough to be implemented into any existing position measurement ASICs. The layout of the sensor cell is shown in Fig. 3. The sensor described has additional advantage over single Hall element, as the offset reduction is for many applications good enough not requiring the spinning technique normally used for ASICs with single Hall element as a ''virtual'' spinning of Hall elements in the array is employed. When additional spinning is employed, the resulted offset is further minimized.

3. Results

The measured results are in detail presented and discussed. Some results are summarized in Table 1.

Table 1. Hall sensor comparison.

	Sensitivity (V/T)	**Power Consumption (mA)**	**Noise (μT/√Hz)**	**Offset (μT)**
Novel Sensor	400	0.2	0.22	4
Single Hall	100	2	0.4	10

The design of the sensor allows higher frequency range compared to conventional Hall element. Fig. 4 shows the response to a magnetic field at 200 kHz demonstrating the bandwidth better than 200 kHz when using the clock frequency of 30 MHz.

The sensor has been implemented in 0.35 μm CMOS technology. The realized sensor cell has been measured and the results show the expected improvement in terms of reduction of power consumption, sensitivity and noise as shown in Table 1.

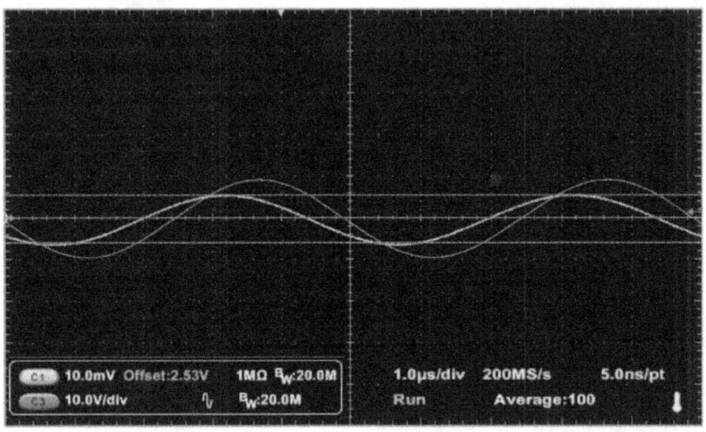

Fig. 4. Resulted signal of applied magnetic field – yellow and magnetic field of 200 kHz –magenta.

4. Conclusion

The proposed magnetic sensor proved to significant improvement of most parameters compared to a conventional Hall element. The improvements range up to 10 times less power consumption for the same sensitivity and signal to noise ratio and is scalable to improve sensitivity and noise for an order of magnitude at the same power consumption. It is simpler to design and occupies nearly the same silicon area.

References

[1]. R. S. Popović, Hall effect devices, 2nd Ed., *Institute of Physics Pub.*, Bristol, Philadelphia, 2004.

[2]. J. Trontelj, Hallov senzor, senzorski sklop, ki vključuje Hallove senzorje in načini krmiljenja senzorrskega sklopa: P 201900120, *Urad RS za Intelektualno Lastnino*, Ljubljana, 2019.

(016)

Mobile Optical Sensor System for DOM Monitoring in Stream Ecosystems

T. Posnicek [1], **R. Preuer** [1], **G. Weigelhofer** [2], **A. Eder** [3] **and M. Brandl** [1]

[1] Danube University Krems, Department for Integrated Sensor Systems, Dr. Karl Dorrek Str. 30, 3500 Krems, Austria

[2] Water Cluster Lunz, BioFrames, Dr. Carl Kupelwieser Promenade 5, 3293 Lunz, Austria

[3] The Federal Agency for Water Management, Institute for Land and Water Management Research, Pollnbergstraße 1, 3252 Petzenkirchen, Austria

Email: Thomas.posnicek@donau-uni.ac.at

Summary: The impact of agricultural land use on the composition of dissolved organic matter (DOM) and its effects on the aquatic carbon cycle are still largely unknown. Fluorescence spectroscopy is the standard method for identifying DOM fluorophores via excitation–emission matrices. The instruments for this method are unhandy and can only be used in the laboratory. A sensor for dissolved organic matter in stream ecosystems based on fluorescence measurement was developed. It was designed as an easy to use mobile optical system for online monitoring of DOM under field-conditions. For the determination of DOM two indices are used, namely the freshness index (BIX) and the fluorescence index (FIX).

Keywords: Dissolved organic matter, DOM, Agriculture, Freshness index, Fluorescence index.

1. Introduction

Agriculture is the dominant form of land use in Lower Austria with a share of more than 46 % of the total area. Agriculture supplies considerable amounts of dissolved organic matter (DOM) to stream ecosystems, changing basic metabolic processes at the water-sediment interface and influences the ecological state and the health of water systems [1, 2]. The amount and composition of the terrestrial DOM influences the growth and respiration of benthic microbes as well as CO2 emission from streams. However, the effect of agricultural land use on the composition of DOM and its effects on the aquatic carbon cycle are still largely unknown [3].

Recent advances in fluorescence spectrophotometry have provided a new tool for the rapid identification of DOM fluorophores via excitation and emission matrices (EEM). An EEM shows fluorescence centers assigned to different DOM components, such as humic, fulvic or protein-like fluorophores [4]. This can be used to identify anthropogenic DOM sources in streams [5] and to distinguish bioavailable from refractory DOM components that determine microbial activity and organic matter processing.

DOM quality can change rapidly due to microbial processing of unstable compounds or exposure to sunlight during transport and handling [2], which affects data quality. Therefore, an easy-to-use real-time fluorescence sensor is required to provide DOM data in high spatial and temporal resolution without such interferences [6].

2. Methods

The laboratory instruments (fluorescence spectrometers) for measuring DOM are unwieldy, heavy and cannot be used outdoors. The common measuring principle is based on fluorescence spectroscopy. To avoid scanning the entire spectrum of the sample, different spectroscopic single wavelength indices were analyzed to determining DOM in aquatic systems. Two prominent indices were considered, namely the freshness index (BIX) and the fluorescence index (FIX). These indices are ratios and thus the most important disturbance parameter (amount of dissolved organic carbon DOC) can be compensated.

The measuring system consists of several LEDs that excite the fluorescence emission of the water sample. The fluorescence light is filtered at the peak emission wavelength with various optical filters mounted on a filter wheel for easy switching. The excitation light for the BIX index is generated by a 310 nm UV LED. To calculate the BIX index, the fluorescence emission was measured at the wavelength 380 nm and 430 nm by adjusting the filter wheel to the correct filter position and then amplifying it with a high gain photo multiplier (PMT). To avoid any influence of ambient stray light on the detector signal, the light source is pulsed by 1 kHz and the photomultiplier output signal is amplified phase synchronously by a lock-in amplifier. The measurement of the FIX index is based on the same principle. For this purpose, a 370 nm UV LED and two optical filters for the fluorescence emission light at 450 nm and 500 nm are used. Fig. 1 show the 3D printed prototype of the measuring chamber and the filter wheel containing different excitation light filters.

Fig. 2 show the complete sensor system. For outdoor use the sensor is housed in a robust and waterproof case. A rechargeable battery supplies the system with power. To perform a measurement, place the sample in the measuring chamber and press the "Measure" button on the touch screen. A flow cell can be used in combination with an external pump system for online monitoring. An integrated GPS module offers the possibility to record the location of the measurement outdoor. The measurement data can be sent to a PC via USB or to a mobile phone via Bluetooth.

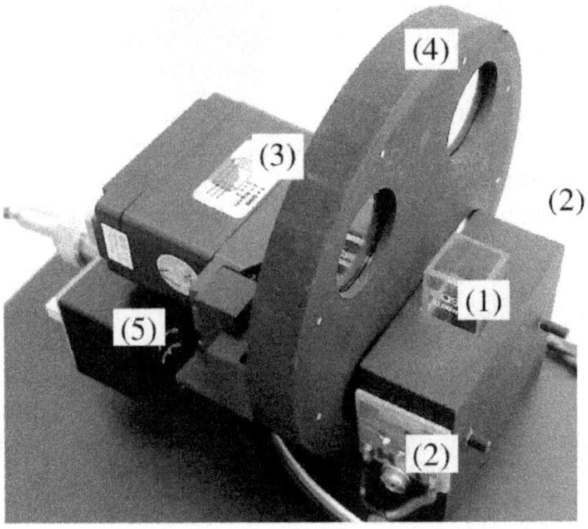

Fig. 1. 3D printed prototype of the measurement chamber with cuvette (1), LED's (2), servo (3) for rotating the optical filter wheel (4) and the photomultiplier (5).

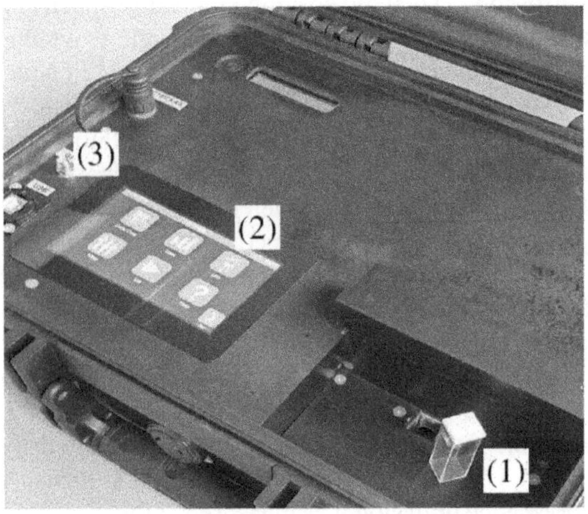

Fig. 2. Sensor system ready to use with measurement chamber (1), touch display (2) and external power connector, USB and GPS (3).

3. Results

The functional verification of the sensor system was carried out on the basis of laboratory and field tests. The sensor was compared with the fluorescence spectrometer LS55 (Perkin Elmer, USA). Samples were taken from a nearby river, see Fig. 3.

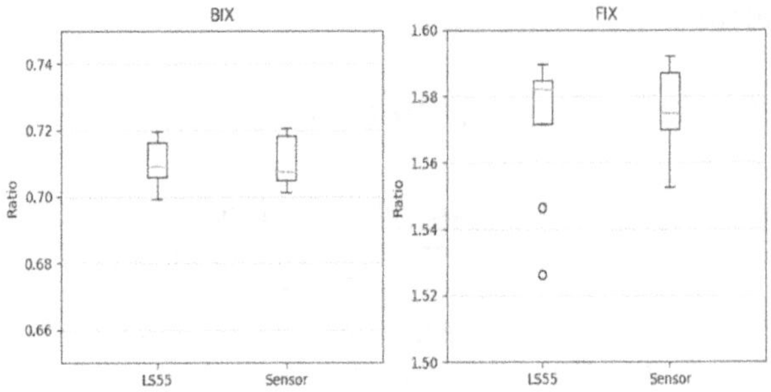

Fig. 3. Comparison of BIX and FIX between the LS55 spectrometer and the sensor (n = 9).

4. Conclusions

The developed sensor offers a simple possibility of real-time measurements of DOM in the free field.

Acknowledgements

The authors thank the NÖ Forschungs- und Bildungsges.m.b.H. (NFB) for financially supporting the project (SC15-002).

References

[1]. A. Eder, G. Blöschl, F. Feichtinger, Indirect nitrogen losses of managed soils contributing to greenhouse emissions of agricultural areas in Austria: Results from lysimeter studies, *Nutr. Cycl. Agroecosyst*, Vol. 101, 2015, pp. 351-364.

[2]. Ch. Fasching, T. J. Battin, Exposure of dissolved organic matter to UV-radiation increases bacterial growth efficiency in a clear-water Alpine stream and its adjacent groundwater, *Aquatic Sciences*, Vol. 74, 2011, pp. 143-153.

[3]. H. F. Wilson, M. A. Xenopoulos. Effects of agricultural land use on the composition of fluvial dissolved organic matter, *Nature Geoscience*, Vol. 2, 2009, pp. 37-41.

[4]. A. Baker, R. Inverarity, M. Charlton, Detecting river pollution using fluorescence spectrophotometry: Case studies from the Ouseburn, *NE England Environ. Poll.*, Vol. 124, 2003, pp. 57-70.

[5]. N. Hudson, A. Baker, D. Reynolds, Fluorescence analysis of dissolved organic matter in natural, waste, and polluted waters-a review, *River Research and Applications*, Vol. 23, 2007, pp. 631-649.

[6]. M. Brandl, K. Kellner, T. Posnicek, I. Bado, D. Falkenhagen, Spectroscopic hemoglobin and bilirubin measurement on optically opaque particulate fluids, *Sensors & Actuators: B. Chemical*, Vol. 182, 2013, pp. 711-717.

(017)

On the Fly Soil Classification Using Impedance Spectroscopy

Olga Chambers [1], Janez Trontelj [1], Jurij F. Tasič [2]
[1] University of Ljubljana, Faculty of Electrical Engineering, Laboratory for Microelectronics, Trzaska 25, 1000 Ljubljana, Slovenia
[2] Innovation technologies d.o.o., 8290 Sevnica, Slovenia
E-mail: janez.trontelj1@guest.arnes.si

Summary: The moisture influence on the soil impedance measurements is presented, as it can be a critical problem during real-time analysis. Comparative analysis was performed using a portable sensor developed at the LMFE laboratory to estimate the influence of the moisture. The measurements indicate that a larger difference between measurements corresponding fertilized soil samples when the moisture level is high.

Keywords: Soil classification, Soil impedance spectroscopy, Real-time soil sensor.

1. Introduction

Soil analysis is an essential problem that plays an important role in a nowadays agriculture [1, 2] to provide complex information about its physical and chemical properties.

The literature indicates a common use of impedance methods for soil quantitative and qualitative analysis that may be used to design low-cost portable sensors for real-time application, i.e. on-the-go soil sensors. In the paper [3] it has been shown that it is possible to perform the classification of soil according to soil impedance spectral data. However, environmental factors, such as the amount of water, ambient light, bulk density and temperature make it challenging to fully understand and provide a connection between measurements and soil properties.

This study investigates the water influence as is a most common factor that influence the soil characterisation using electrical sensors in agriculture.

The goal is to perform the spectral analysis of the soil sample within less than few seconds time that includes the soil sample preparation, measurement and classification using portable impedance sensor designed in LMFE laboratory.

2. Materials and Method

2.1. Materials

Table 1. Chemical analysis of the soil samples.

ID	Ph in KCl mg/100 g	P2O5 mg/100 g	K2O mg/100 g	Mg mg/100 g
P001	6.6	87	48	22
P002	7.0	39	39	21
P003	4.4	6.2	18	9.3

Soil samples were collected from the soil at three different agriculture fields with different properties. To investigate the influence of the water on the soil measurement, an appropriate amount of deionized (DI) water was mixed with air-dried soil. Finally, the soil chemical characterisation was performed at certified laboratory [4] and shown in Table 1.

2.2. Impedance Measurement Using Portable Sensor

Laboratory set-up (see Fig. 1) consists of portable sensor developed at LMFE laboratory that enables fast data capture in less than 1 second. The sensor is connected to personal computer for data storage and processing.

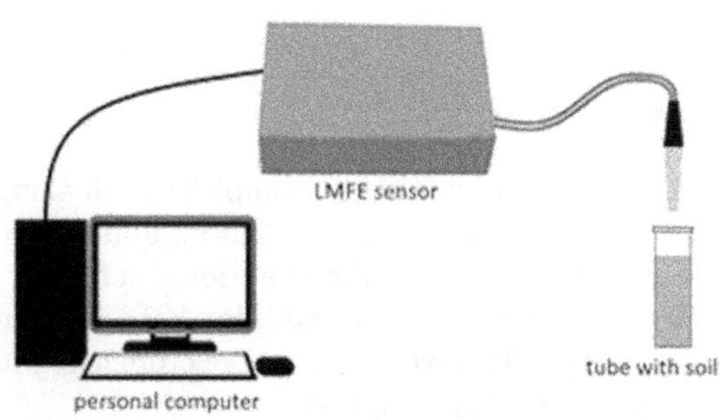

Fig. 1. Set-up scheme for soil impedance measurement.

Measurements were performed for a fixed sample volume (1 ml) placed in a cylindrical plastic holder.

In Figs. 2 and 3 we present the experimental results corresponding to different soils classes with respect to the level of the nutrients, where P003 contains the lowest level of nutrients, P001 with P002 have higher level of the nutrients. Fig. 2a shows the results of the dry soil without any preparation but sieving. Fig. 2b presents the results of the 15 % soil moisture measurements. It can be seen that information about soil classes is completely lost in dry soil, and in soil with low-level of moisture some information is detectable, however, with low-level degree of the probability.

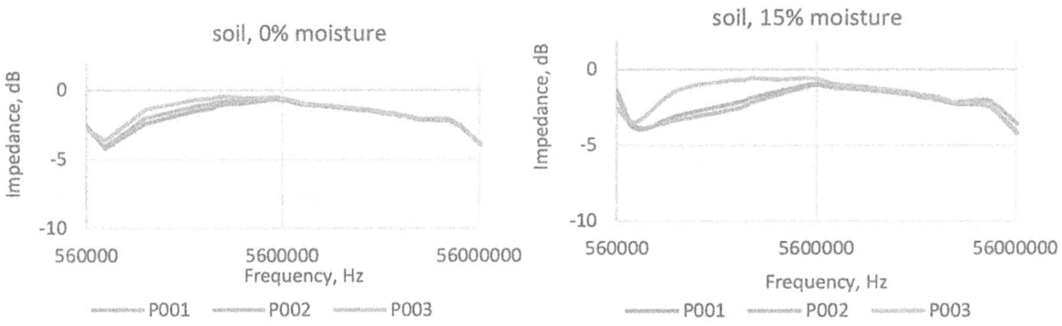

Fig. 2. Plots of the impedance spectra obtained for soil samples having different moisture level: dry soil and 15 % soil moisture. P001, P002 and P003 are names for soil samples under analysis, where P003 contains the lowest level of nutrients, P001 with P002 have higher level of the nutria.

The situation is different if the soil contains a higher degree of the humidity. Fig. 3 demonstrates a clear distinction of three different soils. Top-image shows results corresponding 30 % soil moisture where the difference between soils is notable. Image below shows the reference results obtained from laboratory prepared soil samples and verified by a certified chemical procedure.

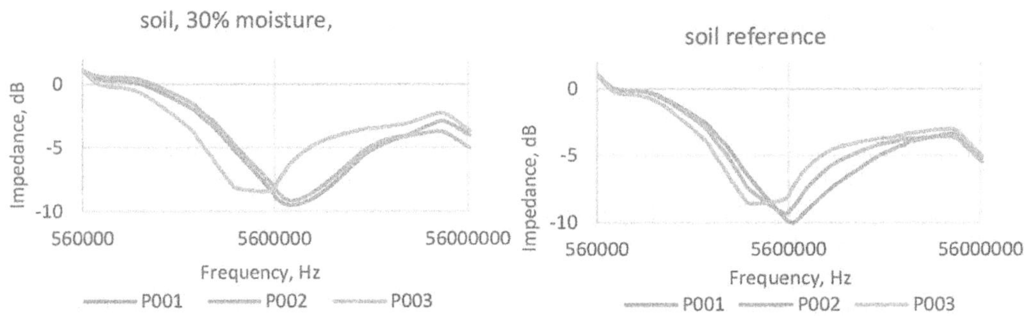

Fig. 3. Plots of the impedance spectra obtained for soil samples having different moisture level: 30 % soil moisture and soil laboratory prepared reference. P001, P002 and P003 are names for soil samples under analysis, where P003 contains the lowest level of nutrients, P001 with P002 have higher level of the nutrients.

4. Conclusion

In this paper the results of a simplified soil sample preparation are presented, which can be implemented in an agriculture machine like vibrational subsoiler or fertiliser spreader with an addition of automatic fertiliser dosage. This will increase the crop, using less fertilisers and less human effort.

Soil samples for analysis were collected from an agricultural field having similar texture characteristics, however with different chemical content.

Experiments performed in our research indicates that dry soil as well as 15 % moisture soil are not suitable for nutrients measurements. Good correlation between nutrition level and soil impedance magnitudes was observed for 30 % moisture of the soil.

References

[1]. R. Albrektiene, M. Rimeika, E. Zalieckiene, V. Saulys, A. Zagorskis, Determination of organic matter by uv absorption in the ground water, *J. Environ. Eng. Landscape Manage*, Vol. 20, 2012, pp. 163-167.

[2]. S. Wang, Z. Li, J. Zhang, J. Wang, L. Cheng, T. Yuan, B. Zhu, Experimental study on frequency-dependent properties of soil electrical parameters, *Electric Power Syst. Res.*, Vol. 139, 2016, pp. 116-120.

[3]. O. Chambers A. Sešek, R. Ražman, J. F. Tasič, J. Trontelj, Fertiliser characterisation using optical and electrical impedance methods, *Computers and Electronics in Agriculture*, Vol. 155, 2018, pp. 69-75.

[4]. Agrecalture Institute of Slovenia, http://arhiv.kis.si

(019)

Microfabricated Thermal Flow Rate Sensor

B. Neji [1]
[1] College of Engineering and Technology, American University of the Middle East, Kuwait
Tel.: +965 22251400, fax: +965 26548484
E-mail: bilel.neji@aum.edu.kw

Summary: Flow rate sensors have been widely used in biomedical engineering and research. Miniaturization of such critical devices has become essential to operate in micro scale circuits, including microfluidic devices that can be used in applications such as preclinical diagnosis and drug delivery. This paper presents the design, microfabrication and characterization of a flow rate sensor. The device is capable of measuring fluid flow rates starting from $10\ \mu L/h$ up to $1\ mL/h$. The sensing part of the device, excluding the power supply, measures $1\ mm$ by $4\ mm$. The channel width and height are both equal to $100\ \mu m$. Experiment results demonstrate that measured output voltages ranging from $0.1\ mV$ to $0.28\ mV$, correspond to flow rates between $100\ \mu L/h$ and $700\ \mu L/h$. The device response time is less than 3 seconds.

Keywords: Flow rate measurement, Heat transfer, Temperature sensing, Microfluidics, Microfabrication.

1. Introduction

With the recent advances in biology and medical fields, nano and micro sensors became essential for accurate measurements and diagnosis. Flow rate sensors are key components used in a wide range of applications, including drug delivery, blood flow monitoring, and Lab on a Chip devices [1-3]. Several techniques have been used to measure fluid flow rate. These techniques use either thermal or non-thermal methods. Here, the focus is on thermal based sensors, with no moving parts. Thermal flow sensors have several advantages such as their small size, low power consumption, and high sensitivity at different flow rates. Several thermal flow sensing methods have been employed in the field, including hot-wire and hot-film, time of flight, and calorimetric method, which is mainly based on measuring the difference between the fluid's initial temperature and its temperature after being in contact with a heating element [4, 5].

In this paper, a calorimetric microfabricated thermal flow rate sensor is proposed for microfluidic applications. Experimental results demonstrate that an applied heat generated by an applied voltage V_{in} to the heating element can be transduced into an electrical signal V_{out}, measured at the outlet temperature sensor, to determine the fluid flow rate.

2. Thermal Flow Sensors Overview

The working principle of thermal flow rate sensors is based on the heat transfer between the heating element, also called heater, and temperature sensors through the microfluidic channel. The output temperature is proportional to the flow rate change. There are different forms of thermal flow sensors, and the most known ones are hot-wire and hot-film, time-of-flight, and calorimetric techniques. Hot-wire and hot-film method relies on the measurement of the heat transfer from the heating element to the fluid: The sensor uses a resistive wire in the case of hot-wire technique, and a resistive thin film is used in the case of hot-film technique. When the liquid flows over the sensor, carrying some of the heat from the heating element, the temperature sensor resistance changes accordingly, and that change corresponds to a specific flow rate change. On the other hand, time-of-flight technique is based on the concept of measuring the transition time of a thermal pulse generated by the heating element to reach the temperature sensing element. The measured time is proportional to the properties of the fluid, including its thermal conductivity and its diffusivity, the average velocity of the flow, and the distance between the heater and the temperature sensor. Calorimetric sensors use a heating element and two temperature sensors: one at the inlet of the channel, and the other at the outlet of the channel. The outlet temperature sensor detects the thermal profile caused by the liquid flowing by. The difference in the thermal profile will be proportional to the liquid flow rate [6]. The flow direction can be deduced from the difference between the two temperature sensors' measurement. It should be mentioned that the type of the fluid and its proprieties have to be well known for an accurate measurement of the flow rate. The presented thermal flow rate measurement techniques are illustrated in the diagrams of Fig 1.

Now a day, the miniaturization of flow rate sensors based on the described techniques has become essential in the field of micro and nano technology. Thermal flow sensors have a great utility in industrial, biomedical and research fields for monitoring and control operations [7]. Several research projects have been focusing on micro systems development that can accurately measure fluid flow rates. A MEMS based flexible flow sensor for online monitoring of body fluid and blood is presented by the authors in [8]. The authors analyzed the effect of different geometries of the resistive heating element placed on a substrate of different materials. Thermal flow rate sensors have also been used to monitor the accuracy and life time of micro pumps. Drug supply micro pumps have been widely used as a medical treatment of diabetes to supply insulin. These pumps have to be small and accurate. The research in [9] presents an osmotic pressure pump, and a magnetic-driven pump, using miniaturized actuators based on a semi-permeable membrane and a magnet. In the recent years, with the research and technology advances, it became possible to fabricate complete systems combining a variety of micro sensors. PDMS (Polydimethylsiloxane) has been widely used to create microfluidic systems made of different modules that are either microfabricated using lithography and metal deposition, or using advanced 3D printing [10, 11].

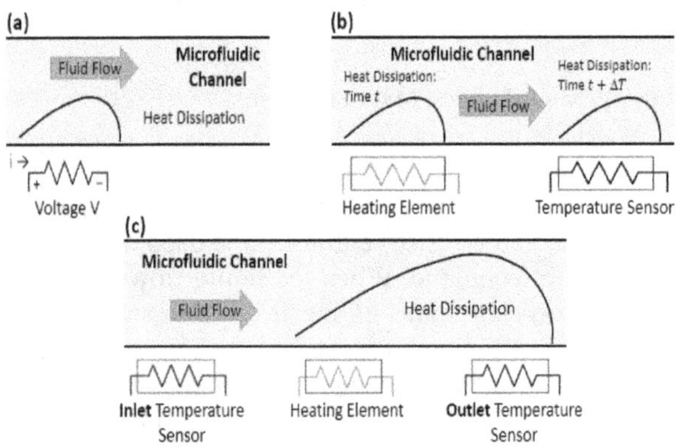

Fig. 1. (a) Hot-wire and thin-film method; (b) Time of flight method; (c) Calorimetric method [4].

3. Flow Rate Micro Sensor Design

3.1. Flow Rate Sensor Principle

The proposed thermal flow rate sensor uses the calorimetric technique at the micro scale. Energy cannot be created nor destroyed; it can only be converted from one form to the other. The designed sensor uses this concept to transduce an applied voltage into thermal energy, to be transferred to a fluid in contact with a heating element. The fluid will carry the heat and transfer it to the outlet temperature sensor. An inlet temperature sensor is used to determine the fluid's initial temperature. Each measured temperature raise at a specific applied voltage will correspond to a specific flow rate of the fluid. A block diagram of the proposed flow rate sensor is shown in Fig. 2.

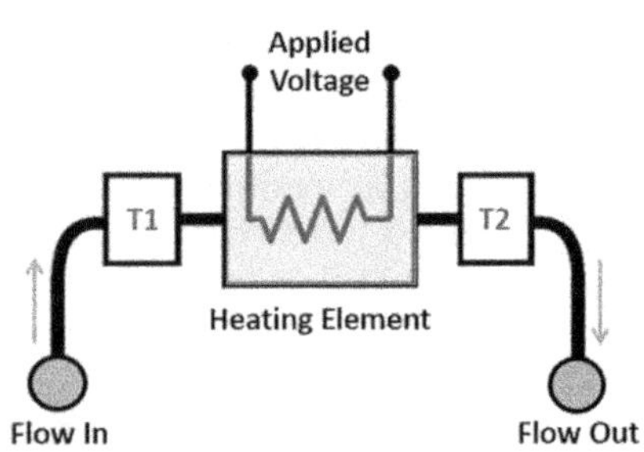

Fig. 2. Flow rate sensor block diagram.

3.2. Sensor Simulation

The simulation of the thermal flow rate sensor has been performed using two different fluids: oil, and water. The proposed design of the thermal flow rate is shown in Fig. 3. Different structures with different dimension have been simulated.

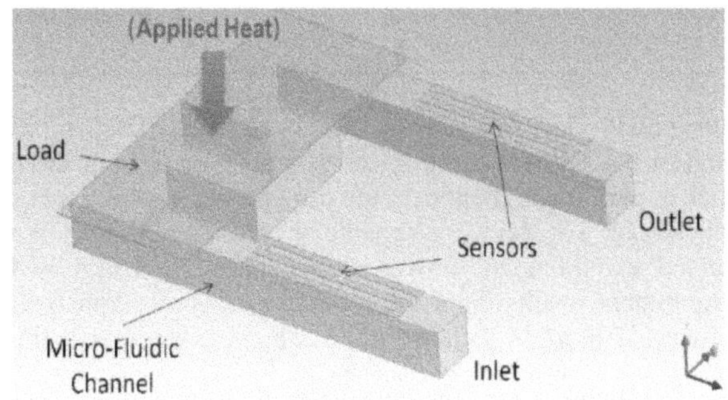

Fig. 3. Proposed micro calorimeter design for simulation.

The simulation results confirm the working principle of the calorimetric method for flow rate sensing. This can be illustrated in Fig. 4.

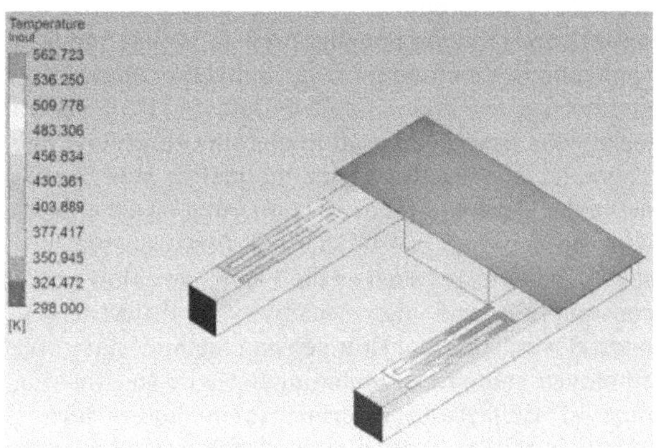

Fig. 4. Example of temperature profile of the heating element and the inlet & outlet sensors.

The simulation results in Table 1 correspond to a heating element structure that has the following specifications: Thickness is 1 μm; length is 160 μm, and width is 400 μm. The used flow rate is 1000 $\mu l/h$ and the applied power to the load is equal to 100 μW.

Table 1. Temperature variation between the input sensor and output sensor using oil then water as fluids.

	Inlet Sensor's $T1$	Outlet Sensor's $T2$	ΔT (°K)
Oil	352	366	14
Water	298	320	22

It can be concluded that water is better than oil for heat transferring from the heating element to the outlet temperature sensor. Water transfers almost double the amount of heat transferred by oil. Therefore, water will be used to simulate the real prototype.

3.3. Flow Rate Sensor Physical Design

The physical design of the microfabricated flow rate sensor is shown in Fig. 5a. It consists of a microfluidic channel with a serpentine structure bonded on the heating element to be used to transfer the heat to the fluid. PDMS is used to make the microfluidic channels. The heat is generated by the applied voltage V_{in}, the equivalent of an applied power P_a. The heating element is a thin film made of tantalum nitride. Tantalum nitride has been chosen due to its low temperature coefficient of resistance (TCR). Low TCR is critical for the heating element to maintain uniform resistance in the large range of operating temperatures. The temperature sensors are made of platinum due to its linearity and relatively high TCR [12]. All metals and PDMS are deposited on a glass substrate. Glass has a very low thermal conductivity, which is 100 times smaller than that of silicon. Lower thermal conductivity is required to limit unwanted heat loss to the substrate and the surroundings. Fig. 5b presents a zoom-in of the microfluidic channel, the temperature sensors and the heating element. The heating element is shown in red; the microfluidic channel is in blue; and the inlet & outlet temperatures sensors and the coplanar waveguide, used to transfer the applied voltage to the heating element, are in grey.

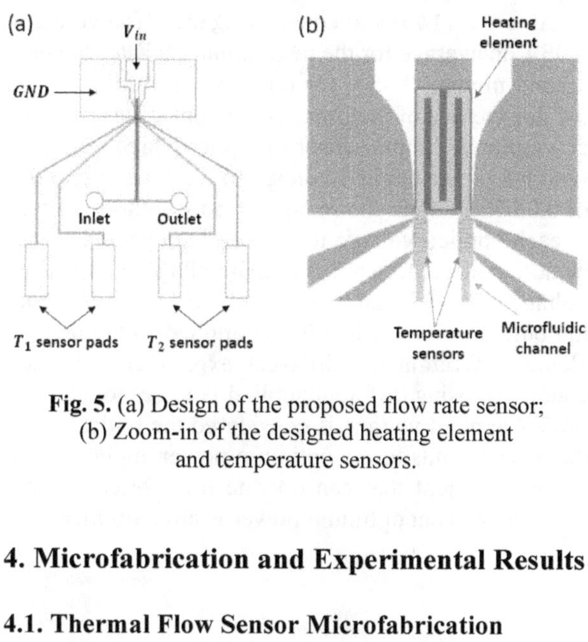

Fig. 5. (a) Design of the proposed flow rate sensor; (b) Zoom-in of the designed heating element and temperature sensors.

4. Microfabrication and Experimental Results

4.1. Thermal Flow Sensor Microfabrication

The proposed flow rate sensor design is microfabricated in a 1000 class cleanroom environment. Photolithography and e-beam evaporation are employed for tantalum nitride and platinum deposition on glass. In the process of microfluidics fabrication, a mold is first produced. The mold is prepared on a silicon substrate and using the concept of photolithography to get the required design onto the mold. Then, soft lithography is employed to make the PDMS fluidic layer. It should be noted that different devices are fabricated while optimizing several parameters of the design, including the width of the sensors' patterns, the spacing between the patterns, the microfluidic channel width, and the minimum spacing between channels. In addition, metal deposition process is optimized by varying several parameters such as the deposition rate, temperature, and the chamber pressure. Fig. 6 represents a picture of the microfabricated flow sensor with PDMS channels and metal deposited on the glass substrate, after the different mentioned optimizations.

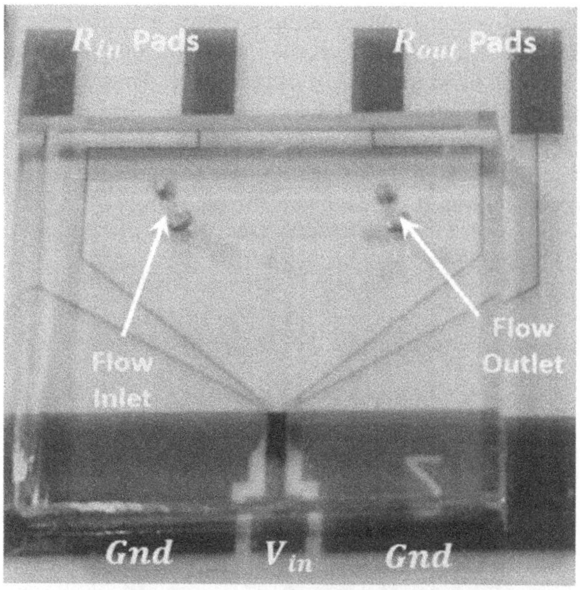

Fig. 6. Microfabricated flow rate sensor. The total device size is $1.5\ cm \times 1.5\ cm$.

A microscopic image of the sensing part of the device, including the heating element and the temperature sensors (Microfluidic channels not bonded) is presented in Fig. 7a. A zoom-in of the microfabricated microfluidic channel, with a serpentine shape, in contact with the heating element is shown in Fig. 7b.

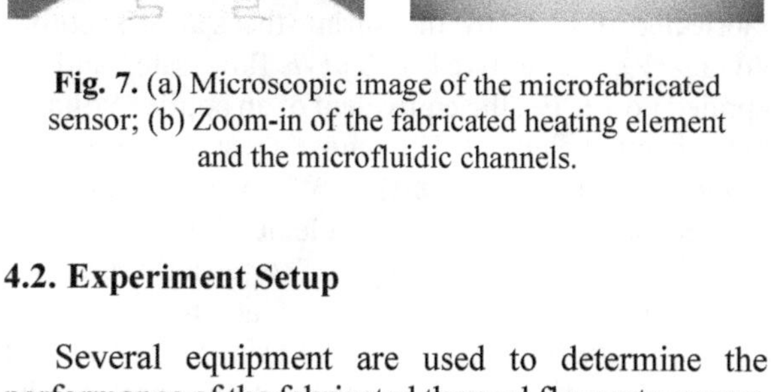

Fig. 7. (a) Microscopic image of the microfabricated sensor; (b) Zoom-in of the fabricated heating element and the microfluidic channels.

4.2. Experiment Setup

Several equipment are used to determine the performance of the fabricated thermal flow rate sensor. The measurement setup is shown in Fig. 8a. A "Fisher Scientific" mechanical syringe pump is used to pump the fluid in the micro channel. In addition, a nano volt meter, Agilent 34420A, is used to measure the output voltages of the temperature sensors. A Signatone probe station with six probes and a microscope is used to setup the connections between the device and the other equipment: two probes are used to apply DC power to the heating element, and four probes are used to measure the voltage across the temperature sensors. The applied DC power to the heating element is performed using a Tetronix ps2521 power supply. An applied power to the heating element will generate a heat that is transferred via the fluid to the outlet sensor, resulting in an increase in its resistance. Therefore, the sensor's output voltage increases accordingly. Deionized water is used for the experiment. The microfabricated flow sensor, while being tested on the probe station, is shown inside the blue circle of Fig. 8b.

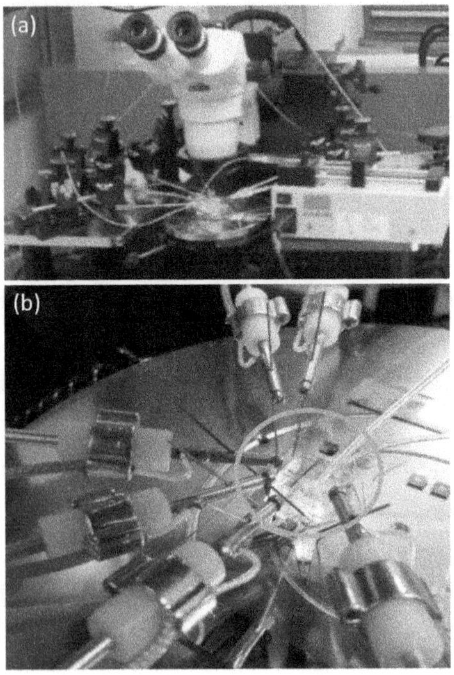

Fig. 8. a. DC measurement initial setup; b. Probes setup for DC power measurement.

4.3. Experimental Results

Several microfabricated devices are used for the characterization of the proposed flow sensor. The calorimetric principle is tested using one of the fabricated devices by measuring the sensor's output voltage V_{out} at a fixed 100 $\mu L/h$ flow rate, and an applied power P_a, the equivalent of an applied voltage, to the heating element. The senosr's response is around 0.9 mV when P_a is 2 mW. When increasing the applied power to the heating element, V_{out} increases accordingly as shown in Fig. 9. The sensor's response V_{out} increses to around 0.23 mV at an applied P_a equal to 24 mW. It can be noted from the figure that the response time of the sensor while incresing the applied power to the heating element is in the order of 2 to 3 seconds.

The optimization of the device's power consumption is essential in order to increase the battery life time of the complete system. Therefore, it is very important to determine the value of the smallest typical amount of P_a, whithout disturbing the function of the microfabricated device. Therefore, several experiments with different applied power P_a to the heating element are conducted separately to define the typical value of P_a to be used to operate the flow rate sensor.

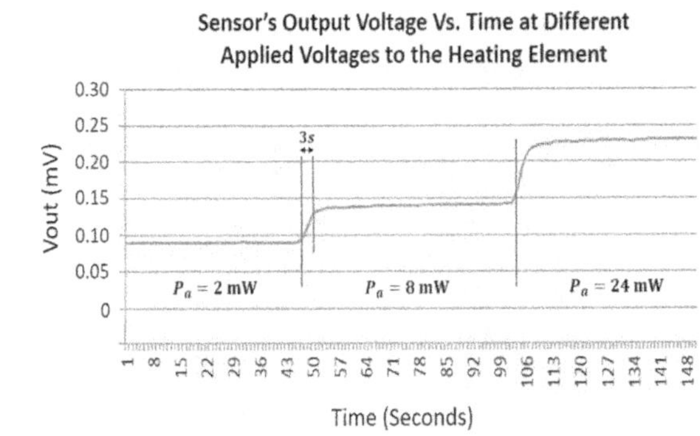

Fig. 9. Temperature sensor output voltage vs. time at different applied powers at the heating element.

The blue curve in the chart of Fig. 10 shows the applied voltage to the load over time. It is initially equal to 0 V (14 mV if considering the offset voltage), 0.05 V in average for the next minute, 0 V for the next second minute, 0.05 V for the next third minute, and 0 V for the rest of the time. On the other hand, the red curve represents the sensor's output voltage over time function of the applied voltage to the load, either 0 V or 0.05 V. Despite the relatively noisy output voltage at each applied voltage to the heating element, it can be noticed that the sensor's output voltage at an applied voltage of 0.05 V can be relatively differentiated from the output voltage when 0 V is applied to the heating element. Whithin the different experiments that are conducted using different applied voltages to the load, 0.05 V applied voltage is considered the typical value that corresponds to the optimum power applied to the heating element that can operate the fabricated flow rate sensor. That optimum power is around6 mW.

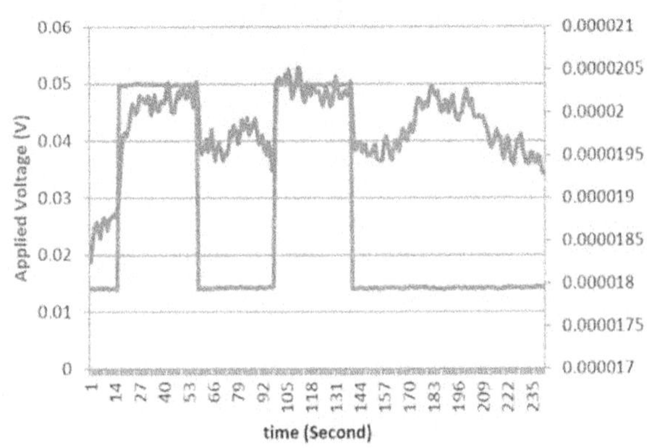

Fig. 10. Temperature sensor output voltage vs. applied voltage to the heating element.

The described experiment is conducted for different microfabricated devices to verify the consistence of the operation. Fig. 11 shows the results summary of two different devices, different from the device used for the results in Fig. 10. The blue diamonds represent the average of sensor's output voltage when no power is applied to the heating element, 0 V, and red squares represent the average of sensor's output voltage when 0.05 V is applied to the heating element. It can be confirmed that there is a noticeable difference between the case where power is applied to the heating element and the case where no power is applied to the heating element, and this confirms the consistency of the device operation for an applied power $P_a = 6\,mW$.

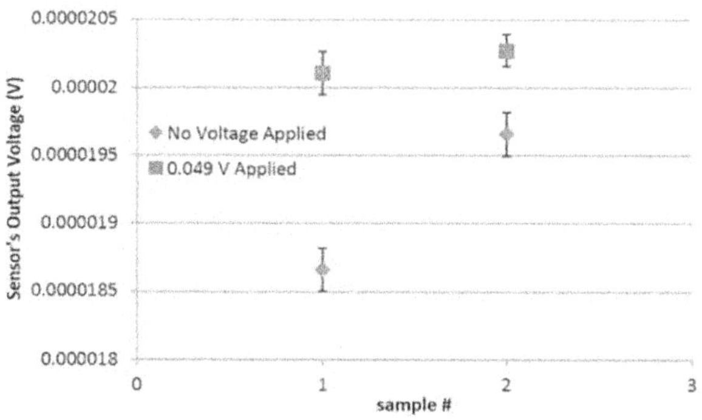

Fig. 11. Temperature sensor output voltage vs. applied voltage to the heating element using two different devices.

A set of microfabricated devices is tested to determine the relationship between the output voltage V_{out} of the temperature sensor at the outlet and the fluid flow rate. It should be noted that each device has to be initially calibrated in order to be used in a complete system. V_{out} vs. the flow rate is presented in the graph of Fig. 12. Different applied powers to the heating element are experimented: 2.5 mW (black curve), 6 mW (orange curve), and 12 mW (blue curve). It can be concluded that the resolution of the sensor increases as the applied power to the heating element is bigger.

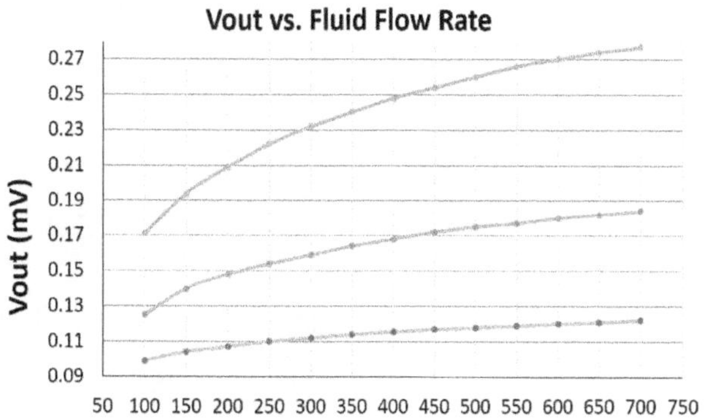

Fig. 12. Outlet temperature sensor voltage vs. fluid flow rate at different applied powers to the heating element.

Results indicate that at a fixed applied voltage to the heating element, the flow rate increases, while less heat is being transferred to the output temperature sensor, and V_{out} variation is reduced.

5. Conclusion

A microfabricated flow rate sensor is demonstrated. Experimental results show that an applied voltage V_{in} to the heating element can generate heat that is transduced into an electrical signal V_{out} to determine the corresponding flow rate of the fluid in use. In the future, the device will be tested in real life biomedical applications. This invasive approch will be much easier to install, use, and manipulate, compared to non-invasive methods such as ultrasound and Doppler measurement [13].

References

[1]. B. Khan, S. Ahmed, V. A Kakkar, Comparative analysis of thermal flow sensing in biomedical applications, *International Journal of Biomedical Engineering and Science,* Vol. 3, 2016, pp. 2016-3301.

[2]. S. Silvestri, E. Schena, Micromachined flow sensors in biomedical applications, *Micromachines*, Vol. 3, 2012, pp. 225-243.

[3]. S. Cerimovic, F. Keplinger, R. Beigelbeck, A. Jachimowicz, H. Antlinger, B. Jakoby, Monitoring the glycerol concentration in aqueous glycerol solutions using a micromachined flow sensor, in *Proceedings of the Microelectronic Systems Symposium*, 2014, pp. 1-6.

[4]. J. T. W. Kuo, L. Yu, E. Meng, Micromachined thermal flow sensors – A review, *Micromachines*, Vol. 3, 2012, pp. 550-573.

[5]. J. Tyson, Micro-thermal Flow-rate Detection for Lab on-a-chip Technologies, Project Description, 2018.

[6]. B. Neji, J. Xu, A. H. Titus, J. Meltzer, Micro-fabricated dc comparison calorimeter for RF power measurement, *Sensors,* Vol. 14, 2014, pp. 20245-20261.

[7]. F. Shaun, et al., On the co-integration of a thermo-resistive flow-rate sensor in a multi-parameter sensing chip for water network monitoring, in *Proceedings of the 19th International Conference on Solid-State Sensors, Actuators and Microsystems (TRANSDUCERS'17)*, 2017, pp. 1069-1072.

[8]. Y. Hayashida, et al., C208 development of micro pressure sensor for applying to MEMS blood-flow meter, in *Proceedings of the JSME Conference on Frontiers in Bioengineering,* Vol. 26, 2015, pp. 145-146.

[9]. Y. Mukouyama, et al., PDMS balloon pump with a microfluidic regulator for the continuous drug supply in low flow rate, in *Proceedings of the 28th IEEE International Conference on Micro Electro Mechanical Systems (MEMS'15)*, Estoril, 2015, pp. 666-669.

[10]. C. M. Ho, S. H. Ng, K. H. H. Li, Y. Yoon, 3D printed microfluidics for biological applications, *Lab on a Chip*, Vol. 15, 2015, pp. 3627-3637.

[11]. A. K. Au, W. Huynh, L. F. Horowitz, A. Folch, 3D printed microfluidics, *Angew. Chem. Int. Ed.*, Vol. 55, 2016, pp. 3862-3881.

[12]. K. Hamaguchi, T. Tsuchiyama, J. Matsushita, Oxidation of tantalum nitride, *Materials Science Forum*, Vol. 761, 2013, pp. 125-129.

[13]. L. Ma, J. Liu, J. Wang, Study of the accuracy of ultrasonic flowmeters for liquid, *AASRI Procedia*, Vol. 3, 2012, pp. 14-20.

(020)

Low Frequency Hydrophone for Marine Seismic Exploration Systems

E. V. Egorov [1], A. S. Shabalina [1], D. L. Zaytsev [1] and G. Velichko [2]
[1] Moscow Institute of Physics and Technology, 9 Institutsky Per., 141701 Dolgoprudny, Russia
[2] «Yuzhmorgeologiya», JSC, 20 Krymskaya St., 353461 Gelendzhik, Russia
Tel.: + 74987446995, fax: + 74987446995
E-mail: EgorovEV@mail.ru

Summary: The low frequency hydrophone with the frequency range 1-300 Hz for marine seismic exploration systems was developed. The principle of operation of the hydrophone is based on molecular electron transfer [1] that allows you to achieve high sensitivity in the low frequency region. The paper presents the stabilization method of the frequency response in all frequency range with depth up to 30 m. Laboratory and marine tests confirmed the stated characteristics and the possibility of using this sensor in marine seismic systems. An experimental sample of a hydrophone has successfully passed a comparative marine test in Gelendzhik Bay (Black Sea) with technical support from JSC 'Yuzhmorgeologiya'.

Keywords: Hydrophone, Acoustic sensor, Molecular-electronic technology, Negative electrodynamic feedback, Seismic exploration.

1. Introduction

Inertial motion sensors are the main elements of all mineral exploration systems. Currently, in addition to widely used electromechanical geophones, electrochemical geophones operating on the principle of molecular-electronic transfer (MET) [1] have begun to be used in such land systems [2-4].

Hydrophones are one of the main components of marine seismic exploration systems. An experimental sample of the hydrophone was also made and laboratory studies of its main characteristics, such as amplitude-frequency response and self-noise, were carried out.

Most marine operations to search for mineral deposits are held shelf, in coastal and transition zones. Laboratory experiments have shown that in order to stabilize the amplitude-frequency characteristic with the increasing depth, it is necessary to increase the rigidity of the bubble in the upper cover.

2. Stabilization Method of the Frequency Response of the Hydrophone

The detailed scheme and principle of operation of the MET hydrophone (Fig. 1) is described in the paper [5].

If there is air under the cover during assembly, it is at the pressure of 1 atm. In this case, if the constant pressure on the outer membrane increases (for example, when a hydrophone is immersed), the inner membrane will flex until the pressure levels out. Moreover, for the volume V of the bubble under the cover, the following expression will be valid:

$$P_0 V_0 = P_1 V_1, \tag{1}$$

where P_0 = 1 atm., V_0 is the initial bubble volume at the pressure of P_0, V_1 is the bubble volume at the pressure P_1.

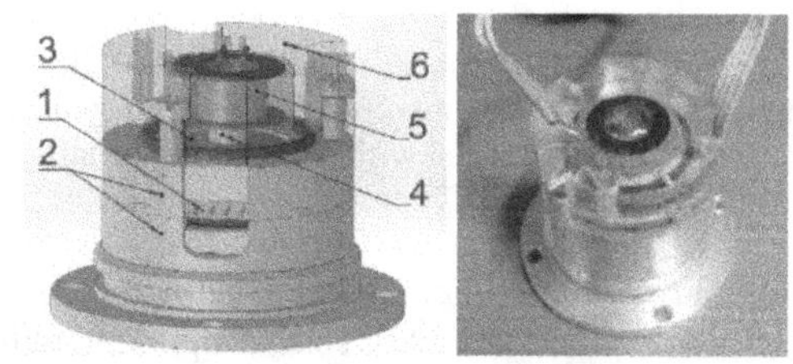

Fig. 1. The scheme and experimental sample hydrophone. 1 - sensing element; 2 - metal flanges; 3 - elastic rubber membranes; 4 – magnet; 5 - electromagnetic coil; 6 - upper cover.

The volume of the internal bubble will decrease inversely to the external pressure. This leads to the fact that the membrane begins to mechanically rest against the coil and thereby distort the output signal. Secondly, the rigidity of the air bubble increases, which leads to a decrease in sensitivity at high frequencies with increasing external pressure. It should also be noted that the feedback option used cannot compensate for the effect of the constant pressure.

Thus, stabilization of the frequency response is reduced in order to increase the rigidity of the isolated volume, which will change slightly with increasing pressure. For this, it was proposed to fill the volume with a compressible liquid, and polymethylsiloxane (PMS) liquid was chosen for that purpose.

3. Laboratory Tests

Two experimental model MET hydrophones were assembled. Their sensitivity was adjusted to 0.8 mV/Pa ±1 dB in the range of 1-300 Hz. The isolated volume of one of them was completely filled with PMS liquid, other second was partially completely filled. Calibration results for different pressure are shown in Fig. 2.

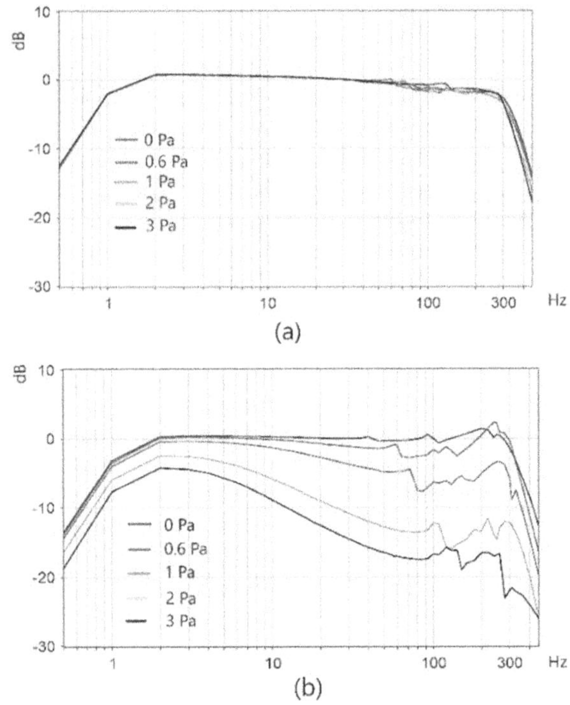

Fig. 2. Frequency response (0.8 mV/Pa) of MET hydrophones with the volume completely filled with liquid under the upper cover (a) and with the volume partially filled under different pressure differences.

Laboratory experiments have shown that the frequency response of MET hydrophone with a completely filled volume under the upper cover varies by -0.1 dB at low frequencies and -0.5 dB at high frequencies with increasing pressure difference on the membranes to 3 atm. While the frequency response of the second changes much more (-22 dB).

Further experiments were carried out with first type of MET hydrophone.

4. Marine Experiments

Marine experiments were held in Gelendzhik Bay (Black Sea) with technical support from JSC «Yuzhmorgeologiya». The experiment was to compare the response of the reference hydrophone (MP-25-250, ARAM ARIES II system) and the MET hydrophone to an external disturbing signal generated by a single pneumatic source of 40 cubic inches. The spectra of one of the elastic signals are shown in Fig. 3.

We see a good match (up to 20 %) of signals in the range from 15 to 100 Hz. The discrepancy at low (1-15 Hz) frequencies is caused by maintaining the flat characteristic of the MET hydrophone in this range and the lower boundary frequency by 10 Hz (-3 dB) at the reference hydrophone. The discrepancy in the range from 100 to 200 Hz can be caused by inaccuracy of placing the hydrophones at one point relative to the signal source. The discrepancy in the spectra at high frequencies (from 200 Hz) is due to the fact that the upper frequency of the working range of the MET hydrophone is 300 Hz, and the frequency response of the hydrophone remains flat.

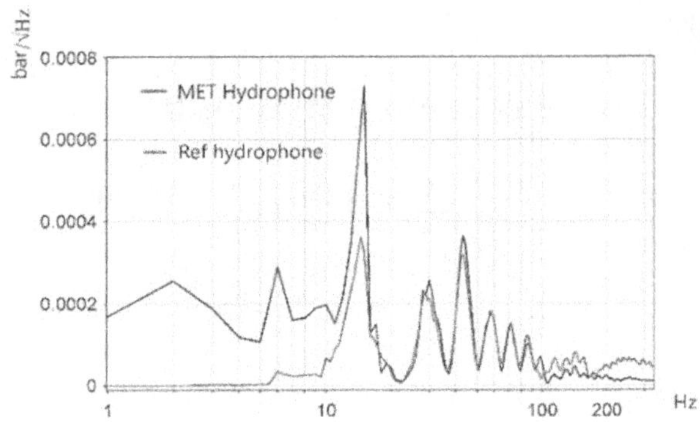

Fig. 3. Spectra of hydrophone signals of the same disturbing action in marine experiments.

5. Conclusions

As a result, the method of stabilizing the frequency response of MET hydrophone in range 1-300 Hz was proposed, tested and confirmed by laboratory and marine tests. Proven working depths are up to 30 m.

Comparative marine tests of the developed model of the MET hydrophone were carried out, which showed the possibility of using this type of hydrophone in marine seismic exploration systems.

Acknowledgements

This work was supported by the Russian Ministry of Education and Science, Project ID RFMEFI57817X0243; Russian Foundation for Fundamental Research, Project ID 17-07-01334-a.

References

[1]. A. S. Bugaev, A. N. Antonov, V. M. Agafonov, K. S., et al., Molecular electronic transducers for measuring instruments, *Journal of Communications Technology and Electronics*, Vol. 5, Issue 12, 2018, pp. 1339-1351.

[2]. I. R. Sharapov, et al., Innovative passive microseismic methods in the oil and gas industry – Experience of use in Russia, in *Proceedings of the IV International Geological Conference (AtyrauGeo'17)*, Atyrau, Kazakhstan, May 24-25.

[3]. G. N. Antonovskaya, N. K. Kapustian, et al., New seismic array solution for earthquake observations and hydropower plant health monitoring, *Journal of Seismology*, Vol. 21, Issue 5, pp. 1039-1053.

[4]. E. V. Egorov, V. M. Agafonov, et al., Angular molecular-electronic sensor with negative magnetohydrodynamic feedback, *Sensors*, Vol. 18, Issue 1, 245.

[5]. D. L. Zaitsev, V. M. Agafonov, S. Yu. Avdukhina, E. V. Egorov, Broadband MET hydrophone, in *Proceedings of the 80th EAGE Conference and Exhibition 2018*, Copenhagen, Denmark, 11-14 June 2018, pp. 4757-4761.

(021)

Optical Method to Diagnose Circulatory Disorders

V. Shapar [1], **V. Lysenko** [1], **A. Savchuk** [2]
[1] V. Lashkaryov Institute of Semiconductor Physics, NAS of Ukraine, 41 Nauky Ave., 03028 Kyiv, Ukraine
[2] International Center "Institute for Applied Optics", NAS of Ukraine, 10-g Kudryavska St., 04053 Kyiv, Ukraine
E-mail: v.shapar51@gmail.com

Summary: We propose a new optical method and arrangement to diagnose microcirculatory bed of the peripheral blood circulation in the subcutis of human soft tissues. This diagnostic method is based on simultaneous measurement of two indices related to blood circulation: total hemoglobin (THb) and degree of hemoglobin saturation with oxygen in the tissue area considered (SaO_2).

Keywords: Peripheral blood circulation, Microcirculatory bed, Optic diffusion spectroscopy, Hemoglobin.

1. Introduction

The system of peripheral blood circulation is one of the most important systems to provide vital activity. The blood vessels of microcirculatory bed are in direct contact with the tissues and serve the metabolic, regulatory, heat exchange, signal and many other functions. Metabolism between blood and extravascular medium as well as tissue saturation with oxygen occur through the thin walls of capillaries formed by a single layer of endothelium cells. Many physiological processes occurring in a human organism (such as regeneration and healing of defective tissues, inflammatory processes, biochemical reactions, etc.) depend on blood microcirculation.

Thereby the diagnostics of local disorders in the blood microcirculation system is an important medical problem. Those sick with peripheral arterial disease especially need such diagnostics [1].

2. The Methods of Blood Flow Study

Now there are several methods to study vascular tone and blood flow in small size vessels. Plethysmography is the most widely applied. This method is based on graphic registration of sphygmic (and slower) oscillations of a volume of some body part related to the dynamics of vessels blood filling. To this end, the studied body part is put into a hermetic compartment filled with air or water. The oscillations induced by variation of vessels blood filling are transmitted to the sensors of a measuring device.

High labor input for hermetization of some body parts to perform measurements under clinical conditions became one of the reasons to develop new methods that do not require a transmission medium to register blood filling variations.

The optical diagnostic methods are free from the mentioned drawbacks. Their advantages are that they have no adverse effect on human organism and make it possible to create compact inexpensive measuring devices to perform screening examinations of people. This is of considerable significance bearing in mind abundance of vascular diseases.

Among the existing optical methods of microvascular bed diagnostics, those based on investigation and analysis of variations of biotissue optical properties are distinguished by their simplicity. Their realization became possible because the progress in development of the theory of optical diffusion spectroscopy (ODS) that makes it possible to investigate the biotissue optical properties through human skin.

The modern pulse oximeters that measure such important blood index as the SaO_2 parameter (the ratio between oxidized hemoglobin HbO_2 and reduced hemoglobin Hb) operate using the ODS methods. The disturbances of peripheral blood circulation and damages in cutaneous covering are often accompanied by violation of the ratio between HbO_2 and Hb. According to the studies performed in [2], the SaO_2 parameter can be taken as one of the diagnostic indices of microvascular bed state.

At present there is no direct optical method to measure blood filling. Therefore, in [1] it was proposed to measure the amount of hematocrit instead of blood filling, and in [3] it was proposed to measure total hemoglobin (THb) (placed in the studied tissue area) as the measured parameter responsible for blood filling.

However, it should be noted that the problem of measuring hematocrit and total hemoglobin in a tissue using the optical method still remains unsolved up to now. Among the known methods of measuring total hemoglobin, only those based on the photon transport equation and Monte-Carlo calculations have got practical application. Unfortunately, the measuring equipment using the above methods requires powerful computer aids and does not provide the desired accuracy.

3. A Novel Method of Hb and HbO_2 Measuring

In the present work we propose a simple and yet sufficiently exact method (as shown experimentally) for noninvasive measurement of different chromophores placed in soft human tissues. The computational algorithms for concentrations of different chromophores (in particular, THb and SaO_2 parameter) were obtained using the physical-mathematical model of beams propagation in a tissue. This model is described by the equation similar to the modified *Booger-Lambert-Ber equation proposed in* [4] to solve the pulse oximetry problem.

A substantial difference from the equation in [4] is that now the effective path length L of optical beams within a biotissue (depending on the light wavelength) is replaced by parameter r of the distance between the radiation source and radiation receiver that are located on the surface of the human cutaneous covering studied. Taking this into account, the equation for optical density D of the biotissue under investigation is:

$$D = \lg\frac{I_0}{I} = \sum_i \varepsilon_i C_i \cdot r \cdot B + G \qquad (1)$$

Here r is the radius of the circular optical channel in a fiber optic measuring head; ε_i is the extinction coefficient of the ith chromophore; C_i is concentration of the ith chromophore; G is some geometric factor related to the construction of the device measuring head; B is the length factor related to the scattering process (taken as a constant).

To technically realize the proposed measuring method, a special multichannel fiber optic head [5] with circular interface structure and small (100-200 µm) width of circular channels was developed. When studying signals reflected from the bulk of human biotissue, the experimental check of Eq. (1) with the mentioned measuring head in the wavelength range $\lambda = 0.45 \div 1.3$ µm showed that Eq. (1) is accurate to no worse than 3 % if the distance between the circular receiving channels does not exceed 1.6 mm.

Fig. 1 shows typical dependences of the logarithm of the normalized reflection coefficient of light (ln(R/S) by the subcutaneous layer of an adult's finger from the radius of the receiving channel for different wavelengths λ. (Here S is the ring area).

Following are the ratios to calculate the total hemoglobin C_{Hb} and SaO_2 parameter obtained from the solution of Eq. (1) (neglecting the methemoglobin component and carboxyhemoglobin HbCO):

$$C_{Hb} = \frac{A_1(\varepsilon_{HbO2}^{\lambda1} - \varepsilon_{Hb}^{\lambda1}) - A_2(\varepsilon_{HbO2}^{\lambda2} - \varepsilon_{Hb}^{\lambda2})}{\left[\varepsilon_{Hb}^{\lambda1}(\varepsilon_{HbO2}^{\lambda2} - \varepsilon_{Hb}^{\lambda2}) - \varepsilon_{Hb}^{\lambda2}(\varepsilon_{HbO2}^{\lambda1} - \varepsilon_{Hb}^{\lambda1})\right]} \times \frac{1}{r_2 - r_1} \qquad (2)$$

$$SaO_2 = \frac{(A_1 / A_2) \cdot \varepsilon_{HbO2}^{\lambda2} - \varepsilon_{Hb}^{\lambda1}}{(\varepsilon_{HbO2}^{\lambda1} - \varepsilon_{Hb}^{\lambda1})} \qquad (3)$$

Here A_1 and A_2 are the device parameters, $I_1^{\lambda1}$ and $I_1^{\lambda2}$ are the optical signal intensities in the first circular channel for wavelengths λ_1 and λ_2, $I_2^{\lambda1}$ and $I_2^{\lambda2}$ are the optical signal intensities in the second circular channel at wavelengths λ_1 and λ_2, $\varepsilon_{HbO2}^{\lambda1}$ and $\varepsilon_{Hb}^{\lambda1}$ are the extinction coefficients of the corresponding hemoglobin forms at wavelength λ_1, $\varepsilon_{HbO2}^{\lambda2}$ and $\varepsilon_{Hb}^{\lambda2}$ are the extinction coefficients of the corresponding hemoglobin forms at wavelength λ_2.

The $\varepsilon_{HbO2}^{\lambda1}$, $\varepsilon_{Hb}^{\lambda1}$ and $\varepsilon_{HbO2}^{\lambda2}$, $\varepsilon_{HbO2}^{\lambda2}$ were obtained from the published sources [6].

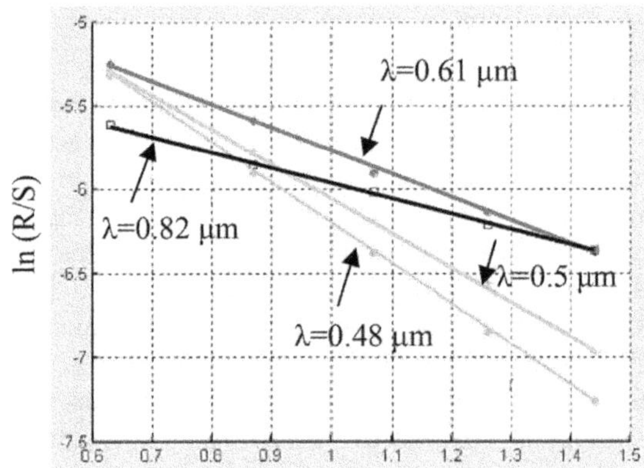

Fig. 1. The radius of the annular channel, mm.

The above Eqs. (2) and (3) are applied in the fabricated mockup sample of the portable diagnostic device. Now it is subjected to medical tests.

Using the four-wave measurement method and the corresponding calculation algorithm Oximetry readings are accurate in patients with carbon monoxide poisoning or methemoglobinemia.

References

[1]. M. A. Khalil, H. K. Kim, I. K. Kim, M. Flexman, R. Dayal, G. Shrikhande, A. H. Hielscher, Dynamic diffuse optical tomography imaging of peripheral arterial disease, *Biomedical Optics Express*, Vol. 3, 2012, pp. 2288-2298.

[2]. G. Yu, G. Lech, C. Zhou, B. Chance, et al., Time-dependent blood flow and oxygenation in human skeletal muscles measured with noninvasive near-infrared diffuse optical spectroscopies, *J. of Biomedical Optics*, Vol. 10, Issue 2, 2005, 024027.

[3]. U. Wolf, M. Wolf, J. H. Choi, et al., Localized irregularities in hemoglobin flow and oxygenation in calf muscle in patients with peripheral vascular disease detected with near-infrared spectrophotometry, *J. Vasc. Surg.*, Vol. 37, Issue 5, 2003, pp. 1017-1026.

[4]. A. Zourabian, et al., Trans-abdominal monitoring of fetal arterial blood oxygenation using pulse oximetry, *J. of Biomedical Optics*, Vol. 5, Issue 4, 2000, pp. 391-405.

[5]. V. Shapar, V. Lysenko, A. V. Bondarenko, Multichannel fiber-optic ring converters for optical sensors of physical quantities, *J. of Nano Research*, Vol. 39, 2016, pp. 221-227.

[6]. S. Prahl, Tabulated Molar Extinction Coefficient for Hemoglobin in Water, Oregon Medical Laser Center, http://omlc.ogi.edu/spectra/hemoglobin/summary.html

[7]. E. D. Chan, M. M. Chan, M. M. Chan, Pulse oximetry: Understanding its basic principles facilitates appreciation of its limitations, *Respiratory Medicine*, Vol. 107, 2013, pp. 789-799.

(022)

Invisible Sensors for Early Prediction of Discontinuous Bed-leaving Behavior Patterns

H. Madokoro [1], K. Nakasho [2] N. Shimoi [1], H. Woo [1] and K. Sato [2]
[1] Akita Prefectural University, 84-4 Tsuchiya, Yurihonjo City, Akita, 015-0055, Japan
[2] Yamaguchi University, 1677-1 Yoshida, Yamaguchi City, Yamaguchi, 753-8511, Japan
Tel.: + 810184272081, fax: + 810184272000
E-mail: madokoro@akita-pu.ac.jp

Summary: This paper presents a novel bed-leaving sensor system for real-time recognition of bed-leaving behavior patterns. The proposed system comprises five pad sensors installed on a bed, a rail sensor inserted in a safety rail, and a behavior pattern recognizer based on machine learning. We obtained benchmark datasets of continuous and discontinuous behavior patterns from 10 subjects. We compared machine learning algorithms of five types to recognize five behavior patterns. The experimentally obtained results revealed that the proposed sensor system improved recognition accuracy for both datasets. Moreover, we achieved improved recognition accuracy after integration of learning datasets as a general discriminator.

Keywords: Ambient sensors, Home agent, Life monitoring, Machine learning, Quality of life, Random forest.

---

1. Introduction

Recently, various bed-leaving sensors are commercially available from manufacturers. For example, clip sensors, mat sensors, infrared (IR) sensors are widely used at hospitals and nursing-care facilities. Although, clip sensors are used easily because they are the most reasonable means available, care recipients are restrained by sensor wires because they are attached directly to the patient's nightwear. Moreover, a risk exists that a sensor wire might wrap around the neck of a care recipient. Therefore, the use of clip sensors has been discouraged recently. Mat sensors, which are inexpensive even compared with clip sensors, are used widely at clinical sites because they entail no restraint. One shortcoming of mat sensors is their slow detection and response from the position where a care recipient sits at the bed terminal while putting their feet on it. Another shortcoming is the excessive reaction even when a caregiver or a family member passes through while stepping on it. Furthermore, a care recipient might attempt to leave while consciously avoiding stepping on a mat sensor because it is a visible sensor. IR sensors present similar shortcomings to those of mat sensors. Moreover, caregivers must check the sensor installation status because care recipients touch them occasionally.

To prevent fall accidents, sensor systems with bed-leaving behavior predictions at an early stage have been studied. Asano et al. [1] proposed a detection system using a depth camera. They employed support vector machines (SVMs) to recognize bed-leaving behavior patterns after optimizing parameters and motion features combined with the body size, position, and orientation of respective subjects. Their experimentally obtained result achieved 92.65 % recall after 68 iterations. However, the precision was insufficient for practical application because false detection occurred in within 24 iterations. Moreover, they used a depth camera for capturing images. Although it is difficult to identify profiles solely from depth images, a challenging task remains: eliminating unpleasantness felt by a patient who is monitored by a camera.

Kawamura et al. [2] proposed a wearable sensor system using a three-axis accelerometer. For their system, unrestrained measurements are actualized using a lightweight sensor module of 13 g. With consideration of clinical applications, they obtained not only metaparameters that contributed to recognition, but also experimentally obtained results with bed-up and wheelchair locomotion. No recognition accuracy was provided as detailed sensor characteristics. Moreover, they provided neither recognition accuracy nor detailed sensor characteristics. Existing bed-leaving sensors entail persistent difficulties related to quality of life (QoL), detection speed, convenience, and cost. No sensors that satisfy these requirements have been put to practical use. In modern society, the declining birthrate and aging population are progressing rapidly. We therefore regard development of sensor systems that overcome these problems as an urgent task to.

This study was conducted to develop a bed-leaving recognition sensor system that is inexpensive, convenient, and maintainable with advanced QoL for care recipients. For improving recognition accuracy and reliability compared with our earlier bed-leaving sensor system [3], our novel sensor prototype comprises pad sensors and a rail sensor installed respectively on a bed frame and a bed-side safety rail. For evaluating our sensor system, we obtained original benchmark datasets of two types: continuous datasets with behavior transitions from sleeping to bed leaving in a predefined interval and discontinuous datasets with free and random movements obtained from 10 subjects. We compared machine learning algorithms of five types to recognize five behavior patterns. Our earlier study [3] provided respective classifiers because bed-leaving behavior patterns have unique characteristics along with their subjects. Nevertheless, for this learning strategy, the recognition accuracy was insufficient to ensure reliability for discontinuous datasets. Therefore, we developed a single classifier using all continuous datasets. The experimentally obtained results revealed that the proposed sensor system improved recognition accuracy for both datasets.

2. Sensor System

2.1. System Structure

Fig. 1 depicts the whole system structure of our novel sensor prototype, which comprises pad sensors, a rail sensor, sensor boards, a wireless router, and a monitoring computer. Output signals are collected to the sensor boards with a wired connection. The sensor boards convert analog signals to digital signals in real time. Digital signals are sent to a monitoring computer with wireless connection. We used ZigBee, a short distance wireless communication protocol, to provide cost-effective implementation and low power consumption. The transmitted measurement signals are displayed on the monitoring computer in real time. Behavior recognition algorithms based on machine learning are incorporated in the monitoring computer.

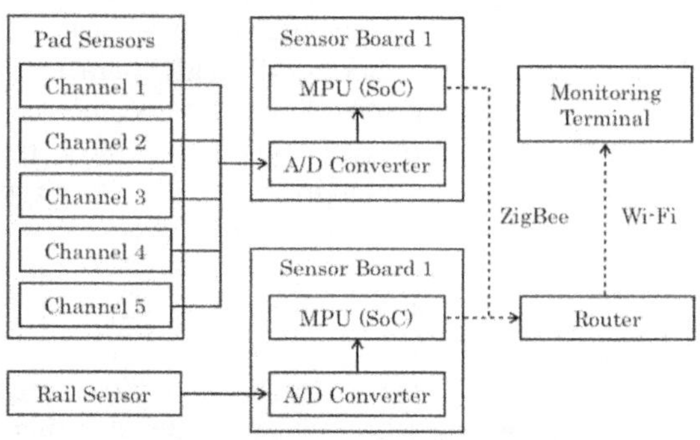

Fig. 1. Whole system structure.

2.2. Pad Sensor

Fig. 2 depicts our originally developed pad sensor prototype. The pad sensors are non-restrictive, invisible, cost effective, and independent of sensor driven power. We used a piezoelectric film, which generates a potential difference when distortion occurs in an arbitrary direction against external forces such as receiving vibration. A commercial piezoelectric film (DT2-028K/L; Tokyo Sensor Co., Ltd.) is sandwiched by 1-mm-thickurethane sheets with 50 deg hardness. The durability and elasticity of the piezoelectric film are improved using urethane sheets. Moreover, polyethylene terephthalate (PET) boards with a larger size than the urethane sheets provide extended sensing ranges. The sensor core is protected using PET plates that have 200 mm diameter and 0.5 mm thickness. We used ultraviolet stiffened resin to adhere the piezoelectric film and the urethane sheets.

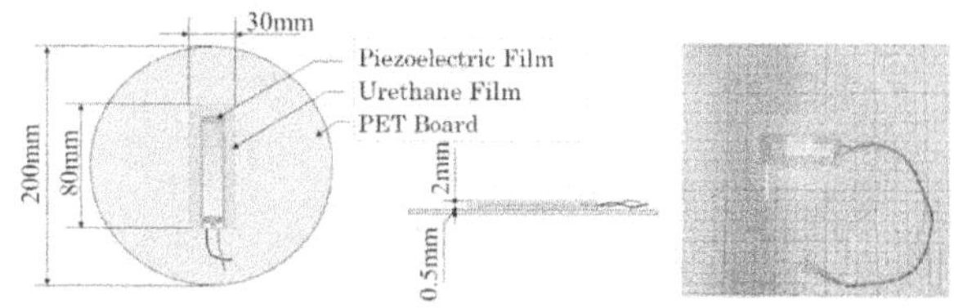

Fig. 2. Interior architecture and pad sensor appearance.

2.3. Rail Sensor

Care recipients whose legs and body are weak sometimes grip a safety rail beside a bed when they try to stand up. Motegi et al. [4] reported that approximately 82 % of care recipients gripped a safety rail when they left from their bed. Therefore, we specifically examined a bed side safety rail for the prevention of a fall. We consider that the recognition accuracy is improved if gripping of a safety rail is detected using a dedicated sensor.

Fig. 3 depicts our originally developed rail sensor prototype. We inserted a piezoelectric film, DT2-028K/L, into a silicon tube with 50 mm length, 10 mm outer diameter, and 5 mm inner diameter. As a stopper and a protector, a metal cap is stuffed to the sensor upside.

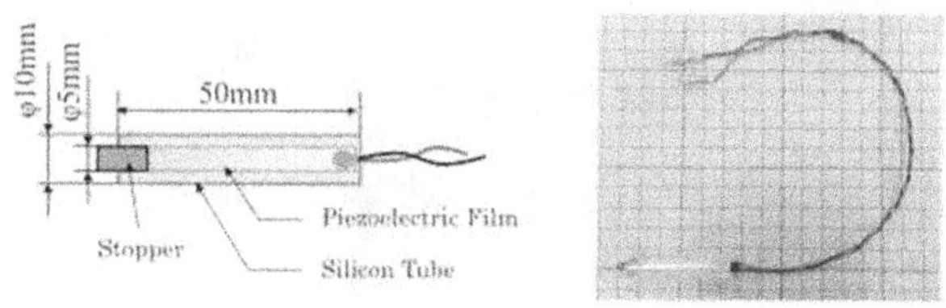

Fig. 3. Interior architecture and appearance of the rail sensor.

2.4. Sensor Installation

To sense the distributed weight of a sleeping body on a bed, pad pressure sensors are installed to five areas between a mattress and a bed frame. Fig. 4 depicts the sensor configuration on a bed. The approximate measurement ranges of each sensor are the upper body for channels 1 and 2 (CH1 and CH2), legs for the channels 3 and 4 (CH3 and CH4), and the hip for channel 5 (CH5). A rail sensor is installed to a safety bed rail of one side. Fig. 5 depicts photographs of installed sensors. Compared with the pre-installed sensor bed [5], our sensors can be installed on various beds as a post-installation system. Fig. 6 depicts a sensor measurement board with a ZigBee module for wireless communication. Sensor signals are transmitted to a monitoring computer via the board. Our system provides easy and simple monitoring using only six sensors of two types and two measurement boards.

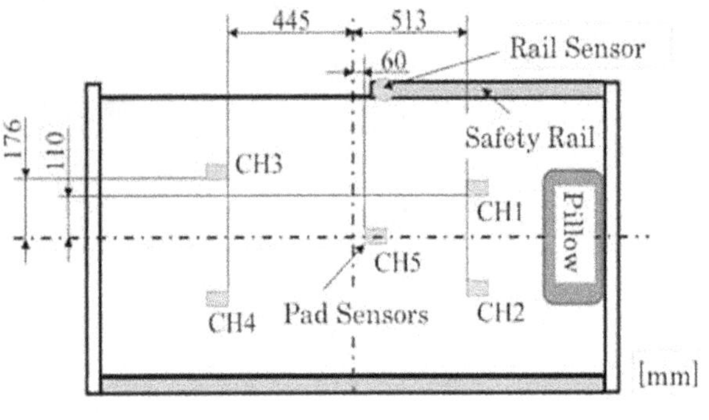

Fig. 4. Sensor configuration.

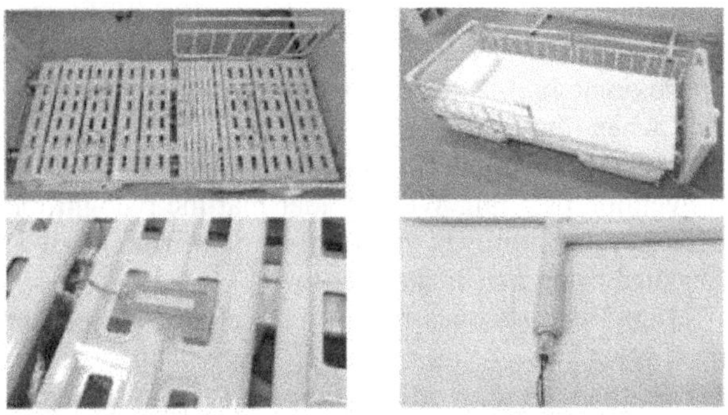

(a) Pad Sensors (b) Rail Sensor

Fig. 5. Sensor installation.

Fig. 6. Sensor measurement board with ZigBee module.

2. Datasets

2.1. Target Behavior Patterns

The following are features and estimated sensor responses for the respective patterns:

SLP – Sleeping: a subject is sleeping on a bed normally.

LOS – Longitudinal sitting: a subject is sitting Longitudinally on a bed after rising.

LAS – Lateral sitting: a subject is sitting laterally on a bed after turning the body from longitudinal sitting.

TES – Terminal sitting: a subject is sitting in the terminal position on a bed trying to leave a bed. Rapid and correct detection are necessary because of the terminal situation for leaving a bed.

LEB – Left a bed: a subject has left the bed. Herein, sensor responses disappear in the status of losing consciousness or a life crisis. For such circumstances, monitoring devices such as electrocardiographs are used. Such circumstances are beyond our prediction targets.

Table 1. Profile of subjects.

Subject	**A**	**B**	**C**	**D**	E
Height [cm]	161	169	177	168	170
Weight [kg]	51	66	91	51	60
Sex	F	M	M	M	M
Subject	**F**	**G**	**H**	**I**	**J**
Height [cm]	167	177	178	170	165
Weight [kg]	61	84	80	78	70
Sex	M	M	M	M	M

3.2. Datasets Obtained for Conditions

We obtained bed-leaving behavior pattern datasets at a simulated experimental room that resembled a clinical site. We used an electro-actuation bed (KA-36121R; Paramount Bed Co., Ltd.) equipped with three actuators for reclining the back and feet panels and for adjusting its height. We obtained datasets without using the back plate reclining function for avoiding load pattern changes on the bed. The route for a subject to leave from the bed is restricted to one side with two attached safety rails.

The subjects were 10 persons: nine men and one woman. Table 1 presents profiles of all subjects. We set two protocols to obtain different characteristic datasets. The first protocol comprises the same procedures as those of our earlier study [3]. Each subject switched their behavior patterns of five types with 20 s intervals. For this study, we call them continuous datasets (CDS). We obtained 10 sets of CDS from each subject. Herein, the data sampling rate was set to 50 Hz.

The second protocol comprises behavior patterns as discontinuous datasets. The order and duration of bed-leaving behavior patterns are various along with body parameters and health conditions in each subject. For example, behavior patterns from LOS to SLP without changing to LAS are occurred frequently. We consider that the employment of our sensor system at a clinical site gives rise to dramatically lower recognition accuracy. As a basic consideration for aiming a practical application, we obtained datasets without fixed sequences or time intervals for subjects of their basic behavior pattern transitions. We designate them as discontinuous datasets (DDS). The data acquisition period was set to 15 min per person.

For calculating recognition accuracy, ground truth (GT) labels are indispensable to DDS. However, the burden to allocate GT labels is excessively high because each subject moved freely for 15 min. Therefore, we used a depth camera to record video images for annotation. We allocated GT labels to DDS manually from video image observation.

4. Evaluation Experiment

4.1. Evaluation Criteria

Let T_{num} and G_{num} respectively be the total numbers of test signals and GT labels. For evaluation criteria, the recognition accuracy R for a test dataset is defined as

$$R = \frac{T_{num}}{G_{num}} \times 100$$

Herein, we define mean R as R_{mean}. Moreover, we define R of SLP, LOS, LAS, TES, and LEB as R_{SLP}, R_{LOS}, R_{LAS}, R_{TES}, and R_{LEB}.

We used K-Fold cross validation for evaluating results along with machine-learning and evolutional-learning approaches. Herein, we set K = 5 based on results of earlier studies [6, 7].

We conducted four evaluation experiments using behavior pattern datasets of two types. Table 2 summarizes experimental details.

Table 2. Experimental conditions.

Section	Learning	Test Dataset	Discriminator
V-B	CDS	CDS	Each Subject
V-C	CDS	CDS	Each Subject
V-D	CDS	DDS	Each Subject
V-E	CDS	DDS	All Subject

4.2. Comparison Results of Learning Algorithms

We evaluated recognition accuracy of machine learning algorithms using CDS. Table 3 depicts comparison results. We used CPNs as a discriminator for our earlier study [3]. As a comparison result, R_{mean} using CPNs was 75.4 %, which is the second lowest of six algorithms. In contrast, R_{mean} using RF was 91.1 % that the highest. For all behavior patterns except of LEB, R_{mean} using RF were higher than those of other algorithms. For LEB, L-SVM provided the highest recognition accuracy.

Table 3. Comparison results of learning algorithms.

Algorith	R_{SLP}	R_{LOS}	R_{LAS}	R_{TES}	R_{LEB}	R_{mean}
NB	98.7	32.2	27,4	25.8	0.9	53.2
kNN	98.5	85.0	79.5	80.8	86.1	89.0
L-SVMs	96.2	63.8	64.4	51.1	89.4	78.2
R-SVMs	98.0	62.4	64.6	67.9	88.8	81.0
RF	98.8	88.6	81.8	84.5	88.5	91.1
CPNs	88.1	67.9	59.7	62.5	75.9	75.4

As a commonplace tendency for all algorithms, R_{SLP} achieved the highest. In contrast, R_{LOS}, R_{LAS} and R_{TES} are smaller than 90.0 %. We consider that the recognition accuracies for these three behavior patterns must be high because our system intends to predict bed-leaving behavior. Therefore, we examine measures to improve recognition accuracies for these three behavior patterns from the viewpoint of datasets and discriminators.

4.3. Experimental Results for CDS

Using CDS, we evaluated the capabilities of our originally developed sensors of two types. Fig. 7 depicts comparison results of *Rmean* for the solely used pad sensors and the combined sensors with pad sensors and a rail sensor. Comparison of the results shows that *Rmean* of the combined sensors is 4.0 % higher than that of the pad sensors. Recognition accuracies in respective behavior patterns are improved from 0.4 % up to 12.3 %. Particularly, R_{TES} exhibits the maximum improvement. Moreover, the experiment result demonstrates that the rail sensor, which detects the grasp of a safety rail in TES intensively, contributes to improvement of the overall recognition accuracy.

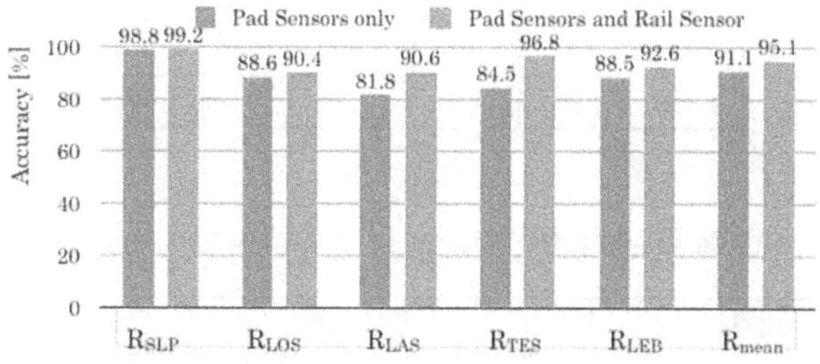

Fig. 7. Comparison results of recognition accuracies.

The output voltage from the piezoelectric film in the rail sensor occasionally exceeded 0.1 V when a subject was in TES. We infer that sufficient recognition accuracy is obtainable using the rail sensor solely if we set the target-only TES. However, we consider that false recognition occurred between TES and LEB, especially for the immediate transition from TES to LEB. Although sensor signals should be approximately 0 V in LEB, sharp signals are outputted. This tendency occurs when a subject tries to leave from the bed with holding or shaking of a safety rail, which enhances vibration. Therefore, we infer that the combination between the rail sensor and the pad sensors is the best for a practical use because of avoiding the problem that is occurred in the case of a solely used rail sensor. Moreover, false recognition to LEB is avoidable accurately because the recognition accuracy of the combined sensors obtained relative superior improvement for LAS, TES, and LEB.

We examine detailed recognition results obtained using a confusion matrix. Tables 4 and 5 respectively present confusion matrixes for the pad sensors and the combined sensors. Specifically examining TES with the highest recognition accuracy, the number of false recognition instances was reduced in all behavior patterns after appending the rail sensor. Particularly, the number of false recognitions to LEB was reduced from 30 signals to 4 signals. We infer that the correct discrimination between TES and LEB engenders improved recognition accuracy.

Table 4. Confusion matrixes for the results of pad sensors.

	SLP	LOS	LAS	TES	LEB
SLP	1047	5	3	3	2
LOS	21	443	14	3	19
LAS	12	9	319	36	14
TES	3	8	18	321	30
LEB	2	6	12	18	292

Table 5. Confusion matrixes for the results of pad sensors and rail sensors.

	SLP	LOS	LAS	TES	LEB
SLP	1052	3	3	0	2
LOS	19	449	14	1	17
LAS	13	3	354	10	10
TES	1	2	8	365	4
LEB	2	10	12	3	303

Table 6. Recognition accuracies in each subject for DDS.

Subject	R_{SLP}	R_{LOS}	R_{LAS}	R_{TES}	R_{LEB}	R_{mean}
A	67.1	65.8	65.6	69.4	84.5	68.9
B	97.8	51.4	22.8	79.6	78.8	82.4
C	95.2	66.0	8.8	10.9	1.1	43.3
D	98.9	84.8	42.2	89.6	94.3	84.6
E	93.9	78.1	60.3	88.1	82.9	83.7
F	99.8	89.2	95.1	90.3	95.2	95.3
G	51.3	43.9	47.0	89.1	15.0	51.6
H	98.1	64.0	51.1	78.3	79.7	73.5
I	95.0	71.0	57.7	61.7	88.9	86.5
J	95.9	55.7	64.6	74.0	30.3	82.1
Average	89.3	67.0	51.5	73.1	65.1	75.2

Although the use of the rail sensor was aimed at improved R_{TES}, *RLAS*) was improved from 81.8 % – 90.6 % as a subsidiary contribution. The number of false recognition instances was reduced, except for SLP. Particularly, the number of false recognition instances for TES was decreased to 26 times. The improved R_{TES} after appending the rail sensor produces decreased false recognition instances of LAS. As a result, correct recognition instances, except for TES, produces an improved R_{mean}. We demonstrated that the addition of a sensor that can reliably recognize a single posture engendering improvement of R_{mean}. The combined sensors have clear benefits for bed-leaving behavior recognition when compared to other configurations.

4.4. Experiment Results for DDS

Table 6 denotes R_{mean} for each subject for DDS. Compared with that of CDS as depicted in Fig. 11, recognition accuracies were lower in all five behavior patterns. Especially, R_{LAS} was drastically lower. This experimentally obtained result revealed that recognition accuracy, except for R_{SLP}, which was the highest accuracy among five behavior patterns, was strongly affected by randomness in DDS.

The disparity of detailed recognition accuracies in respective subjects was from 43.3 % as the lowest to 95.3 % as the highest. Compared with the mean accuracy of 75.2 %, the recognition accuracies were above for seven subjects and below for the remaining three subjects. Therefore, significant lower accuracy for specific subjects decreased the overall recognition accuracy. We infer that this tendency is influenced by individual differences in behavior patterns. Each subject played predetermined behavior patterns in CDS and free behavior patterns in DDS. We consider that the recognition accuracy was significantly lower because behavior patterns were various among the subjects in DDS.

Table 7 depicts the confusion matrix for all subjects. Table 8 presents the confusion matrix for Subject C, with the lowest recognition accuracy. Numerous signals were falsely recognized to SLP. Particularly, correct recognition in LEB was merely 2 of 179 signals. Other signals were falsely recognized to SLP. This trend demonstrated that false recognition occurred in the state that SLP and other behavior patterns were not distinguished. In DDS, *Rmean* was dramatically lower in particular subjects. We consider that this is attributable to behavior variations of subjects between learning datasets and test datasets. Therefore, we consider that *Rmean* improves if learning datasets contain diversity.

Table 7. Confusion matrix of all subjects.

	SLP	LOS	LAS	TES	LEB
SLB	**4,097**	356	142	5	26
LOS	312	**1,283**	162	11	96
LAS	179	201	**795**	116	164
TES	295	21	72	**974**	135
LEB	183	55	26	58	**641**

4.5. Integration of Learning Datasets

For applying our system at a clinical site, learning dataset preparation for each subject might be troublesome and time-consuming. In addition, preserving accuracy has come to be a problem of system reliability because generalization was dropped in DDS. Therefore, we attempt to construct a generic classifier combined with learning datasets for all subjects as depicted in Table 1. For this experiment, we used CDS for learning and DDS for validation.

Table 8. Confusion matrix of Subject C.

	SLP	LOS	LAS	TES	LEB
SLP	**335**	17	0	0	0
LOS	52	**171**	1	1	34
LAS	141	4	**16**	9	11
TES	286	0	0	**35**	0
LEB	177	0	0	0	**2**

Table 9 depicts recognition accuracies obtained before and after integration of learning datasets. Recognition accuracies in six of ten subjects were improved. Particularly, recognition accuracies of Subjects C and G were improved notably. Although recognition accuracies of four subjects dropped, three of them remained dropped percentages up to 2.0 %. However, with regard to Subject E, R_{mean} decreased 5.6 percentage points, especially for steep drops in R_{LOS} and R_{LAS}).

Table 9. Recognition accuracies before and after integration of learning Datasets.

Subject	Before	After	Difference
A	68.9	67.1	-1.8
B	82.4	86.9	4.5
C	43.3	77.9	34.6
D	84.6	84.7	0.1
E	83.7	78.1	-5.6
F	95.3	93.5	-1.8
G	51.6	66.4	14.8
H	73.5	77.4	3.9
I	86.5	87.0	0.5
J	82.1	80.8	-1.3
Average	75.2	80.0	4.8

Fig. 8 depicts the comparison result of recognition accuracy for each behavior pattern. All recognition accuracies were improved in the integrated datasets. Particularly, R_{LEB} improved 15.8 percentage points. Table X denotes the confusion matrix for all subjects after the integration of validation datasets. Compared with results depicted in Table 7, the numbers of false recognition instances were lower in all behavior patterns. Although the false recognition instances for SLP were numerous in Table 7, these results were improved in Table 10. Moreover, false recognition instances appeared frequently in behavior patterns that were close to GT labels except of LAS. We consider that it is a challenging task to recognize intermediate states of two neighbor behavior patterns which change their body among respective behavior patterns. To reduce false recognition instances, we infer that it is necessary to use a method to maintain the previous status until the recognition becomes stable if a present recognition result differs from the previous one. In contrast, false recognition instances of LAS were divided to the other four behavior patterns evenly. We infer that this tendency makes it difficult to distinguish LAS from other behavior patterns.

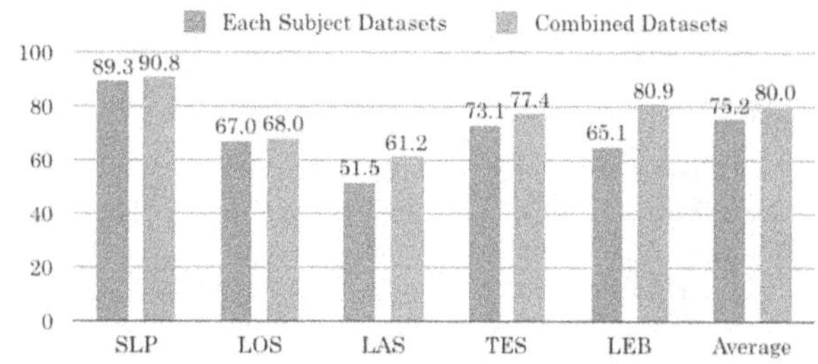

Fig. 8. Comparison result of recognition accuracy in each behavior pattern.

Table 10. Confusion matrix.

	SLP	LOS	LAS	TES	LEB
SLP	**4,190**	310	53	6	67
LOS	315	**1,284**	162	13	90
LAS	150	131	**909**	117	148
TES	48	45	141	**1,166**	97
LEB	21	30	37	74	**801**

5. Conclusion

This paper presented the bed-leaving behavior recognition system that comprises pad sensors installed on a bed, a rail sensor inserted in a safety rail, and a behavior pattern recognizer based on machine learning algorithms. We obtained benchmark datasets of continuous and discontinuous behavior patterns from 10 subjects. The experimentally obtained results revealed that RF obtained the highest recognition accuracy in our benchmark datasets. Compared with our earlier study, results obtained using CPNs, the recognition accuracies were improved by 20.7 % for LOS and 21.9 % for TES. After appending the rail sensor to the pad sensors, the mean recognition accuracy improved 4.0 percentage points, including a 12.3 percentage point improvement for TES. Regarding the difference of behavior pattern transitions, the mean recognition accuracy decreased 22.9 percentage points in discontinuous datasets.

For our future work, we aim to apply our proposed sensor system to a clinical site. We must acquire datasets from numerous subjects. Additionally, we must demonstrate the system reliability for conducting long-term monitoring.

Acknowledgments

This study was conducted with the support of the Ministry of Internal Affairs and Communications,

Strategic Information and Communications R&D Promotion Program (SCOPE 152302001) in Japan. We are grateful to Prof. Katsumi Wasaki and Assoc. Prof. Masaaki Niimura of Shinshu University for helpful discussions related to this study as our joint study project.

References

[1]. H. Asano, T. Suzuki, J. Okamoto, Y. Muragaki, H. Iseki, Bed exit detection using depth image sensor, *Journal of Tokyo Women's Medical College*, Vol. 84, Issue 2, 2014, pp. 45-53.

[2]. K. Kawamura, Y. Okuno, Y. Hirose, K. Ozone, K. Tomita, Detection of various postures and gait using a wearable triaxial accelerometer, *Journal of Physical Therapy Science*, Vol. 32, Issue 3, 2017, pp. 435-438.

[3]. H. Madokoro, N. Shimoi, K. Sato, Unrestrained multiple-sensor system for bed-leaving detection and prediction, *Nursing and Health*, Vol. 3, Issue 3, 2015, pp. 58-68.

[4]. M. Motegi, N. Matsumura, T. Yamada, N. Muto, N. Kanamaru, K. Shimokura, K. Abe, Y. Morita, K. Katsunishi, Analyzing rising patterns of patients to prevent bed-related falls (Second report), *Trans. Japan Society for Health Care Management*, Vol. 12, Issue 1, 2011, pp. 25-29.

[5]. T. Hatsukari, T. Shiino, S. Murai, The reduction of tumbling and falling accidents based on a built-in patient alert system in the hospital bed, *Journal of Science of Labour*, Vol. 88, Issue 3, 2012, pp. 94-102.

[6]. Y. Mitadera, K. Akazawa, Analysis of incidents occurring in long-term care insurance facilities, *Bulletin of Social Medicine*, Vol. 30, Issue 2, 2013, pp. 123-134.

[7]. H. Asano, T. Suzuki, J. Okamoto, Y. Muragaki, H. Iseki, Bed exit detection using depth image sensor, *Journal of Tokyo Women's Medical College*, Vol. 84, Issue 2, 2014, pp. 45-53.

(023)

Development of a Sensitive and Selective Mixed-potential Ammonia Sensor for Automotive Exhausts

G. Nematbakhsh Abkenar [1], **J. P. Viricelle** [1], **M. Rieu and P. Breuil** [1]
[1] Mines Saint-Etienne, Univ. Lyon, CNRS, UMR 5307 LGF, Centre SPIN, F – 42023 Saint-Etienne France
Tel.: +33477426659
E-mail: gita.nematbakhsh@emse.fr

Summary: Selective catalyst reduction (SCR) system using ammonia gas as a reductant, is one the most reliable ways to control NO_x emissions from diesel engine vehicles and trucks. In order to optimize the conversion rates of NO_x and to prevent inducing excessive NH_3 to the air, an NH_3 sensor is required to control the SCR system for maintaining NH_3 slip at an acceptable level. In this study, a highly selective ammonia sensor was developed to detect ammonia emissions from automotive exhaust. The effects of operating temperature and electrode composition on NH_3 response were investigated. The results indicate that the selectivity and stability of sensors are extremely dependent on previous parameters. These sensors also show remarkable sensing properties even for low concentrations of ammonia with logarithmic dependence on the gas concentration.

Keywords: Ammonia, Mixed-potential sensors, Selectivity, Vanadium oxide, Gas sensor.

---

1. Introduction

One of the most effective technologies in decreasing large-scale NO_x emission produced by diesel engine vehicles is Urea-SCR (selective catalytic reduction) system. However, an excess of urea injection can cause NH_3 emissions to the environment [1, 2].

So, many studies have been carried out in order to develop an effective ammonia sensor for harsh environments of automobiles exhausts. Nevertheless, most of existing sensors suffers from cross-sensitivity to other toxic gases, low sensitivity and stability [3].

V_2O_5 has been proved to be one of the interesting metal-oxides suitable for gas-sensing applications due to its catalytic properties [4].

In the present study, low-cost highly selective YSZ-based mixed potential sensors with Au-V_2O_5 working electrode and platinum reference electrode were fabricated.

2. Experimental

Sensors were fabricated using Al_2O_3 (2 cm × 5cm) substrates. An electrolyte layer of YSZ (8 mol% Y_2O_3-doped) was screen-printed followed by 2 hours sintering at 1380 °C. Afterwards, the reference electrode (platinum) and working electrode (Au-V_2O_5, 3 different mass ratio 85-15, 70-30 and 50-50) were printed on YSZ layer using commercial inks of platinum and gold, and home-made ink V_2O_5. A platinum resistor was also printed at the backside of the alumina support to control the sensor temperature. The sintering temperature of reference and working electrodes were 850 °C and 600 °C respectively. The measured sensors' responses (ΔV) are the potential difference between reference and working electrode.

3. Results and Discussion

Fig. 1 shows the responses of sensors with Au-15 %V_2O_5 working electrode, to 100 ppm CO, NO_2, NO and 20 ppm of NH_3 at four different temperatures of 450, 500, 550 and 600 °C. It can be seen that the sensors respond to all gases at lower temperatures while by increasing temperature to 600 °C the selectively toward NH_3 is greatly improved.

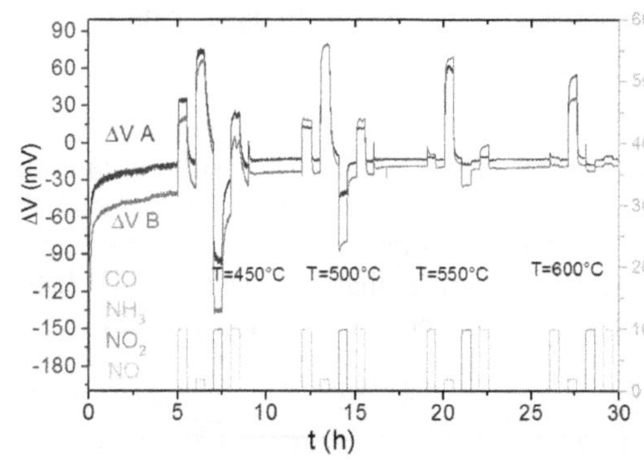

Fig. 1. Responses of Au-15 %V_2O_5 sensors to 100 ppm CO, NO_2, NO and 20 ppm of NH_3 at temperatures of 450, 500, 550 and 600 °C.

The stability of sensors was tested by exposing them at 550 °C and 600 °C for long duration. Fig. 2 clearly shows that working electrode is completely destroyed after 17 h working at 600 °C while it is stable after 48 h working at 550 °C. Hence, investigate were made to decrease the selective working temperature of sensors.

After testing different mass percentages of V_2O_5 in working electrode, we concluded that by increasing this value to 50 %, the selective temperature of ammonia sensors could be decreased to 550 °C. The selectivity of sensors were also confirmed by testing other possible gases such as hydrocarbons and hydrogen. Fig. 3 reveals that these sensors shows almost no response to 20 ppm of H_2 and 100 ppm of C_xH_y.

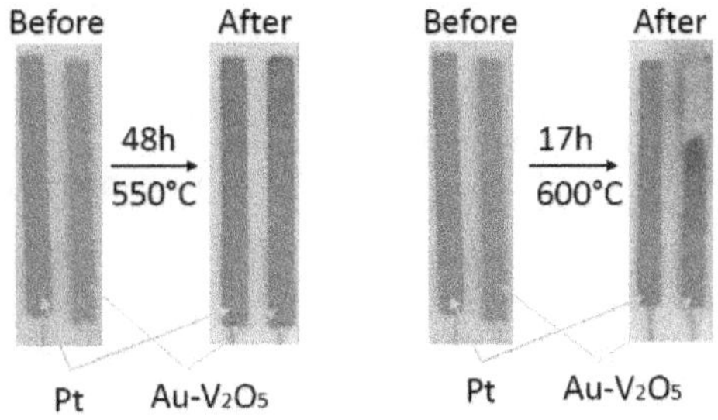

Fig. 2. Electrodes' appearances for new-fabricated Au -15 %V_2O_5 sensors sintered 1 h at 600 °C before and after tests for 48h at 550 °C and 17 h at 600 °C.

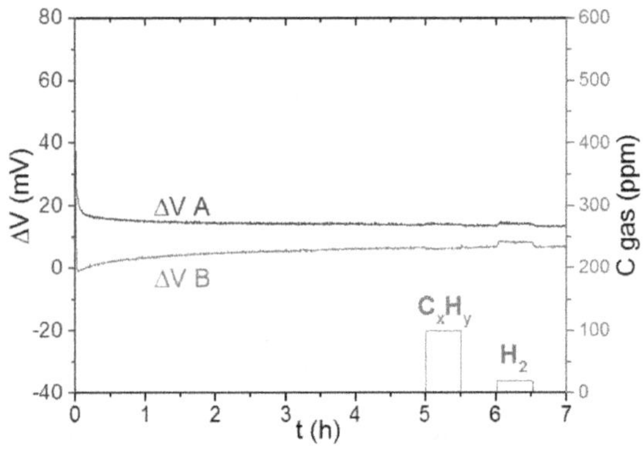

Fig. 3. Responses of Au-50 %V_2O_5 sensors to 20 ppm of H_2 and 100 ppm of C_xH_y (a mixture of 56ppm propene (C_3H_6), 28 ppm n-butane (C_4H_{10}) and 16 ppm methane (CH_4)) at 550 °C.

The sensing performance of resulting ammonia sensors at 550 °C was investigated by stepwise change in the NH_3 concentration from 1 to 40 ppm (Fig. 4). These results indicate that the sensors have high sensitivity to NH_3, even at low concentrations.

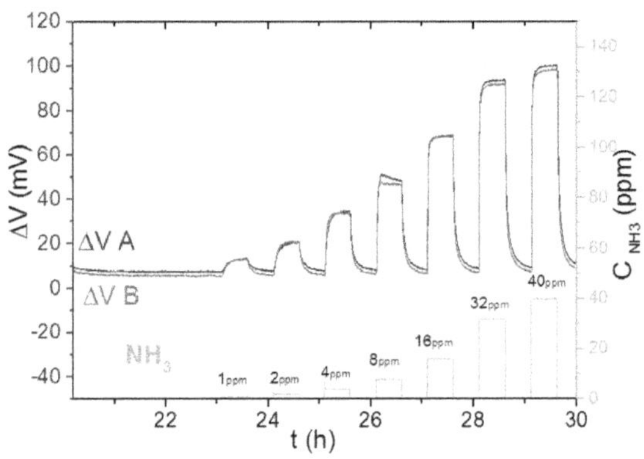

Fig. 4. Responses of Au-50 % V_2O_5 sensors to different concentration of NH_3 at 550 °C.

As illustrated in Fig. 5, the measured sensors' responses have logarithmic characteristic as a function of NH_3 concentration with sensitivity of around 55 mV/decade.

Further investigations are going on to gain deeper insight in sensing mechanism of V_2O_5 based working electrode which governs the sensor's performance.

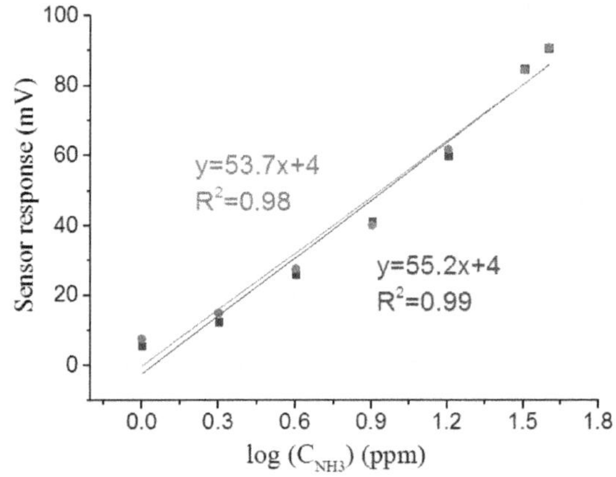

Fig. 5. Logarithmic dependence of sensors' response to NH_3 concentration at 550 °C.

4. Conclusions

Selective ammonia sensors were developed using V_2O_5-Au as working electrode, and platinum reference electrode. Selectivity of sensors were found to be so temperature dependent. The durability of these sensors were increased by increasing the mass of percentage of V_2O_5 to 50 % in working electrode with optimum operating temperature of 550 °C. A logarithmic trend of sensor responses as a function of NH_3 concentration were observed with sensitivity of around 24 mV/decade.

References

[1]. K. Shimizu, I. Chinzei, et al., Doped-vanadium oxides as sensing materials for high temperature operative selective ammonia gas sensors, *Sensors and Actuators B*, Vol. 141, 2009, pp. 410-416.

[2]. M. van Nieuwstadt, I. Dpadhyay, et al., Control of urea SCR systems for US diesel applications, in *Proceedings of the IFP Energies Nouvelles International Conference*, Dearborn, USA, 2011, pp. 655-665.

[3]. A. Muzamil, F. Qazi, et al., Recent trends in electrochemical detection of NH_3, H_2S and NO_x gases, *International Journal of Electrochemical Science*, Vol. 12, 2017, pp. 1711-1733.

[4]. V. Modafferi, S. Trocino, et al., Electrospun V_2O_5 composite fibers: Synthesis, characterization and ammonia sensing properties, *Thin Solid Films*, Vol. 548, 2013, pp. 689-694.

(025)

Optical Back-reflecting Sensor Based on Vertical Grating Coupler Configuration

A. Demeter-Finzi [1] and S. Ruschin [1]
[1] Tel-Aviv University, Faculty of Engineering, School of Electrical Engineering, Tel-Aviv University, 69978 Tel-Aviv, Israel
E-mail: anatdemeter@gmail.com

Summary: We present a one-port sensing configuration based on vertical grating coupler and waveguide reflector arrangement. Using scattering matrix model the sensitivity and limit of detection of the sensor are optimized. Absolute sensitivity value is provided, including a theoretical limit of detection better than 10^{-8} RIU.

Keywords: Optical sensors, Optical waveguides, Grating couplers, Photonic devices.

1. Introduction

Waveguide sensors combining grating couplers attracted considerable attention for many years owing to their high sensitivity and capability of rapid detection of analytes [1, 2]. There are two types of sensors based on grating couplers. The first is based on the dependence of the grating couplers' functionality on the specific properties of the underlying waveguide (e.g. the effective refractive index). The second is based on the interference effect between two waveguide channels. Since interferometers are sensitive to phase changes, they are considered to be among the most sensitive sensors. Resolutions of the order of 10^{-4}-10^{-6} RIU were reported, with potential of attaining even higher resolving powers when dispersion effects are suitable harnessed [3].

Here, we suggest a new sensing configuration based on vertical grating coupling arrangement aided by back-reflection. The use of vertical grating coupler is very attractive option, due to its high performance, easy fabrication, and reduction of the system's tuning complexity.

For optimizing the sensor sensitivity and limit of detection (LOD), a simple analytical model based on scattering matrix (S-matrix) formalism that was developed by us is used [4]. Using this model, an upper limit for the sensitivity and lower limit for the LOD of the sensor are determined. By locating the simulating sensor on a contour map calculated by the S-matrix model and adjusting its location, the device sensitivity can be farther improved. This new sensing configuration together with our S-matrix optimization model provides highly sensitive optical sensor which can be farther optimize based on user constrains.

2. Sensor Design

2.1. Sensing Configuration

The suggested sensor is presented in Fig. 1. It is a one-port sensor composed of two gratings: one for coupling the light vertically from a single mode fiber into the optical waveguide (right grating in Fig. 1) and the other for directing the light to a specific direction (left grating in Fig. 1). Between both of the gratings a resonant cavity is formed. Our model assumes that the cover refractive index stays constant during the sensing process. This situation is suitable for label-free optical bio-sensing where the effective mode index is changed by a linking process of the analyte to an adsorbed layer in the interacting region [2].

During the sensing process the refractive index in the film-cover region changes as a result of the presence of an analyte. While the light is guided in the film layer, in the cover and the substrate layers it decays exponentially as a function of the distance from the interface of the film-cover or film-substrate. The analyte interacts with the decaying field in the cover region, as a result of the interaction of the wave with the analyte and the effective refractive index of the guided mode changes. This change is measured by locating a photodetector at the input source direction and measuring the back-reflected intensity, I_R.

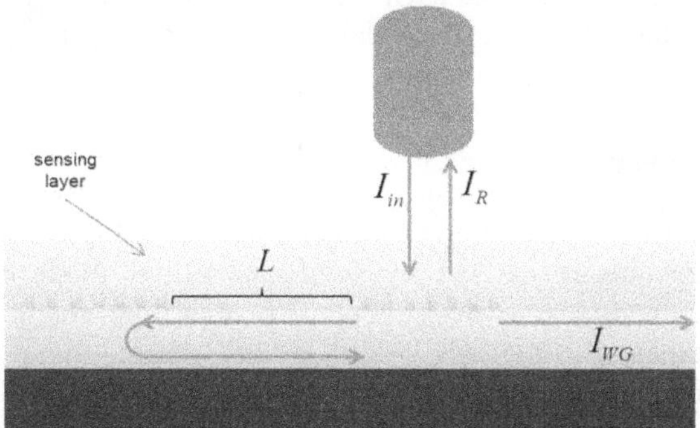

Fig. 1. Schematic drawing of the suggested sensor.

2.2. Sensor Optimization

Our optimization method is based on S-matrix formalism [4, 5]. It is a general model that can be applied to any 3-port coupler based device in order to achieve optimal sensitivity.

The device sensitivity, S, and LOD are calculated by the following equations [6]:

$$S = \max\left|\frac{\partial(I_R / I_{in})}{\partial n}\right|,$$
$$LOD = 0.2\% \cdot (\frac{I_R / I_{in}}{S}), \qquad (1)$$

where I_{in} is the source intensity, I_R is the back-reflected intensity and n is the waveguide effective refractive index.

A systematic search of the grating's diffraction efficiency values allows the determination of the sensor optimal grating parameters. By assuming a steady-state regime, the following expression is attained for the field back-reflected into the optical fiber:

$$E_{out} = a_1 \sigma \exp(i\phi_\sigma) + a_2 \rho \exp(i\phi_\rho), \qquad (2)$$

where a_1 is the complex amplitude of the right propagating field towards the coupling grating and $a_2 = 1$ is the input field. Both fields are related by:

$$a_1 = \frac{\sigma \exp(i\phi_\sigma) r_G \exp(-2i\beta L)}{1 - r_G(\xi \exp(i\phi_\xi))\exp(-2i\beta L)} a_2, \qquad (3)$$

where L is the waveguide length, β is the light propagation constant inside the waveguide, and r_G is the waveguide back reflectivity, the coefficients: ρ, σ, ξ and $\phi_{\rho,\sigma,\xi}$ were defined in [5].

By assuming power conservation and reciprocity in the system, the optimization problem can be farther reduced to two independent variables, namely the electrical field reflecion and transmission coefficients, which were chosen to be ρ and η [4]. global optimal values were extracted by allowing these two parameters all possible values in their definition range.

3. Results

Referring back to the sensing scheme in Fig. 1, the structure is excited by fundamental TE mode at $\lambda = 1.55$ µm wavelength. The light is coupled vertically to the waveguide via the input grating. An ideal waveguide reflector, $r_G = 1$, and a waveguide length of $L = 100$ µm are assumed. The waveguide effective refractive index is changed during the sensing process in the range of $3 \le n \le 3.01$. Contours plot for the sensor sensitivity as a function of the two independent parameters ρ and η is displayed in Fig. 2. Absolute optimal sensitivity of 71039 $[\% \cdot RIU^{-1}]$ and LOD of 3×10^{-8} RIU were determined for the sensor device.

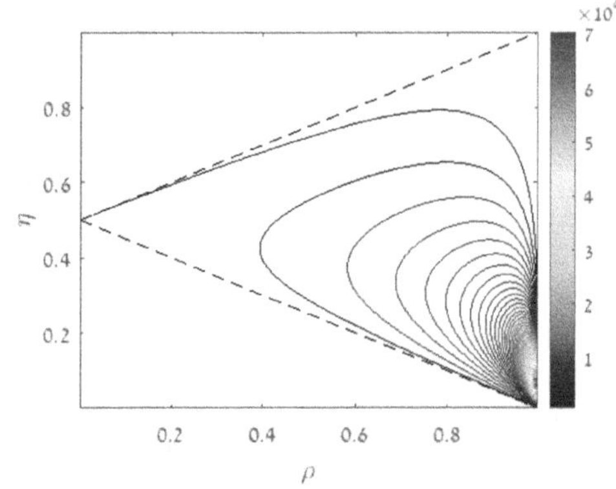

Fig. 2. Contours plot of the sensor sensitivity as a function of the two independent parameters ρ and η.

4. Conclusions

We presented a new sensing configuration based on vertical grating and resonant cavity arrangement. S-matrix optimization model was used in order to achieve maximum sensitivity. The formalism suggested here should provide guidelines for the design and fabrication of an optimized practical device for implementation in optical sensing or communication [7].

References

[1]. D. Paul, M. Textor, I. Szendr, J. Ramsden, G. Cs, D. Spencer, Optical grating coupler biosensors, *Biomaterials*, Vol. 23, 2002, pp. 3699-3710.

[2]. M. C. Estevez, M. Alvarez, L. M. Lechuga, Integrated optical devices for lab-on-a-chip biosensing applications, *Laser Photonics Review*, Vol. 6, Issue 4, 2012, pp. 463-487.

[3]. R. Levy, S. Ruschin, Critical sensitivity in hetero-modal interferometric sensor using spectral interrogation, *Optics Express*, Vol. 16, Issue 25, 2008, pp. 20516-20521.

[4]. A. Demeter-Finzi, S. Ruschin, S-matrix absolute optimization method for a perfect vertical waveguide grating coupler, *Optics Express*, Vol. 27, Issue 12, 2019, pp. 16713-16718.

[5]. A. Bunkowski, O. Burmeister, K. Danzmann, R. Schnabel, Input-output relations for a three-port grating coupled Fabry-Perot cavity, *Optics Letters*, Vol. 30, Issue 10, 2005, pp. 1183-1185.

[6]. J. Homola, S. S. Yee, G. Gauglitz, Surface plasmon resonance sensors, *Sensors and Actuators: B Chemical*, Vol. 54, 1999, pp. 3-15.

[7]. A. Demeter-Finzi, S. Ruschin, Back-reflecting interferometeric sensor based on a single grating on a planar waveguide, *Journal of Optics*, Vol. 18, Issue 1, 2015, 15801.

5th International Conference on Sensors Engineering and Electronics Instrumentation Advances (SEIA' 2019), 25-27 September 2019, Canary Islands (Tenerife), Spain

(026)

Sensors Based on Ion Transfer Voltammetry across the Polarized Ionic Liquid/Water Interface

J. Langmaier and Z. Samec
J. Heyrovský Institute of Physical Chemistry of the Czech Academy of Sciences, Dolejškova 3, 182 23 Prague 8, Czech Republic
Tel.: (+420) 266 053 895, fax: (+420) 266053910
E-mail: jan.langmaier@jh-inst.cas.cz

Summary: Some electrochemical techniques such as voltammetry at the polarized interfaces between two immiscible electrolyte solutions (ITIES) represent interesting alternatives to classical electrochemistry. The benefit of the ion transfer voltammetry lies in the possibility of detection of ionic species which are not otherwise redox active. The methodology enables apart of direct determination of ionic samples (including pharmaceutical and clinical ones) also to monitor reaction processes (acidobasic, enzymatic, etc.), determination of reaction substrates and products in one experimental step, evaluation of reaction and transport kinetics, and estimation of liphilicity of involved species.

Keywords: Ion transfer voltammetry, Ionic liquid, Liquid-liquid interface, Electrochemical detection.

1. Introduction

In the past two decades, electrochemistry at ITIES, alternatively referred to as the polarized liquid-liquid or water-organic solvent interfaces, has proved to be exceptionally useful tool for electrochemical investigations from both theoretical and practical points of view [1-3]. Lately, organic solvents such as nitrobenzene, 1,2-dichloroethane or 2-nitrophenyl octyl ether were replaced by room-temperature ionic liquids (RTILs) composed of highly hydrophobic cations and anions, which served as both the solvent and the electrolyte [4]. Compared to the classical molecular solvents, RTILs are extremely stable in a wide range of experimental conditions including elevated temperature, pH range 2-13; they are also non-volatile and environmentally friendly.

2. Experimental

RTIL is composed of tetrakis[3,5-bis(trifluoro-methyl)phenyl]borate ($TFPB^-$) or tetrakis(pentafluoro phenyl)borate ($TPFPB^-$) and tridodecylmethyl ammonium ($TDMA^+$), shown in Fig. 1.

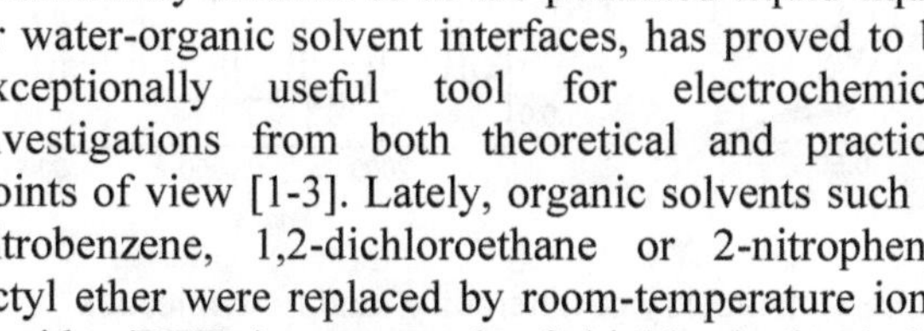

Fig. 1. Examples of hydrophobic components of RTIL components including fluorinated anions and quaternary ammonium cations.

The experimental setup involves a polyvinyliden fluoride microporous membrane impregnated by RTIL separating two aqueous electrolyte solutions [5]. Measurements are carried out in a four-electrode homemade cell schematically depicted in Scheme 1.

Scheme 1. Four-electrode cell layout, RE1, RE2 and WE1, WE2 stand for reference and working electrodes, respectively.

This arrangement enables to perform the classical electrochemical and electroanalytical experiments, in particular cyclic voltammetry, (differential) pulse voltammetry and impedance spectroscopy, and to obtain analytical and physico-chemical data needed for the model considerations, or analytical applications. Ionic compounds of interest are detected and determined directly, and their concentration profile is monitored *in situ*, i.e., in the reaction mixtures, or separately in batch or flow-through style.

3. Results

ITV is used to investigate the transfer of alkali metal cations, protons, and ammonium ions facilitated by the complex formation with valinomycin at the interface between an aqueous electrolyte solution and RTIL, and enables to evaluate the stability constants of their 1:1 complexes with valinomycin [6]. Based on these results, an amperometric ion-selective electrode for alkali metal cations is designed and the addition of water-soluble crown ethers allows to distinguish the voltammetric responses of alkali and alkaline earth metal cations [7].

ITV approach is next used for elucidation of kinetics and mechanism of methylglyoxal reaction with biguanides metformin, phenformin and 1-phenylbiguanide under the physiological conditions

[8]. Kinetic analysis is based on the time profiles of concentrations of the protonated biguanides by measured by ITV, which further enables an evaluation of the acid dissociation constants of these biguanides [9].

Evaluation of the lipophilicity of ionic and neutral drug forms, estimation of distribution coefficients (log D), partition coefficients (log P) and dissociation constant (p*K*) using ITV contribute to the deeper understanding of the behaviour of drugs during their controlled releases from the non- or semi-polar drug carriers. ITV at RTIL membrane makes it possible to compare the lipophilicity of several biologically active ions such as neurotransmitter acetylcholine and several related ions including choline, muscarine, protonated atropine, protonated scopolamine [10].

In a related experiment [11], ITV is used for an evaluation of the diffusion coefficients and the standard Gibbs energies of ion transfer of the protonated antimuscarinic agents tolterodine, fesoterodine, and their common metabolite 5-hydroxymethyl tolterodine, as well as for their determination in the aqueous samples and urine. An analysis of the pH effect provides the parameters characterizing their lipophilicity both in their ionic and neutral forms. The application of ITV for a monitoring of the enzymatic hydrolysis of fesoterodine is demonstrated. Differential pulse mode of ITV provides linear concentration dependences of the peak current for all these protonated drugs with detection limits below μmol L^{-1} and the flow-through injection arrangement makes the limit significantly lower. Fig. 2 presents some examples of the amperometric detection offered by ITV at the polarized RTIL membrane.

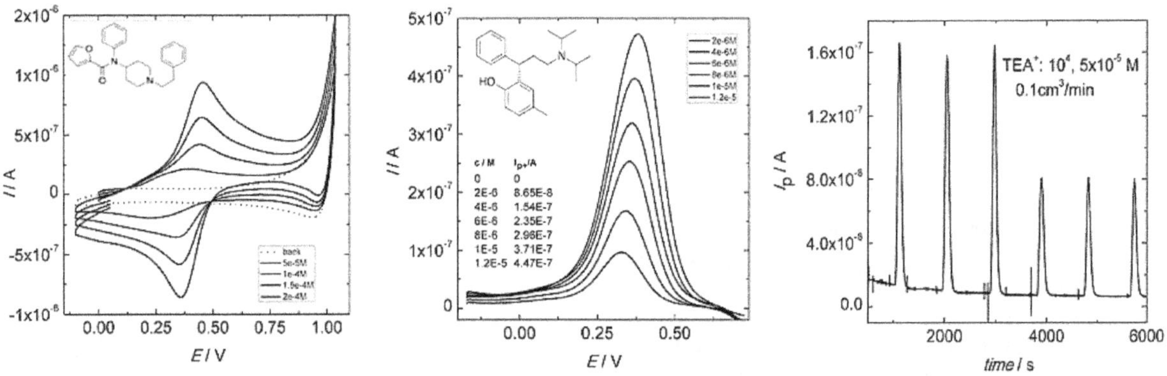

Fig. 2. Cyclic voltammetry of furanylfentanyl$^+$, differential pulse voltammetry of tolterodine$^+$ and flow-injection amperometry of TEA^+.

4. Conclusions

ITV enables to perform both the classical electrochemical and electroanalytical experiments, such as cyclic voltammetry, (differential) pulse voltammetry and impedance spectroscopy, and to obtain analytical and physico-chemical data. Ionic compounds of interest are detected directly, and/or their concentration profile is monitored *in situ*, i.e., in the reaction mixtures, or separately in batch or flow-through style.

Acknowledgements

The authors gratefully acknowledge financial support by Czech Science Foundation project No. 17-05387S.

References

[1]. V. Marecek, Z. Samec., et al., Charge-transfer across the interface of 2 immiscible electrolyte-solutions, *Advances in Colloid and Interface Science*, Vol. 29, Issue 1-2, 1988, pp. 1-78.

[2]. M. Senda, T. Kakiuchi, et al., Electrochemistry at the interface between 2 immiscible electrolyte-solutions, *Electrochimica Acta*, Vol. 36, Issue 2, 1991, pp. 253-262.

[3]. O. Shirai, S. Kihara, et al., Voltammetry for the ion transfer through a membrane, *Analytical Science*, Vol. 7, 1991, pp. 607-610.

[4]. Z. Samec, J. Langmaier, et al., Charge-transfer processes at the interface between hydrophobic ionic liquid and water, *Pure and Applied Chemistry*, Vol. 81, Issue 8, 2009, pp. 1473-1488.

[5]. J. Langmaier, Z. Samec, Cyclic voltammetry of ion transfer across a room temperature ionic liquid membrane supported by a microporous filter, *Electrochemistry Communication*, Vol. 9, 2007, Issue 10, pp. 2633-2638.

[6]. J. Langmaier, Z. Samec, Voltammetry of ion transfer across a polarized room-temperature ionic liquid membrane facilitated by valinomycin: Theoretical aspects and application, *Analytical Chemistry*, Vol. 81, Issue 15, 2009, pp. 6382-6389.

[7]. J. Langmaier, A. Trojanek, et al., Amperometric ion-selective electrode for alkali metal cations based on a room-temperature ionic liquid membrane, *Electroanalysis*, Vol. 21, Issue 17-18, 2009, pp. 1977-1983.

[8]. J. Langmaier, Z. Samec, et al., Voltammetric and capillary electrophoretic study of scavenger kinetics of methylglyoxal by antidiabetic biguanide drugs, *Journal of Electroanalytical Chemmistry*, Vol. 777, 2016, pp. 26-32.

[9]. J. Langmaier, M. Pizl, et al., Extreme basicity of biguanide drugs in aqueous solutions: Ion transfer voltammetry and dft calculations, *Journal of Physical Chemmistry A*, Vol. 120, Issue 37, 2016, pp. 7344-7350.

[10]. J. Langmaier, S. Zalis, et al., Lipophilicity of acetylcholine and related ions examined by ion transfer voltammetry at a polarized room-temperature ionic liquid membrane, *Journal of Electroanalytical Chememistry*, Vol. 815, 2018, pp. 183-188.

[11]. J. Langmaier, J. Skopalova, et al., Detection of antimuscarinic agents tolterodine and fesoterodine and their metabolite 5-hydroxymethyl tolterodine by ion transfer voltammetry at a polarized room-temperature ionic liquid membrane, *Electrochimica Acta*, Vol. 304, 2019, pp. 54-61.

(027)

Simulation of Crystallization and Mechanical Properties of Ge Thin Film for Flexible Sensor in Communication Devices

Y. Kogure [1], T. Funayama [1] and Y. Uchida [1]
[1] Teikyo University of Science
Adachi-ku, Tokyo, Japan
Tel.: + 81-47-411-6265
E-mail: kogure@ntu.ac.jp

Summary: Materials for blood flow sensor that operate in close contact with the skin has been investigated The sensor maintain the quality of life and the communication device for disabled peaple. Low-temperature crystallization and mechanical properties of germanium thin films have been simulated by means of molecular dynamics under free boundary condition. Tersoff bond order potential was adopted for calculating the interactions between germanium atoms. Amorphous model was prepared by quenching the molten syatem at 1800 K. The model was consisted of 10,500 atoms. By annealing the amorphous state at 400 K, partial crystallization was realized. Crystallization was verified by the radial distribution function of the model. Then the film was deformed and stress-strain relation was calculated. The change of atomic structure during the deformation was investigated from the cross-sectional views. The mechanical strength and flexibility are discussed from microscopic point of view. Simulation plays a complementally role with the experimental research and contributes to the development of the flexible sensors.

Keywords: Germanium thin film, Molecular dynamics, Tersoff potential, Crystallization, Stress-strain relation, Flexible sensor, Communication device.

1. Introduction

A flexible sensor that functions closely to the skin is an indispensable tool in the field of care and rehabilitation. Recently, it has become possible to measure the vital signs for a long time. In order to be able to measure in daily life, it is necessary to have a flexible sensor that can be deformed according to the shape of skin. Resetting is required when a gap appears between sensor and skin. Stable sensor operation is really required. We have been studying hemoglobin measurement as a communication device for disabled people. Such a device employs near-infrared spectroscopy (NIRS) [1]. The fabrication of sensors that adhere to the skin is essential to accurately measure the oxyhemoglobin in the blood flow. The present authors have been engaged in the study to fabricate the germanium thin films on the flexible substrates in the project to improve the yes/no device [2, 3]. The low-temperature metal-induced poly-crystallization of electrodeposited germanium thin film on flexible substrate has been realized experimentally. Germanium has a high carrier mobility and low processing temperature. In our experimental study germanium films are deposited at 60 °C and crystallized at 150 °C (423 K). However, the microscopic process of the mechanism of crystallization is not well understood in many cases [4, 5]. Molecular dynamics simulation using reliable interatomic potential is a powerful way to visualize microscopic processes of crystallization [6]. One of the present authors has been applied the molecular dynamics simulation on the studies of crystal defects and mechanical properties of metals [7]. The program code developed in the studies can be utilized in the present investigation.

2. Method of Simulation

2.1. Atomic Potential

The bond order potential function developed by Tersoff is used in the simulation [8, 9]. The potential energy depends on the bond length and bond angle. The potential energy is described as

$$E = \sum_i E_i, E_i = \frac{1}{2}\sum_{j \neq i} V_{i,j}, \tag{1}$$

$$V_{ij}(r) = f_C(r)[f_R(r) + b_{ij} f_A(r)], \tag{2}$$

$$\begin{aligned} f_R &= A\exp(-\lambda r), \\ f_A &= -B\exp(-\mu r) \end{aligned} \tag{3}$$

where $f_c(r)$ is the truncation function, b_{ij} is related to the coordination number and represent the many body nature of the potential. Constants A, B, λ, μ are determined to reproduce the experimental values of physical property.

The elementary structure of germanium is shown in Fig. 1(a). The potential energy depends on the bond length r and the bond angle θ. The contour plot for the potential energy of central atom is shown in Fig. 1(b). The energy shows a minimum at $r = 2.43$ angstrom and $\theta = 116$ deg. The force acting on an atom is calculated by differentiating the energy by distance r.

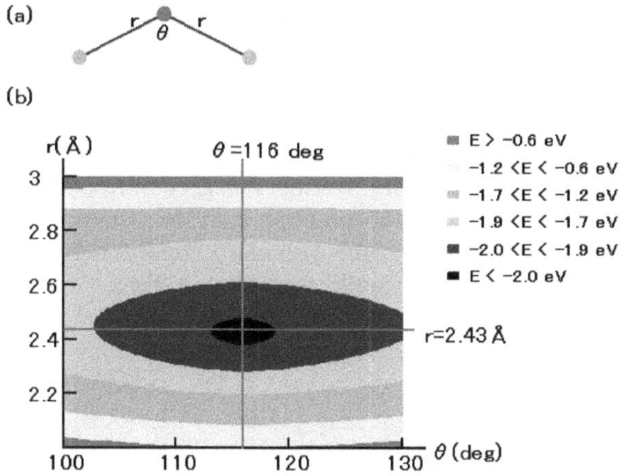

Fig. 1. Elementary structure (a), and contour plot of potential energy of Tersoff potential for the central atom in the elementary structure (b).

2.2. Molecular Dynamics

The equation of motion for an atom is solved by integrating the potential numerically by the Verlet algorithm. The time interval for the numerical integration is chosen to be $\Delta t = 1.0 \times 10^{-15}$ sec, which is much smaller than the period of the thermal vibration in germanium.

The temperature of the atomic system can be controlled by rescaling the mean velocity $< v^2 >$ of atoms through the relation $m < v^2 >= 3kT$, where m is the atomic mass and k is the Boltzmann constant. The simulations are performed by using a FORTRAN code developed by the present authors and results are visualized by Visual Basic (Microsoft).

3. Results and Discusion

3.1. Formation of Amorphous State

Thin film samples for the simulation is consisted of 10,500 atoms. In the initial state, atoms are arranged in a perfect crystal of diamond structure of rectangular shape with {100} faces. The size is 9 nm × 9 nm × 3 nm, as shown in Fig. 2(a), where the nearest neighbor atoms are connected by straight lines. Then the temperature is raised to 1900 K or more to bring it into a molten state (Fig. 2(b)). The atoms in the system of molten state are in completely disordered state. After keeping in the molten state for 30,000 MD steps the system is suddenly quenched to 0 K, and then the amorphous state is realized as shown in Fig. 2. (c)). The temperature variation during the amorphization process is shown as a function of MD step in Fig. 3. The temperature in molten state is seen to be controled flat and suddenly drops to 0 K.

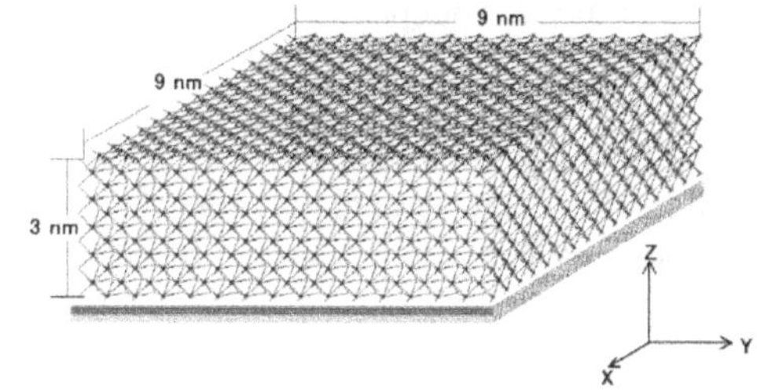

(a) Initial crystal.

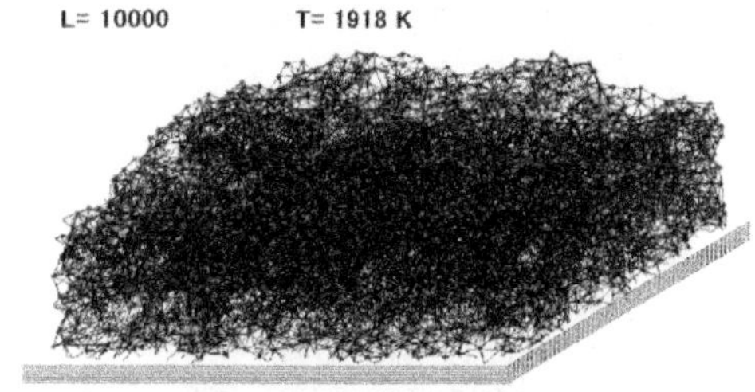

(b) Molten state.

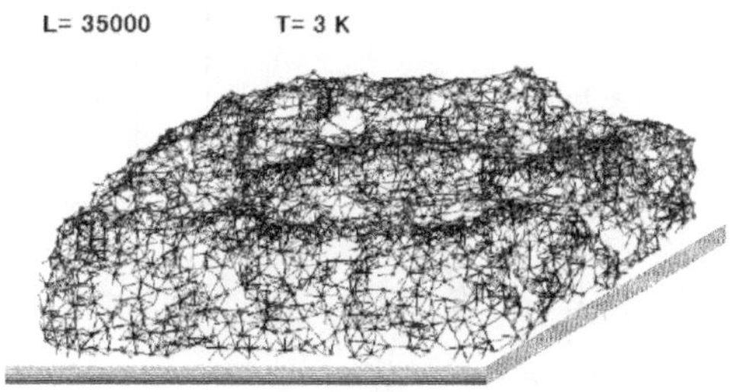

(c) Amorphous state.

Fig. 2. Preparation of amorphous sample.

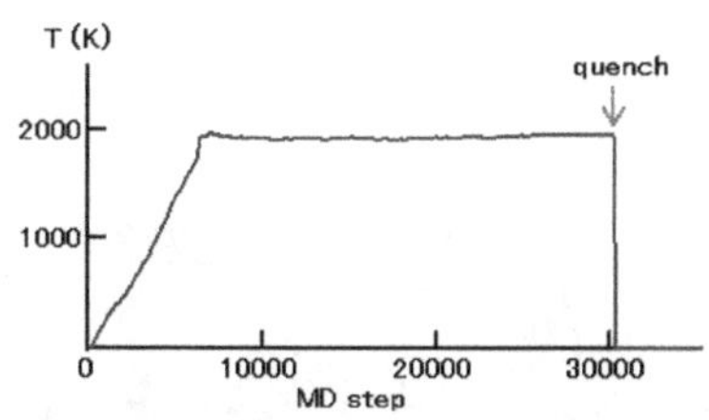

Fig. 3. Temperature variation during amorphorization.

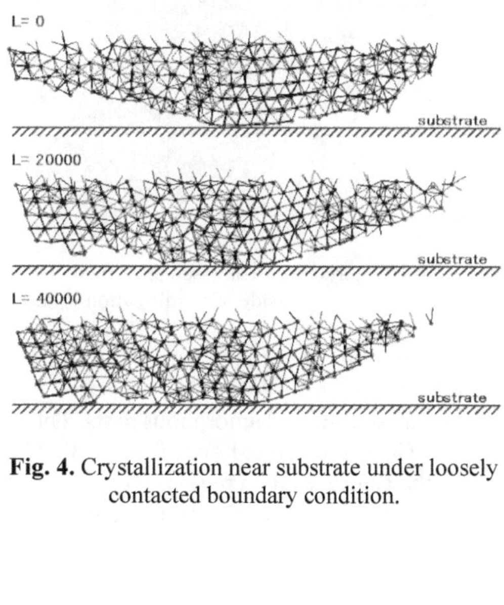

Fig. 4. Crystallization near substrate under loosely contacted boundary condition.

3.2. Crystallization at 400 K

In the experimental study, Cu-induced polycrystallization of electrodeposited films were observed at 150 °C (423 K). The molecular dynamics simulations for annealing and crystallization in large thin film (10,500 atoms) samples were performed at 400 K, where the metal element was not involved. Two types of boundary conditions between the sample and the substrate were tried:

1) The substrate and the sample are in closely contacted and atoms can slide on the surface.
2) The sample is loosely contacted with the substrate allowing a gap within 10 angstrom.

In the boundary condition 1), crystallization has hardly progressed because atoms are mostly fixed at the lower end. On the contrary, in the boundary, condition 2), it can be seen that polycrystallization is more advanced because of the boundary close to the free surface even at the lower end. An example of crystallization at the substrate is shown in Fig. 4. A progress of crystallization at the left side is seen. In order to occur a crystallization, a degree of freedom in the three-dimensional direction may be necessary at the surface. Namely, the boundary condition is very important for the crystallization of thin films.

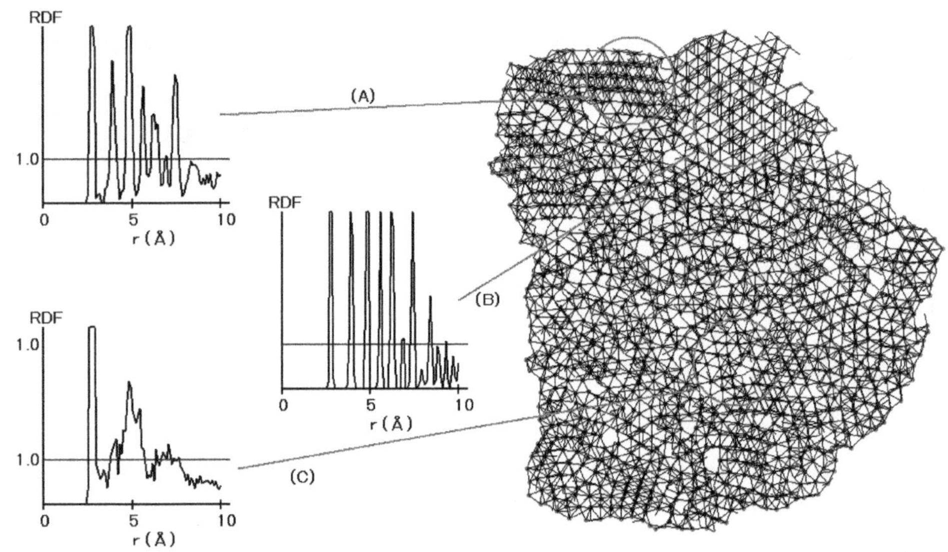

Fig. 5. Annealed thin film sample and radial distribution functions.

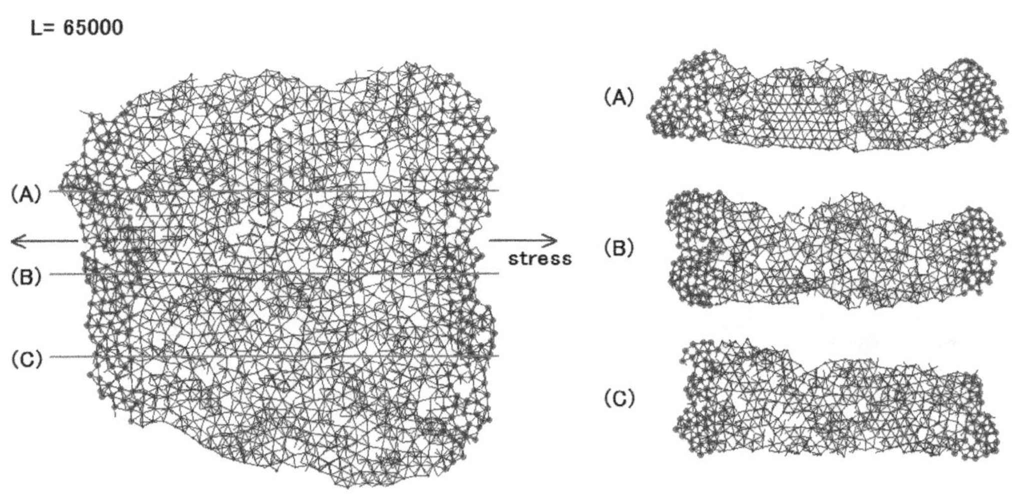

Fig. 6. Atomistic structure in stressed specimen. Typical cross-sectional views cut at (A), (B), (C) are shown in right hand side. Crystallization is seen to be progressing at the cross section (A).

Atomic structure in a plane parallel to the bottom of thin film is shown in Fig. 5. It is a heterogeneous structure of crystalline and amorphous parts. The radial distribution (RDF) at typical zones (A), (B), (C) are inserted in the figure. RDF in regions (A) and (B) show crystalline character, but RDF in (B) is sharper and has shown higher crystallinity than (A). RDF in (C) shows disordered and amorphous nature.

Judging from Fig. 4 and Fig. 5, the characteristic sizes of the crystalline and amorphous regions appear to be comparable to the thickness of thin film. Typical time steps of the molecular dynamics is 100,000.

Meanwhile the outershape and the heterogeneous structure of films were changing. The average potential energy of atoms increased once, and decreased again. At that time, a change in heterogeneous structure was observed. Simulation at a different temperature, 500 K, was performed, and similar behavior was seen. The movement of atoms was more intene when the temperature was higher.

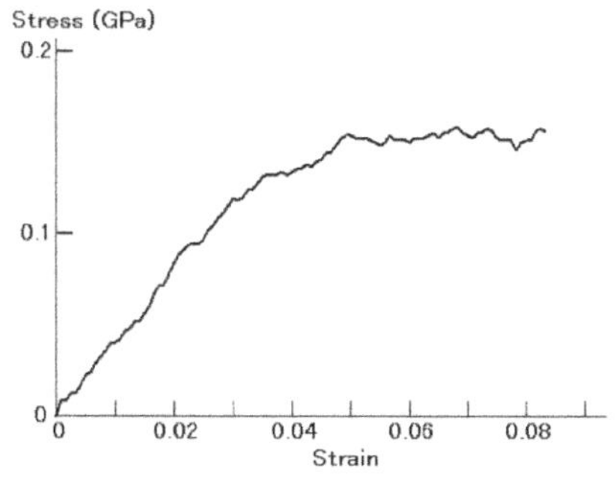

Fig. 7. Stress-Strain relation.in a deformed specimen.

3.3. Mechanical Properties

A tensile stress was applied on the plate-shape sample. The atomic configuration during deformation is shown in Fig. 6, where the atoms shown by red circles are rigidly displaced in opposite directions to apply the tensile strain. The strain rate is 1.6×10^9 / sec and maximum strain is about 8 %. The stress can be calculated from the sum of the forces acting on fixed atoms during deformation. In fact, since the forces acting on atoms inside the fixed region cancel each other. Only the forces at the boundary remain and contribute to the stress. Calculated stress-strain relation is shown in Fig. 7. The stress is seen to be increasing in proportion to the strain in the first part of the stress-strain relation in Fig. 2. Young's modulus may be able to be estimated as 3 GPa, which is much smaller than the perfect crystal of germanium. This may be due to the sparse structure of present sample. The stress approaches a constant value at a strain of 4 % or more. The atomic structure during deformation is a mixture of crystalline parts and amorphous disordered structures as shown in Fig. 6. The mosaic structure changes as the deformation progressed. The mechanism of deformation of this thin film seems to be similar to the deformation of polycrystals by slippage of grain boundaries. The motion of dislocations and other point defects might be concerned.

4. Conclusions

Low temperature crystallization of germanium thin film has been simulated by means of molecular dynamics using Tersoff potential. In particular, the movement of individual atoms at the interface with the substrate was found to be important. This point is really helpful in the experimental study. The heterogeneous structure consist of crystalline and amorphous regions was realized. Basic calculation on the mechanical properties of germanium thin films were also made. The stress-strain relation of heterogeneous film was successfully delivered by the simulation. The analysis of flexible natures of sensors linked to actual skin movement is difficult because the models that can be handled by molecular dynamics are too small. Adoption of multiscale model combining atomic model with continuum model seems to be effective.

Present study explore the communication possibility of disabled people. The accuracy of measurement indoor life has increased. However it may not be used outdoors with heavy activity, because the device still lacks flexibility and stability.

References

[1]. M. Naito, Y. Michioka, K. Ozawa, I. Ito, M. Kiguchi, T. Kanazawa, A communication means for totally locked-in ALS patients based on changes in cerebral blood volume measured with near-infrared light, *IEICE Transactions on Information and Systems*, Vol. E10-D, 2007, pp. 1028-1037.

[2]. Y. Uchida, T. Funayama, Y. Kogure, W. Yeh, Metal-induced low-temperature crystallization of electrodeposited Ge YJIN film, *Japanese Journal of Applied Physics*, Vol. 55, 2016, 031303.

[3]. Y. Uchida, T. Funayama, Y. Kogure, W. Yeh, Low-temperature Cu-induced poly-crystallization of electrodeposited germanium thin film on flexible substrate, *Physica Status Solodi C*, Vol. 13, 2016, pp. 864-867.

[4]. D. Shahrjerdi, B. Hekmatshoar, S. S. Mohajerzadeh, A. Khakifirooz, M. Robertson, High mobility Poly-Ge thin-film transistors fabricated on flexible plastic substrates at temperatures below 130° C, *Journal of Electronic Materials*, Vol. 33, Issue 4, 2004, pp. 353-357.

[5]. W. Knaepen, S. S. Gaudet, C. Detavernir, R. L. Van Meirrhaeghe, J. Jordan Sweet, C. Lavoie, In situ X-ray diffraction study of metal induced crystallization of amorphous germanium, *Journal of Applied Physics*, Vol. 105, 2009, 083532.

[6]. Y. Kogure, T. Funayama, Y. Uchida, Atomistic simulation of structure and dynamics in crystallizing germanium thin films, *Sensors & Transducers*, Vol. 227, Issue 11, November 2018, pp. 21-27.

[7]. Y. Kogure, Y. Hiki, Atomistic simulation of shear mode deformation of nanocrystalline copper with different grain sizes, *Materials Transactions*, Vol. 55-1, 2014, pp. 64-68.

[8]. J. Tersoff, New empirical approach for the structure and energy of covalent systems, *Physical Review*, Vol. B37, Issue 12, 1988, pp. 6991-7000.

[9]. J. Tersoff, Modeling solid-state chemistry: Interatomic potential for multicomponent systems, *Physical Review*, Vol. B39, 1989, pp. 5566-5568.

(028)

Monolithic Integrated PIN Photodiode Study with Backend Stack Optimization

I. Jonak-Auer, F. Roger and O. Synooka
ams AG, Device R&D, Tobelbaderstrasse 30, 8141 Premstaetten, Austria
[2] Tel.: +43-3136-500-32000, fax: +43-3136-500-932000
E-mail: ingrid.jonak-auer@ams.com

Summary: This paper constitutes a systematic analysis of the impact of low-doped intrinsic base material, different bottom anti-reflective coatings and dielectric filter deposition on the electrical and optical performance of Si photodiodes processed in standard CMOS fabrication processes. Photodiode designs cover full area as well as interdigitated variants. Optimization of the photodiode's spectral responsivity for a specific wavelength has been achieved by different bottom anti-reflective coating concepts as well as direct deposition of interference filters on Si-wafer. While standard bottom anti-reflective coating is the most efficient way of optimizing the optical response, embedded bottom anti-reflective coating offers the possibility of additional interference filter deposition directly on the wafer. For full area photodiodes with respective anti-reflective coating in place, the photodiode's quantum efficiency approaches 100 % for λ = 750 nm. For interdigitated photodiodes, the spectral responsivity in the wavelength range of 400 nm to 500 nm can be significantly enhanced compared to full area photodiodes and show values as high as 0.21 A/W @ 400 nm and 0.37 A/W @ 500 nm.

Keywords: Integrated PIN Photodiode, Intrinsic epitaxial Si substrate, Bottom antireflective coating, Dielectric Filter, Spectral responsivity; Leakage current.

1. Introduction

A photodiode (PD) is a device based on a p-n junction collecting photocurrent generated in the substrate. The combination of reverse biasing and large depletion area are necessary for improving the quantum efficiency of the photodiode [1]. The depletion area can be enlarged by using low-doped intrinsic Si, thus forming a p-intrinsic-n (PIN) photodiode. The low-doped intrinsic epitaxial material (iEPI) additionally reduces the recombination rate of the photo carriers, thus improving the efficiency of the device. The ability to detect only a defined portion ("band") of incoming light with respect to wavelength is key for applications such as color sensing, spectral imaging, UV light detection and ambient light sensing. The photodiode can be tuned to be especially sensitive to specific wavelengths by using appropriate p-n junctions as well as adding a variety of add-on filters and coatings like bottom anti reflective coating (BARC), embedded bottom anti reflective coating (eBARC) and/or dielectric filters as part of the CMOS fabrication process, In this paper the impact of the different intrinsic p-type base materials, different photodiode and guard ring designs and different add-on layers are analyzed with respect to optical and electrical performance of the optical sensors.

2. Materials and Methods

For this study we designed full area photodiodes as well as interdigitated photodiodes. For full area photodiodes we designed circular and square high doped shallow n-type layers in p-type iEPI (Fig. 1a), hereafter called nPD, For the interdigitated photodiodes we designed high doped shallow n-type and p-type fingers constituting intermeshing comb structures as depicted in Fig. 1b.

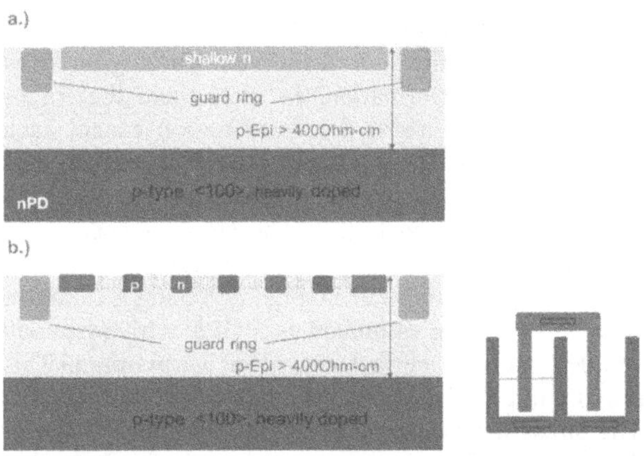

Fig. 1. PIN photodiode schematics of (a) full area nPD and (b) interdigitated PDs of alternate n-type and p-type stripes.

Regarding guard ring design we have processed and analyzed photodiodes with pnp guard rings as well as photodiodes with just p-type substrate contacts of different depths. Low doped p-type Epi of resistivity > 400 Ohm-cm on top of a highly p-doped substrate was used as base material.

All photodiodes have been processed in an industry standard 350 nm high-voltage CMOS technology, using a very cost effective way to integrate PIN photo detectors [2]. Processing of an additional isolation well as host for all circuitry components guarantees full functionality of the standard CMOS logic while the photo detectors highly benefit from the low doping concentrations of the intrinsic material. Special surface protection techniques are performed to maintain the low doping concentrations of the substrate during the complete CMOS processing. Complete modularity of the CMOS process remains uncompromised by the implementation of PIN photodiodes.

To further enhance the photo sensor's quantum efficiency in specified wavelength regimes, different anti-reflective coating concepts were applied. In the standard BARC concept [3], shown in Fig. 2a, the reflected fraction of the optical power is minimized by removing the whole dielectric stack on top of the photodiode before BARC deposition. For photo sensors, which require additional filter deposition or 3D integration by through silicon vias the standard BARC concept is not applicable, because 3D integration as well as filters require planar wafer surfaces for further processing. To make the general concept of an appropriate λ/4 plate also accessible to these sensors, an embedded BARC (eBARC) concept has been developed (Fig. 2b).

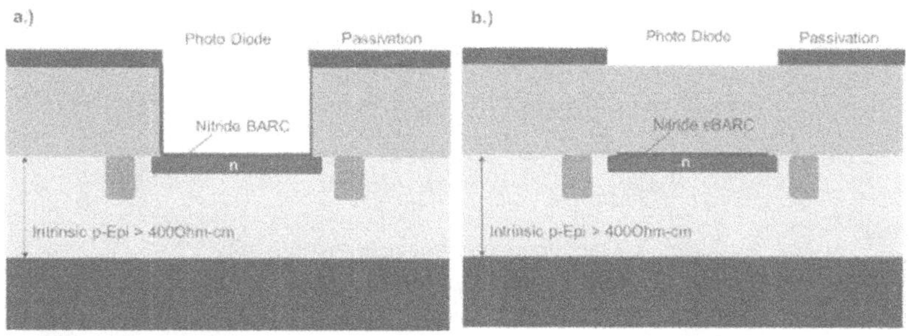

Fig. 2. Schematics of (a) standard BARC and (b) embedded BARC concepts.

Some photodiodes were processed with dielectric filter stacks in addition to eBARC. Interference filter processing uses reactive magnetron sputtering and lift-off patterning, which permits lateral filter dimensions smaller than 100 µm. Band pass filters were designed for transmission of different spectral regimes and processed in-line with the photodiodes. Special care was taken by layout rules and wafer edge treatment to minimize mechanical stress and wafer bow for combinations of up to 6 and more filters integrated on the wafer.

Characterization of the PIN photodiodes covers capacitance and leakage current measurements as well as spectral responsivity measurements. Photodiode layout permits direct pad access to anode and cathode without any signal amplification circuit. Capacitance measurements have been performed with a Cascade Micromanipulator and a HP4284A precision LCR meter, for leakage current measurements we used the Cascade Micromanipulator equipped with thermo-chuck and a Parameter Analyzer HP4155B. All optical measurements were performed on wafer level according to the measurement setup depicted in Fig. 3. A Xenon lamp is used as light source, which is discrete band filtered in 1nm wavelength steps via a spectrometer. The monochromatic light beam is split, one half goes into a reference detector for actual power measurement, while the other half is fed onto the optically active area of the device under test. To determine the responsivity of the device under test, the photocurrent generated by the incident light power is measured with a Parameter Analyzer HP4155B. The biasing of the photodetector device under test is also done by the Parameter Analyzer. Measurement data recordings as well as filter steering, calibration and reference light power measurements are controlled via serial bus interfaces and personal computers.

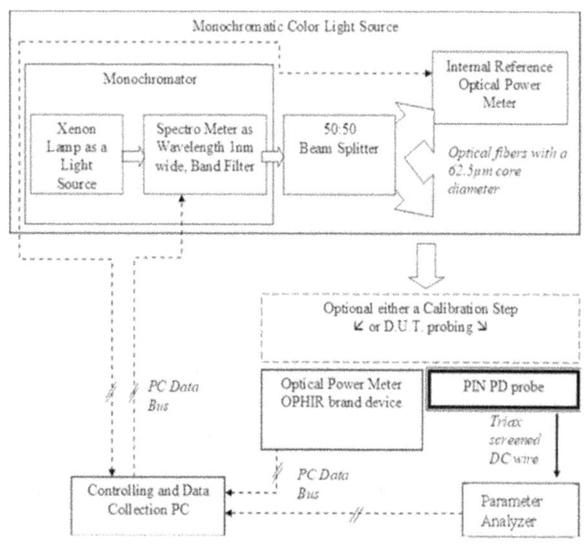

Fig. 3. Experimental setup for spectral responsivity measurements.

3. Results

Fig. 4 presents the impact of the iEPI- and BARC-thickness on spectral responsivity (SR). The spectral responsivity reaches 100 % of the quantum efficiency for BARC layers optimized for $\lambda = 750$ nm and $\lambda = 900$ nm and 82 % for layer optimized for $\lambda = 425$ nm wavelengths.

The leakage current of a circular nPD of 365 µm diameter in dependence on iEPI- and BARC-thickness is shown in Fig. 5.

From a theoretical point of view, thicker iEPI should cause larger space charge width and thus higher leakage current. But TCAD simulations showed no such effect as depicted in Fig. 6, where the leakage current of nPDs has been simulated for 20 µm and 30 µm iEPI.

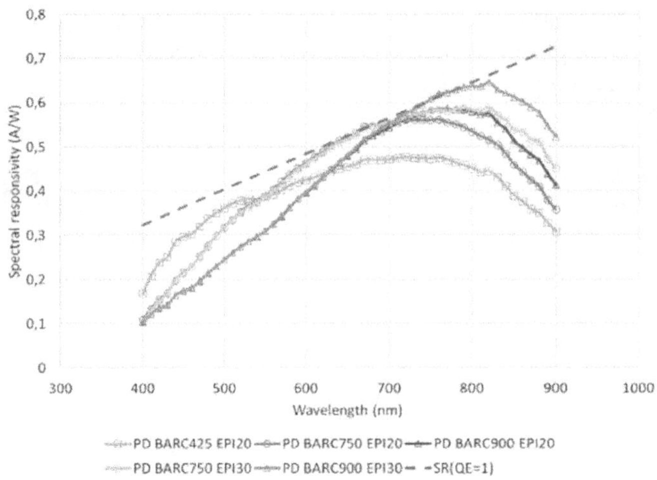

Fig. 4. Dependence of spectral responsivity on iEPI- and BARC-thickness.

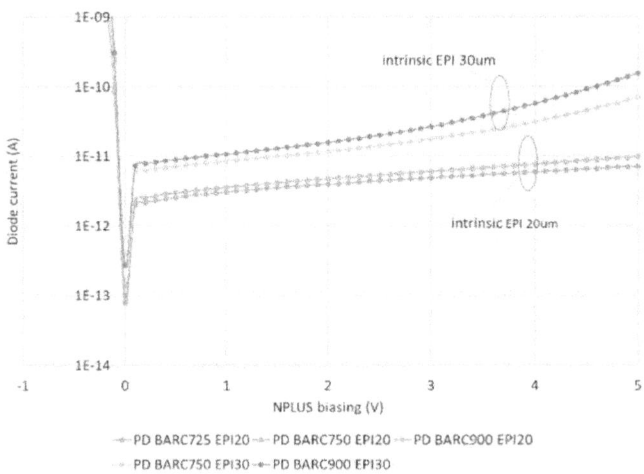

Fig. 5. Leakage current dependence on iEPI- and BARC thickness.

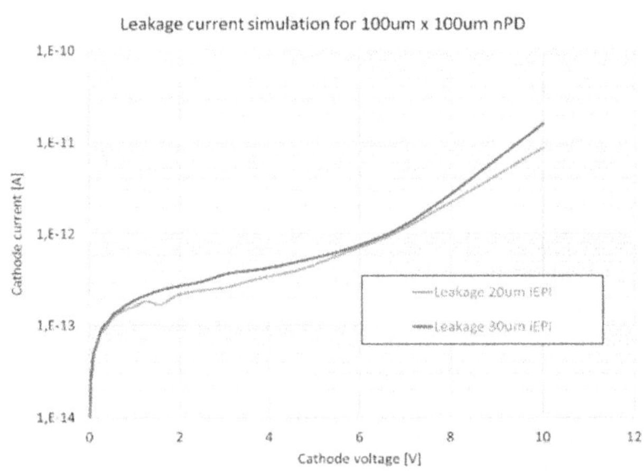

Fig. 6. TCAD simulation of nPD leakage current for two EPI thicknesses.

Essentially we could confirm by TCAD simulations, that the increased leakage of the nPD in 30 µm iEPI was caused by lower doping of the 30 µm iEPI in comparison to the 20 µm iEPI. Figs. 7 and 8 show the extension of the space charge region in dependence on the EPI doping concentration and according leakage current simulations, respectively. The simulation results show the same trend as our measurements. Significant difference in leakage current of photodiodes processed on iEPI of different thickness can be observed only for very lowly doped materials. As depicted in Fig. 9, photodiodes processed on 20 µm and 30 µm iEPI show diverging leakage values only for doping concentrations lower than 1E13/cm³ corresponding to resistivity values higher than 1 kOhm-cm. Lower leakage of photodiodes in thinner iEPI is caused by the limited extension of the space charge region in depth due to the low resistive substrate placed below the iEPI layer. We measured the leakage current of circular nPDs of different size on 20 µm (red curve) and 30 µm iEPI (blue curve) as shown in Fig. 10. Leakage current is linearly dependent on the photodiode diameter.

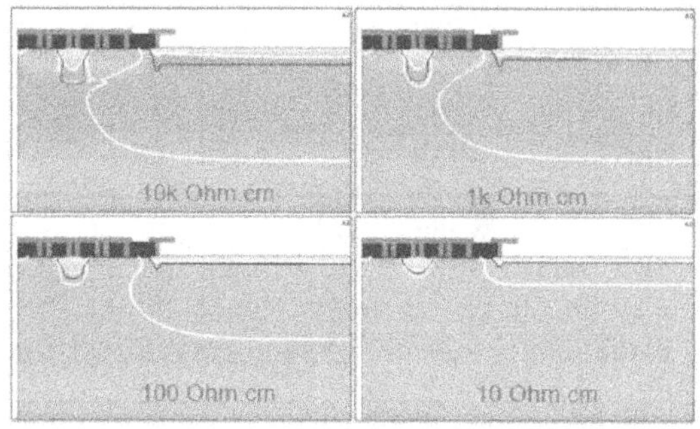

Fig. 7. TCAD simulation of space charge region extension in dependence on EPI doping.

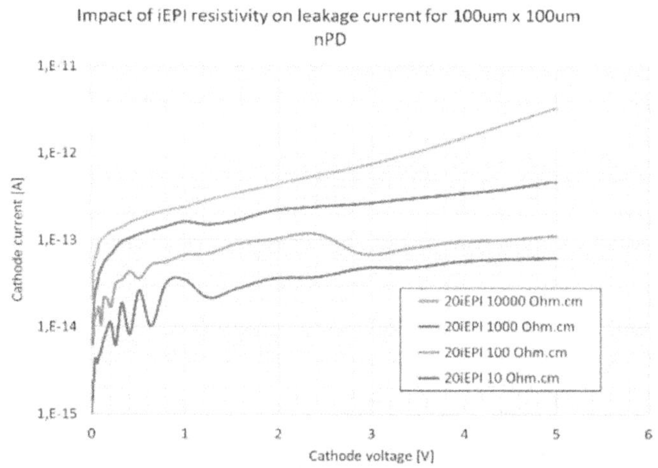

Fig. 8. TCAD simulation of nPD leak current for different EPI resistivity.

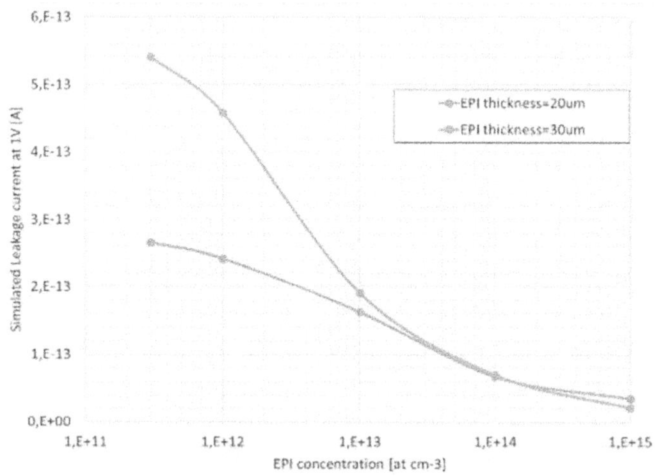

Fig. 9. TCAD simulation of nPD leak current dependence on iEPI concentration for two EPI thicknesses.

When designing photodiodes with pnp type guard rings, where the n type contact layer is used for the reduction of photodiode to photodiode electrical and optical crosstalk, care must be taken with respect to photodiode to guard ring spacing. This configuration generates a lateral npn bipolar with the photodiode which increases the leakage current of the photodiode [4, 5].

The capacitance of the circular nPD is also dependent on the iEPI thickness and photodiode size. The nPD capacitance is limited by the iEPI thickness because the low resistive substrate placed below the iEPI layer is blocking the extension of the space charge width in depth. Fig. 11 shows that the circular photodiodes' capacitance is linearly dependent on the area of the photodiode while the impact of the BARC is limited.

In addition to circular PIN photodiodes, square PIN photodiodes with a fixed single photodiode geometry of 200 µm × 200 µm are placed. We also processed arrays of 16 square photodiodes (4×4, each 200 µm × 200 µm), 49 square photodiodes (7×7) and 121 square photodiodes (11×11). The cathodes of all the photodiodes placed in an array are connected together. For arrays, the pnp guard ring is only placed around the array while each block has only a p- type substrate contact. Fig. 12 shows the capacitance of a single square photodiode in 20 µm and 30 µm iEPI, as well as the capacitance of different arrays. As observed for circular photodiodes, the capacitance is lowered when the iEPI thickness is increased. Fig. 13 shows the evolution of capacitance of the square PIN photodiode versus the number of photodiodes placed in an array. The capacitance is varying linearly with the number of blocks. As each block has a fixed geometry, the capacitance is varying linearly with the area of the array.

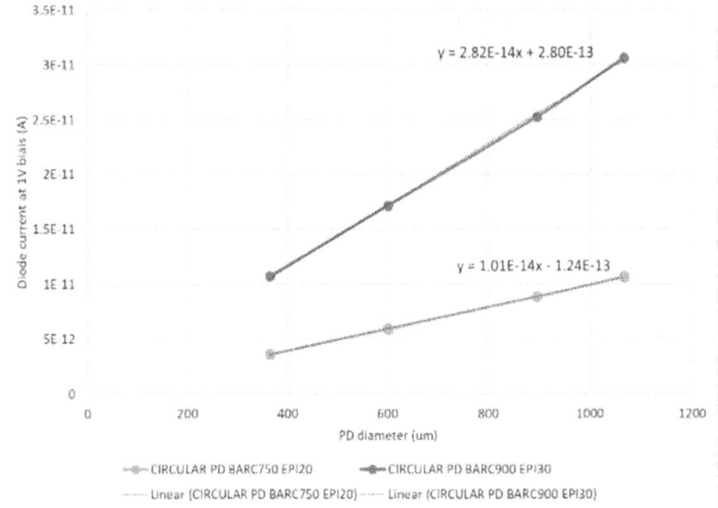

Fig. 10. Measured leakage current magnitude as a function of photodiode diameter.

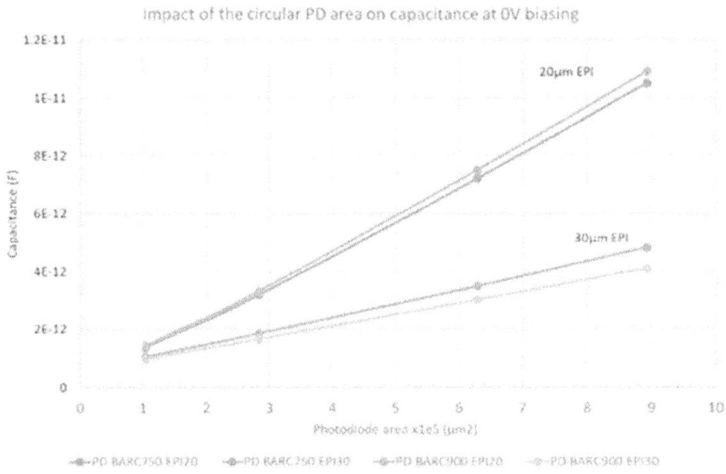

Fig. 11. Evolution of capacitance at 0 V with PD area for different iEPI thickness and different BARC thickness.

For photodiodes with p-type substrate contact rings instead of pnp guard rings, increasing depth of the p-type substrate contact ring causes decreasing leakage current and increasing capacitance of the photodiodes [2]. The dependence of photodiode leakage and capacitance on the distance between deep p-type substrate contact ring and photodiode is shown in Figs. 14 and 15, respectively. nPDs with 2 µm, 5 µm and 10 µm distance between substrate contact ring and photodiode were analyzed. The base material was 20 µm iEPI. While leakage current shows hardly any dependence on distance, capacitance values vary significantly.

Interdigitated photodiodes (Fig. 1b) target for short wavelength light (UV – blue), which is absorbed very close to the Si surface [6]. The lateral electric field between the stripes collects the generated charge, thus significantly improving the optical response in the respective wavelength range. Fig. 16 shows the influence of finger spacing on spectral responsivity and compares interdigitated photodiodes to nPDs (red curve). For 1.2 µm finger spacing spectral responsivity values as high as 0.21 A/W @ 400 nm and 0.37 A/W @ 500 nm could be achieved. Unfortunately, because of the internal npn bipolar action between the n-type/ p-type fingers, the leakage current is much higher with respect to nPD leakage current.

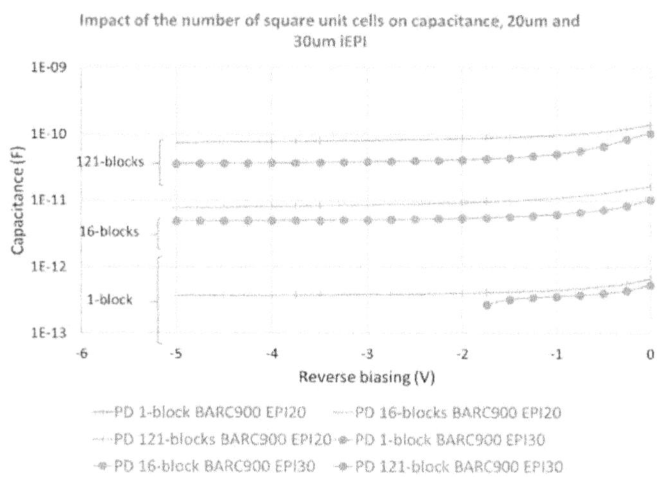

Fig. 12. Impact of scaling of square nPDs in 20 μm and 30 μm iEPI on the capacitance.

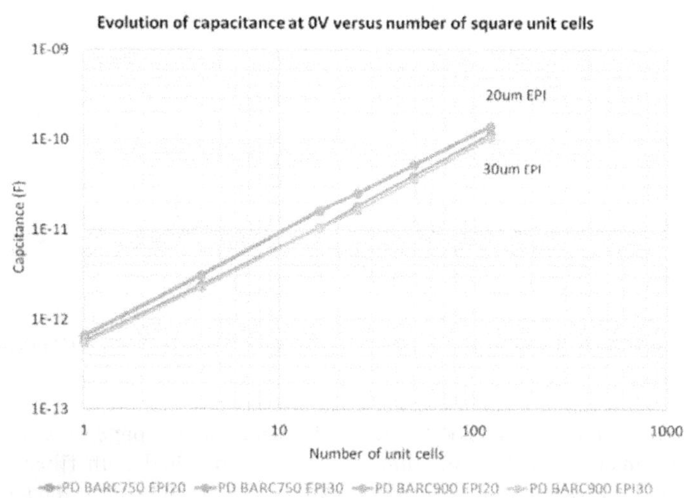

Fig. 13. Evolution of capacitance at 0 V versus number of unit cells for square nPDs.

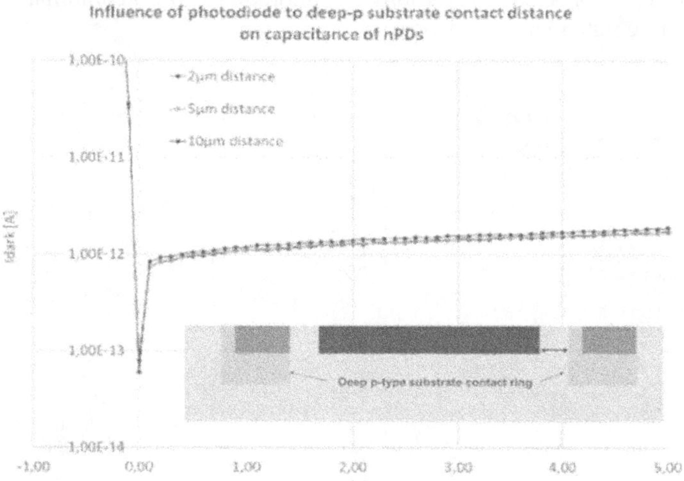

Fig. 14. Impact of nPD distance to deep-p substrate contact on leakage current.

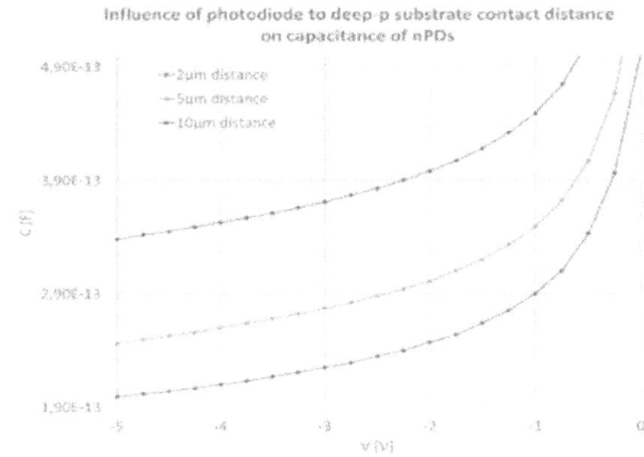

Fig. 15. Impact of nPD distance to deep-p substrate contact on capacitance.

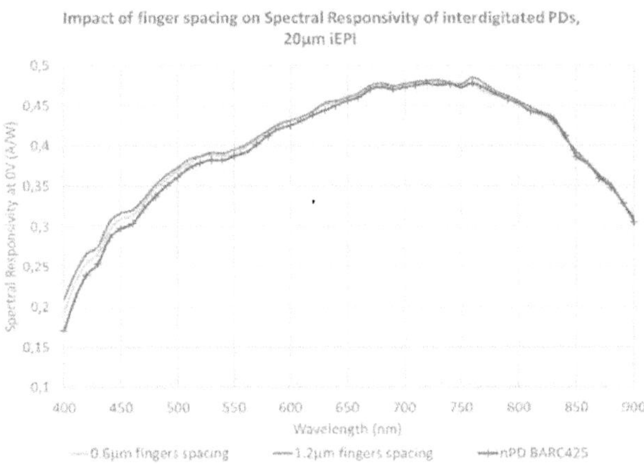

Fig. 16. Influence of finger spacing of interdigitated PDs on spectral responsivity.

Implementation of filter technologies is key for applications such as color sensing, spectral imaging, and ambient light sensing. The required spectral transmission characteristics of the filters is frequently determined by industry standards and customer requests. State of the art and commercially available interference filters deposited on glass or specifically absorbing glasses can be used for such applications. However, for monolithic integrated smart sensors, as described in this paper, the required integration density can't be reached with filters on glass substrates. The filters have to be part of the CMOS fabrication process, taking advantage of the available lithography steps. ams AG offers a large variety of inorganic filters as part of their CMOS fabrication processes, which can be utilized as stand-alone filters together with a photodiode, or in combination as stacked filters.

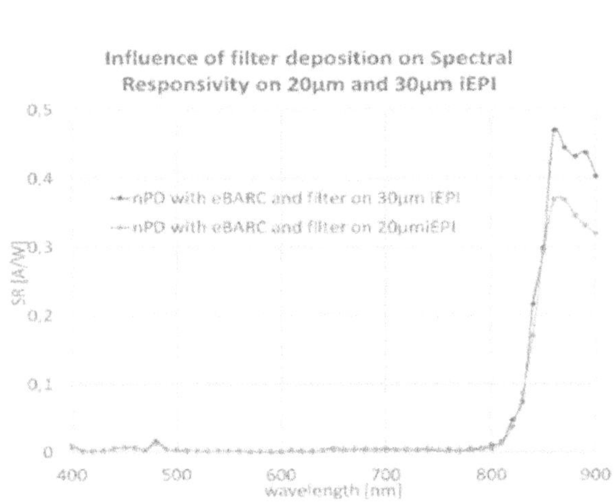

Fig. 17. Influence of long pass filter deposition on spectral responsivity.

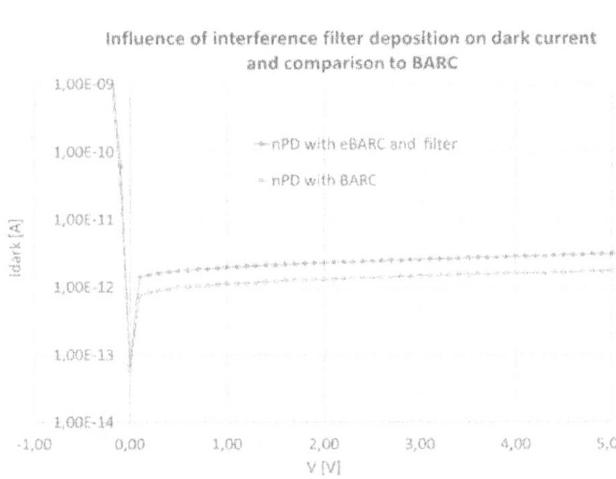

Fig. 18. Influence of long pass filter deposition on leakage current.

4. Conclusions

In this paper, we present the impact of low-doped intrinsic Si Epi material and various backend coatings (BARC, eBARC and filter) on the optical and electrical performance of various PIN photodiodes. Full area PIN photodiodes (high doped n-cathode in p- epitaxial layer) reach a spectral responsivity of 0.63 A/W and 100 % quantum efficiency with the deposition of BARC optimized for $\lambda > 750$ nm. The leakage current is larger with thicker iEPI thickness and varies from 3.5 pA (20 µm iEPI) to 10 pA (30 µm iEPI) for photodiodes with 365, diameter at 1 V reverse biasing. Their capacitance is reduced for larger iEPI thickness and varies from 0.97 pF (30 µm iEPI) to 1.43 pF (20 µm iEPI) at 0 V biasing. Further tuning of capacitance is possible by distance variations to the surrounding guard rings and substrate contacts. For interdigitated photodiodes, the spectral responsivity in the wavelength range 400 nm to 500 nm can be significantly enhanced compared to full area photodiodes and show values as high as 0.21 A/W @ 400 nm and 0.37 A/W @ 500 nm.

References

[1]. M. S. Simon, K. N. Kwok, Physics of Semiconductor Devices, 3rd Ed., *Wiley*, Hoboken, NJ, USA, 2007, pp. 671-682.

[2]. I. Jonak-Auer, et al., New integration concept of PIN photodiodes in 0.35 µm CMOS technologies, *Proceedings of SPIE*, Vol. 8431, 2012.

[3]. I. Jonak-Auer, et al., PIN photodiodes with significantly improved responsivities implemented in a 0.35 µm CMOS/BiCMOS technology, *Proceedings of SPIE*, Vol. 7719, 2010, 77190U.

[4]. J. Teva, et al., Dark current study for CMOS fully integrated-PIN-photodiodes, *Proceedings of SPIE*, Vol. 8073, 2011, 80731P.

[5]. J. Teva, et al., Gathering effect on dark current for CMOS fully integrated-, PIN- photodiodes, *Proceedings of SPIE*, Vol. 7605, 2010, 76050A.

[6]. H. Zimmermann, et al., Blue-enhanced PIN finger photodiodes in a 0.35-µm SiGe BiCMOS technology, *IEEE Photonics Technology Letters*, Vol. 21, Issue 22, 2009, pp. 1656-1658.

(029)

Design and Implementation of the Embedded System for Environmental Variables Measurement

M. Pies [1], **R. Hajovsky** [1] **and J. Velicka** [1]
[1] VSB-Technical University of Ostrava, FEECS, Department of Cybernetics and Biomedical Engineering, 17 listopadu 2172/15, 70800 Ostrava-Poruba, Czech Republic
Tel.: +420 597 325 858
E-mail: martin.pies@vsb.cz

Summary: The paper deals with the description of the carbon dioxide concentration measurement system in the rooms of the Department of Cybernetics and Biomedical Engineering. There is a measuring chain, where the pivotal element is a device for monitoring the carbon dioxide concentration with wireless transmission of the measured quantities via IQRF technology. The paper summarizes the processing of measured data on one example of measured data.

Keywords: Carbon dioxide, IQRF, NDIR, Measurement chain, Wireless sensor network.

1. Introduction

The concentration of carbon dioxide influences air quality. In interior rooms, CO_2 concentration should be lower than 1000 ppm. If this concentration is exceeded psychical performance is reduced, vigilance is infringed and, with more pronounced exceeding, headache, weariness, etc. can occur. If the concentration of 5000 ppm is exceeded severe health risks are imminent.

In school premises, stipulated values for temperature and humidity are essential, as laid down in Annex 3 to Directive of the Ministry of Education, Youth and Sports No. 343/2009 (on hygienic requirements for rooms and operation of institutions and premises for training and education of children and juveniles). In this case, values for school classrooms and office rooms are selected, see [1].

Directive of the Ministry of Regional Development No. 20/2012 (on technical requirements for buildings) lays down, among others, maximum permissible concentration of CO_2 in the rooms of stay at 1500 ppm. Values above this limit can be designated as inconvenient. Based on these regulations, limit values for measured variables have been specified, see Table 1.

Table 1. Limit values of measured variables.

	Min. value	Rec. value	Max. value
Concentration of CO_2 [ppm]	–	800-1000	1500
Temperature [°C]	20 (except. 18)	20-24	28 (except. 30)
Relative humidity [%]	30	30-64	70

This paper summarizes briefly the description of the device for measuring the concentration of carbon dioxide, temperature and relative humidity of the air and atmospheric pressure, which was first introduced in [2] and [3].

The monitoring of carbon dioxide concentration at the Department of Cybernetics and Biomedical Engineering was based on the thesis [5]. At that time, there was a growing need to monitor the concentration of carbon dioxide in view of the fatigue rate of people in these areas. Article [7] describes the effect of the dependence of carbon dioxide concentration on the number of persons in the monitored environment. The authors of this paper were used for experimental measurements an e-nose for indoor odor monitoring and photoacoustic multi-gas instrument for carbon dioxide detection.

Article [8] describes an interesting solution of CO_2 concentration monitoring, which uses a sensor hidden in a table lamp. XBee technology is used to transmit measured carbon dioxide concentration information.

This article describes the measurement of carbon dioxide concentration in the laboratories of the authors' department. For this purpose, a carbon dioxide concentration measuring instrument has been developed, where the transmission of measured CO_2 concentration information is procured by IQRF technology [4] and [6].

2. Measuring Chain

2.1. Room Measurement Carbon Dioxide Device

The measuring chain proposed for monitoring air quality in laboratories of the Department of Cybernetics and Biomedical Engineering is based on a wireless measuring node, hereinafter marked with the abbreviation RMCD (Room Measurement of Carbon Dioxide). This device is permanently powered from the mains via a micro USB power adapter. By using this connector, it is possible to power this device from any commercially available power adapter that meets the requirements of a 5 V supply voltage and a current of at least 500 mA. The photography of one of the RMCD is shown in Fig. 1.

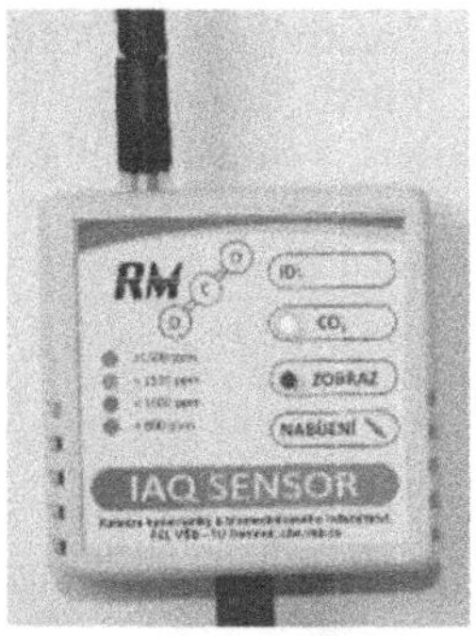

Fig. 1. A photography of installed Room Measurement of Carbon Dioxide device.

The internal structure of the RMCD is shown in Fig. 2. This block diagram of the wireless measurement module shows the links between the blocks.

As already described, the module is powered by a USB power supply. Battery management provides recharging of 3.7 V/1100 mAh Lithium Polymer battery as well as short-circuit and under-charge protection. The battery in this module is capable of operating this wireless sensor module for 24 to 36 hours, depending on the data transmission period of the module. The Power management block provides power adjustment for the TR-76 IQRF module, as well as battery voltage switching for the DC/DC converter and CDM7160 carbon dioxide sensors and the MCP9800 temperature sensor. The DC/DC boost block is a step-up module for powering the carbon dioxide concentration sensor and MCP9800 temperature sensor. The TR-76 block comprises an IQRF transceiver TR-76D. The Voltage Level Translator block is used to modify the voltage on the IIC bus and control signals for the CDM7160 sensor. The BME280 temperature and relative humidity and atmospheric pressure sensor is connected to the IQRF TR-76D directly via the IIC bus.

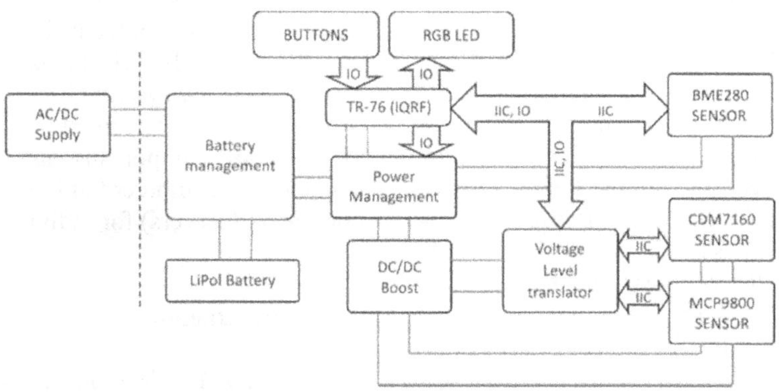

Fig. 2. The internal structure of the Room Measurement of Carbon Dioxide device.

The use of two temperature sensors is optional. It depends on whether the user wants to measure relative humidity and atmospheric pressure. If temperature measurement is sufficient, the cheaper MCP9800 can be used.

The CDM7160 carbon dioxide concentration sensor is one of the Non-Dispersive Infra Red (NDIR) type. It works by reducing the intensity of a certain wavelength of infrared light absorbed by CO_2 molecules in the air. Specifically, it is a differential spectral method. The air naturally (diffuse) flows into the sensor. IR light is used as a light source. IR spectrum selectivity is ensured by a 4.3 μm wavelength filter in front of the detector. The intensity drop is detected by an IR light detector. The FIGARO CDM7160 uses a second detector as the reference detector, which only detects a reference wavelength of 3.8 μm. The concentration is calculated based on the intensity differences. The sensor is equipped with a non-woven particle filter. This filter reduces the effect of light scattering on the particles and contamination of the optical system elements. [9] The advantage of using this sensor is that the carbon dioxide concentration value is directly readable through the IIC bus or via the PWM output in case of simpler wiring. The measurement accuracy is 50 ppm in the range of 300 to 5000 ppm CO_2. The disadvantage is the increased consumption during the measurement, as the current consumption during the measurement reaches up to 60 mA. For this purpose, the sensor is switched off when there is no need to measure the carbon dioxide concentration. The rough value of CO_2 concentration in given intervals is also indicated by RGB LED as it is shown in Fig. 1.

The main part of the RMCD device is an IQRF module from the company Microrisc, in the Fig. 2 denoted as TR-76. IQRF is a platform for low speed, low power, reliable and easy-to-use wireless connectivity e.g. for telemetry, industrial control, and building automation. It can be used with any electronic equipment. It can be used whenever wireless information transfer is needed, e.g. remote control, monitor, alarm, displaying of remotely acquired data or connection of more devices to a wireless network. IQRF is a complete ecosystem from one brand including hardware, software, development support, and services. IQRF is ideal to implement the Internet of Things. The IQRF transceiver can operate in worldwide ISM bands 433 MHz, 868 MHz, and 916 MHz. The data rate is limited to 19.2 kb/s, but for automatic gathering data from sensors or for control the actuators in home automation it is fully sufficient. The IQRF consists of 8-bit microcontroller and RF chip with additional hardware like as EEPROM memory, LED's temperature sensor, etc. There are several types of IQRF modules. The TR-76D module is characterized by a minimum of additional components, which decreases power consumption. The additional advantage is the more input/output ports on it. Communication is packed-oriented and it allows maximum payload 64 B per packet. The maximum amount of nodes in one IQRF network is 240 devices including the coordinator [10].

The Mesh topology consists in connecting each node to all the other ones (Full mesh), thus ensuring high reliability. So, if any of the routes are currently not available, there is an alternative path to deliver the message (usually multiple paths).

In the case of a wireless IQRF network, in the mesh topology, each node has its defined time transmission slot, which enables that transmission does not interfere with other nodes. The node continuously receives messages and forwards them in its transmission slot to all the nearby devices. Thanks to alternative paths, there is a high probability that in the case of damage to multiple nodes, the message arrives at the destination.

IQMESH is a protocol for wireless IQRF networks developed by the company Microrisc. It is based on the mesh topology. IQRF transceivers are deployed in the area, which ensures better coverage and a wider range thanks to mesh topology. The neighbours are those transceivers that are mutually in the transmission range.

In the case of the deployment of, e.g. 15 transceivers in the area in mesh topology, each of them will be transmitting in its time interval; in this case, there are 15 intervals. Transceivers transmit in a synchronized way, therefore they do not interfere mutually when transmitting.

The main control transceiver – the so-called coordinator, sends data when necessary, ambient transceivers receive this data and gradually send it to the neighbourhood in their time interval, where other transceivers receive it; these forward it again gradually in their time interval, and the data spreads within the network in this manner [4, 11].

2.2. Data Processing

The proposed measuring chain is composed of wireless measuring nodes RMCD and a central unit. Measured quantities are: CO_2 concentration, temperature, relative humidity, and atmospheric pressure of surrounding air. Sensors used have digital output with communication via IIC bus. Data measured by these sensors are sent through IQRF transceivers (hereinafter marked as NODE) see [4].

The measured data are processed and visualized on microcomputer Raspberry Pi, see [5]. The coordinator of the IQMESH network can attend up to 239 wireless measuring nodes that can form not only wireless measuring modules RMCD but also other sensors providing for connectivity to IQMESH. The basic structure of the system Raspberry Pi with necessary installed applications is shown in Fig. 3.

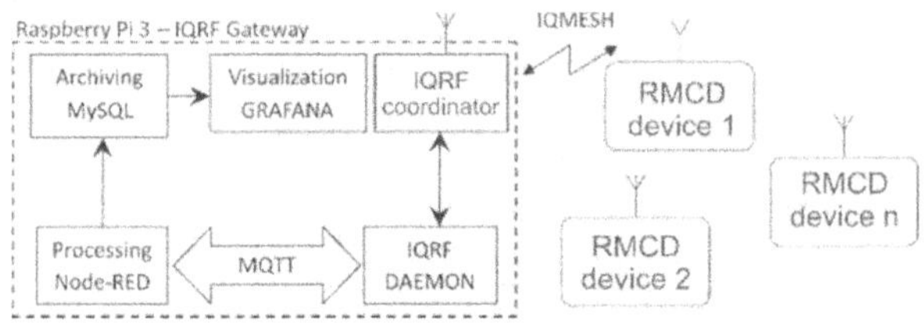

Fig. 3. Flow chart of processing, archiving, and visualization of measured data.

The measured data are accepted form the coordinator of the IQMESH network by the application IQRF-Daemon and subsequently transferred via the service MQTT to application in the environment Node-RED. The data processed in this way are recorded, already in the ready-to-read form, into the local MySQL database. The stored data are visualized in the system Grafana.

In the case presented here, all processing and visualization components run simultaneously on one single board computer but it is possible to use separate computers (servers) for individual services.

2.3. IQRF Daemon

The IQRF daemon provides communication between the IQMESH network coordinator and other systems via UDP, MQTT or WebSocket. The application can be connected to the IQRF IDE development environment through a UDP channel. The IQRF daemon application has a superstructure with a web interface for configuring and controlling the IQMESH network directly from a web browser. The application is designed as a modular open-source.

2.4. Node-RED

Node-RED is a graphical programming environment based on JavaScript. Measured data are received via MQTT protocol, then processed and stored in the database. The advantage of this solution is the high speed of implementation and easy modification or simple extension.

The measured data are received in the raw form via MQTT as JSON text. The data thus obtained are then processed. Processing includes conversion of CO_2 concentration. Temperature, relative humidity, and atmospheric pressure are recalculated using calibration constants of individual sensors. For simplicity, a custom data processing block "RMCD" was created. The values obtained in this way are then used for the compilation of the SQL query, through which the

MySQL database is stored. Each data processed in this way has a unique identifier in the database and contains a timestamp and a NODE identifier.

Calibration constants for the BME280 sensor are separately acquired and stored for each NODE.

2.5. Grafana

Grafana is a visualization tool for the web interface. Visualization of measured data was created in Grafana environment. The user can browse the measured data through historical graphs or he can monitor the current or last measured values.

3. Evaluation of Measurement

The purpose of evaluating the pilot operation was finding out of a part of teaching or working time that students and the staff spend in the environment that can influence their working performance. Based on thus obtained data, modification of the mode of ventilation or recuperation can be recommended.

Values according to the schedule or working time have been selected from the measured values. These data can be evaluated by means of the graph of cumulative frequencies, see Fig. 4. This graph shows in a synoptic form in which part of the selected time period is concentration of CO_2 lower than the chosen boundary values. The same procedure is also possible for other evaluated variables [2].

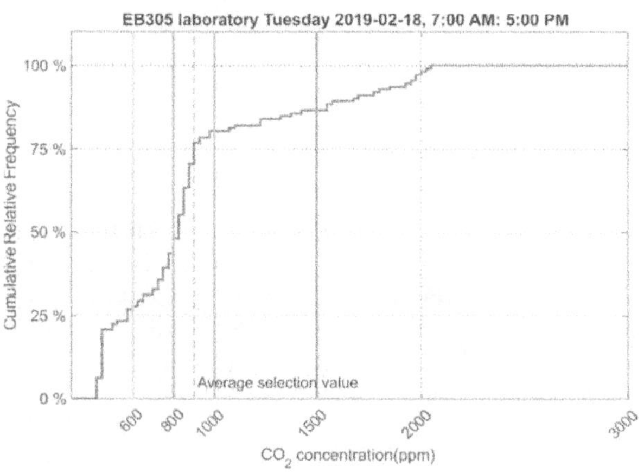

Fig. 4. Graph of cumulative frequency of data selection; Concentration of CO_2 in laboratory EB305, VSB-TUO.

The measured data have been evaluated based on the elaborated methodology of measurement and evaluation. In the charts, the limit values are marked in red/blue colour.

The values measured in laboratory EB305 on February 18, 2019, during the time of teaching (7:00 AM – 5:00 PM) have been selected for evaluation. Approximately 20 persons are moving in this laboratory during the teaching hours.

In the chart of relative cumulative frequency of CO_2 concentration (see Fig. 4), it is possible to read out that the values did not exceed the limit value of 1500 ppm in less than 86 % of the time of stay in the room.

The chart of the carbon dioxide concentration during the day is shown in Fig. 5.

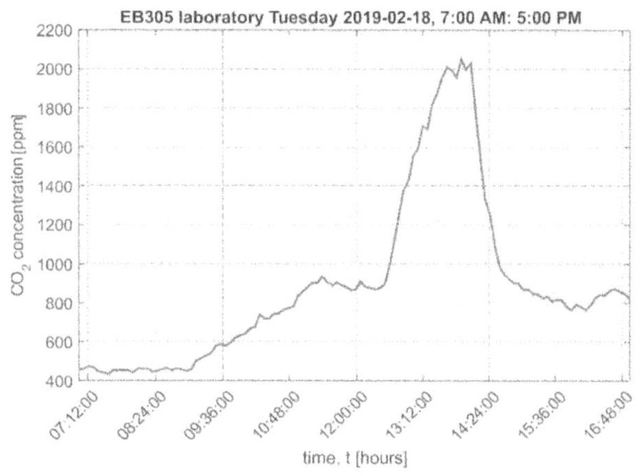

Fig. 5. The chart of the carbon dioxide concentration in laboratory EB305, VSB-TUO.

As can be seen from the curve, the carbon dioxide concentration in the room increased up to around 2200 ppm. This was due to non-ventilation in the laboratory. In the first part of this chart, one window from three ones windows was opened. The increasing of the CO_2 concentration value means, that ventilation was not sufficient. In the second part, after lunch, the windows were closed and the CO_2 concentration has increased rapidly. After lessons, all three windows have been opened for 20 minutes.

For comparison, the environment in the laboratory EB305 to the standard office for 2 workers were chosen. The office EA341 has the same ventilation system as in the laboratory. A graph of relative cumulative frequency of CO_2 concentration for the EA341 office is presented in Fig. 6.

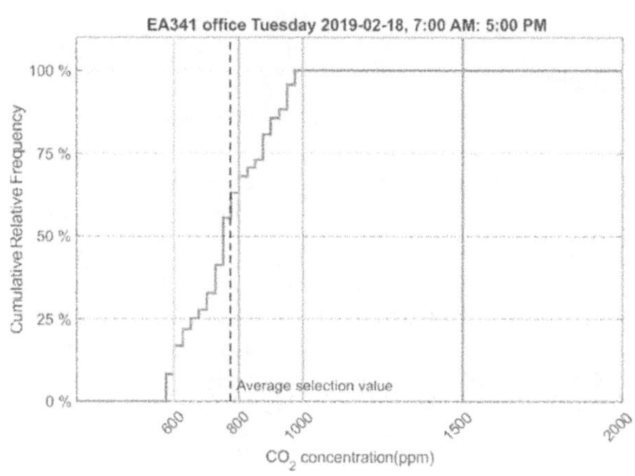

Fig. 6. Graph of cumulative frequency of data selection; Concentration of CO_2 in office EA341, VSB-TUO.

The values measured in office EA341 in the same date and time interval as in the laboratory EB305 have been selected for evaluation. Two workers were there during the working day.

In the graph of relative cumulative frequency of CO_2 concentration (see Fig. 6), it is possible to read out that the values did not exceed the limit value of 800 ppm in less than 68 % of the time of stay in the room so it means more comfortable environment in opposite to the EB305 laboratory.

The chart of the carbon dioxide concentration during the day in the EA341 office is shown in Fig. 7.

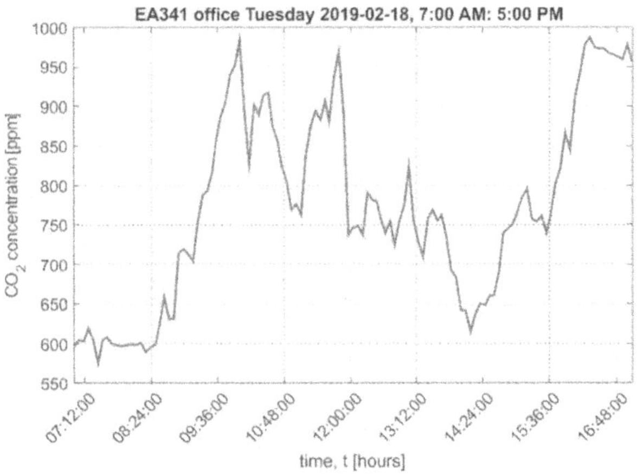

Fig. 7. The chart of the carbon dioxide concentration in office EA341, VSB-TUO.

As can be seen from this chart, the maximum concentration of CO_2 did not exceed the 1000 ppm value. It was caused by the opening the window during the day. At the end of the working day, it is obvious, that decreasing of CO_2 concentration is slow.

4. Conclusion

The proposed and realized monitoring system is intended for monitoring quality of air in interior rooms, predominantly in classrooms and laboratories of the Department of Cybernetics and Biomedical Engineering, VSB-TU Ostrava. However, it can also be applied in other school premises, offices or households.

The measured values have been evaluated within the time period in which stay of persons in the given room is expected. Based on the values obtained in this way, the part of time of stay that the persons spend in an inconvenient environment could be determined.

The building in which both rooms are located is equipped with air-conditioning units, but without air recovery, so it is necessary to ventilate. The installed carbon dioxide monitoring system makes it possible to measure the CO_2 concentration, temperature and relative humidity of the ambient air and possibly atmospheric pressure in the interval from 1 minute to several hours. A period of 5 minutes was chosen as a sufficient period of measurement and data transmission.

The data obtained from the measuring modules have been evaluated according to the limits laid down by the legislation. It follows from the measured values that the limits of CO_2 concentration are sometimes exceeded. The measured values show that the limits of temperature and relative humidity are not exceeded. The evaluated data were gathered during the winter period (February 2019).

One office and one laboratory were selected to compare two environments. The evaluated data were from February 18th, 2019, i.e. in winter when outdoor temperatures do not exceed 0 °C and therefore the ventilation time is minimized. Mean value of CO_2 concentration in the EB305 laboratory was 897 ppm. Mean value of CO_2 concentration in the EA341 office was 773 ppm. It is only 124 ppm difference, but when the data were evaluated as a relative cumulative frequency of CO_2 concentration, the results are more significant. In the EB305 laboratory students stays in the 1500 ppm of CO_2 concentration.

In the EB305 laboratory, students spent in an environment with a concentration of 1500 ppm of CO_2 and more 14 % of teaching time. Worse results occur when evaluating the time spent by students in environments above 800 ppm of CO_2. In this environment, the students were more than 52 % of their teaching time. In the EA341 office for 2 workers, they spent 32 % of their working time in an environment with CO_2 concentrations above 800 ppm. The comparison for 1500 ppm of CO_2 concentration could not be made because the CO_2 concentration at EA341 did not exceed 1000 ppm CO_2 over the evaluated period. The time and intensity of ventilation and the number of people in the room had a significant impact on this situation. Another significant influence is the indication of the interval of CO_2 concentration directly on the RMCD device, which enables immediate evaluation of the ambient air quality in the monitored room.

Acknowledgements

This work was supported by the projects of the Ministry of Industry and Trade - MPO FV10706, MPO FV20519 and MPO FV40132, next by the project No. CZ.01.1.02/0.0/0.0/15_019/0004694 of European structural funds and by the project SP2019/107 of Student Grant System, VSB-TU Ostrava.

References

[1]. K. Gecova, D. Vala, Z. Slanina, W. Walendziuk, Air condition sensor on KNX network, *Proceedings of SPIE*, Vol. 10445, 2017, 104455U.

[2]. J. Velicka, M. Pies, R. Hajovsky, Monitoring of environmental variables in rooms of the department of cybernetics and biomedical engineering, in *Proceedings of the IEEE 20th International Conference on e-Health Networking, Applications and Services (Healthcom'18)*, 2018.

[3]. J. Velicka, M. Pies, R. Hajovsky, Wireless measurement of carbon dioxide by use of IQRF technology, *IFAC-PapersOnLine*, Vol. 51, Issue 6, 2018, pp. 78-83.

[4]. IQRF, Tech. s.r.o., IQRF – Technology for Wireless, 2019, https://www.iqrf.org

[5]. J. Velicka, Measuring and visualization system for measuring environmental quantities using IQRF, PhD Thesis, FEECS, Department of Cybernetics and Biomedical Engineering, *VŠB-TUO*, 2018.

[6]. I. Calvo, J. M. Gil-García, I. Recio, A. López, J. Quesada, Building IoT applications with Raspberry PI and low power IQRF communication modules, *Electronics (Switzerland)*, Vol. 5, Issue 3, 2016, 54.

[7]. L. Eusebio, M. Derudi, L. Capelli, G. Nano, S. Sironi, Assessment of the indoor odour impact in a naturally ventilated room, *Sensors (Switzerland)*, Vol. 17, Issue 4, 2017, 778.

[8]. F. Salamone, L. Belussi, L. Danza, T. Galanos, M. Ghellere, I. Meroni, Design and development of a nearablewireless system to control indoor air quality and indoor lighting quality, *Sensors (Switzerland)*, Vol. 17, Issue 5, 2017, 1021.

[9]. Figaro Japan, CDM7160 Non Dispersive Infrared (NDIR) CO_2 Sensor, 2019, https://www.figaro.co.jp/en/product/feature/cdm7160.html

[10]. R. Hajovsky, M. Pies, Use of IQRF technology for large monitoring systems, *IFAC-PapersOnLine*, Vol. 28, Issue 4, 2015, pp. 486-491.

[11]. M. Pies, R. Hajovsky, Monitoring environmental variables through intelligent lamps, *Lecture Notes in Electrical Engineering*. Vol. 425, 2018, pp. 148-156.

(030)

Measurement System Development for Geotechnical Monitoring

M. Pies [1], **R. Hajovsky** [1]
[1] VSB-Technical University of Ostrava, FEECS, Department of Cybernetics and Biomedical Engineering, 17 listopadu 2172/15, 70800 Ostrava-Poruba, Czech Republic
Tel.: +420 597 325 858
E-mail: martin.pies@vsb.cz

Summary: The paper deals with the design of measuring systems for monitoring geotechnical quantities. There are presented design and implementation of measuring systems for measurement of retaining steel networks and barrier states and also for slope shift measurement by using a load anchor cell. For transferring data from accelerometers to a superior system, IQRF technology is used to communicate between wireless nodes and network coordinator in the MESH topology. An integral part of the article is a description of two measurement systems. The first one is for monitoring the state of retaining steel networks, the second one is for monitoring the slope shift of the gabion walls.

Keywords: Accelerometer sensor, Data processing, IQRF, Load anchor cell, Monitoring system, Retaining steel network, Rockfall barriers.

1. Introduction

Not only in the Czech Republic but also abroad there is a large number of rock formations whose damage directly threatens objects or traffic on the roads located under these formations. For these reasons, the stability of many such rock formations is ensured by means of retaining steel networks and/or dynamic barriers.

Another equally important element of roads protection is the use of gabion walls, which are special iron cages filled with stones. These cages are tied together. Their primary purpose is to prevent slopes from sliding. Even these slopes need to be monitored. For this purpose, special sensors called load anchor cells are used.

Various examples of rehabilitation using barriers and protection networks are de-scribed in the article [11], which deals with the rehabilitation of a rock massif in Germany. Another example is the use of a retaining wall and retaining networks in Albania, see [7] and in the USA, see [6] and [8].

Generally, two basic protection systems are used to rehabilitate rock formations. The first one is the use of high-strength steel network, where the damaged rock mass is covered with this network.

The second method is based on installing dynamic barriers (fences). Mainly this is a special construction of the fence, where a special twin-screw mesh is drawn between the fence rods. This is further reinforced by the use of steel ropes at their edges.

At present, there is no comprehensive monitoring system that informs the network administrator about whether they have been broken, or just capturing a certain amount of fallen rocks. Checking these retaining networks and barriers is done only at a given time. The need for the above-mentioned predictions already starts to be relevant, see [5].

2. Components of the Measuring Chains

This paper describes experimental setups of monitoring the state of retaining steel network and monitoring the load of the gabion walls. An approach closest to our requirement on the solution is presented in the publication [10], where the researchers measured deflection on the network in a laboratory model. Other publications [4, 12] are focused on applications of mathematic models of retaining networks and restraining walls on unstable slopes.

For that purpose, special measuring chains were proposed. Two main parts of these measuring chain are the same. They comprise an IQRF/GSM gateway, which collects data from the wireless nodes and a database system with visualization. The third part of the measuring chain is different according to the type of measurement [1]. The block scheme of the system is shown in Fig. 1.

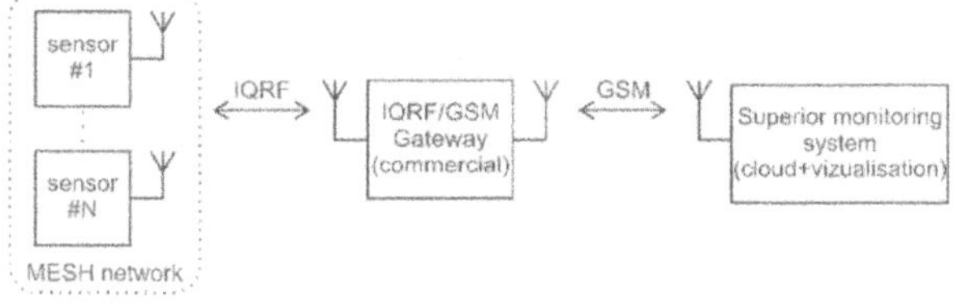

Fig. 1. Block scheme of the monitoring system measurement chain.

Two own nodes with wireless data transfer were developed for the purpose of monitoring the safety on the roads. First one is a simple accelerometer wireless node, which is installed on the retaining steel networks and/or barrier. The second one is the special node, which measures an electrical resistance at the load anchor cell. Its electronic is more complex in comparison to the accelerometer sensor. Each part of the own developed wireless node will be described shortly below.

2.1. Own Developed Accelerometer Wireless Node

The internal structure of the accelerometer node is shown in Fig. 2. This block diagram of the wireless measurement module shows the links between the blocks.

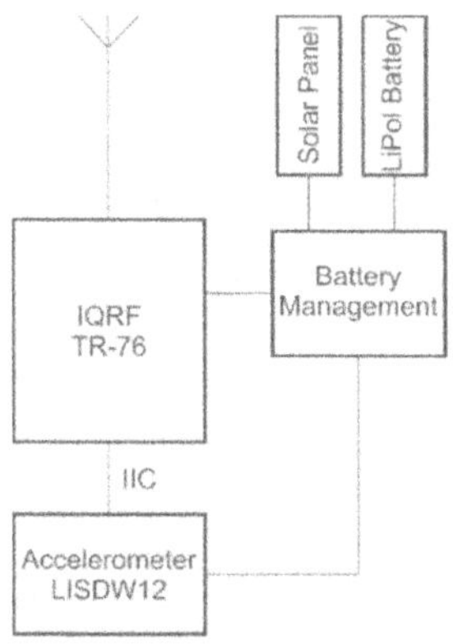

Fig. 2. Internal structure of the own developed accelerometer wireless node with sensor LIS2DW12.

Two types of accelerometer node with wireless transfer has been developed during the last half-year. First one used accelerometer sensor LIS2DS12, the new one used accelerometer sensor LIS2DW12. The differences between them is minor, thereinafter the LIS2DW12 sensor will be described.

The STM LIS2DW12 is an ultra-low-power three-axis MEMS accelerometer. It is very accurate, as it contains 16-bit analog to digital conversion hardware for each channel. Therefore it captures the *x*, *y*, and *z* channels at the same time. This chip also contains an internal temperature sensor, which could be used for temperature compensation of measured data. Measured data are sent through IIC bus to the master, which represents the IQRF transceiver module TR-76D. This module will be described at another sensor closely.

Fig. 3 shows two types of the sensor constructions.

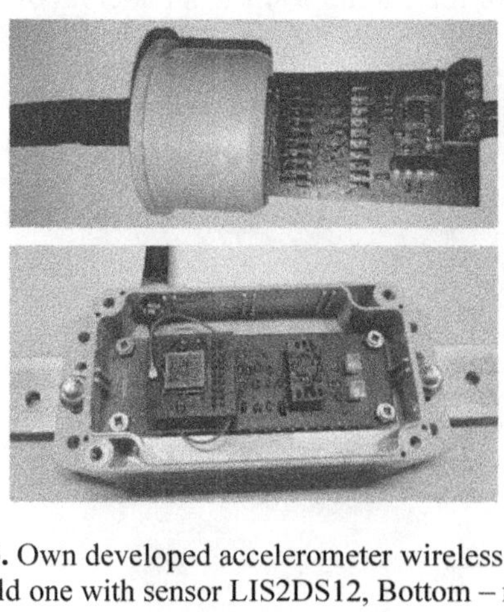

Fig. 3. Own developed accelerometer wireless node; Top – old one with sensor LIS2DS12, Bottom – new one with sensor LIS2DW12.

Photography on the top shows accelerometer sensor LIS2DS12 and its electronic, which are installed to the water-tight plastic tube. Photography on the bottom shows accelerometer sensor LIS2DW12 and its electronic components, which are installed to the plastic box. The main difference between these constructions is in the using of the solar panel. Older version of the sensor uses the outside solar panel with over-dimensioned power of 5 W and $LiFePO_4$ battery with 3.2 Ah capacity. The newer version uses a smaller solar panel with 0.5 W power and LiPol battery with 2 Ah capacity. Different chemistries of batteries must be considered in the design of Battery Management block, see Fig. 2. For that reason, the LT3652 charger with multi-chemistry support was used.

2.2. Own Developed Load Anchor Cell Node

Each load anchor cell works as a Wheatstone bridge that measures the load at the point where the anchor into the rock mass is installed. The pressure of this massif on this measuring element causes a change in the piezoresistive resistances and this results in the bridge unbalancing. The internal structure of the load anchor cell node is shown in Fig. 4. This block diagram of the wireless measurement module shows the links between the blocks.

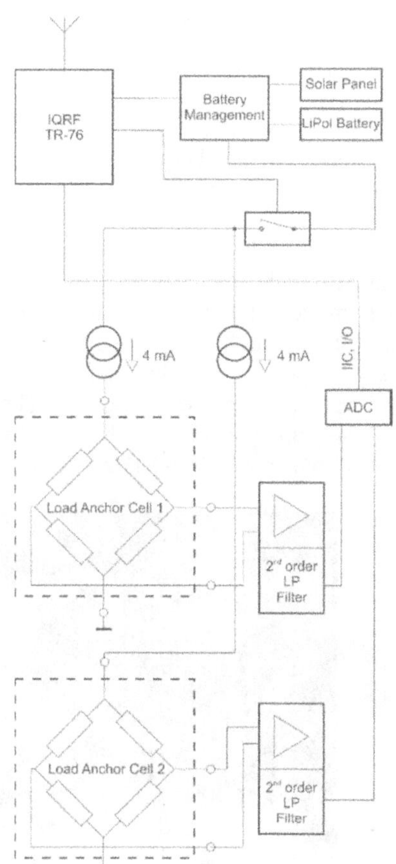

Fig. 4. Internal structure of the own developed load anchor cell node.

For measuring the voltage at the load anchor cell, the constant current of 4 mA must be provided into load anchor cell gauge. For that purpose, a precise current source was used. Each load anchor cell has an input impedance of about 3700 Ω on the power terminals, so to achieve a constant 4 mA current flowing through this bridge, this sensor needs to be powered by at least 14.8 V. Since the whole device is battery-powered, the voltage for these power sources needs to be increased by the boost DC / DC inverter. However, this inverter is a source of interference that needs to be filtered.

The measured voltage at the output of the bridge is up to 1000 mV. This voltage is gained by instrument operational amplifier and filtered by 2nd order low pass active filter. Instrument operational amplifier must accept higher common voltage on its differential input pins due to high voltage from the current source. Measured voltage is converted by Σ-Δ 16-bit AD converter. AD converter communicates with master module through IIC bus. The master module is an IQRF transceiver, which combines an 8-bit microcontroller with RF chip into one small module. The photography of the module is in the Fig. 5 on the left part.

Fig. 5. Own developed load anchor cell node with the IQRF transceiver.

The IQRF transceiver module is the main part of the load anchor cell node. There is a product from the company Microrisc. In the Fig. 2 and the Fig. 4 is denoted as IQRF TR-76. IQRF is a platform for low speed, low power, reliable and easy-to-use wireless connectivity for telemetry, industrial control, and building automation. IQRF is a complete ecosystem from one brand including hardware, software, development support, and services. IQRF is ideal to implement the Internet of Things. The IQRF transceiver can operate in worldwide ISM bands 433 MHz, 868 MHz, and 916 MHz. The data rate is limited to 19.2 kb/s, but for automatic gathering data from sensors or for control the actuators in home automation it is fully sufficient. The TR-76D module is characterized by a minimum of additional components, which decreases power consumption. The additional advantage is the more input/output ports on it. Communication is packed-oriented and it allows maximum payload 64 B per packet. The maximum amount of nodes in one IQRF network is 240 devices including the coordinator [1].

The measured data from load anchor cell node are transmitted by IQRF transceiver in small mesh topology to the IQRF/GSM gateway. Load anchor cell node can measure a voltage on two load anchor cells separately.

2.3. IQRF/GSM Gateway and Cloud

For both experiments, a commercial IQRF/GSM Gateway GW-GSM-02A was used, see [2]. The gateway sends the measured data to the cloud, which offers a communication interface for further data processing. This commercial gateway could be replaced by Raspberry Pi [3].

For the purposes of monitoring protective steel networks and barriers as well as gabion walls, the strategy of autonomous data gathering from wireless sensors that communicate with each other in mesh topology was chosen.

Data collection is periodic with a measurement interval from minutes to hours. This measurement time can be set via the IQRF cloud, where by writing the appropriate command the new value of the polling period is written into the EEPROM memory of the network coordinator. The network coordinator is part of the IQRF / GSM gateway. When collecting data, it is necessary to take into account the energy consumption of measurement and data transmission. Although IQRF technology is characterized by significantly lower power consumption, communication in the GSM network requires much greater transmit power. For this reason, the measurement of the state of the protective steel nets is set to a period of 30 minutes and the measurement of the state of the gabion walls to an interval of 4 hours.

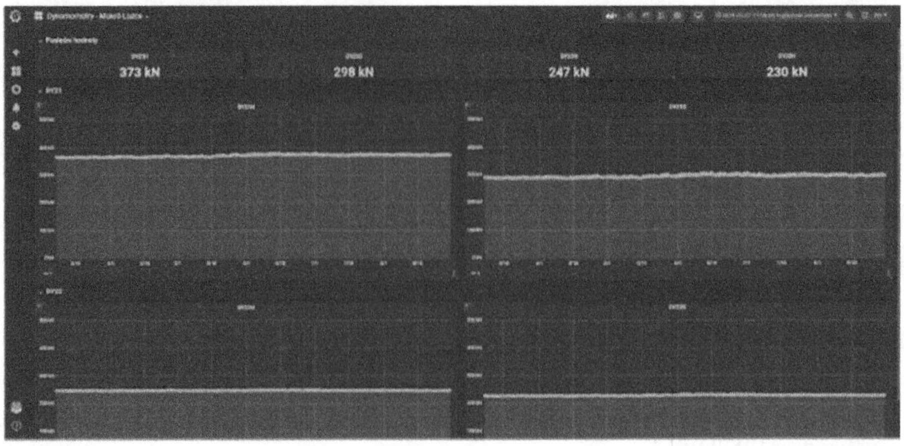

Fig. 6. The Grafana system shows the loads on the gabion walls.

Data from the IQRF cloud are further processed using scripts running on the server of our university. These scripts recalculate raw data from wireless sensors and store it in a MySQL database. This data can be further exported to various formats, such as CSV and further evaluated in MATLAB. For data visualization is used software Grafana, which draws data from MySQL database. An example of the Grafana tool is shown in Fig. 6. This figure shows the load on gabion walls between March and mid-August 2019.

3. Monitoring the State of Retaining Steel Networks

3.1. Installation of the Own Developed Accelerometer Wireless Nodes

To test the developed sensors, a test polygon was created in Málkov, Northern Bohemia. It is made of 3 fields of protective fences, as they are then installed on rock massifs. Photography of these fences is shown in Fig. 7.

Fig. 7. A physical model of dynamic barriers, where the accelerometer nodes are installed.

In the first version, 4 accelerometer sensors were installed on the test polygon for these barriers, which are placed in plastic tubes. The deployment of the two sensors is shown in Fig. 8.

Fig. 8. Detail view of installed accelerometer sensors, placed in plastic tubes.

On the left side of the picture is a detail of the LIS2DS12 accelerometer sensor located in a plastic tube. On the right side of the picture can be seen two of these sensors, which are powered by one 5 W solar panel.

The initial installation was carried out in order to verify the operation of the sensors in real conditions and also in extreme weather conditions, especially in summer and winter. It was also tested the use of solar panels for charging batteries, their activity throughout the year, and what effect on their efficiency, for example, snow cover, etc. Such limit states can be achieved within decades, depending on the state of weathering processes, type and dimensioning of protective measures. In principle, however, it can be safely stated that there is a high level of probability of achieving such a condition [9].

The sensitivity of individual sensors to the change of the protective fence status was also tested depending on the location of the change. Initial fence stretching tests were described in [13]. Long-term tests are briefly described in the following section.

The next step of the installation was the placement of newer types of sensors using the LIS2DW12 accelerometer and its own solar panel with lower power. Fig. 9 shows the location of this sensor on a real installation.

Fig. 9. View of the newer type of LIS2DW12 accelerometer sensor placed in the plastic box with solar panel.

3.2. Installation of the Own Developed Load Anchor Cell Nodes

The reason for the load anchor cells installation is the monitoring of the slope condition. If the slope began to move, it would be very dangerous for the entire road construction and passing cars. Therefore, the measured data on load anchor cells are monitored remotely and, should a sudden change in the measured values (long-term trend) be observed, the dispatching center is able to provide immediate remediation of the site.

These slopes are often supported by gabion walls. In order to monitor the condition of these walls, according to the recommendations of geologists, special load anchor cells have been installed.

An example of insufficient slope condition monitoring is shown in Fig. 10, where the collapsed gabion wall is visible.

Fig. 10. View of the newer type of LIS2DW12 accelerometer sensor placed in the plastic box with solar panel.

For these reasons, the slope in the vicinity of this accident was provided with monitoring of the slope condition on gabion walls. Fig. 11 shows the installed load anchor cell sensor (left). On the right side of the picture is an electronics of own developed load anchor cell node with wireless communication.

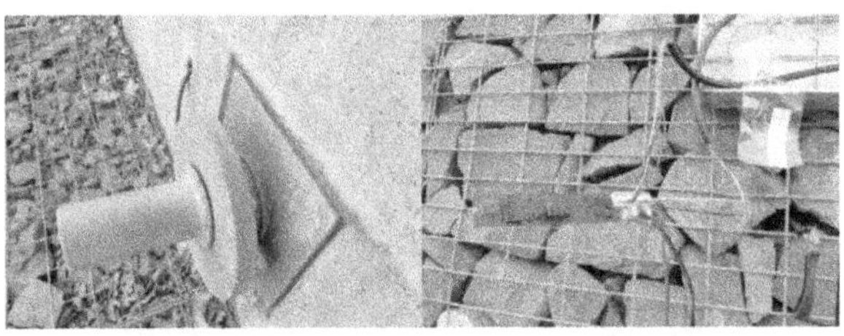

Fig. 11. Left – load anchor cell sensor; Right – own developed load anchor cell node outside the plastic tube.

A total of 4 load anchor cell sensors were installed in this locality. Since the load anchor cell nodes we have developed are 2-channel, it was enough to install 2 of these nodes to measure the state of the gabion walls.

The measured data from these sensors are again sent to the cloud via a commercial IQRF/GSM gateway and then stored in the MySQL database in the same way described above. An example of the load on this wall was shown in Fig. 6, which shows the course of the load on the gabion walls between March and mid-August 2019.

The system is formed as a grid-off, i.e. without external power supply. All components are powered by accumulators, which are charged by solar panels, see Fig. 12.

The IQRF / GSM gateway is recharged by a 40 W solar panel, each own developed load anchor cell node with wireless data transmission is powered by a 5 W solar panel.

As was already mentioned, each node measures the load of the gabion wall at two locations using two sensors.

Fig. 12. Photography of IQRF / GSM gate and solar panel installed on gabion wall.

4. Monitoring the State of Retaining Steel Networks

The testing of the retaining system is carried out from September 2018 continuously in the locality of Málkov, North Bohemia, Czech Republic. There were tested 4 pieces of the accelerometer wireless nodes on the testing model of the barrier. This barrier was randomly deflected by local workers. This simulated the possible states of the barrier system. The data shown in Fig. 13 shows the gravity acceleration of one sensor from the beginning of 2019 to the March.

Measurement of the condition of protective nets was performed with a period of 30 minutes. At the time of attempts to change the load of these fences, the measurement period was set to 1 minute.

As mentioned above, the accelerometer node is based on a 3-axis accelerometer. In reality, the sensor deflection is measured in three axes. The vector sum of the individual deflections should be 1. As can be seen from the fig. 8, the sensors are placed on the protective fences vertically. Therefore, the displacement at steady state in the y-axis is maximal (approx. 1). It should therefore be zero in the remaining axes.

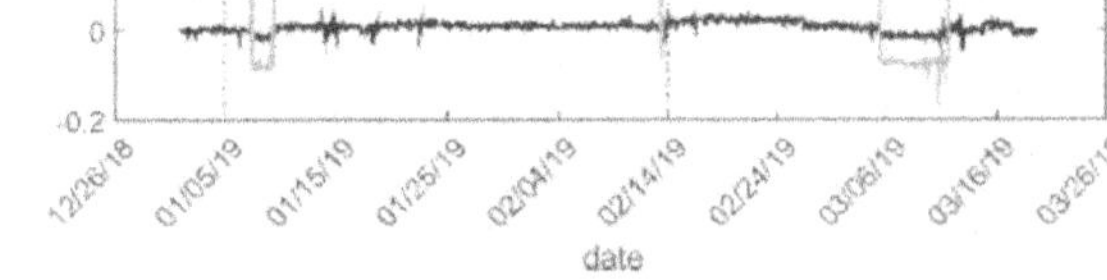

Fig. 13. Deflecting of the dynamic barrier at the location of the accelerometer node No. 1.

As can be seen in Fig. 13, this is not the case. Minor inaccuracies from the ideal state are caused by the installation of the sensor when in real conditions it is not possible to guarantee absolutely accurate mounting of the sensor perpendicularly in individual planes.

When changing the status of the protective fence, i.e. if it is moved due to e.g. falling stones or rock fragments, the sensor will be deflected and thus the measured value of static acceleration in individual axes will change. Changes are also evident from the visualization graphs, where a decrease in values in one axis means an increase in values in another axis while maintaining the condition of the sum in individual axes equals to 1.

Fig. 14 shows the load at the anchor cell, which is located at the bottom of the gabion wall. The load should be constant. Any deflection would indicate an alarm condition indicating slope movement.

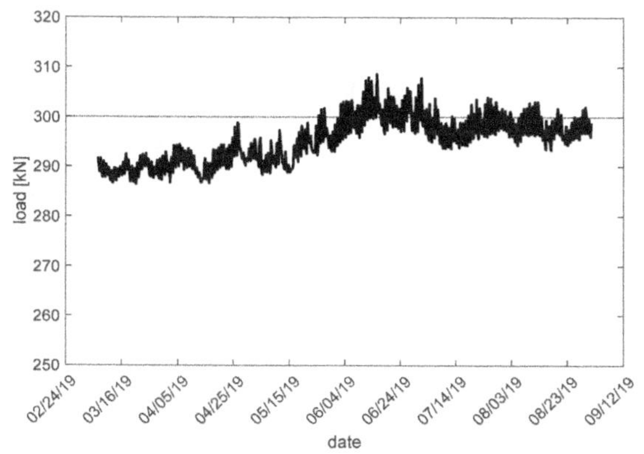

Fig. 14. Load of the gabion wall at its bottom.

As can be seen from the graph, there has been a slight increase in the load on the gabion walls since the beginning of April, but the load trend does not increase or decrease in the following period. This increase in load can be caused by the heating of the rock massif and the gabion wall itself. More detailed conclusions can be evaluated after a year of monitoring.

5. Conclusion

The paper dealt with the description of monitoring systems for measuring geotechnical quantities. There was a proposal and implementation of the system for measuring the status of protective retaining steel networks and barriers and also a system for measurement using load anchor cells. Both systems were practically deployed in-situ under operating conditions. The acquired data are archived in the database and visualized using the Grafana system.

Based on experience with long-term testing of energy performance measurements, two designs of accelerometer sensors with wireless transmission have been developed. Other installations already use a newer type of accelerometer sensor that uses a much smaller solar panel that can reliably power the sensor.

The main advantage of the designed sensors is that they are able to send alarm states themselves by means of asynchronous packets when the given limit values of measured quantities are exceeded. In the past, measurements were made at discrete times when the worker had to drive to the sensor and measure the values. The implementation of wireless value transfer will greatly improve the entire monitoring chain. In case of a sudden occurrence of an anomaly, the sensor sends alarm information and the dispatching center is capable of immediate reaction and intervention at the given location. This will minimize potential damage to property and potentially endanger human health.

Acknowledgements

This work was supported by the projects of the Ministry of Industry and Trade - MPO FV10706, MPO FV20519 and MPO FV40132, next by the project No. CZ.01.1.02/0.0/0.0/15_019/0004694 of European structural funds and by the project SP2019/107 of Student Grant System, VSB-TU Ostrava.

References

[1]. R. Hajovsky, M. Pies, Use of IQRF technology for large monitoring systems, *IFAC-PapersOnLine*, Vol. 28, Issue 4, 2015, pp. 486-491.

[2]. IQRF, Tech. s.r.o., IQRF – Technology for Wireless, 2019, https://www.iqrf.org/

[3]. I. Calvo, J. M. Gil-García, I. Recio, A. López, J. Quesada, Building IoT applications with Raspberry PI and low power IQRF communication modules, *Electronics (Switzerland)*, Vol. 5, Issue 3, 2016, 54.

[4]. H. Lin, H. I. Ling, L. Li, J. G. Collin, D. Leshchinsky, P. Rimoldi, Centrifuge modeling of gabion walls reinforced with geosynthetics, *Design and Practice of Geosynthetic-Reinforced Soil Structures*, 2013, pp. 93-102.

[5]. D. Talend, Going to extremes: Expertise and slope stabilization systems handle some of the toughest sites imaginable, *Erosion Control*, Vol. 20, Issue 4, 2013, pp. 28-35.

[6]. S. Mandavkar, A. Stralla, N. V. Kanati, Construction of MSE walls at St. Anthony Falls (I-35W) bridge, *Geotechnical Special Publication*, Issue 225 GSP, 2012, pp. 3634-3643.

[7]. P. V. Jayakrishnan, Suntharathevan, M. Vicari, G. Uguccioni, R. Budhbhatti, Composite soil reinforcement system in construction of high retaining walls for Albania Highway Project, in *Proceedings of the 5th Asian Regional Conference on Geosynthetics: Geosynthetics for Sustainable Adaptation to Climate Change (GA'12)*, 2012, pp. 939-944.

[8]. S. Mandavkar, A. Stralla, Maccaferri terramesh system, gabion face MSE walls at St. Anthony Falls (I-35W) bridge, *Geotechnical Special Publication*, Issue 211 GSP, 2011, pp. 3449-3458.

[9]. M. F. Benessiuti, G. Pg. Bernardes, E. J. Ananias, Segmental retaining wall: Comparison between predicted and observed slip surface, in *Proceedings of the 9th International Conference on Geosynthetics – Geosynthetics: Advanced Solutions for a Challenging World (ICG'10)*, 2010, pp. 1773-1776.

[10]. D. Castro-Fresno, Q. L. López, E. Bianco-Fernandez, D. Zamora-Barraza, Design and evaluation of two

laboratory tests for the nets of a flexible anchored slope stabilization system, *Geotechnical Testing Journal*, Vol. 32, Issue 4, 2009, pp. 315-324.

[11]. M. Züger, B. Haller, High-energy rockfall barriers for 3,000 KJ impact energies to protect B4 highway at Ilfeld, Germany, *Polish Geological Institute Special Papers*, Vol. 20, 2005, pp. 125-130.

[12]. C. Wendeler, Y. Bühler, P. Bartelt, J. Glover, R. Luis, Application of three-dimensional rockfall modeling to rock-face engineering, *Rock Mechanics and Rock Engineering: From the Past to the Future*, Vol. 2, 2016, pp. 1303-1306.

[13]. M. Pies, R. Hajovsky, Use of accelerometer sensors to measure the states of retaining steel networks and dynamic barriers, in *Proceedings of the 19th International Carpathian Control Conference (ICCC'18)*, 2018, pp. 416-421.

(032)

Concept for Detection of Device Failures Using Active Grid Analysis

A. Faschingbauer [1], C. Sigl [1]
[1] Deggendorf Institute of Technology,
Institute for Applied Informatics, Grafenauer Str. 22, 94078 Freyung, Germany,
Tel.: + 49 8551 91764 22, fax: + 49 8551 91764 69
E-mail: alexander.faschingbauer@th-deg.de

Summary: Typically, device recognition and device failure detection in electrical micro grids use passive measuring, whereby the sensors have to be applied locally, thus requiring an intrusion in the existing wiring. Therefore, a concept is proposed, that uses active grid analysis to detect wire and device faults as well as their location in the grid topology. Instead of the passive grid analysis, the presented approach applies the measurement hardware at socket outlets. This results in an easier, more flexible and less time-consuming installation. The underlying approach for the active grid analysis is based on reflectometry. Due to some benefits compared to other reflectometry types, spread spectrum time domain reflectometry methods is used and implemented, using software defined radio. Moreover, the monitored signals are processed using neural networks.

Keywords: Artificial intelligence, Condition monitoring, Software defined radio, Device recognition, Spread spectrum time domain reflectometry, Device signature.

1. Introduction

Common approaches for detection of device failures are based on passive current sensors, fast analog to digital converters, a kind of preprocessing algorithm and an approach using artificial intelligence. The current sensors must be applied at specific locations in a power grid system. This approach is complicated because a specialist is required to install the system, considering high voltage and installation regulations. Installation is complicated and thus costs are high. With passive fault detection only device failures can be detected, which is a big disadvantage compared to active grid analysis. The presented concept shows an approach which can be easily applied to existing grid installations, using one or more socket outlets and an active analysis method. It is assumed that the system will be capable of learning the existing electrical grid and device topology. Thus, the proposed method is less invasive.

The paper presents a short overview about the state of the art as well as a comparison of methods for passive and active signal analysis. Afterwards, the proposed concept for active grid and device analysis is described in more detail. Finally, the paper is concluded by pointing out the expected outcome and discussing the next steps of planned experiments.

2. Related Work

Roos et al. [1] proposed a non-intrusive load monitoring system (NILM) for device recognition based on fingerprints of devices that can be measured in the power grid. As neural networks are used for pattern recognition, only a single metering point is required for passive grid analysis of several devices.

Smith et al. [2] combine time domain reflectometry (TDR) and spread spectrum techniques to the spread spectrum time domain reflectometry (SSTDR). It is used for fault localization of intermittenet wiring failures (temporary open circuits, short circuits, arcs) on aircrafts by detecting a significant impedance mismatch. Roy et al. [3] show that SSTDR can be used for photovoltaic (PV) ground-fault detection and state, that it could be exploited to detect aging-related impedance variations in a PV system. Roy et al. [4] also used SSTDR for real-time health monitoring of high-power semiconductor switches.

3. Comparison of Passive and Active Grid Analysis

Passive grid analysis is often based on current measurements at one or more measurement points in the grid. The signal of the current sensor is sampled with a relatively high sample rate. The data is often preprocessed and used by a neural network as training data. Afterwards the trained model is used for device recognition and fault detection.

Active grid analysis performs a signal injection into the electric micro grid using a coupling network. Then, the reflected signal is compared to the injected signal. Both signals are used for analysis. The differences between them imply the grid/device topology and faults in the monitored system. Different reflectometry types exist for injection of certain test signals into the system. These reflectometry methods are capable of recognising the topology of the network. One SSTDR-sensor can only measure a certain amount of junctions [6]. This could be enough to enable fault detection in the proposed use case. A short overview about active grid analysis methods is given in the following [2, 5]:

- Time Domain Reflectometry (TDR) uses a voltage step function as test signal. Live testing is only feasible with low frequency signals.
- Frequency Domain Reflectometry (FDR) injects a set of stepped sine waves into the wire and is limited to low frequency circuits.
- Sequence Time Domain Reflectometry (STDR) injects a pseudo noise (PN) sequence and enables real-time testing.
- Spread Spectrum Time Domain Reflectometry (SSTDR) uses a sine wave modulated PN code and is capable of live testing.

For the active measurement of grid signals, SSTDR provides a more suitable method than TDR or FDR because of the capability for real-time testing and the high frequency noise immunity shown in [6]. For this reason, the SSTDR approach is used for active grid analysis.

4. Proposed Concept

4.1. Lab Setup

Fig. 1 presents a simple lab setup, consisting of a electric micro grid, a coupling network and a connected data processing unit (DPU). Furthermore, one or more devices under test (DUTs) can be attached to the microgrid. The following three devices are selected as DUTs, based on [1] and [7]: Power drill, light dimmer and a heater.

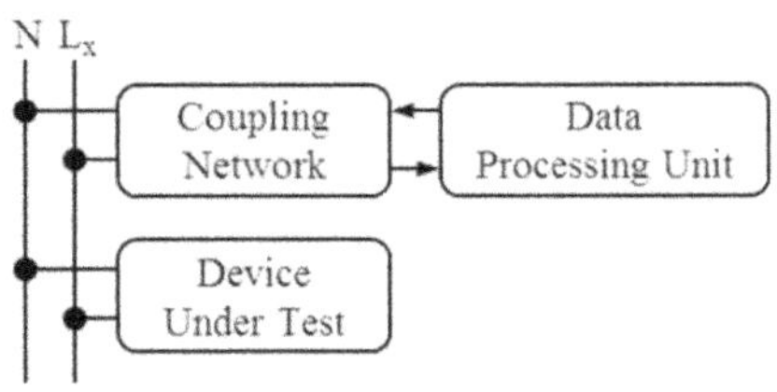

Fig. 1. Lab Setup.

A line impedance stabilization network (R&S ENV216) is used as a coupling network between the grid/DUT and the DPU. The DPU is a software defined radio (SDR). The SDR (USRP1) consists of fast analog to digital converters and digital to analog converters with appropriate signal conditioning circuitry as well as a field programmable gate array (FPGA). The SDR is connected to a Lab PC over USB. Signal processing and calculations are done on the Lab PC using GNURadio.

4.2. Current Work

The first step is to implement SSTDR using a SDR approach. Fig. 2 shows the data and signal flow graph which is used in the setup.

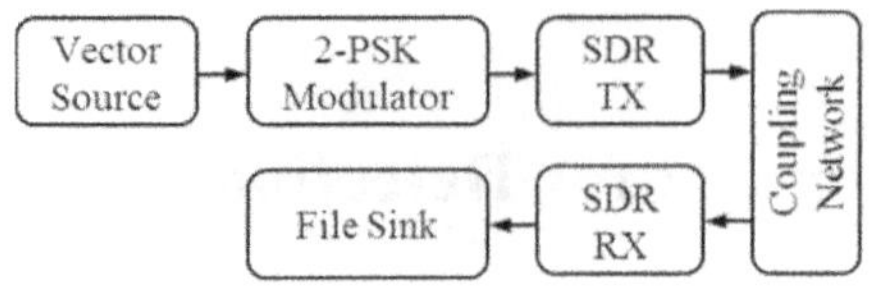

Fig. 2. Data/Signal flow graph.

A specific stimulation signal consisting of a previously generated 64 bit Gold code is used for the active injection in the grid. A vector source sends this bit sequence into a binary phase shift keying (2-PSK) modulator. The output of the 2-PSK modulator is composed of in-phase and quadrature components (I/Q) which are directly feed into the SDR system. The output signal of the SDR system is then feed into the micro grid through the coupling network. The reflected signals from the coupling network are feed back to the Lab PC. A file sink stores the reflected signal represented as I/Q data for postprocessing. This procedure is done to get a well-known signal which can be detected in the received signal stream easily, using cross-correlation.

Equation 1 is used to calculate the cross-correlation and Equation 2 is used to calculate the correlation factor *cf*. The I/Q sample is represented by *n* and thus represents the absolute position in the signal data stream.

$$(t * r)[n] \triangleq \sum_{m=-\infty}^{\infty} \overline{t[m]} \cdot r[m+n], \qquad (1)$$

$$cf[n] = |(t * r)[n]| \qquad (2)$$

A static sequence is required for cross-correlation together with the received/reflected signal from the SDR system. Fig. 3 shows a GNURadio setup to generate this static I/Q data which is stored into a file once.

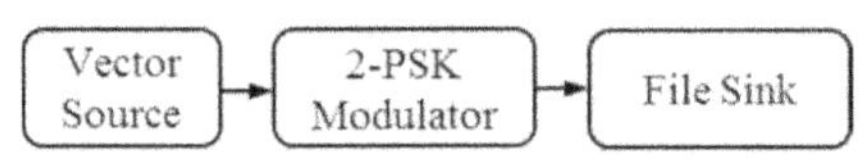

Fig. 3. Template I/Q stream generation.

During static generation as well as during the experiments, the I/Q data is generated by a PSK-Mod block in GNURadio with following parameters. The number of constellation points is two. Bits are gray coded but not encoded differentially. Two samples per symbol are used. Therefore, a frame with a length of 64 bits generates 128 I/Q samples.

There are two ways to calculate the cross-correlation results required for analysis. The received signal can be demodulated and cross-correlated with the originally sent Gold code. Another option is to use the I/Q template (see Fig. 3) for correlation with the reflected I/Q signal directly.

Fig. 4 and Fig. 5 show the cross-correlation results for bit and I/Q streams of a defined segment. Looking at the bitstream, correlation peaks occur in a periodic pattern of frames with 64 bit. Correlation peaks of the I/Q data occur in a periodic pattern of 128 samples per frame. In summary, correlation peaks of bit and I/Q data are equal, because the bitstream size is only half of the I/Q data due to demodulation.

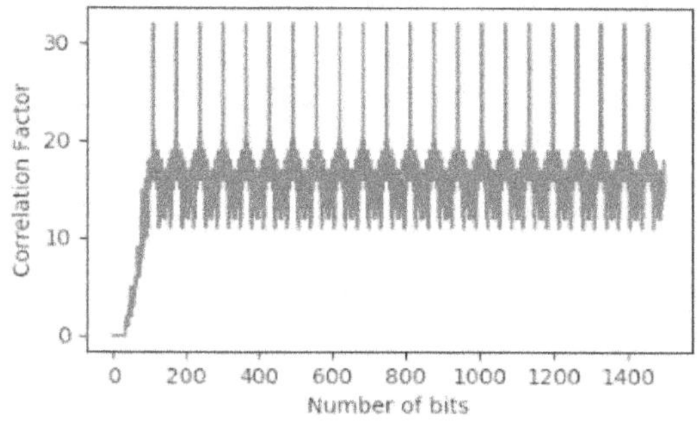

Fig. 4. Cross-correlation of bit streams.

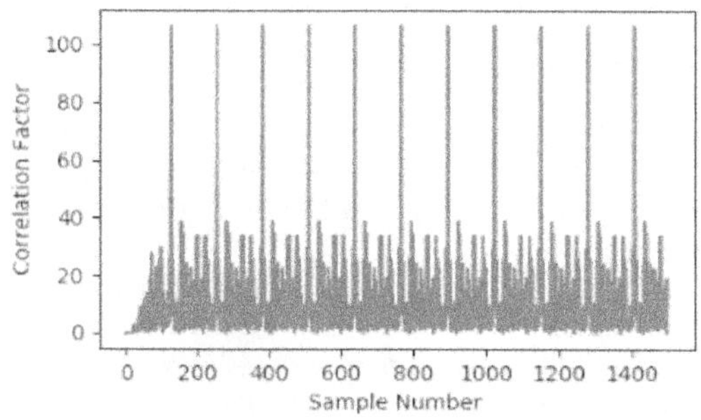

Fig. 5. Cross-correlation of I/Q streams.

Calculating the cross-correlation of bit streams is easier and faster compared to the calculation of the cross-correlation of I/Q data. However, during demodulation of the received 2-PSK signal some bit errors may occur. This makes it more complicated to find a certain frame in the I/Q data. Further, it would cause additional computational expense, too.

If the modulated I/Q representation of the Gold code sequence is used for cross-correlation instead of a pure bitstream, demodulation of the received 2-PSK signal is obsolete and detection robustness increases. Further, the signal information required for analysis is no further included in the bitstream after demodulation. Thus, cross-correlation is applied to the I/Q data stream and the results are used for further analysis.

Fig. 4 and Fig. 5 show a test condition with an ideal reflected signal without disturbance to ensure the signal processing is working properly. Fig. 6 was captured in another test condition where the feedback of the reflected signal was disrupted several times. This results in areas with and without correlation.

A test of the overall setup shows, that a recovery of the 2-PSK modulated Gold code sequence is possible periodically (Fig. 7), which was assumed at the beginning of the test. The position of each cross-correlation spike marks the beginning of a frame directly. Futher signal analysis can be done much easier, when the exact starting position of a frame/sequence is known.

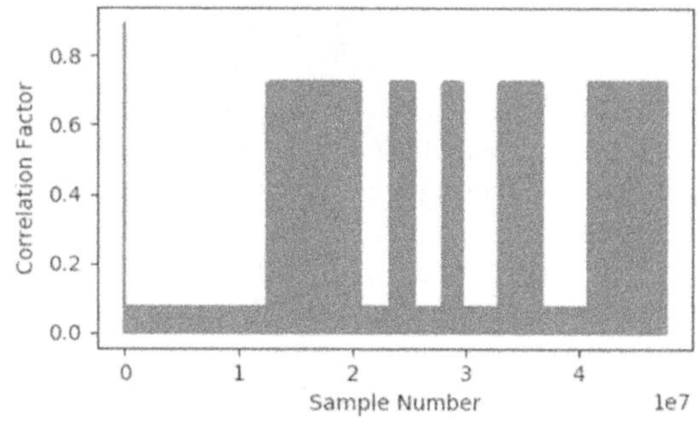

Fig. 6. Cross-correlation of I/Q streams with and without reflection.

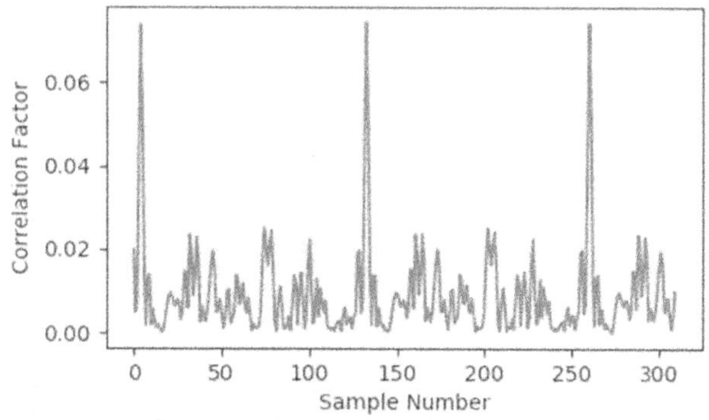

Fig. 7. Cross-correlation of template I/Q and received I/Q data.

4.3. Feature Extraction and Data Analysis

Due to the active signal injection, several features can be extracted from the reflected signal. Using SSTDR, the propagation time of a frame can be measured from its injection until it is received back. Therefore, a very high sample rate is required. In this approach, the sample rate is currently very low (2 MS/s) to get only a small amount of data to facilitate postprocessing. During the experiments it is planned to increase the sample rate.

Also, the FFT of the I/Q data can be calculated on both sides, the transmitting path as well as the receiving path. A comparison of both may be another feature which can be used for analysis. The received I/Q data in combination with its computed cross-correlation peaks can be considered, too. In this case, no further feature extraction is required. Moreover, the data could be analysed using machine learning methods.

4.4. Expected Outcome and Future Work

The first proposed show case consists of only one data acquisition point as depicted in Fig. 1. Further experiments may consist of two or three data acquisition points to get even more information about the grid and attached devices. This show case is illustrated in Fig. 8.

In all show cases, each DUT is switched on and off one by one. Afterwards, combinations of DUTs are switched with overlapping patterns. In the next analysis step, some selected DUTs are artificially aged to manually cause defects. During all test scenarios all inputs, outputs, switch transitions and fault conditions are logged, together with a timestamp. The data is then processed using different machine learning techniques. Long short term memory (LSTM), genetic algorithms and deep neural networks are the first approaches which will be evaluated. It is expected, that the approach gives answers about the type of fault (wire and/or device), as well as the corresponding location of the fault in the micro grid. The determination of the health status of a device, a wire or an electrical connection is another expectation.

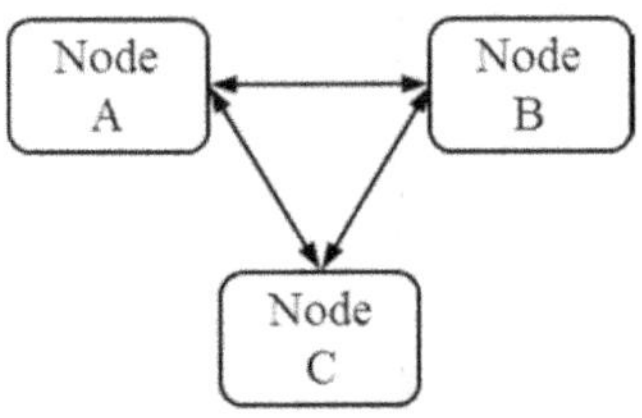

Fig. 8. Satellite architecture.

Using the satellite architecture from Fig. 8 may allow to investigate the grid topology (e.g. wire length, junctions). This can be done by using multipath propagation in combination with a superposition principle. It may also result in a more precise localization of a fault.

Acknowledgement

The research concept presented in this paper is part of the project MobiCM which is funded by the German Federal Ministry for Economic Affairs and Energy under the Central Innovation Programme for SMEs (ZIM).

References

[1]. J. G. Roos, I. E. Lane, E. C. Botha, G. P. Hancke, Using neural networks for non-intrusive monitoring of industrial electrical loads, in *Proceedings of the IEEE Instrumentation and Measurement Technology Conference (IMTC'94)*, Vol. 3, 1994, pp. 1115-1118.

[2]. P. Smith, C. Furse, J. Gunther, Analysis of spread spectrum time domain reflectometry for wire fault location, *IEEE Sensors Journal*, Vol. 5, Issue 6, 2005, pp. 1469-1478.

[3]. S. Roy, M. K. Alam, F. Khan, J. Johnson, J. Flicker, An Irradiance-independent, robust ground-fault detection scheme for PV arrays based on Spread Spectrum Time-Domain Reflectometry (SSTDR), *IEEE Transactions on Power Electronics*, Vol. 33, Issue 8, 2018, pp. 7046-7057.

[4]. S. Roy, F. Khan, Live condition monitoring of switching devices using SSTDR embedded PWM sequence: A platform for intelligent gate-driver architecture, in *Proceedings of the IEEE Energy Conversion Congress and Exposition (ECCE'17)*, 2017, pp. 3502-3507.

[5]. C. Furse, Y. Chung, C. Lo, P. Pendayala, A critical comparison of reflectometry methods for location of wiring faults, *Smart Structures and Systems*, Vol. 2, Jan. 2006, pp. 25-46.

[6]. C. Furse, P. Smith, C. Lo, Y. C. Chung, P. Pendayala, K. Nagoti, Spread spectrum sensors for critical fault location on live wire networks, *Structural Control and Health Monitoring*, Vol. 12, Issue 3-4, 2005, pp. 257-267.

[7]. A. Zoha, A. Gluhak, M. Imran, S. Rajasegarar, Non-intrusive load monitoring approaches for disaggregated energy sensing: A survey, *Sensors (Basel, Switzerland)*, Vol. 12, 2012, pp. 1683-1686.

(033)

SVM Classification of Data Obtained from a Health Condition Monitoring System Using Flexible Force Sensing Resistors

Yasutaka Uchida [1], **Tomoko Funayama** [2], **and Yoshiaki Kogure** [1]
[1] Teikyo Univ. of Science, 2-2-1 Senjyusakuragi, Adachi-ku., Tokyo, Japan
[2] Teikyo Univ. of Science, 2525 Yatsusawa, Uenohara-shi, Yamanashi, Japan
Tel.: + 81369101010, fax: + 81369103800
E-mail: uchida@ntu.ac.jp

Summary: Support vector machine (SVM) learning is applied to the proposed monitoring system which captures changes in health conditions using flexible force sensing resistors. The system consists of eight flexible force sensing resistors, a data acquisition device and a personal computer (PC). Feature quantities are defined by the time difference between the output signal from sensor 1, which specifies the start of measurement, and that from the other sensors. The measurement conditions are normal range of motion, simulated limited shoulder and knee joint. The measurement data are then divided into 30 for learning and 15 test data. The SVM module is used for analysis. Comparing the difference between the linear function kernel and the radial basis function kernel, there is no big difference based on the learning data. However, we observe an 83 % accuracy rate using radial basis function kernel. In the case of test data, the highest accuracy was when t2 and t7 used as the feature quantities.

Keywords: SVM, Machine learning, Classification, Health condition monitoring system, LabVIEW, Force sensing resistors.

1. Introduction

Japan is purported to have the highest proportion of elderly citizens in the world. Its society is experiencing super-aging; according to 2014 estimates, 25.9 % of the Japanese population is aged 65 or above. This proportion is estimated to reach 40 % by 2060 [1]. There are also many elderly people live alone. The self-management of health is necessary to aid those living alone, not only for elderly but also other frail people such as patients [2, 3]. The important thing is to notice signs of poor health as early as possible and to be able to connect to the medical treatment. Therefore, it is necessary to introduce a system that monitors their health conditions daily and issues a warning when it deviates from normal range.

Various kinds of health and welfare equipment, which rely on Internet of Things technology, have been developed and distributed to the market. These includes watching sensors, leaving-the-bed sensors, simple motion analyzing devices, and wearable devices that measure biological information [4-6]. However, they are not being used effectively in the health and welfare field. It is necessary to coordinate the performance of the equipment and on-site needs and privacy issues. Although there is an increasing number of publications on the application of deep learning for health care monitoring, the area is still developing [7, 8]. In this paper, we propose a physical condition monitoring system that uses force sensing resistors [9-11]. We present the preliminary results of 3 class classification using support vector machine (SVM).

This research was conducted with the approval of the Teikyo University of Science's Ethics Committee.

2. Experiment

2.1. Survey Method

The measurement was performed for man in his 60 s. There were three measurement conditions: normal Range of Motion, Simulated limited right knee, and right shoulder. In the shown figures of the classification result, the normal Range of Motion is represented by a black circle, the simulated limited right knee is represented by the green circle, while limited right shoulder is represented by the red circles, respectively.

2.2. Health Condition Monitoring System and Data

The developed system consists of eight flexible force sensing resistors, FSR®408, connected to a LabVIEW Data Acquisition (DAQ) NI-6210 device, which is also connected to a personal computer (PC) through a USB cable, as shown in Fig. 1. The voltage data are saved on the PC as a file in the CSV format. The first sensor is on the pillow and its output signal is the movement detection point as a reference. Three of sensors are spaced 0.1 m apart at the edge of bed, and the remaining four sensors which are spaced about 0.2 m apart and are placed on the floor besides the bed [9]. A folding type bed was used in this experiment. The bed height of 0.4 m is lower than that of a typical home bed. The second and subsequent sensors are located below the mat for protection. Thus, very small output voltage is observed even when the weight is not added.

Fig. 2 shows details of the detecting circuit used in this experiment. The resistance value of the flexible force sensor changes in the range of 100 to 10 MΩ or more. Each flexible force sensor and a 1 kΩ-resistor are connected in series. The voltage drop across the 1 kΩ-resistor is used as the input voltage of the DAQ device. The sampling time is 10 ms, i.e. the output from each of the eight sensors is taken every 10 ms.

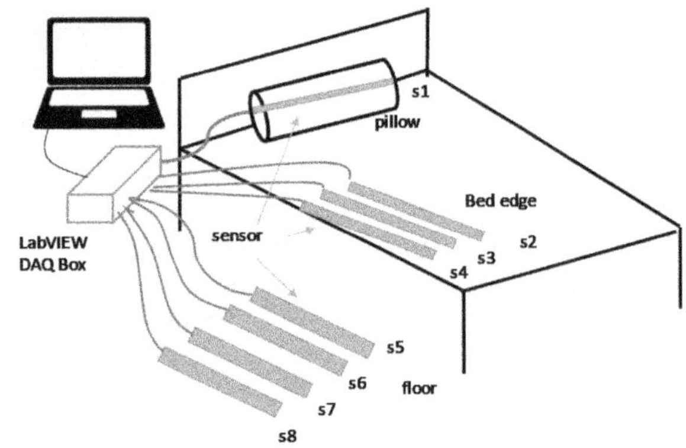

Fig. 1. Health Condition Monitoring System.

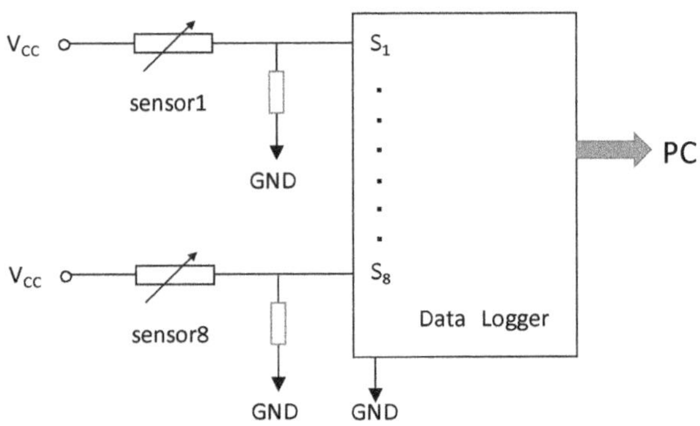

Fig. 2. Details of detecting circuit.

Fig. 3 shows the voltage output obtained from the DAQ. Each of the signals from the flexible force sensors is smoothed by averaging adjacent data.

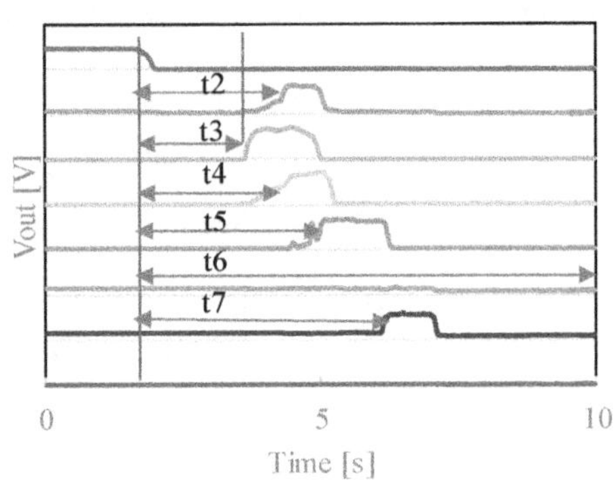

Fig. 3. The voltage output obtained by DAQ.

The quantification of feature quantities for SVM analysis is defined by the time difference as follows: The reference time is the fall time of sensor s1 on the pillow. The difference between the reference time and the rise time of other sensors (s2-s7) signals is defined as the response time, such as t2, t3,..,t7. For waveforms in which the rise time of the signal could not be detected, the final 10 seconds is used as the rise time of the sensor.

2.3. Analysis Method

The library for machine learning uses scikit-learn, and the classification module uses the SVM module. The measurement data were divided into two: 3×10 data for learning and 3×5 data for testing.

First, we consider that the rise time of each sensor reflects the difference between the measurement condition. Then a 2-D map of the rise time is used for comparison. Among the 2-D combinations, three combinations which are considered to be possible to classify are selected as feature quantities. Classification data used for the analysis is a combination of 2 kinds of 3 feature quantities with high possibility of high accuracy rate, as a result of comparing the selection of feature quantities of machine learning.

3. Results and Discussion

Fig. 4(a), (b) and (c) show the classification results using the pair t2 and t7 in the learning data, and the kernels used in classifications are the linear, the radial basis function (C = 1.0, γ = 0.00001) and the radial basis function (C = 1.0, γ = 0.000001), respectively. The value of t2 and t7 is 10 ms. In the case of radial basis function, as the γ value increases, the boundary line changes for a straight line to a curve and more accurate classification is achieved. The accuracy rate of classification obtained by the linear and radial basis function (C = 1.0, γ = 0.00001) using the data for learning are 80 %. In the case of the radial basis function (C = 1.0, γ = 0.000001) is 70 %. After performing SVM learning using data for learning, the 15 test data samples are applied to the learning result to obtain an accuracy rate. The accuracy rates in the linear and radial basis function (C = 1.0, γ = 0.00001) are 93.0 % and 86.7 %, respectively. The accuracy rate of radial function kernel was lower than that of linear function kernel. This result will be improved by optimizing parameters of C and γ.

Fig. 5(a), (b) and (c) show classification results using a pair of t3 and t7 in the learning data, and the kernel used in classifications were linear and radial basis function (C = 1.0, γ = 0.000001), and (C = 1.0, γ = 0.5) respectively. The value of t3 and t7 are also 10 ms. In the case of radial basis function, as the γ value increased, the boundary line changed from straight line to curve and more accurate classification is achieved.

The accuracy rates of classification obtained by linear and radial basis function (C = 1.0, γ = 0.000001) kernel obtained using date for learning are 76.7 % and 56.6 %, respectively. After performing SVM learning using data for learning, 15 data for test were applied to the learning result to obtain an accuracy rate. The accuracy rates in the linear and radial basis function (C = 1.0, γ = 0.000001) kernel are 80.0 % and 86.7 %, respectively. The accuracy rate of radial function kernel is lower than that of linear function kernel. This result will be improved by optimizing parameters of C and γ.

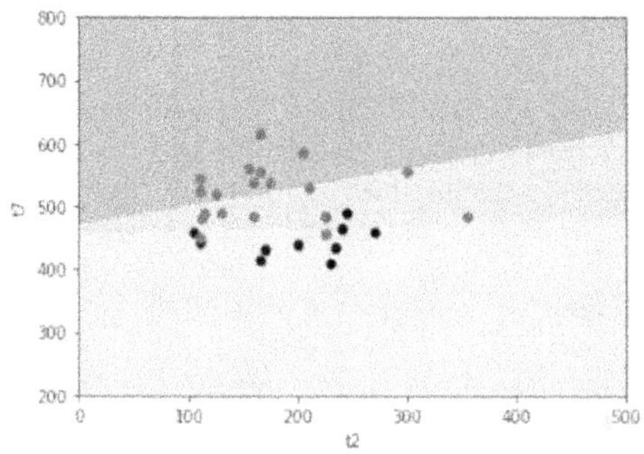

(a) Linear kernel.

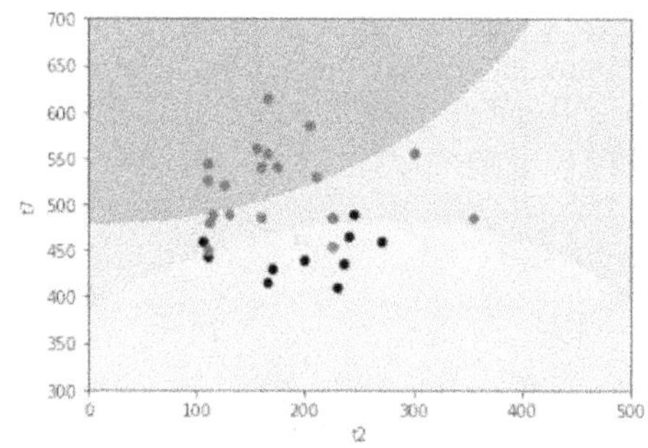

(b) Radial basis function kernel (C = 1.0, γ = 0.00001).

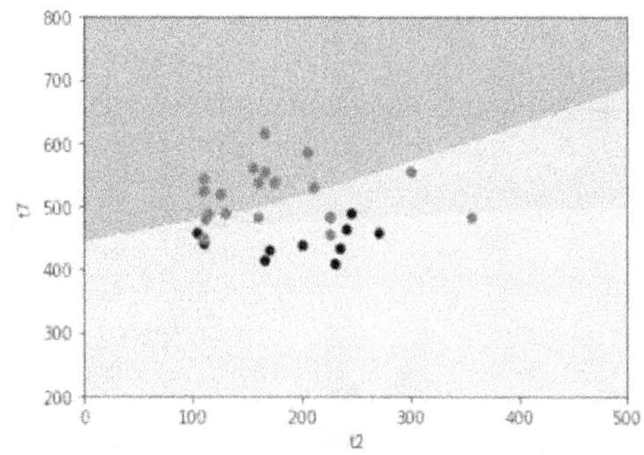

(c) Radial basis function kernel (C = 1.0, γ = 0.000001).

Fig. 4. Classification results for t2-t7.

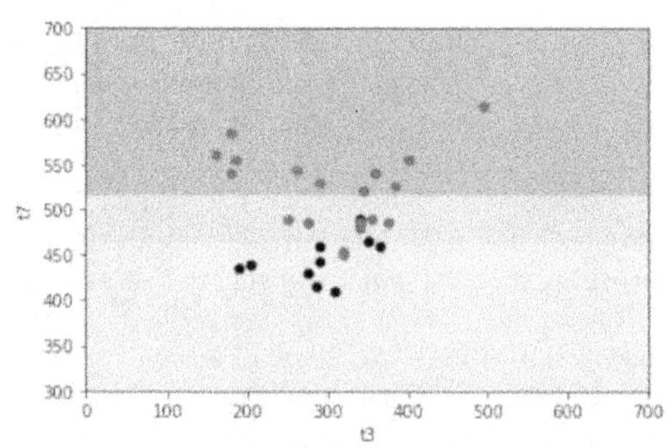

(a) Linear kernel.

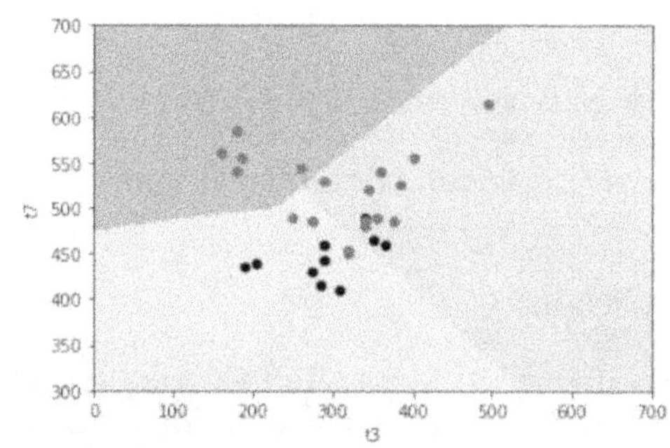

(b) Radial basis function kernel (C = 1.0, γ = 0.000001).

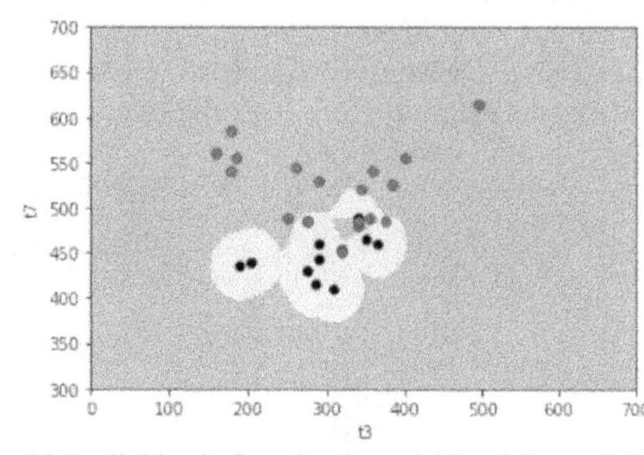

(c) Radial basis function kernel (C = 1.0, γ = 0.5).

Fig. 5. Classification results for t3-t7.

Fig. 6(a) and (b) show classification results using a pair of t4 and t7 in the learning data, and the kernel used in classifications are linear and radial basis function (C = 50.0, γ = 0.000001), respectively. The value of t4 and t7 were also 10 ms. The accuracy rates are 83.3 %. After performing SVM learning using data for learning, 15 test data are applied to the learning result to obtain an accuracy rate. The accuracy rates in the linear and radial basis function kernels are 73 % and 86.7 %, respectively.

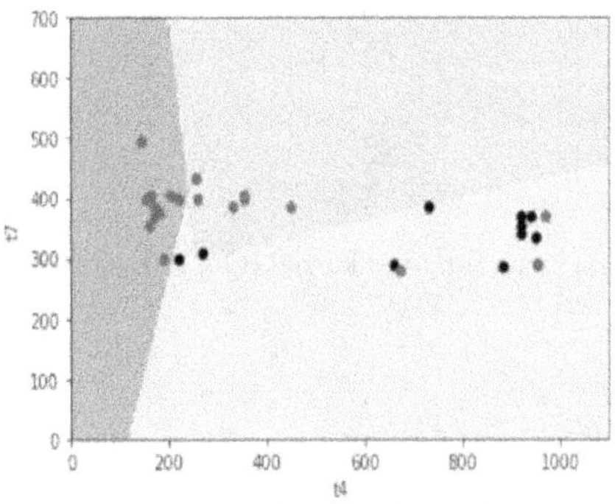

(a) Linear function kernel.

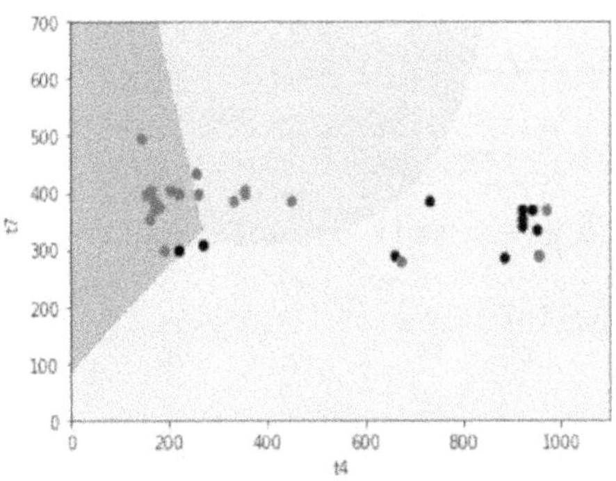

(b) Radial basis function kernel (C = 50.0, γ = 0.000001).

Fig. 6. Classification results for t4-t7.

4. Conclusions

We performed SVM classification using linear and radial basis functions on the data from the health condition monitoring system. As a result of using learning data, the accuracy rate from the classification system using a pair of t4 and t7 as feature quantities was highest at 83.3 %. In classification using test data, only one set of data was used, but the highest accuracy rate of 93.0 % was obtained when the pair of t2 and t7 was used.

Acknowledgements

This work was supported by JAPS KAKENHI Grant Number JP17K01590.

References

[1]. White Paper of Cabinet office, Government of Japan, http://www8.cao.go.jp/kourei/english/annualreport/2017/2017pdf_e.html.

[2]. K. Sudo, J. Kobayashi, S. Noda, Y. Fukuda, K. Takahashi, Japan's health care policy for the elderly through the concepts of self-help(Ji-jo), mutual aid (Go-jo), social solidarity care (Kyo-jo), and governmental care (Ko-jo), *BioScience Trends*, Vol. 12, 2018, pp. 7-11.

[3]. N. Akiyama, T. Shiroiwa, T. Fukuda, S. Murashima, K. Hayashida, Health care cost for the elderly in Japan: Analysis of medical care and long-term care claim records, *PLoS One*, Vol. 13, 2018, e0190392.

[4]. M. M. Alam, E. B. Hamida, Surveying wearable human assistive technology for life and safety critical applications: standards, challenges and opportunities, *Sensors*, Vol. 14, 2017, pp. 9153-9209.

[5]. M. Frydrysiak, L. Tesioroski, Wearable care system for elderly people, *International J. Pharma Medicine and Biological Science*, Vol. 5, 2016, pp. 171-177.

[6]. Y. L. Zheng, X. R. Ding, C. C. Y. Poon, B. P. L. Lo, H. Zhang, X. L. Zhou, G. Z. Yang, Unobtrusive sensing and wearable devices for health informatics, *IEEE Trans. on Biomedical Engineering*, Vol. 61, 2014, pp. 1538-1554.

[7]. M. H. Alvaro, G. Z. Begonya, M. Z. Amaia, Gait analysis methods: an overview of wearable and non-wearable systems, highlighting clinical applications, *Sensors*, Vol. 14, 2014, pp. 3362-3394.

[8]. S. Lee, J. Chang, Oscillometric blood ressure estimation based on deep learning, *IEEE Trans. Industrial Informatics*, Vol. 13, 2017, pp. 461-472.

[9]. Y. Uchida, T. Funayama, Y. Asano, Y. Kogure, Proposal for a simple care system using pressure-sensitive resistors at nursing site, in *Proceedings of the International Conference on Service Science and Innovation (ICSSI'18)*, 2018, p. 41.

[10]. Y. Uchida, T. Funayama, Y. Asano, Y. Kogure, Possibility for application of the machine learning to simple care system, *Human Interface Society SIG-ACI-22*, Vol. 20, Issue 11, 2018, pp. 59-62 (in Japanese).

[11]. Y. Uchida, Y. Kogure, T. Funayama, Y. Asano, A study on data classification method from healthcare monitoring system using pressure sensor, *Human Interface Society SIG-ACI-23*, Vol. 21, Issue 2, 2019, pp. 15-18 (in Japanese).

(034)

Silver Nanoparticle-modified Polypyrrole Love Wave Sensors Sensitive to Acetone

M. Setka [1], **F. A. Bahos** [2], **D. Matatagui** [2], **I. Gràcia** [3], **E. Figueras** [3], **J. Drbohlavová** [1], **and S. Vallejos** [1,3]

[1] CEITEC - Central European Institute of Technology, Brno University of Technology, 61200 Brno, Czech Republic

[2] ICAT, UNAM, 04510 CD MX, Mexico

[3] Instituto de Microelectronica de Barcelona (IMB-CNM, CSIC), Campus UAB, 08193 Bellaterra, Spain

Tel.: + 420-5-4114-6153

E-mail: vargas@feec.vutbr.cz, stella.vallejos@imb-cnm.csic.es

---

Summary: Love-wave gas sensing devices based on polypyrrole (PPy) modified with silver (Ag) nanoparticles (NPs) are developed and validated towards relevant volatile organic biomarkers (e.g., ethanol, acetone and toluene) involved in early disease diagnosis. Results demonstrate the functionality of the sensors with improved limits of detection, for instance 40 ppb for ethanol, 45 ppb for acetone and 89 ppb for toluene, compared to other gravimetric sensors.

Keywords: Gas sensors, Love-wave, Polypyrrole, Silver, Acetone.

---

1. Introduction

Recently, our studies on Love-wave sensors (a type of surface acoustic wave (SAW) sensors) based on a multi-guiding layer consisting of SiO_2 and modified PPy showed suitable towards the detection of low ammonia concentrations at room temperature, due to not only high sensitivity of sensors caused by multi-guiding structure (SiO_2/PPy), but also the modification of PPy with gold catalytic NPs [1]. Further recent studies also indicate that the use of other chemically active particles such as Ag NPs are prone to improve the gas sensing properties of PPy towards vapour organic compounds (VOCs) and in turn the properties (response, sensitivity, selectivity) of Love-wave sensors as compared to other SAW systems in the literature. In this context, this work focusses in the fabrication of multi-guiding Love-wave sensors based on PPy modified with Ag NPs and their validation towards various relevant VOCs (including ethanol, acetone and toluene) found to be involved in the exhaled breath of patients with diseases such as diabetes and cancer.

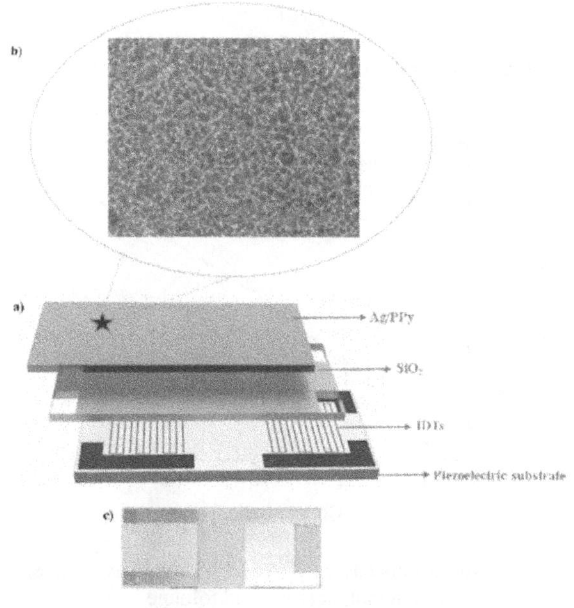

Fig. 1. a) schematic view of the multi-guiding Love-wave device; b) optical image of Ag/PPy layer deposited by spin coating; c) photograph of Love-wave device before Ag/PPy deposition.

2. Experimental

Love-wave based sensors consisting of a piezoelectric substrate (ST-Quartz) with two ports of aluminium interdigital transducers (IDTs) and two guiding layers were used in this work. The first guiding layer is based on SiO_2, while the second guiding/sensitive layer is based on Ag NPs (10-15 nm) modified PPy NPs (35-55 nm) [1]. The fabrication consisted of various microfabrication steps, including metallization, oxide deposition, lithography, and spin-coating of the Ag/PPy layer on the top (Fig. 1).

3. Results

The gas sensing tests of Ag/PPy Love-wave sensors were performed at room temperature using different concentrations (from 0.5 to 5 ppm) of ethanol, acetone or toluene. Fig. 2 shows the response (frequency shift) of the sensors to various concentrations of acetone. Generally, the sensors showed similar reproducible and stable responses for all analytes.

The calibration curves for each analyte are showed in Fig. 3 and the box plots of the sensors sensitivity is displayed in Fig. 4. These results indicate a low-cross sensitivity for acetone, ethanol and toluene, with higher responses for the polar VOCs (i.e., ethanol and acetone) in comparison to the non-polar molecule (i.e., toluene), due to possible formation of hydrogen bond [2]. Moreover, the higher sensitivity registered to acetone compared to ethanol could be related to the higher molecular weight of the first, as the frequency change of the sensor is significantly affected by the mass change of the sorbent (i.e. Ag/PPy layer) upon gas adsorption [3].

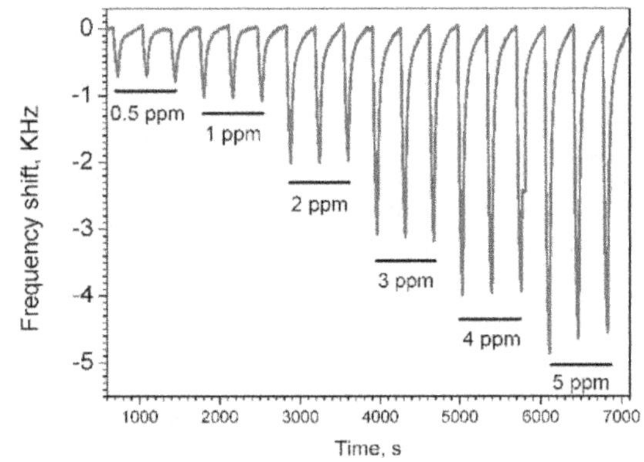

Fig. 2. Response of the Ag/PPy Love-wave sensor to 0.5, 1, 2, 3, 4 and 5 ppm of acetone.

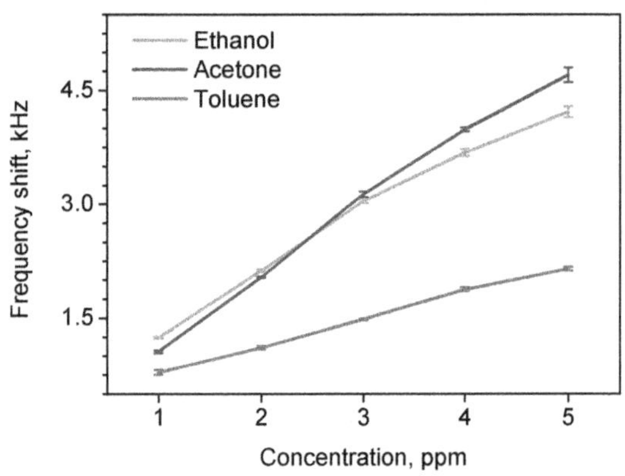

Fig. 3. Calibrations curves of Ag/PPy Love-wave sensors to ethanol, acetone and toluene.

The limit of detection (LOD) of the modified sensors to acetone was estimated as 0.045 ppm, consistent with the range of acetone concentrations determined by spectroscopy techniques in the exhaled breath of healthy (0.04-2.7 ppm) and diabetic patients (1.7-21 ppm). These sensors show remarkable lower LOD towards ethanol, acetone and toluene compared to other gravimetric sensors (Table 1).

4. Conclusions

Love-wave sensors based on silver modified polyprrole were developed in this work. Test of the sensors to acetone, ethanol and toluene demonstrated improved performances with lower LODs towards polar VOCs (acetone, ethanol) as compared to other gravimetric sensors in the literature.

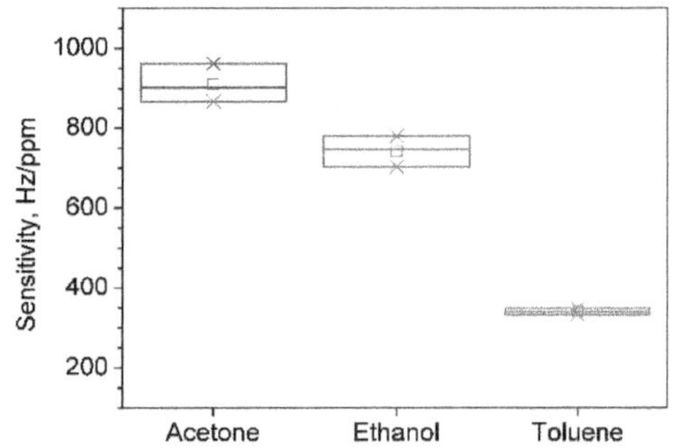

Fig. 4. Box plots of the sensors sensitivity to ethanol, acetone and toluene. Each box displays the median, mean, and upper and lower quartiles (first and third) of the respective distribution.

Table 1. Comparative table showing the limits of detection reported in the literature for other SAW sensors.

Sensor	Gas	LOD (ppm)	Ref.
Ag/PPy	Ethanol Acetone Toluene	0.040 0.045 0.089	This work
PEI/Fe_3O_4	Ethanol Toluene	65 54	[4]
ZIF-8 + AuNPs	Acetone Ethanol	1.1 0.5	[5]

Acknowledgements

This work has been supported by the Czech Science Foundation (GAČR) via Grant no. 17-16531S.

References

[1]. M. Šetka, F. A. Bahos, Polypyrrole based love-wave gas sensor devices with enhanced properties to ammonia, *MDPI Proceedings*, Vol 2, 2018, 786.

[2]. S. Hamilton, M. J. Hepher, Polypyrrole materials for detection and discrimination of volatile organic compounds, *Sensors and Actuators B: Chemical*, Vol. 107, 2005, pp. 424-432.

[3]. J. Devkota, K. J. Kim, Zeolitic imidazolate framework-coated acoustic sensors for room temperature detection of carbon dioxide and methane, *Nanoscale*, Vol. 10, 2018, pp. 8075-8087.

[4]. C. Viespe, D. Miu, Characteristics of surface acoustic wave sensors with nanoparticles embedded in polymer sensitive layers for VOC detection, *Sensors*, Vol. 18, 2018, 9.

[5]. F. A. Bahos, A. Sainz-Vidal, ZIF nanocrystal-based Surface Acoustic Wave (SAW) electronic nose to detect diabetes in human breath, *MDPI Biosensors*, Vol. 9, 2018, 13.

(035)

A Multivariate Emissivity Database for Industrial Infrared Radiation Thermometry

E. Chalkley
Advanced Forming Research Centre, University of Strathclyde, UK
E-mail: eleanor.chalkley@strath.ac.uk

Summary: The Advanced Forming Research Centre's Emissivity Calibration Furnace is an instrument designed to provide high volumes of multivariate emissivity data to improve accuracy in infrared radiation thermometry used in metal forming processes. The initial instrument was used to create a database of emissivity measurements for commonly used materials.

Keywords: Pyrometry, Metals processing, Infrared thermometry, Emissivity.

1. Introduction

In metal forming industries, the key process variable is temperature. This controls the outcome and needs to be measured and controlled throughout. Advances in infrared radiation thermometry mean that it is possible to track the surface temperature of products in metal forming processes from start to finish. However, take up of these technologies is not universal and implementation of infrared temperature measurements can be imperfect.

A leading cause of poor take up and imperfect implementation is the complexity of accounting for the finite emissivity of metal surfaces. While ratio pyrometer [1] is a mature technology, many sites still use single wavelength or wideband pyrometers, sometimes without taking emissivity into account, even though specific technologies exist [2]. This leads to internally consistent temperature readings which can be used to repeat a working process, but not to a traceable measurement of temperature which can be used to draw metallurgical conclusions. An emissivity database containing information specific to high temperature metal forming can be used to correct.

1.1. A Multivariate Emissivity Database

As a centre concentrating on improving the supply chain for technical metal products, the AFRC seeks to find ways to bring improved thermal metrology into the forge. One approach is to construct a large database of emissivity measurements, varying in wavelength and temperature.

The equation which forms the basis for ratio pyrometry gives luminance of a body as a function of wavelength and temperature

$$L(\lambda, T) = \frac{c_1 \varepsilon(\lambda, T)}{\lambda^5 \left(e^{\frac{c_2}{\lambda T}} - 1\right)} \tag{1}$$

Equation (1) informs the design of an instrument for measuring the emissivity of samples of as-processed metals and high temperature materials. The instrument consists of, a regulated tube furnace, a water-cooled freely radiating zone and a freely sliding silicon carbide platform that can accommodate samples of up to 35×35×10 mm.

The instrument was designed to use dual wavelength ratio pyrometers from the Land Instruments SPOT range to measure the radiance from the sample. The pyrometer remains at a fixed distance from the sample as the sample holder is withdrawn from the effective black body of the tube furnace to the freely radiating water-cooled end zone. The ratio between radiance measured in the hot zone and the water-cooled end zone corresponds to the emissivity of the sample.

Ratio pyrometers function by measuring $L(T)$ for two wavelengths. If it can be assumed that emissivity of the surface at two wavelengths is equal, then by dividing $L(T)_{\lambda_2}$ by $L(T)_{\lambda_1}$ gives the radiance temperature

$$T_R = \left(\frac{1}{T} + \frac{\ln(\varepsilon_1/\varepsilon_2)}{c_2(\lambda_2^{-1} - \lambda_1^{-1})}\right) \tag{2}$$

The assumption that $\varepsilon_1 = \varepsilon_2$ is not always appropriate, especially for surfaces which undergo chemical and physical change during the process. The emissivity database identifies these materials and provides the data required to produce high quality pyrometric temperature measurements.

1.2. Materials Surveyed

As well as surveying the range of materials that are processed within the AFRC, a range of refractory and transitional materials were surveyed, including the brick which forms the interior walls of the AFRC's major furnaces, alpha case grown on Ti during heating and silicon carbide.

This database aims to extend and improve upon the widely used reference data in Touloukian [3]. Fig. 1 shows the data collected for Inconel 718 samples during 2018.

2. Spectroscopic Emissivity Measurement

With only measurements at the wavelengths that correspond to the specific optics setup of the pyrometers, the emissivity dataset is limited in scope and also in speed of acquisition. Acquiring continuous spectra in addition to the pyrometry wavelengths allows greater data-gathering speed and examination of wavelength-dependent features.

Existing pyrometer optics were adapted to accommodate a ZrF optical fibre which feeds a spectrometer with measurement range from 900nm to 2500 nm. The short wavelength accuracy limit of the spectrometer measurement is related to the cut-on wavelength of the black body curve.

Table 1. Scope of initial pyrometer-based emissivity data collection.

Material	Wavelengths	Temperature Range
Inconel	1 µm, 1.5 µm, 2.2 µm, 2.4 µm	300 °C to 1100 °C
Titanium	1 µm, 1.5 µm, 2.2 µm, 2.4 µm, 8-14 µm	250 °C to 1100 °C
Aluminium	1 µm, 1.5 µm, 2.2 µm, 2.4 µm, 8-14 µm	250 °C to 550 °C
Waspalloy	1 µm, 1.5 µm, 2.2 µm, 2.4 µm, 8-14 µm	300 °C to 1100 °C
Silicon Carbide	1 µm, 1.5 µm	600 °C to 900 °C
Refractory brick	1 µm, 1.5 µm, 2.2 µm, 2.4 µm, 8-14 µm	200 °C to 1050 °C

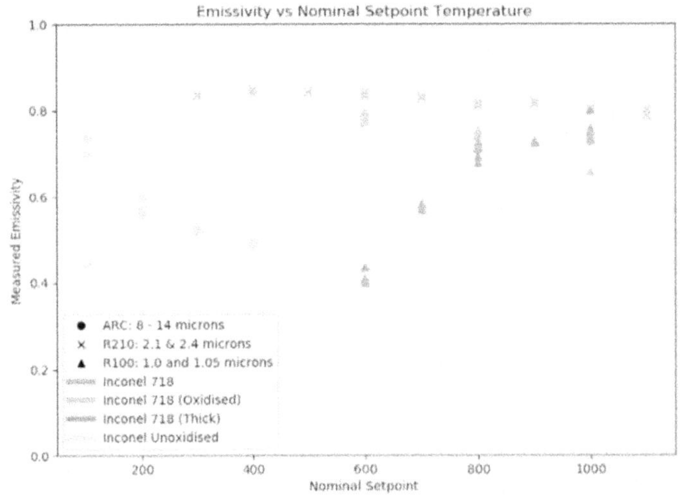

Fig. 1. Collected emissivity measurements for Inconel 718 as a function of furnace set point, wavelength and sample quality.

The comparison plot in Fig. 2 shows the emissivity measurements for the 4 pyrometer wavelengths plotted alongside the spectrometer data for the same material at the same temperature. The spectrometer method of emissivity measurement records consistently lower values than the band limited and historical data. Further analysis of systematic differences in radiance measurements between the pyrometer and spectrometer systems is planned.

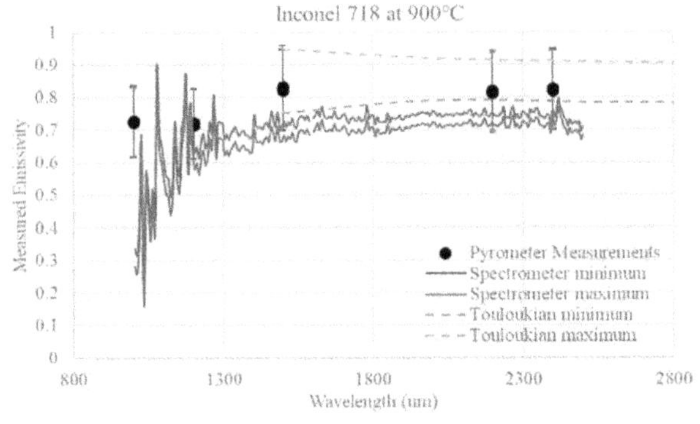

Fig. 2. Comparison of pyrometer and spectrometer measurements of the emissivity of heavily oxidised Inconel 718 at 900 °C with reference data from [3].

4. Conclusions

Measurements produced during the initial phase of narrowband measurement of emissivity using the Emissivity Calibration furnace are in line with historical measurements. This data is now being supplemented by spectroscopic multiwavelength data to support thermal measurements in heavy industry.

Acknowledgements

The author would like to acknowledge Ametek Land Instruments for the donation of instruments used in this work.

References

[1]. A. F. Gibson, A two-colour infra-red radiation pyrometer, *J. Sci. Instrum.*, Vol. 28, 1951, pp. 153-155.

[2]. F. Turner, Smart aluminium temperature sensors for Industry 4.0, *Aluminium International Today*, 2018, Vol. 31, Issue 3, pp. 25-27.

[3]. Y. S. Touloukian, D. P. DeWitt, Thermophysical Properties of Matter, Technical Report, Vol. 7, *Data Book, United States*, 1970.

(040)

Prediction of Temperature in WSN Using Artificial Intelligence

L. Formanek [1], M. Chochul [1] and O. Karpiš [1]
[1] University of Žilina, Faculty of Management Science and Informatics, Department of Technical Cybernetics, Univerzitná 8215/1, 01026 Žilina, Slovakia
Tel.: +421 41 513 4357
E-mail: lukas.formanek@fri.uniza.sk

Summary: This article describes the use of an artificial neural network (ANN) to predict the temperature at a certain point in wireless sensor network based on measurements of other nodes in WSN. The aim is to show that a neural network is capable to learn the properties of this nonlinear physical system with certain accuracy. The theoretical part describes the introduction to WSN and feedforward neural network. The experimental part consists of data collection, preprocessing and comparison of different simple neural network structures that have been applied to solve this problem.

Keywords: Wireless sensor network, Artificial intelligence, Neural network, Regression, Internet of Things.

1. Introduction

Nowadays, wireless technologies are part of our daily lives. They are used in various areas of industry, automation and environmental monitoring. These networks are a collection of different devices that use some kind of wireless technology to communicate with each other. For the transmission of information on the network is used for example radio, infrared or microwave radiation as a transmission medium. Most of the current technologies use a central node to collect sensor data, where the measured values are processed and further evaluated.

Typically, each node implements one or more sensors, a control unit, a communication module, a memory and a power source [1]. The block diagram of the wireless node can be seen in the Fig. 1.

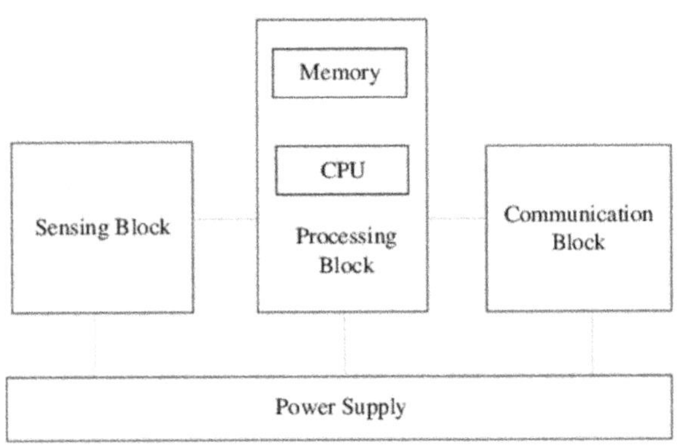

Fig. 1. Block diagram of wireless sensor node.

Increasingly, we come across the term smart sensor. This sensor not only senses a given quantity but also pre-processes the measured values and sends only the necessary data. This ensures saving of transmission channel capacity. The most important parameters in wireless sensor networks are a signal range, network throughput and power consumption. We include the issue of WSN networks in the ever-growing field of embedded systems.

Wireless sensor networks (WSN) are often used to measure different physical quantities, for example, temperature, humidity or gas concentrations over large geographic areas [2].

The problem is that sometimes it is not possible to place the sensor at any place in a given location, or long-term operation of this critical sensor is impossible in several aspects such as energy consumption or economic reasons. On the other hand in some cases, it is necessary to know if the sensor in a given location is measuring correctly or whether there was any crisis situation like a forest fire.

For all these cases artificial neural network (ANN) can be used to predict measured physical quantity at a specific location based on measurements in surrounding places.

This article describes data collection from 4 nodes that measure outside temperature and light intensity. These nodes were deployed in various peripheral parts of the Žilina city, several kilometers apart. The next chapters briefly describe the introduction to the wireless sensor network and feedforward neural network. The practical part is devoted to data collection, preprocessing and comparison of different neural network structures that have been used to predict the temperature in the location, where the fourth node is located. At the same time, the data from this sensor represents the teacher´s response during the learning of the neural network.

In this case, we will deal with regression. In a regression problem, the goal is to predict the output of a continuous value, such as temperature or light intensity. In contrast, there is a classification problem, where the aim is to predict a discrete label, for example, if the picture contains a dog, or if the voice record contains the given word [3].

1.1. Hardware Platform

A wireless sensor network consists of spatially distributed sensor nodes. In WSN, each sensor is capable of independently performing data capture, preprocessing, and data processing tasks. Besides, nodes communicate with each other or with a central node depending on the network topology.

As shown in the Fig. 1, each node consists of a control unit, a sensing unit, a communication unit and a power supply. The main task of the control unit is to process the data and control the functions of other hardware components such as sensors or communication module. Most often, the control unit is a microcontroller, because it implements many different peripherals. They serve mainly for communication with various sensors and communication unit. Another advantage of using a microcontroller is its relatively low cost and low power consumption. This fact has a positive effect on keeping the node in operation for a longer time.

The largest amount of energy is used for communication within the WSN, especially for sending data. On the other hand, energy is also consumed when measuring and processing data from sensors. Therefore, it is important to design proper algorithms for data sensing and data processing. Power can be supplied from batteries or supercapacitors. Batteries are usually the main power source. Due to the limited power options, great emphasis is placed on minimizing node consumption. Energy harvesting can also be used to convert energy from the environment into electrical energy [4].

A sensor is a hardware device designed to measure ambient characteristics. Sensors allow the transformation of non-electrical quantities into electrical signals. As an example, we can mention temperature, CO2 emissions, humidity, speed, pressure or electromagnetic field. The basic parameters of the sensors are resolution, stability, sensitivity, inertia, non-linearity and hysteresis, permissible error, noise, response time and frequency response. All these features are important when selecting the appropriate sensor for our application. The continuous analogue signal measured by the sensor is digitized using an analogue-to-digital converter and further processed by the node control unit.

1.2. Feedforward Neural Network

Generally, the feedforward neural network consists of an input layer, may include one or more hidden layers and an output layer. Each node of the input layer represents an input variable. The nodes in this layer are associated with nodes from the hidden layer. The hidden layer nodes (neurons) may be connected to the nodes from the next hidden layer or may be connected to the output layer. The output layer may be represented by one or more variables. In addition to the nodes that are located in the input layer, each node has several inputs. The individual inputs are multiplied by the weights of the W_{xy} link and then added together. The nonlinear activation function is applied to the resulting value, which is being considered as the output value of the node. This value then represents the node entry of the next layer [5]. A simplified example of the feedforward neural network model is shown in the Fig. 2.

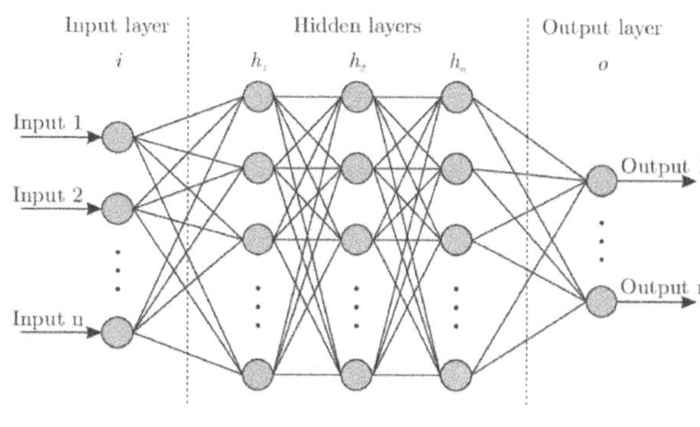

Fig. 2. Model of artificial neural network.

The setting of individual interconnection weights between nodes takes place in the neural network training process. There are many algorithms for training a neural network, however backpropagation algorithm is usually used. In the feedforward network, the information moves in only one direction -forward. From the input nodes, information moves through the hidden nodes (if any) to the output nodes. The commonly used neuron transfer function is: [6]

$$y(n) = f\left(\sum_{i=0}^{N-1} x_i(n)w_i(n)\right), \tag{1}$$

where x is the input vector, w is the weight vector and f is the activation function, usually used tanh, ReLU function, sigmoid or linear function.

2. Data Collection and Preprocessing

Our nodes measure outside temperature and light intensity in one-minute interval. Example of measured data is shown in the Fig. 3. Nowadays, there are many different temperature sensors. Some of them provide analogue output, others digital. We created a low-cost solution of the node, so we chose a digital sensor DS18B20 because of its good availability and price.

This sensor communicates with the control unit via one-wire interface. The temperature is measured in degrees Celsius. The photoresistor PGM5506 is used to measure the light intensity, which is characterized by a positive integer value in the range <0.65535>. The lower the number, the light intensity is higher.

The control unit consists of the STM32F103C8T6 microcontroller with cortex M3 core. Measured data is sent to the access point, which stores it on the SD card and forwards data to the server, where it is stored in the database.

For wireless communication, LoRa technology is used, which is one of the many technologies that are designed for the Internet of Things (IoT) and devices that communicate with each other over a long distance.

They are characterized by low data flow and low power consumption. This technology is based on spread spectrum modulation that has similar characteristics to FSK modulation but with much longer range (up to 15 km in open area) [7].

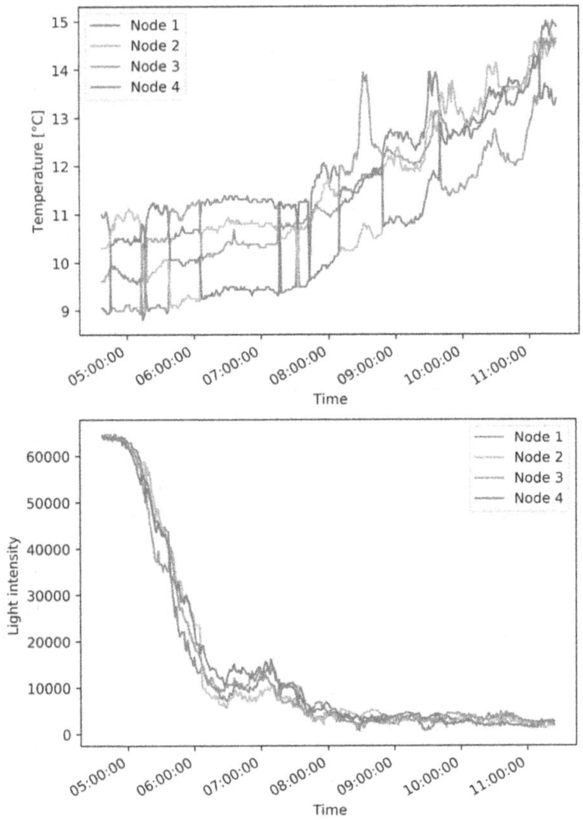

Fig. 3. Example of measured data.

In LoRa modulation the spreading of the spectrum is achieved by generating a chirp signal that continuously varies in frequency. An advantage of this method is that timing and frequency offsets between transmitter and receiver are equivalent, greatly reducing the complexity of the receiver design. The frequency bandwidth of this chirp is equivalent to the spectral bandwidth of the signal. The wanted data signal is chipped at a higher data rate and modulated onto the chirp signal [8].

We used data that was collected in 8 days. Data from the first 7 days were used to train neural networks, last day was used for verification.

3. Experimental Results

This part describes the usage of different neural networks structures that have been used to calculate the node 4 temperature based on the remaining three nodes measurements.

All in all, we created a dataset of approximately 11500 values. First 10000 values were for the training process and the rest for testing. One line in the dataset consists of the time the measurement was made and of the 8 values from the sensors (each node measured temperature and light intensity). We also normalized data into <0.1> interval and mixed the order for training data. The program was written in Python with TensorFlow framework usage.

Neural network input represents the currently measured data from node 1, node 2 and node 3 (six values). The output is the predicted temperature of node 4 at the current time.

Generally, we focused on fully connected feedforward neural networks with backpropagation training algorithm. Four different models of neural networks have been tested. The first is a linear model without a hidden layer. Other models contain one, two and three hidden layers, each containing 8 neurons.

In all cases, constant parameters and learning rate is used $\eta = 0.0002$. ReLU function was selected as the activation function and Xavier method was used for weights initialization. The same number of training epochs (50) was selected for all artificial neural network models. In the Fig. 4 we can see a comparison of target and predicted values for each structure of the neural network.

For each neural network model absolute error, standard deviation (Sigma) and root mean square error (RMSE) were calculated. All these errors can be seen in the following table.

Table 1. Calculated prediction errors.

Model	Absolute error, (°C)	Sigma, (°C)	RMSE, (°C)
1	1.451	1.945	1.984
2	0.838	1.023	1.030
3	0.718	0.844	0.844
4	0.633	0.755	0.757

4. Conclusions

In this paper, we described the possibility of using artificial feedforward neural network to learn the output of node measurement in WSN based on the remaining nodes measurements. We tested and compared four different simple neural networks, each with the different number of hidden layers. Adding additional input parameters could help achieve more accurate results.

For example, one of them is the time when the measurement was made. The next step is to test other types of neural networks like a recurrent neural network. The aim is to find the structure of the neural network as simple as possible to get the most accurate results. This neural network will be implemented in the nodes of the wireless sensor network.

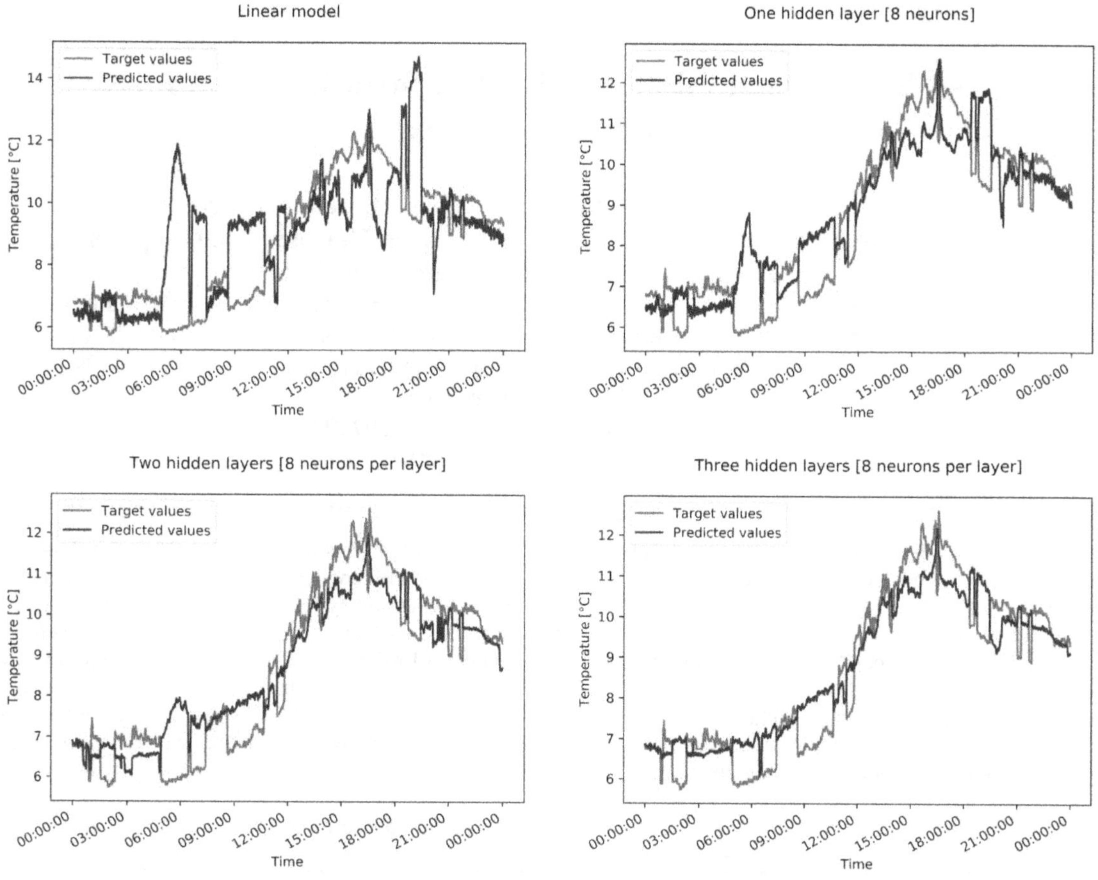

Fig. 4. Graphical results.

References

[1]. R. Berger, Introduction to wireless sensor networks, in *Proceedings of the NI Technical Symposium*, 2009.

[2]. D. Antolin, A. Bayo, N. Medrano, B. Calvo, S. Celma, WubiNet: A flexible WSN for applications in environmental monitoring, in *Proceedings of the IEEE International Instrumentation and Measurement Technology Conference (I2MTC'12),* 2012, pp. 2608-2611.

[3]. M. A. Nielse, Neural Networks and Deep Learning, *Determination Press*, 2015.

[4]. G. Serpen, J. Li, L. Liu, AI-WSN: Adaptative and intelligent wireless sensor network, *Procedia Computer Science*, Vol. 20, 2013, pp. 406-413.

[5]. V. Kvasnička, et al., Introduction to Neural Network Theory, *Bratislava*, 1997, pp. 70-77.

[6]. D. Marček, M. Marček. Neural Networks and Their Applications, *Žilina*, 2006, pp. 35-43.

[7]. K. Mikhaylov, J. Petaejaejaervi, T. Haenninen, Analysis of capacity and scalability of the LoRa low power wide area network technology, in *Proceedings of the European 22th European Wireless Conference*, 2016, pp. 1-6.

[8]. Semtech Technology, https://www.semtech.com

(041)

Label-free Optical Fiber Sensing Platform Based on Lossy Mode Resonances

C. R. Zamarreño [1,2], P. Zubiate [1], A. Ozcariz [1], C. Elosua [1], A.B. Socorro [1], A. Urrutia [1], D. Lopez [1], N. De Acha [1], J. Ascorbe [1], J. M. Corres, M. Hernaez [3], J. Goicoechea [1], F. J. Arregui [1,2], I. R. Matías [1,2] and I. del Villar [1]

[1] Institute of Smart Cities (ISC), Public University of Navarra, Pamplona, 31006, Spain

[2] Electrical and Electronic Engineering Department, Public University of Navarra, Pamplona, 31006, Spain

[3] School of Chemistry, Faculty of Science, University of East Anglia, Norwich Research Park, Norwich, NR4 7TJ, United Kingdom

Tel.: + 34948168445, fax: + 34948169720

E-mail: carlos.ruiz@unavarra.es

Summary: Lossy-mode resonance (LMR)–based optical fiber devices, with a sensitivity comparable to surface plasmon resonance (SPR)-base devices, have emerged as a versatile tool for many applications since the first experimental demonstration in 2009. In contrast to surface plasmon resonances (SPR), generated by transverse magnetic (TM) light, LMRs can be generated with transverse electric (TE) and TM light. Contrary to SPRs that are typically generated by metallic thin-films (gold and silver), semiconducting and dielectric materials can support LMRs, which broadens the range of applications and simplifies some chemistry processes. All in all, the study and optimization of LMR-based optical fiber devices is still at its infancy and presents a promising future for the development of versatile and highly sensitive tools as well as for the utilization of these devices as a label-free sensing platform suitable to be used in many different applications, such as industry, biology, chemical or environmental among others.

Keywords: Lossy mode resonances, Optical fiber sensor, Thin-films, Sensing platform, Refractive index.

1. Introduction

The utilization of resonance-based refractometers for sensing purposes has become a hot topic due to their large wavelength range, rapid response, simple instrumentation and, what is more important, a high sensitivity. A first approach, known as the Kretschmann configuration [1], was described in 1960 by means of surface plasmon resonances (SPR) supported by a gold thin-film fabricated onto a prism and with some technological improvements still remains as the gold standard sensing platform for commercial devices in biochemical applications.

On the other hand, Kretschmann configuration presents several disadvantages, such as bulky size or complexity compared with the proposal presented in 1993 by Jorgenson and Yee [2] with an optical fiber acting as the high-index prism that couples the incident light for the excitation of the SPR (see Fig. 1). Both, Kretschmann and optical fiber configurations, consist of label-free refractometric sensing platforms based on SPR and can be used for chemical and biomedical applications based on monitoring refractive index changes through resonant wavelength shifts, which has motivated the works of many researchers in that field [3].

SPR generation is mainly associated to thin-films of noble metals, such as gold and silver, and conductive materials with a negative value of the real part of the dielectric permittivity. However, a different type of resonance, known as lossy mode resonance (LMR) can be obtained with semiconductors or dielectric materials, such as metal oxides or polymers, with a positive real part of the dielectric permittivity, which extends the range of available materials for resonance generation, with comparable sensitivity and the possibility to be obtained with TE and TM polarized light at the same time [4].

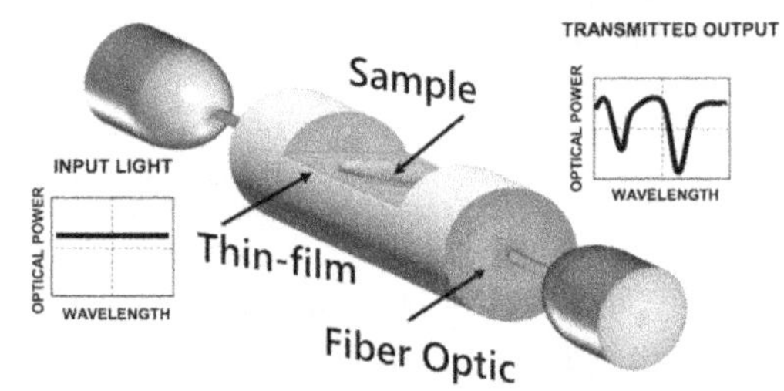

Fig. 1. Schematic representation of the resonance-based optical fiber sensing configuration.

Since the first experimental demonstration in 2009 by the group at Public University of Navarra [5], the utilization of LMR-based devices has emerged as a versatile tool that can compete with SPR and many other authors and groups worldwide have explored the utilization of different structures and materials for the fabrication of highly sensitive LMR-based refractometers as it is described later.

2. LMR-based Refractometers

Apart from the optical fiber geometry, the thin-film properties play an important role in the performance of

LMR-based optical fiber refractometers, such as sensitivity in nanometers (nm) per refractive index unit (RIU), wavelength working range and refractive index measuring range. In this sense, many different fabrication techniques can be explored for the fabrication of the thin-films [6].

Although polymeric thin-films can generate LMRs, the utilization of metallic oxides, such as titanium oxide (TiO_2), indium tin oxide (ITO), aluminum doped zinc oxide (AZO) or indium gallium zinc oxide (IGZO), has coped the fabrication of LMR-based optical fiber refractometers due to the high values of the real part of the refractive index that enhances the sensitivity of the device and low but not zero extinction coefficient, which reduces the losses. Table 1 shows a comparison of the sensitivities of these devices as a function of the resonance generating material, working wavelength and refractive index measuring range. Among them, the highest sensitivity corresponds to tin oxide. Consequently, this material has been selected for the development of many interesting applications, such as environmental monitoring, biological detection or medical diagnosis among others.

Table 1. Comparison of LMR sensitivity as a function of the LMR generating material (Mat).

Mat.	Sensitivity nm/RIU	Working Range (RIU)	λ (μm)	Ref.
ITO	8742	1.365-1.38	1.2-1.6	[7]
In_2O_3	4926	1.33-1.38	1.4-1.6	[8]
SnO_2	14501	1.33-1.334	1.4-1.5	[9]
IGZO	5111	1.33-1.37	1.3-1.6	[10]
TiO_2	2893	1.35-1.41	1.2-1.5	[11]
AZO	2280	1.33-1.45	1.2-1.4	[12]

3. Outlook

LMR-based optical fiber devices are still at an early stage for the development of a fully functional sensing platform and it is necessary to find the most suitable materials for LMR generation that can cover a broad interrogation range with the highest sensitivity. However, a tremendous work has been done in the development of cutting-edge applications and more and more groups have joined this work in the last years, which predicts a promising future for a label free LMR-based optical fiber sensing platform.

Acknowledgements

This work was supported in part by the Spanish Ministry of Education and Science-FEDER TEC2016-78047-R, Government of Navarra 0011-1365-2017-000117 and Public University of Navarra PJUPNA26 research grants.

References

[1]. E. Kretschmann, H. Raether, Radiative decay of non-radiative surface plasmons excited by light, *Zeitschrift fur Naturforsch. – Sect. A J. Phys. Sci.*, Vol. 23, 1968, pp. 2135-2136.

[2]. R. C. Jorgenson, S. S. Yee, A fiber-optic chemical sensor based on surface plasmon resonance, *Sensors Actuators B. Chemical*, Vol. 12, 1993, pp. 213-220.

[3]. C. Caucheteur, T. Guo, J. Albert, Review of plasmonic fiber optic biochemical sensors: improving the limit of detection, *Anal. Bioanal. Chem.*, Vol. 407, 2015, pp. 3883-3897.

[4]. I. D. Villar, F. J. Arregui, C. R. Zamarreño, et. al., Optical sensors based on lossy-mode resonances, *Sensors Actuators B,* Vol. 240, 2017, pp. 174-185.

[5]. I. del Villar, C. R. Zamarreño, M. Hernaez, F. J. Arregui, I. R. Matias, Lossy mode resonance generation with indium-tin-oxide-coated optical fibers for sensing applications, *Journal of Lightwave Technology*, Vol. 28, Issue 1, 2010, pp. 111-117.

[6]. C. R. Zamarreño, I. R. Matias, F. J. Arregui, Nanofabrication techniques applied to the development of novel optical fiber sensors based on nanostructured coatings, *IEEE Sensors Journal*, Vol. 12, Issue 8, 2012, pp. 2699-2710.

[7]. P. Zubiate, C. R. Zamarreño, I. del Villar, I. R. Matias, F. J. Arregui, High sensitive refractometers based on lossy mode resonances (LMRs) supported by ITO coated D-shaped optical fibers, *Optics Express*, Vol. 23, Issue 6, 2015, pp. 8045-8050.

[8]. C. R. Zamarreño, et al., Sensing properties of indium oxide coated optical fiber devices based on lossy mode resonances, *IEEE Sensors Journal*, Vol. 12, Issue 1, 2012, pp. 151-155.

[9]. F. J. Arregui, I. del Villar, C. R. Zamarreño, P. Zubiate, I. R. Matias, Giant sensitivity of optical fiber sensors by means of lossy mode resonance, *Sensors and Actuators B*, Vol. 232, 2016, pp. 660-665.

[10]. A. Ozcariz, M. Dominik, M. Smietana, I. Del Villar, C. R. Zamarreño, F. J. Arregui, I. R. Matias, Lossy mode resonance optical sensors based on indium-gallium-zinc oxide thin film, *Sensors and Actuators A: Physical*, Vol. 290, 2019, pp. 20-27.

[11]. P. Zubiate, C. R. Zamarreño, I. Del Villar, I. R. Matias, F. J. Arregui, Experimental study and sensing applications of polarization-dependent lossy mode resonances generated by D-shape coated optical fibers, *Journal of Lightwave Technology*, Vol. 33, Issue 12, 2012, pp. 2412-2418.

[12]. A. Ozcariz, D. Piña-Azamar, C. R. Zamarreño, F. J. Arregui, Aluminum doped zinc oxide (AZO) coated optical fiber LMR refractometers, *Sensors and Actuators B*, Vol. 181, 2019, pp. 698-704.

(044)

Message Compression Concept for Software Defined IoT Edge Devices

R. Poeschl [1] **and A. Faschingbauer** [1]
[1] Technische Hochschule Deggendorf, Institute for Applied Informatics, Grafenauer Str. 22, 94078 Freyung, Germany
Tel.: + 49 8551 9176421, fax: + 49 8551 9176469
E-mail: rainer.poeschl@th-deg.de

Summary: An efficient data transfer and flexible and reconfigurable edge devices are key elements of the IoT. In this paper a concept is proposed which uses a lightweight protocol and compressed XML messages using efficient XML interchange (EXI). The message system can combine several state of the art communication links. The concept also includes an approach for reconfigurable edge devices. Therefore, configuration data, stored in structured XML format, is transformed to edge specific parameter files or even application code using extensible stylesheet transformation. After transferring it to the edge node, the application code has to be executed by an interpreter. Using EXI shows high compression rates compared to standard XML.

Keywords: Internet of Things, Sensor system, Edge devices, EXI, Message compression, Reconfiguration.

1. Introduction

Breaking IoT down into its parts, communication is an essential aspect. Each device which is directly connected to the internet has enough bandwidth available. When it comes to small sensors, bandwidth is more valuable. Therefore, this paper presents a concept which uses a special type of data compression to enable efficient data transfer. The concept is explained with a communication system, consisting of at least two communication partners.

Frequently, larger IoT-systems, sensor networks or interconnections of smart sensors are based on a heterogeneous hardware interface and software structure. Therefore, a high degree of flexibility regarding the hardware interfaces and the application itself is required to get interoperable and adjustable e.g. in industrial environment [1].

Thus, edge devices often have to be reconfigurable, particularly when they have to deal with changing applications. Therefore, the paper focuses also on an approach for a reconfigurable distributed IoT-system. The concept is especially suitable for low bandwidth datalinks but can be used on any other communication system, too.

2. Related Work

The related work focuses on the main topics of this paper: the efficient data transfer and the flexibility and heterogeneity of IoT-devices.

An early attempt to tackle the problem of heterogeneous hardware interfaces is the smart transducer interface standard set IEEE 1451. The standard IEEE 1451.0 standardizes a transducer electronic datasheet (TEDS) containing data like ID and calibration data, independent of the physical layer [2]. Further standards of this family deal with different communication protocols and TEDS formats. Mainly due to its complexity, the practical importance of IEEE 1451 is not as high as initially anticipated [3]. Furthermore, some of the standardized interfaces and components are outdated and not appropriate for the IoT. The Open Geospatial Consortium (OGC) Sensor Web Enablement is an initiative to build a middleware for the interconnection of sensors, based on web-technologies and XML-sensor models. XML-based SensorML is a core component of the Sensor Web. It makes a framework to model sensors, actors and sensorsystems as processes available. Also the interconnection of process chains can be modelled [4]. Together with all other components of the Sensor Web a large ecosystem of tools and reference implementations mainly with focus on geo-information systems exist [5].

However, the Sensor Web sets up on a higher level than IEEE 1451 and therefore does not solve the problem of the heterogeneous sensor interfaces [6]. The sensor interface descriptor (SID) concept was initially meant to close the gap between the heterogenous sensor interfaces and the Sensor Web protocols by adding SID-adaptors to the sensors [7]. At the moment, SID is described as OGC discussion paper but not further developed for years [8].

The OGC protocol PUCK defines a protocol for Ethernet and RS232 to tackle installation and configuration challenges in sensor nets. Therefore, PUCK standardizes a memory which includes instrument driver and descriptive documents in formats like SensorML or IEEE 1451 [9]. An approach to combine the protocols and concepts SID, SensorML and PUCK to a reconfigurable sensor system with limited flexibility is also shown [9].

Some projects and approaches mainly in the environmental monitoring domain also use and adapt the Sensor Web standards [10, 11], but the standards don't seem to have a high importance in other domains. Although, these Sensor Web extension concepts are attractive, they don't fit the requirements of a highly flexible IoT-system with lightweight edge-nodes completely.

Multi agent based systems provide a high degree of flexibility, but come also with high system complexity [12]. The software defined sensor network concept offers a high degree of flexibility regarding interfaces and applications due to techniques like over-the-air programming and a wireless sensor node built-up on FPGA and software defined radio [13]. However, this concept requires high computing performance on the wireless sensor node respectively the edge node.

3. Concept

The data transfer presented in this paper sets up on a message protocol using lightweight protocol data units (PDU) consisting of a minimal header containing a startbyte, four byte length information of the PDU and the message as payload. The basic structure of the PDU is shown in Table 1.

Table 1. Structure of the PDU.

Header					Payload
Start-byte	Length [3]	Length [2]	Length [1]	Length [0]	Message
0xFA	0x00	0x00	0x00	0x05	EXI

The payload data is based on a structured XML representation. As XML itself has high syntax overhead with efficient XML interchange (EXI) an efficient compression method is used. It encodes XML documents in a binary data format using a grammar to optimize the entropy encoding. The grammar coding of EXI works similar to the Huffman coding. This means, the coding assigns codewords in different length according to the probability of occurrence of the regarding symbol.

There are two technical options to generate the grammer: statically generation at compile time or dynamically at the start of the application. Due to the enhanced flexibility, the dynamical generation is chosen for the concept. This means, after an extension or modification of the communication protocol, the application at the nodes needs only a restart procedure with the new EXI schemas. Fig. 1 shows the generation of the grammar.

The EXI encoding can be executed on XML documents without schema and also on documents using XML schema. The compression results are much better when using schemas, but the programming effort is also increased. The application of EXI combines the benefits of the structured XML data, including easy mapping to a SQL-database, with the efficient data transfer of binary data.

The concept uses the embeddable EXIP implementation in C to do the conversion between the XML and EXI documents [14]. The schema is converted using ExiProcessor [15]. To manage different interfaces, a thread pool based communication server which merges the data streams from different communication links is used. To realize the send and receive queues a producer consumer based model is used. Fig. 2 shows the conceptual system architecture.

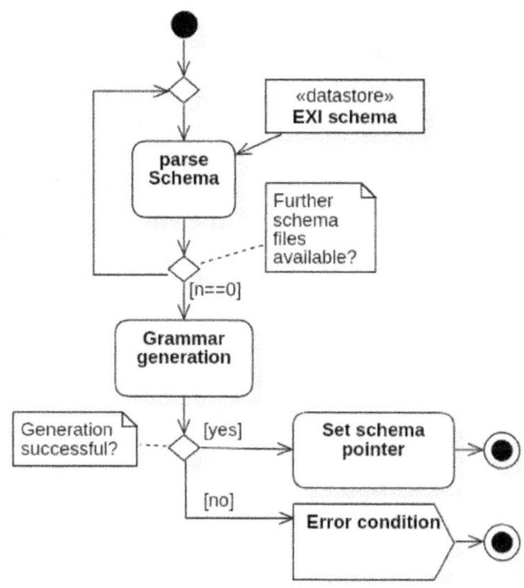

Fig. 1. Generation of the EXI grammar.

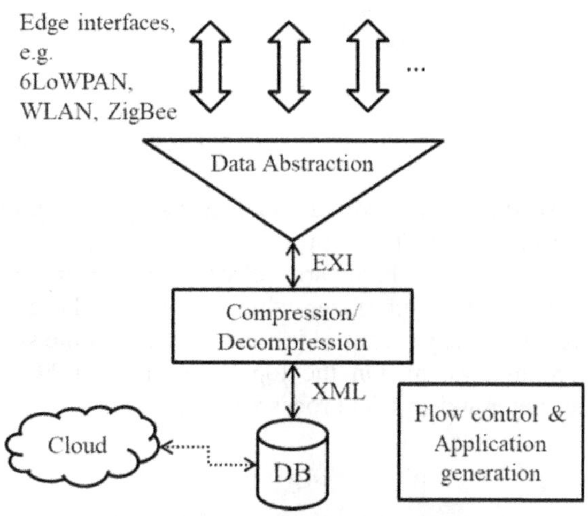

Fig. 2. Conceptual system architecture.

```
<?xml version= " 1.0 " encoding= " UTF-8 " ?>
<Configuration type= " ADC " >
    <ConfigName> SampleConfigADC </ConfigName>
    <Intervall_ms> INTERVALL_100ms </Intervall_ms>
    <ComInterface> Ethernet </ComInterface>
    <SNID>
        <SNID0> 20 </SNID0>
        <SNID1> 78 </SNID1>
        <SNID2> 1 </SNID2>
        <SNID3> 255 </SNID3>
        <SNID4> 2 </SNID4>
        <SNID5> 2 </SNID5>
    </SNID>
    <ConfigData>
        <ChannelNr> 5 </ChannelNr>
        <Resolution_bit> 14 </Resolution_bit>
        <HWRange_mV> 4500 </HWRange_mV>
        <HWSamplerate_Hz> 1000 </HWSamplerate_Hz>
        <Key> Temperature </Key>
        <UnitSystem> MSB </UnitSystem>
    </ConfigData>
</Configuration>
```

Fig. 3. Configuration example for an analog to digital converter.

To get a high degree of flexibility and reconfigurability regarding interfaces and applications, the edge code is generated on a master node. For that, the application information represented also in XML format is transformed using a XSL stylesheet with XSLT to the edge code. A configuration example in XML for an analog to digital converter edge node is shown in Fig. 3. The configuration contains the message interval, the chosen communication interface, six byte sensor node ID and interface specific configuration data, like the sample rate.

The configuration can be an adapted list of parameters or even the whole application code. Both will be transferred as EXI message to the respective edge node. In the case of application code updates, the edge node must be able to interpret the new application code. Therefore, the concept uses application code in C performed by a C-interpreter on a 32-bit microcontroller. Application code in Lua or Python may be better suited alternatives when using processors with higher resources. Fig. 4 shows the generation of new application code as UML activity diagram.

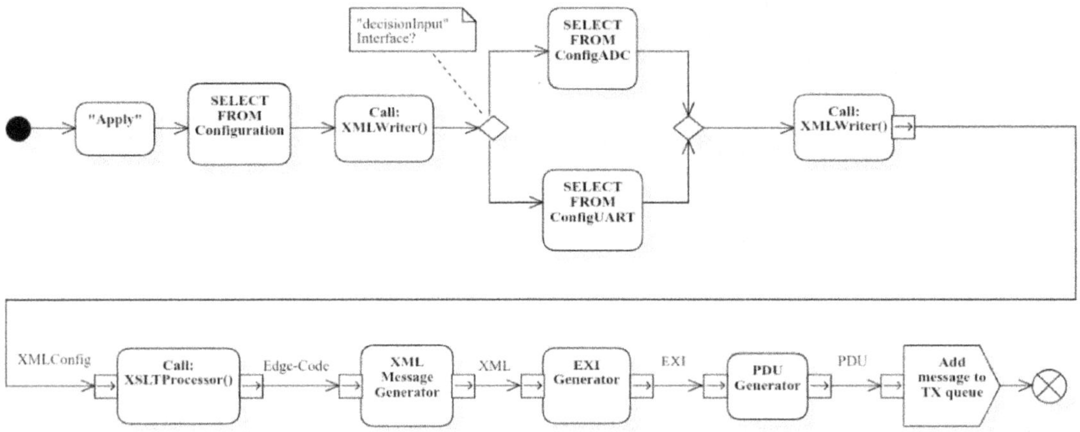

Fig. 4. Example of application code generation.

At the edge node a concept using a runtime environment (RTE) is proposed (Fig. 5). The RTE consists of a hardware abstraction layer and communication interfaces which set up on the edge hardware. The job control is done by a small time slice operating system. On the top layer of the RTE an interpreter and an EXI processor are placed.

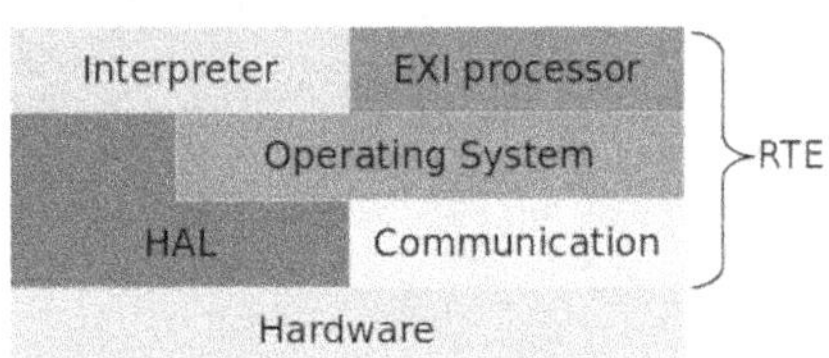

Fig. 5. Concept of the runtime environment.

At the edge node the application code is received, packed as an EXI message which is handled by the EXI processor. At nodes with limited resources EXI processors written in C like EXIP or ws4d-uexi [16] are available. Afterwards, the application code is stored in the code memory of the edge node.

Fig. 6 visualizes the upstart process at the edge node after performing an application update as UML interaction diagram.

The new application stored in the code memory has to be read and interpreted by a C-interpreter like Pic oC [17] at runtime. Then, the application generates, converts to EXI and puts cyclic HeartBeat-messages on the communication link to show the availability of the node. Subsequently also the payload data (e. g. measured values) are being sent with the MeasuredValue message.

4. Results

The result is a flexible concept using state of the art communication interfaces which can be used even with low storage and clock resources such as microcontroller based nodes. The concept merges and unifies data streams without reducing flexibility in application and interfaces. The protocol causes very low overhead by using PDUs and EXI compression. EXI compression is reasonable especially on (wireless) low bandwidth links. Fig. 7 shows the size of some system messages in the formats XML, EXI without schema and schema-based EXI. It shows a large compression effect compared to XML for status or measurement messages (up to 96 %) but lower effect on configuration messages (22 %) as they have lower protocol overhead. Normally, the incidence of the configuration messages is not too high. In contrast, message transfer occurs much more often. Thus, the average compression rate should be sufficient for most IoT applications.

The concept combines the advantage of structured XML-data (after decoding) with the minimized overhead of binary data transfer. This could be a crucial advantage for wireless IoT systems with limited bandwidth.

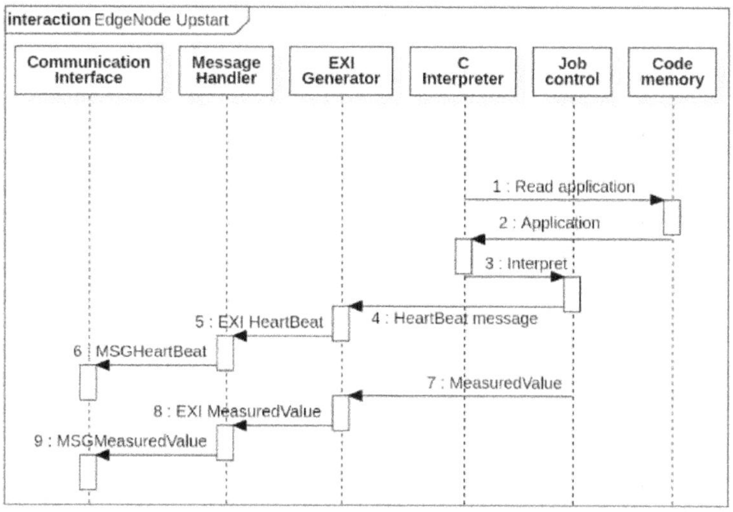

Fig. 6. Upstart process at the edge node after an application update.

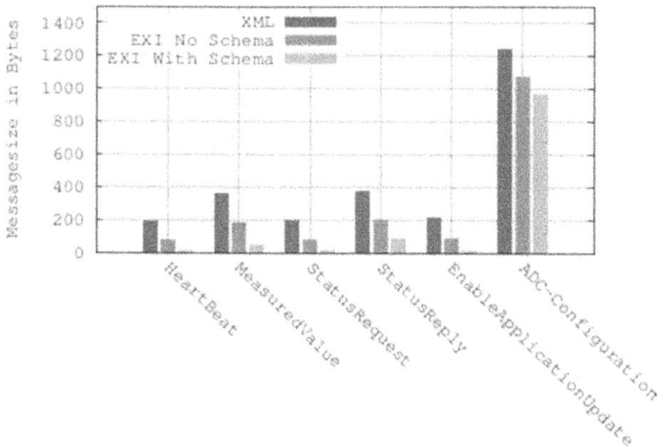

Fig. 7. Size of messages using different data formats.

The generation of edge code using XSL-transformation is promising in terms of flexibility, but needs an interpreter on edge nodes or is limited to adapt parameter lists.

5. Conclusion and Future Work

An efficient communication system as well as a high system flexibility are crucial points of the IoT. The presented concept addresses both points using a lightweight protocol in combination with the binary XML compression EXI and integrate that with a dynamic edge node application generation using XSLT. However, the high flexibility of the edge application has the drawback that an interpreter at edge node side is needed. Yet, this disadvantage seems to be acceptable for some applications.

Further steps are the completion of the prototypical system in order to perform a complete system with the reconfigurable sensor nodes to figure out the performance and system behavior by using different low power wireless IoT links like 6LoWPAN or ZigBee.

References

[1]. J. Cecílio, P. Furtado, Distributed configuration and processing for industrial sensor networks, in *Proceedings of the 6th International Workshop on Middleware Tools, Services and Run-time Support for Networked Embedded Systems (MidSens'11)*, Lisboa, Portugal, 2011.

[2]. IEEE Standard for a Smart Transducer Interface for Sensors and Actuators – Common Functions, Communication Protocols, and Transducer Electronic Data Sheet Formats, *Institute of Electrical and Electronics Engineers*, Jul. 2007.

[3]. F. Hu, Q. Hao, Intelligent Sensor Networks: The Integration of Sensor Networks, Signal Processing and Machine Learning, *CRC Press*, Boca Raton, Florida, 2012.

[4]. M. Botts, A. Robin, OpenGIS ® Sensor Model Language (SensorML) Implementation Specification, *Open Geospatial Consortium Inc.*, 17-Jul-2007.

[5]. M. Botts, SensorML-2.0 Examples, SensorML 2.0, http://www.sensorml.com/sensorML-2.0/examples/

[6]. A. Bröring, *et al.*, New generation sensor web enablement, *Sensors*, Vol. 11, Issue 3, Mar. 201, pp. 2652-26991.

[7]. A. Bröring, S. Below, T. Foerster, Declarative sensor interface descriptors for the sensor web, in *Proceedings of the International Workshop on Pervasive Web Mapping, Geoprocessing and Services (WebMGS'10)*, Como, Italy, 2010.

[8]. A. Bröring, S. Below, OpenGIS ® Sensor Interface Descriptors, *Open Geospatial Consortium Inc.*, Jun. 2010.

[9]. T. O'Reilly, OGC ® PUCK Protocol Standard Version 1.4, *Open Geospatial Consortium*, Jan. 2012.

[10]. E. Martínez, D. Toma, S. Jirka, J. del Río, Middleware for plug and play integration of heterogeneous sensor resources into the sensor web, *Sensors*, Vol. 17, Issue 12, Dec. 2017, 2923.

[11]. S. Lin, F. Lyu, A capable smart sensor interface module for interoperable ocean observatories, *IOP Conference Series: Earth and Environmental Science*, Vol. 171, 2018, 012005.

[12]. T. Suganuma, T. Oide, S. Kitagami, K. Sugawara, N. Shiratori, Multiagent-based flexible edge computing architecture for IoT, *IEEE Network*, Vol. 32, Issue 1, Jan. 2018, pp. 16-23.

[13]. D. Zeng, T. Miyazaki, S. Guo, T. Tsukahara, J. Kitamichi, T. Hayashi, Evolution of software-defined sensor networks, in *Proceedings of the IEEE 9th International Conference on Mobile Ad-hoc and Sensor Networks (MSN'13)*, Dalian, China, 2013, pp. 410-413.

[14]. EXIP Download, SourceForge.net, https://sourceforge.net/projects/exip

[15]. ExiProcessor download, SourceForge.net, https://sourceforge.net/projects/exiprocessor

[16]. WS4D-UEXI – Google Code Archive, https://code.google.com/archive/p/ws4d-uexi

[17]. Z. Saleeba, Picoc GitLab, https://gitlab.com/zsaleeba/picoc

(046)

Laser-based Gas Sensors for Breath Analysis: From Research to the Clinic

S. M. Cristescu
Radboud University, Institute for Molecules and Materials, Heyendaalseweg 135, 6525 AJ, Nijmegen, the Netherlands
Tel.: + 31243652277
E-mail: s.cristescu@science.ru.nl

Summary: Detection and analysis of volatile compounds in exhaled breath is an attractive tool for monitoring the metabolic status of a patient and disease diagnosis, since it is non-invasive and fast. Several breath-analysis methods are currently available, ranging from mass spectrometry to laser-based spectroscopy and sensor arrays. This talk presents the current status of laser-based methods for clinical breath gas analysis and describes recent technological developments and their applications. A SWOT analysis (strengths, weaknesses, opportunities and threats) is placing the laser-based techniques in the clinical framework of breath research.

Keywords: Breath, Gas analysis, Laser spectroscopy, Sensors.

1. Introduction

Humans exhale hundreds of gases arising from normal body metabolism. Other compounds are produced by pathogens and represent infectionspecific biomarkers (e.g. infections in cystic fibrosis). Analysis of these biogenic volatiles can be used for disease diagnosis, fingerprint and prognosis, as well as for drugs selection and therapy monitoring. The most common methods currently employed for breath analysis are based on mass spectrometry; these are analytical methods allowing detection of multiple gas species, although quantitative analysis and/or positive identification of compounds are not always possible. In comparison, laser-based spectroscopy can provide an accurate and precise quantitative analysis result, with an instrument that does not necessarily require an expert user. Depending on the spectroscopic method and the laser source, laser spectroscopy can provide real-time (breath-cycle resolved) detection of a single or few volatile compounds, with detection limits ranging from the part-per-million (ppm) to less than part-per-billion (ppb) [1].

2. Laser-based Spectroscopic Techniques

Several near-infrared (near-IR) and mid-IR laser-based breath analyzers with various spectroscopic methods have shown their suitability for detection of specific molecules of interest for medical community. The detection limit that is achieved is determined by the sensitivity of the chosen technique. To detect compounds in breath at less than ppm levels, the absorption path length can be increased, by either using a multi-pass cell or employing a resonant cavity-enhanced approach (e.g. cavity-enhanced absorption, CEAS). In addition, wavelength- (WMS) and frequency modulation spectroscopic (FMS) methods are also considered for noise reduction. Alternatively, photoacoustic spectroscopy (PA) can be employed; rather than measuring the decrease of laser intensity by absorption, PA measures the result of absorption via an increase in acoustic intensity. Comparable sensitivities to absorption spectroscopy with short optical path length (tens of cm) can be achieved.

2.1. Exhaled Volatiles Detection: Clinical Applications

The exhaled volatiles are present at different concentrations, varying from percentages (CO_2) to ppm (NH_3, CH_4, CO), down to ppb levels (C_2H_4, C_2H_6, NO, HCN).

Therefore, detection of one or more of these analytes requires different detection schemes and laser sources (e.g. distributed feedback (DFB) diode lasers (0.7-3 gm), vertical-cavity surface emitting lasers (VCSELs) in the visible and near-IR, DFB interband cascade lasers (ICLs) (3-6 gm) and quantum-cascade lasers (QCLs) (4-14 gm) in the mid-IR. Few applications are presented.

For exhaled breath biomarkers at concentrations levels around 1 ppm (CO, CH_4, NH_3 and N_2O), real time detection and compact setups are used in practice employing tunable diode laser absorption spectroscopy (TDLAS) in combination with WMS and/or multipass cells. This technique is well-suited for targeted detection of CO_2 and CO for clinical applications [2].

Detection of compounds at low ppbv level or below, requires CEAS or Off Axis-CEAS to be employed. Typical sensitivities that can be achieved are in the 10^{-10}-10^{-11} $cm^{-1}Hz^{-1/2}$ range. Compact sensors have been developed for the determination of oxygen consumption on a breath-by-breath basis in intubated patients [3] or monitoring CO as a biomarker for the selection of lung grafts for transplantation [4]. The latest is performed in the tightly controlled clinical setting of organ transplantation and it requires sensitive detection scheme because an isolated organ delivers CO concentrations that are one order of magnitude smaller than the entire living body.

In laser photoacoustics (PA) the absorption path length is kept very short (mm cm) and that the signal is background free. Using CO_2 laser-based PA, continuous monitoring of ethylene production from exhaled breath of intubated patients during cardiac surgery has revealed for the first time the real-time signatures of the surgical response triggered by oxidative stress including lipid peroxidation [5].

Different isotopic labeled gases (e.g. $^{12}C^{16}O^{16}O/^{13}C^{16}O^{16}O/^{12}C^{16}O^{18}O$, $^{14}N^{14}N/^{14}N^{15}N/^{15}N^{15}N$, etc.) can be measured in exhaled breath after administration of labeled probes using fiber-enhanced Raman multi-gas spectroscopy [6]; these tracer experiments have the potential to make the transition from research to the clinic.

The current field of medical applications demands also portable devices with low power consumption. Quartz-enhanced photoacoustic spectroscopy (QEPAS) can meet these demands to a large extend [7]. Here, the (resonant) PA cell is replaced by a tuning fork, therefore decreasing the sampling volume; in addition, tuning forks are mass produced and inexpensive. Several studies have been shown the potential of this technique for measuring $^{13}CO_2/CO_2$ ratio in relation with identification of Helicobacter pylori infection, liver malfunction, and excessive growth of bacteria in the body, carbon disulfide (CS_2) as indicator of cirrhosis and NH_3 as product of protein metabolism.

2. SWOT Analysis

A brief SWOT analysis (strengths, weaknesses, opportunities and threats) will be presented. Without doubts, the strength of optical methods lies in specific, reliable and quantitative, real-time analysis of single analyte, and often without the need for (frequent) calibration. However, simultaneous detection of multiple species is very often desired. The advances in laser technology and detection schemes allow fast developments of novel design of compact instruments for use outside the labs. Rapidly growing field of nanotechnologies, optoelectronics and photonics integration are becoming competitive with the laser-based spectroscopic techniques. The electronic nose or semiconductor-based sensor array have become popular and are used in clinical applications, in spite of their practical limitations (cross-sensitivities, ageing, need for frequent calibrations).

4. Conclusions

Laser-based spectroscopy has demonstrated its maturity in many clinical tests. At the moment, it needs to gain the trust of the medical community and be promoted in (new) medical applications. Further tests and clinical validation needs to be supported within a multidisciplinary collaboration between the laser spectroscopy researchers and the medical specialists.

References

[1]. B. Henderson, A. Khodabakhsh, M. Metsala, I. Ventrillard, F. M. Schmidt, D. Romanini, G. A. D. Ritchie, S. te Lintel Hekkert, R. Briot, T. Risby, N. Marczin, F. J. M. Harren, S. M. Cristescu, Laser spectroscopy for breath analysis: towards clinical implementation, *Applied Physics B*, Vol. 124, 2018, 161.

[2]. R. Ghorbani, F. M. Schmidt, ICL-based TDLAS sensor for real-time breath gas analysis of carbon monoxide isotopes, *Optics Express*, Vol. 25, Issue 11, 2017, pp. 12743-12752.

[3]. L. Ciaffoni, D. P. O'Neill, J. H. Couper, G. A. Ritchie, Hancock, P. A. Robbins, In-airway molecular flow sensing: A new technology for continuous, noninvasive monitoring of oxygen consumption in critical care, *Science Advances*, Vol. 2, Issue 8, 2016, e1600560.

[4]. M. Maignan, S. Gennai, G. Debaty, D. Romanini, M. Schmidt, V. Brenckmann, A. Brouta, I. Ventrillard, R. Briot, Exhaled carbon monoxide is correlated with ischemia reperfusion injuries during ex vivo lung perfusion in pigs, *Journal of Breath Research,* Vol. 11, Issue 3, 2017, 036004.

[5]. S. M. Cristescu, R. Kiss, S. T. Hekkert, M. Dalby, F. J. M. Harren, T. H. Risby, N. Marczin, B. S. I. Harefield, Real-time monitoring of endogenous lipid peroxidation by exhaled ethylene in patients undergoing cardiac surgery, *American Journal of Physiology – Lung Cellular and Molecular Physiology*, Vol. 307, Issue 7, 2014, pp. L509-L515.

[6]. T. Bogozi, J. Popp, T. Frosch, Fiber-enhanced Raman multi-gas spectroscopy: What is the potential of its application to breath analysis?, *Bioanalysis,* Vol. 7, Issue 3, 2015, pp. 281-284.

[7]. Y. F. Ma, G. Yu, J. B. Zhang, X. Yu, R. Sun, F. K. Tittel, Quartz enhanced photoacoustic spectroscopy based trace gas sensors using different quartz tuning forks. *Sensors*, Vol. 15, Issue 4, 2015, pp. 7596-7604.

(047)

Differential Pressure Sensing Based on Phase Sensitive Detection for Liquid Level Measurement

P. Esmaili [1], **F. Cavedo**[1] **and M. Norgia** [1]
[1] Politecnico di Milano, via Ponzio, 20133, Milan, Italy
Tel.: +39-02-2399.3601, fax: fax.3412
E-mail: parisa.esmaili@polimi.it

Summary: In this paper, a capacitive differential pressure sensor is characterized as a liquid level measurement system. The sensor is configured as a first-order low pass filter with a cut off frequency at 10 kHz, driven by a sinusoidal signal at 100 kHz. The phase-sensitive demodulation of the signal is performed using a single-tone DFT algorithm, acquiring both excitation and output signals. Each signal is acquired at 10 MSPS and the DFT is computed on 8000 samples. The sensitivity of the proposed instrument is 40 ppm/mm with very good linearity and error as low as ±0.3 mm. The measurement characterization of the sensor indicates good performances as low-cost level measurement system.

Keywords: Liquid level measurement, Differential capacitive pressure, Phase sensitive detection, Digital Lock-in amplifier, Single-tone DFT.

1. Introduction

In a wide range of scientific and industrial applications, precise liquid level measurement plays a crucial role. To achieve accurate level detection, request for dedicated techniques that are working to tighter accuracy limits become more severe while production costs are limited due to economic issues. As a result, the requirement for designing an instrument, to be both accurate and cost-effective, becomes ever harder to satisfy. Various techniques have been proposed in this manner. These approaches can be categorized into either direct/indirect or contact/non-contact methods. Typically, direct methods such as floating based level measurement systems are limited by the lack of precision and accuracy. Several approaches in the field of contact-based indirect level measurement such as time domain reflectometry (TDR), capacitive and pressure-based reading systems have been proposed. These approaches eliminate the requirement of moving part as in conventional techniques. However, the capacitive probes suffer from miss-reading not only due to parasitic capacitance but also to fluctuations [1, 2]. Although TDR based approaches are very accurate solutions, but pulse-duration measurements with high resolution requires high costs for the overall data-acquisition system. In addition to contact based methods, different non-contact indirect methods such as ultrasonic [3, 4], vision [5], optical [6-8], microwave [9-12] have been proposed to measure the level. In techniques based on Time of Flight (TOF) principle, the performance becomes poor when they are subjected to measure the quantity of liquid with low dielectric constant due to weak back reflection. In addition, the presence of the foam on the surface of the liquid results into inaccurate level detection during acceleration. In this paper, a low-cost and indirect contact-based level detection instrument is proposed based on differential pressure sensing approach. A Capacitive differential pressure sensor is used as a sensing element in the configuration of first-order low pass filter where a sinusoidal signal at 100 kHz is generated by Digital-to-Analog Converter (DAC) of multi-function instrument (DIGILENT-Analog Discovery) to drive it. In order to eliminate any non-linearity, induced by distortion of the driving signal, the output signal of the proposed instrument is expressed as the ratio between the signal at the output of the operational amplifier and input signal. Then, single-bin DFT (Discrete Fourier Transform) is performed based on evaluating the phase and quadrature component with a digital synchronous receiver, which performs the multiplication of the sampled signal for two quadrature signals to then evaluate the Fourier coefficients. The performance of the proposed instrument has been evaluated experimentally under steady state condition.

2. Liquid Level Measurement Based on Differential Pressure Sensing Principle

Pressure based level sensing technique has been proposed for the analysis and monitoring of a ship's stability [13]. The Pressure transmitter measures level based on the principle that pressure (P) is proportional to the level h of the liquid multiplied by its specific gravity given by (1).

$$P = \rho g h, \qquad (1)$$

where ρ and g are the density and gravitational acceleration respectively. Depending on the technology, different types of pressure transducers are commercially available such as piezoresistive, piezoelectric and capacitive pressure transducers. Fig. 1 represents the working principle of the capacitive differential pressure sensor for measuring liquid level, used in the proposed level instrument. The output of the transducer is an equivalent capacitance which changes its value as a function of the applied pressure.

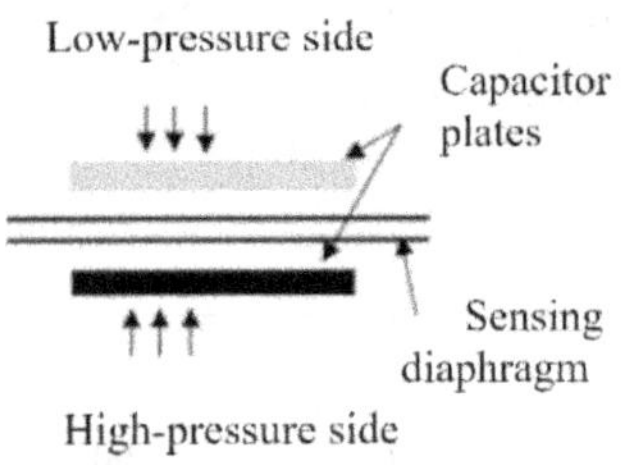

Fig. 1. Capacitance variation in capacitive differential pressure transducer due to applied pressure.

In the frequency-shift-based implementation, the capacitive sensor is used in an RC low pass filter, such that a change in capacitance causes a change in frequency response. The transfer function of a low pass filter is represented in (2) as a function of frequency where f_c is its cut-off frequency.

$$A_v = \frac{V_{out}}{V_{in}} = \frac{1}{\sqrt{1+(\frac{f}{f_c})^2}}, \qquad (2)$$

While working at cut-off frequency provides maximum sensitivity in the phase, working at the frequency 10 times higher guarantees maximum sensitivity in gain and linearity. This is where the excitation signal is generated to derive transducer and measure accurately the level of the liquid.

3. Measuring Circuit

Due to different frequency dependency of various noise sources, as in many electrical and physical systems, there is increasing in noise level as the frequency approaches DC. In case of having flexibility in the choice of the modulation frequency, part of the spectrum can be chosen as the operating zone where the noise level is the lowest. Working at higher frequencies requires a narrowband filter to recover the signal. Alternatively, phase sensitive detection technique is introduced to move the modulated signal back to dc while filtering out other signals that are not synchronized to the reference signal [14]. In this paper, single-bin DFT is performed based on evaluating the phase and quadrature component with a digital synchronous receiver, which performs the multiplication of the sampled signal for two quadrature signals (sine and cosine at the same frequency), to then evaluate the Fourier coefficients. From the two-phase and quadrature signals, amplitude and phase measurements are easily obtained, which becomes the more accurate depending on how many points have been acquired. This is due to the contribution of white noise on every single bin which affects the measurement accuracy in case of dealing with low SNR (Signal to Noise Ratio). However, the frequency-domain SNR can be effectively increased the by increasing the DFT length N. Increase the observation duration or in the other words performing longer DFT implies that each bandpass filter covers a smaller portion of the frequency band resulting less noise power passed by each bin's corresponding filter. Through this technique, very high resolution for amplitude and phase measurements can be achieved. However, there is always tradeoff between time and frequency resolution. Fig. 2 shows the MATLAB/Simulink model of single-bin DFT implementation. Here, Computational specification to extract a single tone (assuming real input) are $2N$ multiplications and $2N + 1$ addition.

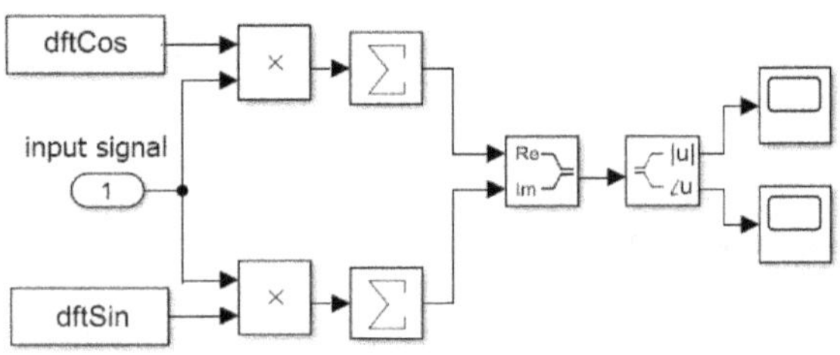

Fig. 2. MTALAB/ Simulink model of Single-bin DFT algorithm.

In this paper, the instrument design is divided into three main parts: the capacitive differential pressure transducer, the analog frontend for signal condition and a digital part to provide sensor operating condition and making the elaboration as represented in Fig. 3. The Excitation signal is generated by DAC. Using two ADCs working in parallel and triggered by the same clock signal, it is possible to acquire the DAC output used to drive the transducer and the output of analog frontend at the same time. In this case any non-linearity induced by distortion of the driving signal is eliminated. Then, Digital Signal Processing unit (DSP) processes the acquired signals using DFT and provides respective phase and quadrature.

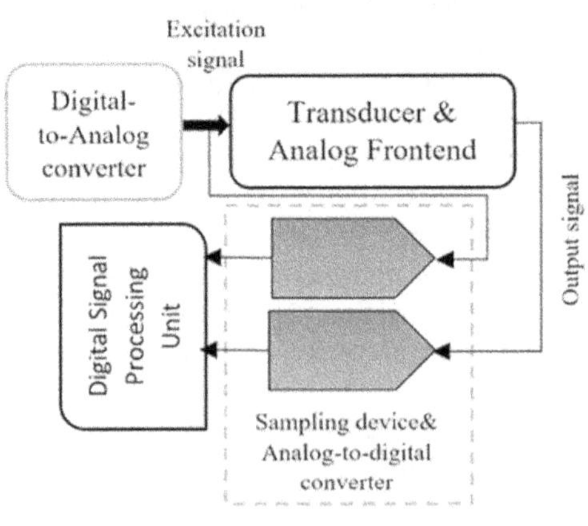

Fig. 3. Block diagram of the proposed instrument including capacitive differential pressure senor, one DAC and two ADCs.

4. Experimental Setup

The prototype of proposed level instrument and experimental setup with the liquid tank have is illustrated in Fig. 4. Graduated cylinder test tank is mounted on high pressure while low pressure side is open to the air.

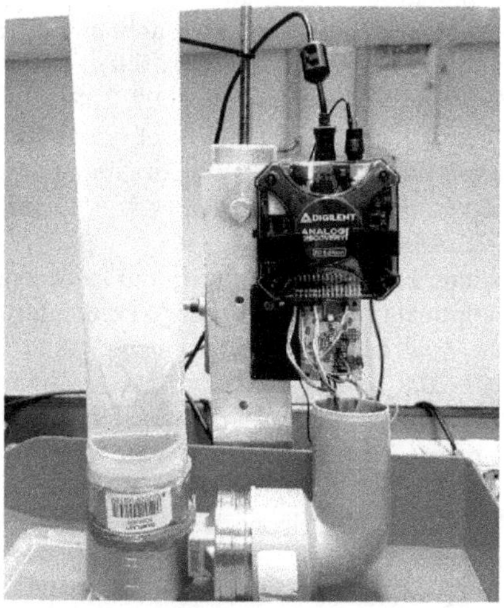

Fig. 4. Prototype of the proposed level instrument.

The Pressure sensing element is a differential capacitive pressure sensor in range of 0~7.5 kPa which is connect to the input operational amplifier in the configuration of first order low pass filter. A multi-function instrument (DIGILENT-Analog Discovery) is used to generate excitation sinusoidal signal at 100 kHz with amplitude of 2 V_{p-p} provided by function generator (±5 V, 14-bit, 100 MS/s). Using Dual channel 14-bit, ADC form Analog Discovery 2, the excitation and output of the operational amplifier are acquired simultaneously. Both signals are processed using DFT and provides respective real and imaginary values. Each signal is acquired at 10 MSPS and the DFT is computed on 8000 samples. Fig. 5 demonstrates measuring circuit with digital signal generator and two synchronous samplers. Further steps to perform DFT algorithm and amplitude detection is done in LabVIEW from National Instrument.

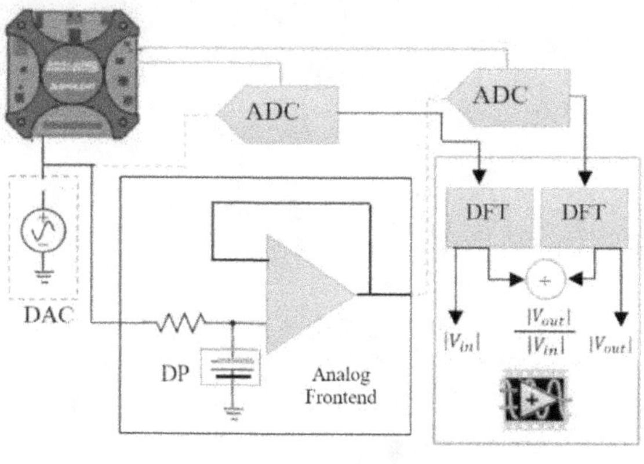

Fig. 5. Measuring circuit block diagram of the proposed instrument with signal generation, acquisition and processing block diagram.

To determine sensitivity curve, the water level in a test tank was increased in steps by adding water. Fig. 6 shows the sensititivty curve of the proposed instrument. As shown in this figure, the sensitivity is 40 ppm/mm.

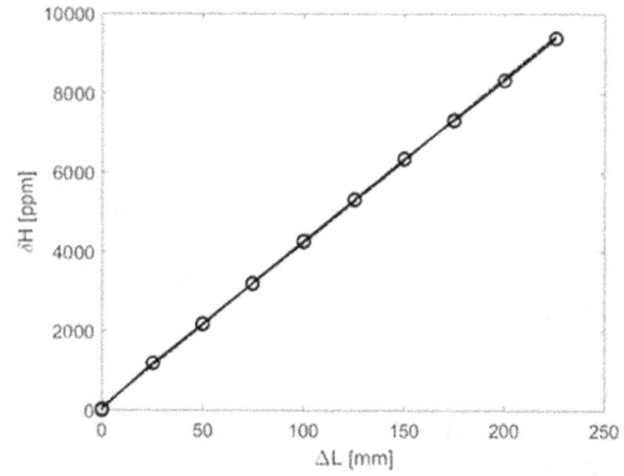

Fig. 6. Sensititivty curve of the proposed instrument.

To evaluate repeatability error, additional experiments have been performed. Fig. 7 shows the measured level increment and decrement using proposed instrument for repeated measurements to evaluated repeatability error of the instrument. Corresponding error is illustrated in Fig. 8. As result shows, the repeatability error for the proposed instrument is about ± 0.3 mm.

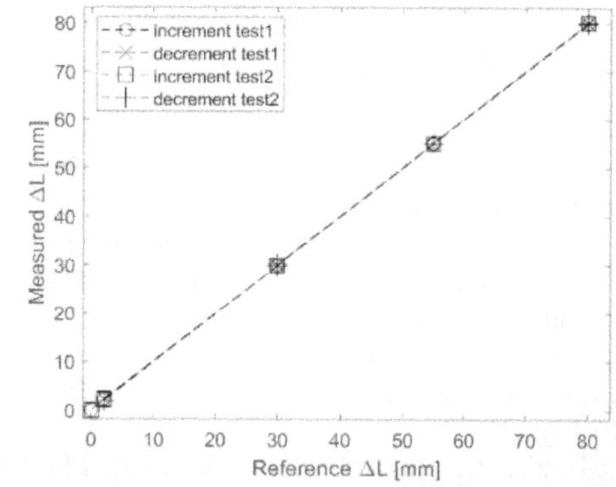

Fig. 7. Measured relative value of the level V.S reference values during repeatability test.

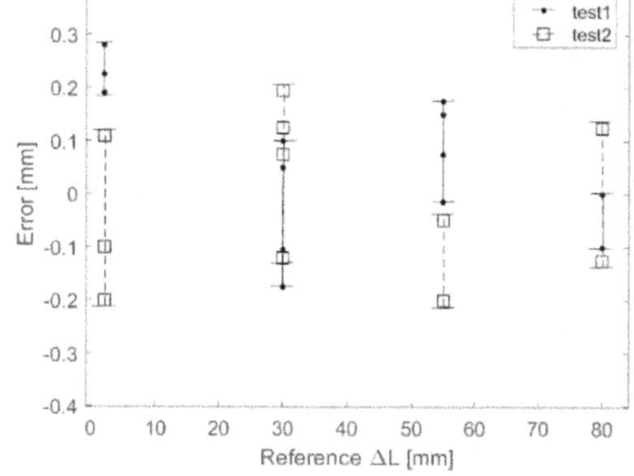

Fig. 8. Absolute values for repeatability error of the proposed instrument.

4. Conclusion

As low-cost level measurement system, a digital-lock in amplifier based liquid level instrument is proposed in this paper, using differential pressure sensing approach. The sinusoidal signal at 100 kHz is used as the excitation signal. The output signal of the proposed instrument is defined as the ratio between the signal at the output of the operational amplifier and input signal to eliminate any non-linearity, induced by distortion of the driving signal. Dual channel ADC form DIGILENT is used to acquire the excitation and output of the operational amplifier simultaneously. Both signals are processed using DFT and provide respective real and imaginary values. Each signal is acquired at 10 MSPS and the DFT is computed on 8000 samples. The sensitivity of the proposed instrument reported here is 40 ppm/mm, with good linearity. In addition, the repeatability test is performed and the error is determined as $\pm$0.3 mm.

Acknowledgements

The authors would like to thank Fondazione Ing. Paolo Foresio for the financial support of the research activity.

References

[1]. H. Canbolat, A Novel Level Measurement Technique Using Three Capacitive Sensors for Liquids, *IEEE Transactions on Instrumentation and Measurement*, Vol. 58, Issue 10, 2009, pp. 3762-3768.

[2]. S. Bera, H. Mandal, S. Saha, A. Dutta, Study of a modified capacitance-type level transducer for any type of liquid, *IEEE Transactions on Instrumentation and Measurement*, Vol. 63, Issue 3, 2014, pp. 641-649.

[3]. D. Zou, M. Yang, X. Zhan, R. He, X. Li, Assessment of air entrainment in stirred tanks using capacitive sensors, *Sensors and Actuators A: Physical*, Vol. 216, 2014, pp. 92-101.

[4]. J. Terzic, C. Nagarajah, M. Alamgir, Fluid level measurement in dynamic environments using a single ultrasonic sensor and Support Vector Machine (SVM), *Sensors and Actuators A: Physical*, Vol. 161, Issue 1-2, 2010, pp. 278-287.

[5]. P. Li, Y. Cai, X. Shen, S. Nabuzaale, J. Yin, J. Li, An accurate detection for dynamic liquid level based on MIMO ultrasonic transducer array, *IEEE Transactions on Instrumentation and Measurement*, Vol. 64, Issue 3, 2015, pp. 582-595.

[6]. L. Nguyen, B. Schaeli, D. Sage, S. Kayal, D. Jeanbourquin, D. Barry, L. Rossi, Vision-based system for the control and measurement of wastewater flow rate in sewer systems, *Water Science & Technology*, Vol. 60, Issue 9, 2009, 2281.

[7]. K. Sohn, J. Shim, Liquid-level monitoring sensor systems using fiber Bragg grating embedded in cantilever, *Sensors and Actuators A: Physical*, Vol. 152, Issue 2, 2009, pp. 248-251.

[8]. W. Wang, F. Li, Large-range liquid level sensor based on an optical fibre extrinsic Fabry-Perot interferometer, *Optics and Lasers in Engineering*, Vol. 52, 2014, pp. 201-205.

[9]. D. Melchionni, M. Norgia, Optical system for liquid level measurements, *Review of Scientific Instruments*, Vol. 85, Issue 7, 2014, 075113.

[10]. A. Cataldo, L. Tarricone, F. Attivissimo, A. Trotta, Simultaneous measurement of dielectric properties and levels of liquids using a TDR method, *Measurement*, Vol. 41, Issue 3, 2008, pp. 307-319.

[11]. A. Cataldo, E. Piuzzi, E. De Benedetto, G. Cannazza, Experimental characterization and performance evaluation of flexible two-wire probes for TDR monitoring of liquid level, *IEEE Transactions on Instrumentation and Measurement*, Vol. 63, Issue 12, 2014, pp. 2779-2788.

[12]. L. Bengtsson, New design ideas for TDR-based liquid level detectors, in *Proceedings of the IEEE International Instrumentation and Measurement Technology Conference (I2MTC)*, 2017, pp. 1-6.

[13]. H. Zheng, Y. Huang, Y. Ye, New level sensor system for ship stability analysis and monitor, *IEEE Transactions on Instrumentation and Measurement*, Vol. 48, Issue 6, 1999, pp. 1014-1017.

[14]. M. L. Meade, Lock-in Amplifiers: Principles and Applications, *Peter Peregrinus Ltd*, London, 1983.

[15]. R. Beck, A. G. Dempster, I. Kale, Finite precision goertzel filters used for single tone detection, *IEEE Trans. On Circuits and Systems – II: Analog and Digital Signal Processing*, Vol. 48, Issue 6, June 2001, pp. 691-700.

(048)

Analysis of Climatic and Oceanic Variables in the Canarian Archipelago in the Last 15 Years through Remote Sensing

Nerea Marrero Betancort [1], **Dionisio Rodríguez Esparragón** [1] **and Javier Marcello Ruiz** [1]
[1] Instituto de Oceanografía y Cambio Global, Universidad de Las Palmas de Gran Canaria, Unidad Asociada ULPGC-CSIC, Campus de Tafira, 35017, Las Palmas de Gran Canaria. España. Tel: +34 928 45 73 65
E-mail: nerea.marrero102@alu.ulpgc.es, dionisio.rodriguez@ulpgc.es, javier.marcello@ulpgc.es

Summary: The oceans regulate our weather and climate from global to regional scales. The prediction of the consequences of Climate Change is based on a detailed diagnosis of its evolution and the identification of its strongest tendencies. It is of special importance to obtain local data, since the manifestations of climatic anomalies depend on the geographical area. Furthermore, islands are especially vulnerable to climate change, reason why this work have a focus on the Canary Islands. The main objective encompasses a multi-temporal study, based on data from various remote sensing sensors to obtain and interpret environmental variables related to Global Change. Specifically, this study focuses on finding the correlation between oceanic and atmospheric parameters (chlorophyll, sea surface temperature, aerosols and wind) occurred in 2003-2017.

Keywords: Climate change, Multi-temporal study, Remote sensing, Atmospheric parameters, Oceanic parameters.

1. Introduction

Over the last 50 years, oceans have absorbed more than 90 % of the excess of heat received by our warming planet [1]. At the same time, they have absorbed nearly 30 % of anthropogenic CO_2 emissions leading to ocean acidification [2]. These human-induced changes interfere with the natural flow of energy in the climate system.

Insular areas, possess a great environmental, social and economic fragility, being able to present significant alterations due to the climatic change. The transformation of these aspects in the case of the Macaronesian region and, especially, of the Canary Islands, implies alterations both in the climatic conditions and in the increase of the territorial risks, derived from a foreseeable growth of the danger before certain meteorological phenomena [3].

Ocean data acquired from satellite imagery have reached the necessary maturity to be reliable and used in several scientific disciplines. Currently, the high level of development of remote sensing techniques and data processing methods have converted the perception of "simple satellite images" into quantitative measurements of variables relevant to marine and environmental studies.

Hence, the objective of this work is to combine remote sensing with oceanographic and meteorological knowledge, to reveal the possible changes produced in the last 15 years, in the geographical area of Macaronesia, particularly in the surrounding waters of the Canary archipelago, by the contributions obtained from the analysis and processing of remote sensing data.

2. Materials and Method

2.1. Study Area

For this study we have chosen an open ocean site in the north of the Canary archipelago (-15° W, 29.5234° N), is windward of the Canary Islands to avoid wake effects of both the major currents and winds.

2.2. Data

The analyze of the following three variables was conducted: sea surface temperature (SST), chlorophyll concentration (CHL-a), and aerosols optical thickness (AOT). The data relative to these variables have a monthly temporal resolution, and were obtained from the MODIS sensor on board of NASA's Aqua satellite

Additionally, the wind data that we have employ have been developed by the National Center for Atmospheric Prediction and the National Center for Atmospheric Research of the United States (NCEP/NCAR). Monthly means of the zonal (u) and meridional (v) wind components measured at 10 m height were processed, to calculate the wind direction (W.D) and wind intensity (W.I).

The spatial resolution of this study covers a time period of 15 year. From January 2003 until December 2017.

3. Results

Due to the large amount of data analyzed, only the most relevant results are presented in this section.

In Fig. 1, the variables are represented. They have been standardized so that their comparison is easier. It can be observed that SST and CHL have an inversely proportional correlation, reaching the maximum peaks of CHL for the lower temperature ones (approx. <18 ºC). This is because the cold waters are rich in nutrients, which favor the growth of phytoplankton [4].

In spite of knowing African dust events are especially intense and more frequent in winter and early spring, the AOT does not have a marked pattern [5].

As shown in Table 1, the trend of the SST is positive, that is, the SST has increased in the waters of the archipelago during the study period, as the Intergovernmental Panel on Climate Change (IPCC) has already announced [6].

Table 1. Trends in the study period (2003-2017).

VARIABLE	SST	CHL	AOT	W.D	W.I
SLOPE	0.0016	4_{10}^{-5}	-3_{10}^{-4}	0.0127	-5_{10}^{-4}

On the other hand, the wind direction trend is positive, this indicates that the wind is rotating clockwise, predominantly from NE-E (trade winds). In addition, the wind intensity trend is negative, corroborating the studies of [7]. The change of direction of the wind and the decrease of the intensity, can be the cause of the negative tendency of AOT in the last 15 years.

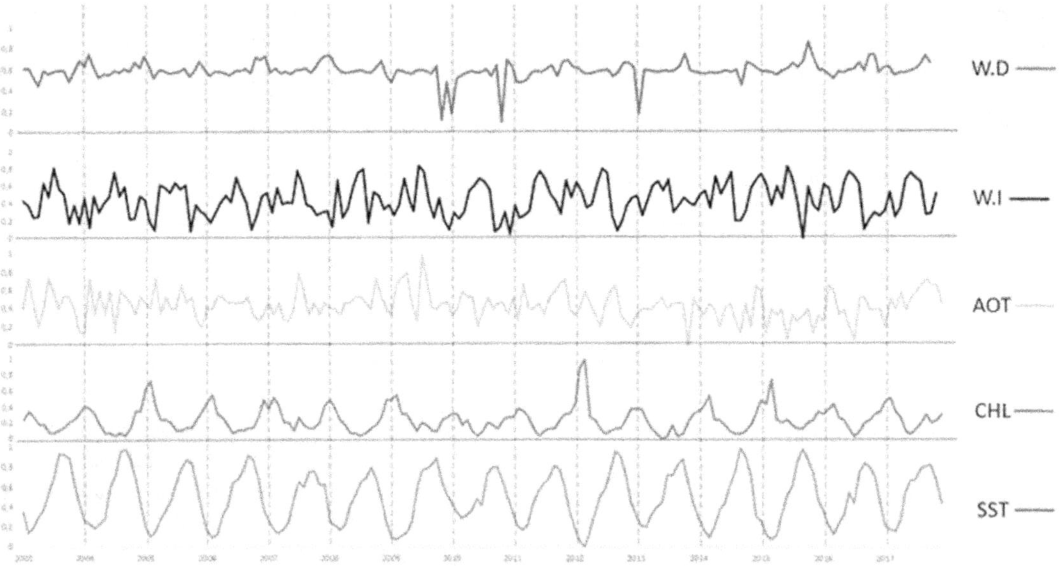

Fig. 1. Time series of standardized variables in the study period (2003-2017).

4. Conclusions

In this current work, different climatic variables have been analyzed in order to find significant changes in the waters of the Canary archipelago, which could be related to climate change.

The average monthly trends in the climatology of 2003-2017 revealed that SST, CHL and wind direction tend to increase. On the other hand, AOT and wind strength tend to decrease.

Another hypothesis that can be drawn from the results is that a slight decrease in wind strength has been observed and changes in its direction could have affected the presence of AOT in the region of the study.

These results could be due to changes in global weather phenomena. The influence of the main synoptic mode of atmospheric circulation and climate variability in the Atlantic Ocean, the North Atlantic Oscillation (NAO), is noticeable across the subtropical Canary Island archipelago [7]. Therefore, it would be interesting to approach the study of correlations between the study variables and the NAO.

This study is part of a research project which aims to understand the correlation between the presence of AOT and respiratory pathologies. In addition, an exhaustive study of AOT with large-scale meteorological phenomena would be of interest.

Acknowledgements

This work has been supported by the PASTOR (ProID2017010072) project, funded by the Department of Economy, Industry, Commerce and Knowledge of the Canary Islands Government, and the European Fondo Europeo de Desarrollo Regional (FEDER).

References

[1]. S. Levitus, J. Antonov, T. Boyer, Warming of the world ocean, 1955-2003, *Geophysical Research Letters*, Vol. 32, Issue 2, 2005, L02604.

[2]. C. Le Quéré, R. Moriarty, R. M. Andrew, J. G. Canadell, S. Sitch, J. I. Korsbakken, R. A. Houghton, Global carbon budget, *Earth System Science Data*, Vol. 7, 2015, pp. 349-396.

[3]. J. Olcina Cantos, Turismo y cambio climático: Una actividad vulnerable que debe adaptarse, *Investigaciones Turisticas*, Vol. 4, 2012, pp. 1-34.

[4]. A. J. Richardson, Plankton and Climate, In Encyclopedia of Ocean Sciences, 2nd Ed., *Academic Press,* 2009.

[5]. S. Engelstaedter, I. Tegen, R. Washington, North African dust emissions and transport, *Earth-Science Reviews,* Vol. 79, Issues 1-2, 2006, pp. 73-100.

[6]. 5th Assessment Report, The Physical Science Basis, Contribution of Working Group, *The Intergovernmental Panel on Climate Change*, 2013.

[7]. C. Azorín-Molina, et al, Wind speed variability over the Canaray Island, 1948-2014: Focusing on trend differences at the land-ocean interface and below-above the trade-winds inversion layer, *Climate Dynamics,* Vol. 50, 2018, pp. 4061-4081.

(050)

A Simple Method for Determination of the Frequency Response of the Constant-temperature Hot-wire Anemometer

S. Takagi [1], A. Inasawa [1] and M. Asai [1]

[1] Tokyo Metropolitan University, Department of Aeronautics and Astronautics, 6-6 Asahigaoka, Hino/Tokyo, Japan

Tel.: + 81 42 585 8666

E-mail: pantaka@tmu.ac.jp

Summary: This paper deals with a simple method for determination of the frequency response of the hot-wire anemometer operated with the constant-temperature mode. Since the sensor-cable reactance in the Wheatstone bridge of the anemometer complicates the frequency response of the anemometer to velocity fluctuations, a small-sized anemometer with an extremely short sensor-cable (of negligible sensor reactance) was newly designed. A simple electric test was conducted to check the transfer function derived theoretically, and the performance of the anemometer was demonstrated. Excellent agreement on the response was found between the test and theory. In addition, the newly-developed anemometer enabled to eliminate inherent electric noise, which improved the dynamic range significantly.

Keywords: Hot-wire anemometer, Constant-temperature mode, Transfer function, Electric test, Frequency response.

---

1. Introduction

Hot-wire anemometry is one of powerful tools to measure velocity fluctuations in turbulent flows. Dynamic characteristics of constant-temperature anemometer (CTA) have been investigated. When the reactance (capacitance and/or inductance) of the sensor cable between hot-wire and servo-amplifier in the CTA bridge is active, the transfer function of the anemometer to velocity fluctuations constitutes the fifth order equation [1]. Whreas if the cable reactance is negligible, it is represented by a quadratic equation [1].

In the present study, a tiny CTA bridge circuit board was newly designed with a sufficiently short sensor-cable to eliminate the reactance effect and was installed inside a metallic slender box to impede electromagnetic noise. This CTA board was attached at the end of the hot-wire support stem, and thus whole anemometer was able to be immersed in the flow.

2. Experimental Arrangement

A single hot-wire sensor was made of tungsten wire whose diameter and length were 5 μm and 1 mm, respectively. The wire except sensing portion was copper-plated with thickness of about 30 μm. Prevalent CTA curcuit[1)] was employed, and the newly designed circuit board was contained in a metallic box with a cross-section of 14×14 mm^2 and length of 120 mm. Thus, the sensor BNC cable in the hot-wire support stem is only 450 mm, whose reactance is negligibly small. The CTA box has a removable pin connector for the CTA output, DC power supply and external electric signals. For the electric test, the square-wave signal with a frequency of 1 kHz from a function generator, or the white-noise signal, which is generated in a laptop computer, is introduced into the CTA bridge. The CTA system was tested in the desk-top calibration wind tunnel with the core region of 50 mm round jet flow.

The CTA output via a 4th order anti-aliasing Filter with a rolloff frequency of 215 kHz was acquired through an A/D converter with a resolution of 24 bits at a sampling rate of 500 kHz.

3. Theory for CTA Frequency Response

Heating current i of the hot-wire deviated from the mean value in the constant-temperature mode is related with velocity fluctuation u and external electric signal e_t as follows.

$$i = \frac{K(\partial I/\partial U)u - A(1+j\omega M)e_t/(\tilde{R}_W + R_A)}{(K+K\delta+1)\{1+2\zeta(j\omega/\omega_r)+(j\omega/\omega_r)^2\}}, \quad (1)$$

where ζ is the damping coefficient, ω_r is the resonance frequency in the feedback loop, δ is the imbalance parameter, K is the feedback rate, M is the thermal inertia of hot-wire sensor and μ is the first order time delay of the servo amplifier.

If the square wave current is introduced into the CTA bridge, the servo amplifier begins to return the bridge to balance. This transient waveform as shown Eq. (2) is described by means of the Laplace trans-formation of Eq. (1) with the Heaviside step $e_t = 0$ at $t < 0$, then $e_t = 1$ at $t \geq 0$.

$$i(t) = \frac{T_r^2}{\mu M}\left[1 + e^{-\frac{\varsigma t}{T_r}}\left(-\cos\frac{t\sqrt{1-\zeta^2}}{T_r} - \frac{\varsigma}{\sqrt{1-\zeta^2}}\sin\frac{t\sqrt{1-\zeta^2}}{T_r} + \frac{M}{T_r}\frac{1}{\sqrt{1-\zeta^2}}\sin\frac{t\sqrt{1-\varsigma^2}}{T_r}\right)\right],$$
$$T_r = 1/\omega_r \quad (2)$$

3. Results and Comparison with Theory

Fig. 1 shows a transient waveform normalized with the peak value against introduction of the square wave current in comparison with Eq. (2) at 20 m/s, where the curve fitting method gives 0.225 ms, 0.752 and 1.77×10^5 for three parameters M, ζ and. ω_r , respectively. Comparison represents excellent agreement.

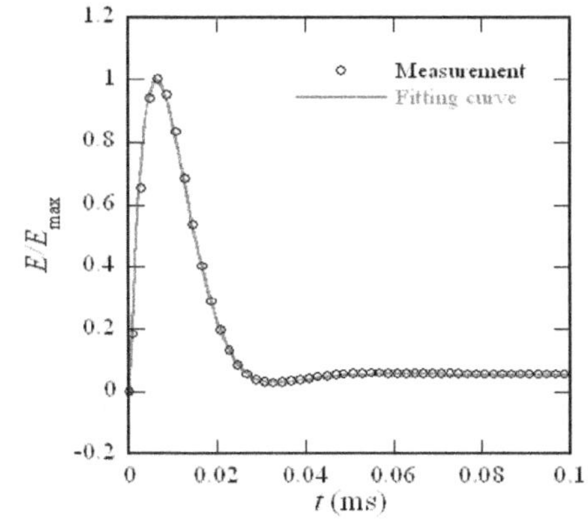

Fig. 1. Response to square wave input, compared to the theoretical curve.

Instead of the square wave, white-noise current was introduced in the bridge. Fig. 2 illustrates power spectrum of the CTA output conducted at 20 m/s. Components at the lower frequency range than 300 Hz are residual turbulence of the wind tunnel, while a large bump in the frequency range exceeding 300 Hz is due to white-noise. The blue curve represents the magnitude of Eq. (1) with u = 0 and three parameters obtained from curve fitting, where the blue curve and spectral component at the bump peak coincides. If we consider the frequency characteristic of the anti-aliasing filter, it is found that the experimental result is consistent with theoretical analysis. Here, the most importance is that introduction of the white-noise yields the ratio of the peak value at the bump to the amplitude at the lower frequency than 300 Hz.

Once the ratio is given as mentioned before, it is possible to eliminate the inherent electric noise from the overall spectral components. A red curve in Fig. 3 depicts the power spectrum of the CTA output at 20 m/s when $e_t = 0$.

A similar bump peaking at 20-30 kHz is noticeable, which is due to electric noise of the servo amplifier.

So, we could estimate the CTA noise characteristic (black curve) using the datasheet opened by servo-amplifier suppliers. The blue curve shows the power spectrum of residual turbulence eliminated the estimated servo amplifier noise, indicating that the mix region at around 1 kHz between electric noise and residual turbulence expands up to 3-4 kHz.

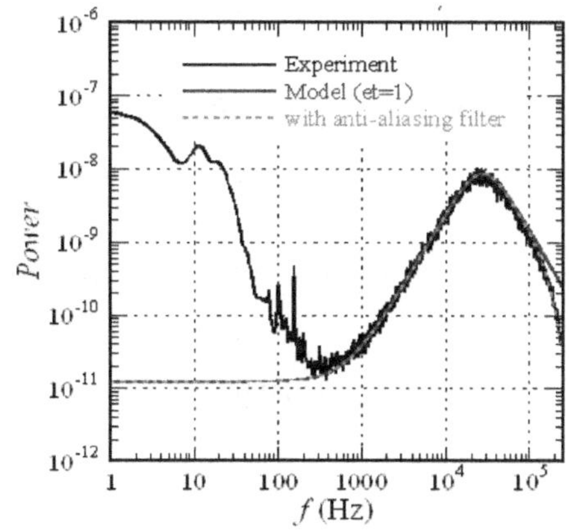

Fig. 2. Power spectrum of CTA bridge output forced by white noise.

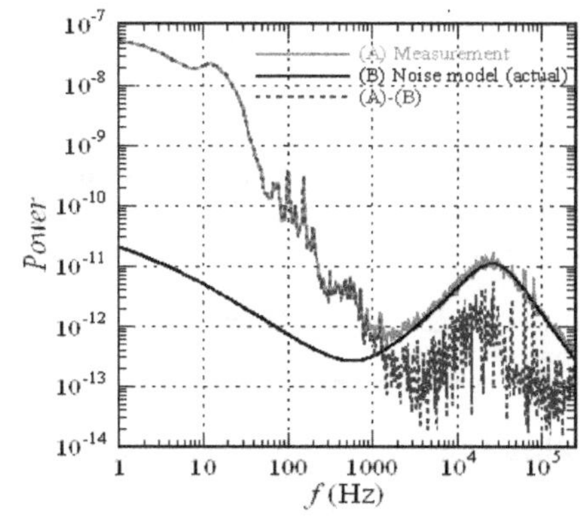

Fig. 3. Power spectra of the CTA bridge with no external signal and of estimated CTA noise.

4. Conclusions

A compact low-noise CTA system, free of the sensor-cable reactance, was newly developed to establish the theoretical model of CTA. Parameters in the model function were successfully determined in a simple manner, and the validity of the model was confirmed by introducing external electric signal to the CTA bridge. The theoretically-described transfer function allowed us to improve significantly the signal-to-noise ratio of the velocity measurements using CTA system.

Acknowledgements

This work was supported by JSPS KAKENHI (Grant Number 17K18939).

References

[1]. N. B. Wood, A method for determination and control of the frequency response of the constant-temperature hot-wire anemometer, *J. Fluid Mech.*, Vol. 67, part 4, 1975, pp. 769-786.

(051)

Mobile Air Quality Monitoring System Using Unmanned Aerial Vehicles & Wireless Sensors Networks

R. M. Camarillo [1], J. Flores [2], J. M. Camarillo [1], E. Hernandez [1], J. A. Ramirez [1], F. J. Castro [1]

[1] Tecnológico Nacional de Mexico, ITLaguna, Blvd. Revolucion & Cuauhtemoc, S/N, Center, 27000, Torreón, Coah. Mexico

[2] Universidad de Guadalajara, CUCEI, Blvd. Gral. Marcelino García Barragán 1421, Olímpica, 44430, Guadalajara, Jal., Mexico

Tel.: + 52 (871) 7051339

E-mail: rosycamarillo01@gmail.com

Summary: In this work, a mobile real-time monitoring system for air quality is proposed. The developed system uses UAV unmanned aerial vehicles and WSN wireless sensor networks, increasing the area of analysis and difficult access. The electronic instrumentation was integrated with an array of sensors, microcontroller, memory and RF modules. The acquired data is stored, processed and transmitted to a computer as a base station through wireless communication, where a developed graphical interface shows the concentrations of polluting gases present in the atmosphere at a minimum height of 10 to 300 m. The weight, size and power consumption of the miniaturized system are 50 g., 50×50×3 cm and 1.5 W respectively. The system has the ability to detect important criteria pollutants such as carbon monoxide (CO), nitrogen dioxide (NO2), nitrogen monoxide (NO), sulfur dioxide (SO_2); and other polluting gases such as Ethanol (C_2H_5OH), Hydrogen (H_2), Ammonia (NH_3) and Propane (C_3H_8).

Keywords: Air quality, WSN, UAV, Criterial pollutans.

1. Introduction

In recent years, deaths caused by respiratory and cardiovascular diseases have increased by 10 %. In October 2018, the World Health Organization (WHO) stated that 93 % of the world's children breathe polluted air every day. According to this information, 1.8 billion children breathe air so polluted that their health and development are in serious danger. The WHO estimates that 600,000 children died in 2016 from acute respiratory infections caused by air pollution. This increase in atmospherically active gases leads to effects such as soil and water acidification [1]. In 2018, the city of Torreón, Mexico, was ranked fourth in 248, as one of the most contaminated areas of ammonia on the planet [2].

In the absence of a national regulatory for the design and installation of air quality monitoring networks, the networks that currently operate in Mexico have different operational criteria and have strong limitations in terms of availability of financial, human and material resources. For this reason, the design of new networks to measure air quality, as well as the redesign of current networks and the expansion of their coverage, are necessary to improve their operation, by homogenizing the criteria of their establishment, operation and maintenance [3].

Previously, the measurement of air quality was carried out by non-continuous methods known as "wet chemistry" for gases and high volumes for suspended particles using complex and very expensive stationary equipment [4] to "collect" data. This paradigm is changing, new low-cost and easy-to-use sensors have been developed, providing high resolution and real-time data that can be "accessed" [5]. These attributes provide opportunities to improve a range of existing air pollution monitoring capabilities and provide avenues for new applications. The range of contaminants to be detected by sensors has been increasing. The appearance of new technologies such as UAV, WSN, the internet of things, advances in computing and communications achieves availability and accessibility of data in real time [5].

A key factor in establishing effective air quality monitoring programs is to assign the optimal location for monitoring stations based on five criteria (population, wind direction and spatial proximity to roads, industries and high traffic areas) that are considered the most important criteria for assigning optimal locations for station monitoring [7]. A large number of developments have been reported to measure environmental parameters present in the air as well as the use of technologies. An investigation was development to measuring the PM10 fugitive dust emission factors for a range of vehicles types and examines the influence of vehicle and wake characteristics on the strength of emissions from an unpaved road. This study was based on the determination of the vertical mass concentration profile, the speed and direction of the ambient wind, and the time it took to generate a trail of dust when passing through instrumented towers [8]. Another work, determined that the contribution of methane CH_4 comes from road traffic and domestic heating, revealing that heating accounts for more than 80 % of the observed CO_2 flows. CH_4 flows expressed as CO_2 equivalents represent approximately 8 % of CO_2 emissions, ranging from 16 % in summer to 4 % in winter and, therefore, cannot be neglected when assessing the impact of cities on the greenhouse effect

[9]. On the other hand, by developing of 3D computational model predicting extent of gas leakage was development based on the building location from Google Earth. It development was based on the concentration differences at the near wall and far wall, and the velocity field change with the leakage time, to provide reliable data for the city gas safe operation and emergency management [10]. Using on a dedicated Unmanned Aerial Vehicle (UAV) was development a new instrument to mapping of NO_2 columns at high spatial resolution allowing to subsample satellite measurements within the extent of a typical ground pixel. The instrument was based on a compact ultra-violet visible spectrometer and a scanner to achieve whiskbroom imaging of the trace gases fields. Including the housing and the electronics [11]. A system online GPRS-Sensors Array was developing integrated a Mobile Data-Acquisition Unit (Mobile-DAQ, based a microcontroller, sensors arrays and GPS) and a fixed Internet-Enabled Pollution Monitoring Server (Pollution-Server and Google Maps to display real-time pollutants), to monitoring air pollutants levels (CO, NO_2, and SO_2) [12]. The development of these devices allow emissions to the atmosphere to be measured, establish permissible limit values, which do not present risks to the health of the population and adverse effects to the environment. Other results of an important investigation, shows a study for 10 years where the presence of ammonia affects air quality with consequences on human health, the environment and climate change. Anthropogenic emissions of ammonia dominate the natural ones and originate mainly from agricultural, domestic and industrial activities. The critical ammonia emission points of the world were identified and categorized and quantified using a high resolution map of atmospheric ammonia observed by IASI satellite. In the study, 248 critical points with diameters less than 50 kilometers were located, associating them with specific agricultural and industrial sources. This will lead to better environmental and health impact assessments of atmospheric ammonia and the implementation of appropriate nitrogen management strategies [13]. One of this critical points was located in Torreon, Coahuila, Mexico which is the fourth place of concentration of amaniaco in the world. This was the reason the engineering academy started this research project. In this work, a mobile real-time monitoring system for air quality is proposed. The developed system uses UAV unmanned aerial vehicles and WSN wireless sensor networks. The system has the ability to detect important criteria pollutants such as carbon monoxide (CO), nitrogen dioxide (NO_2), nitrogen monoxide (NO), sulfur dioxide (SO_2); and other polluting gases such as Ethanol (C_2H_5OH), Hydrogen (H_2), Ammonia (NH_3) and Propane (C_3H_8).

2. Experimental Methodology

The mobile system for air quality monitoring uses emerging technologies such as UAV and WSN. The electronic instrumentation was integrated with high-range microdevices. A small wireless network of 3 nodes was implemented as an application example. Robust MEMS smart sensor nodes have the ability to detect up to 6 pollutant gases and communicate up to a distance of 750 m. between nodes and towards the base station by a PC. A storage system to protect acquired data was integrated, it is shown in Fig. 1. For the acquisition, storage, processing and transmission of data, an electronic control was implemented, which is mounted in each UAV, making the mobile monitoring system an autonomous device. A small UAV (drone) Spark dji was used.

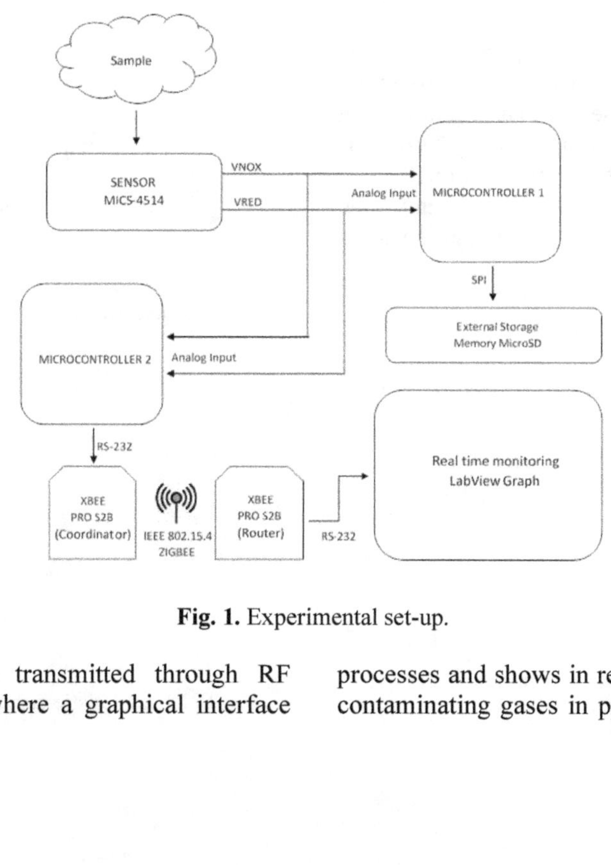

Fig. 1. Experimental set-up.

The acquired data is transmitted through RF modules to a computer, where a graphical interface processes and shows in real time the concentration of contaminating gases in ppm. The graphical interface was programmed in LabVIEW and it is show in Fig. 2. The interface has other characteristics such as regionalization, organization and storage of files and reading of files obtained from memories of the devices.

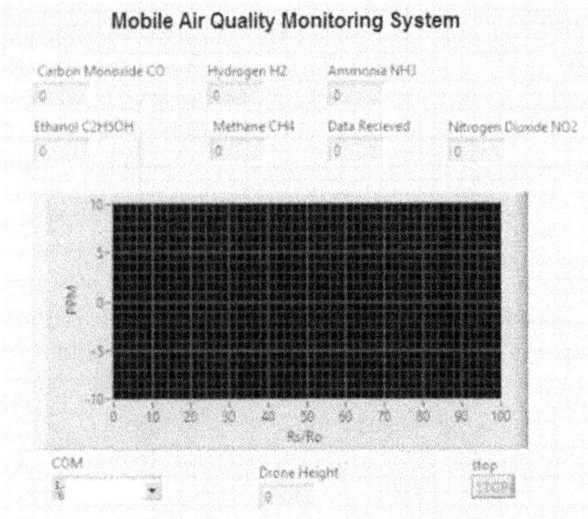

Fig. 2. Graphical interface for Mobile Air Quality Monitoring System.

3. Results and Discussion

By means of a resistance it is possible to obtain a dimensionless relation of resistances that determines the ppm of each element. The concentration of contaminating gases was determined by the relationship:

$$x = \frac{R_S}{R_O}, \tag{1}$$

where x is the gas to be measured in ppm, R_s is the resistance of the sensor in Ω, R_0 the adjustment resistance. R_0 = 22 kΩ for NO_x and R_0 = 45 kΩ for residual gases.

Tests were performed and were performed in the chemistry laboratory under a controlled environment, in order to characterize the sensor. The equation that describes the concentration of NO_2 is $y = 0.1491\,(NO_2) + 0.01$, this is shown in Fig. 3.

Other measurements were made and the characteristic equation was found. The results shown in Figs. 4 to 9.

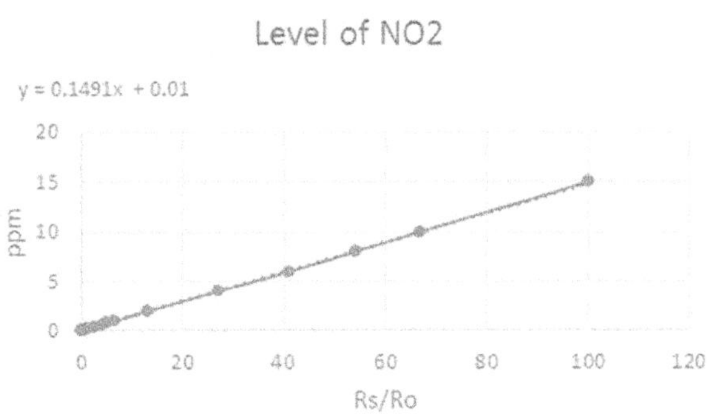

Fig. 3. Characteristic curve for concentrations of NO_2 in the air.

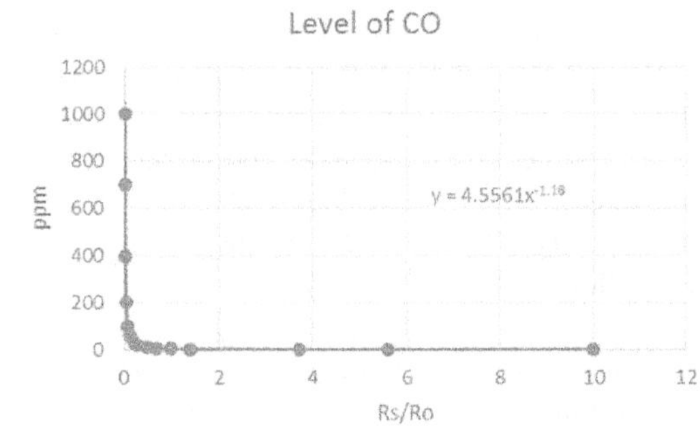

Fig. 4. Characteristic curve for concentrations of CO.

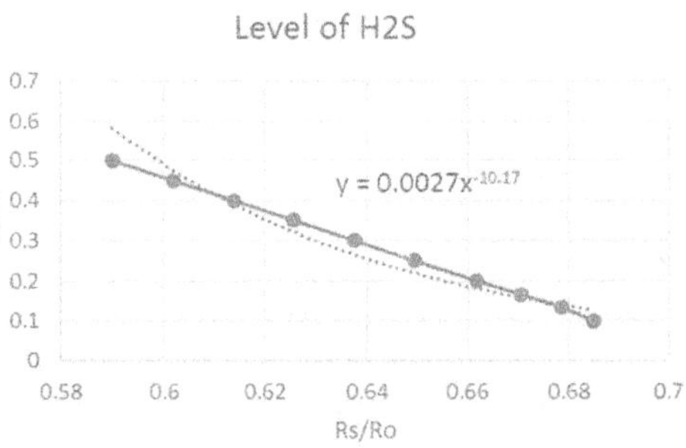

Fig. 5. Characteristic curve for concentrations of H_2S.

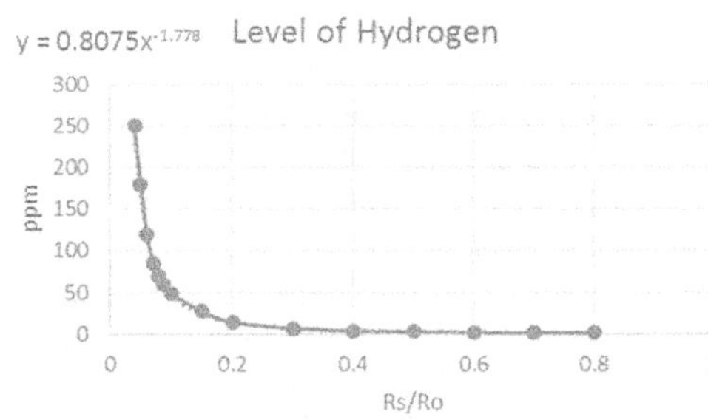

Fig. 6. Characteristic curve for concentrations of Hydrogen.

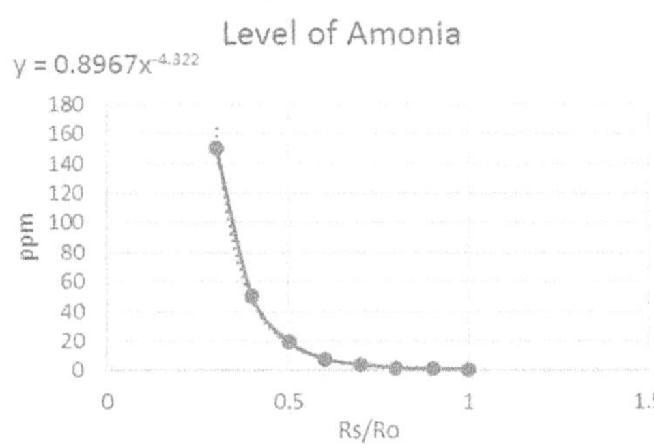

Fig. 7. Characteristic curve for concentrations of Ammonia.

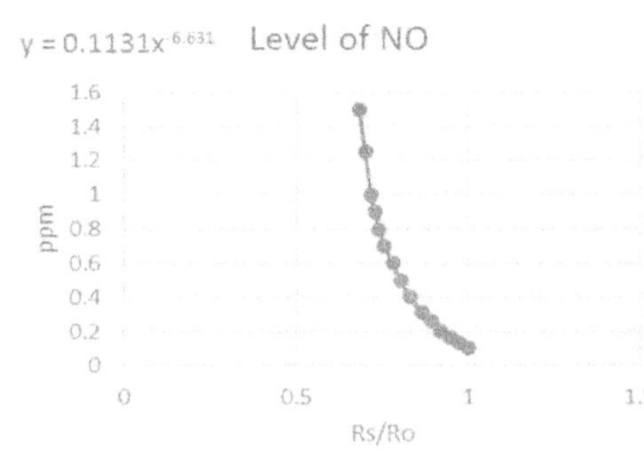

Fig. 8. Characteristic curve for concentrations of NO.

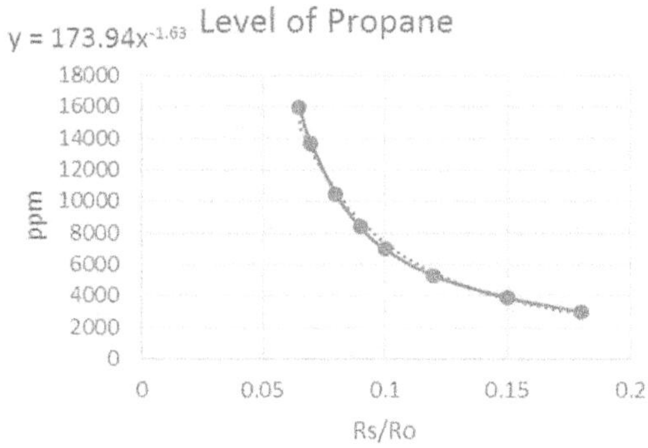

Fig. 9. Characteristic curve for concentrations of Propane.

Environmental tests of air quality were performed in the city of Torreon, Coah. Fig. 10 shown aerial view of city. The system was probed to measure traffic pollution at peak hours and industrial operations at 100 % (13:00 hrs) at a flight height of 15 – 30 meters. The results of this preliminary test shown in Fig. 11.

The results show four critical pollutants including ammonia. We recommended other an exaust probes to made a decisive conclusions about ammonia.

Fig. 10. Aerial view of Torreon, Coah [14].

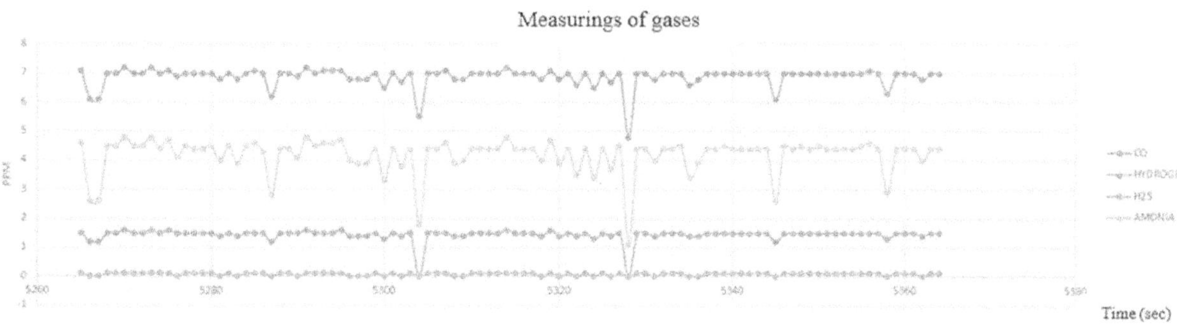

Fig. 11. Measurement of exhaust gases in rush hour in the air.

4. Conclusions

A wireless distributed mobile air pollution monitoring system was designed, implemented and tested using the UAV and WSN. The system measured concentration of pollutant gases such as CO, NO_2, NO, Hydrogen, Ammonia, Propane, and H_2S. The pollution data from various mobile sensor arrays is transmitted to a central server in PC. The data shows the pollutant levels and should be considered by the local air quality regulatory.

The system is based in MICS-4514 sensor and it is focused on the detection of residual gases (RED) and nitrogen gases (NOX). It is mainly dedicated to CO and NO_2, but it is also capable of detecting ammonia, propagation, ethanol, hydrogen and H_2S values with certain limits. A lot of work is required to complete a robust system.

References

[1]. Tecnologia Ambiental, 2017, https://tecnologiaambiental.mx/2017/03/09/aire-puro

[2]. M. Vann Dame, et al., Industrial and agricultural ammonia ponit sources exposed, *Nature,* Vol. 564, 2018, pp. 99-103.

[3]. SINAICA-SEMARNAP-INECC, Manual Redes Estaciones y Equipos de Medicion de la Caliad del Aire, *Instituto Nacional de Ecologia*, Cd. de Mexico, 2016.

[4]. J. C. Chow, Measurement methods to determine compliance with ambient air quality standards for suspended particles, *J. Air Waste Manage Assoc.*, Vol. 45, Issue 5, 1995, pp. 320-382.

[5]. E. G. Snyder, et al., The changing paradigm of air pollution monitoring, *Environmental Sci. & Technol.*, Vol. 47, 2019, pp. 11369-1177.

[6]. R. M. White, et al., Sensors and Apps for community-based atmospheric monitoring, *Air Waste Manage. Ass. Magazine for Envronmental Manager,* Vol. 5, 2012, pp. 36-40.

[7]. M. M. Alsahli, M. Al-Harbi, Allocating optimum sites for air quality monitoring stations using, *Urban Climate*, Vol. 24, 2018, p. 875-886.

[8]. J. A. Gillies, V. Etyemezian, H. Kuhns, D. Nikolic, D. A. Gillette, Effect of vehicle characteristics on unpaved, *Atmospheric Environment*, Vol. 39, 2005, pp. 2341-2347.

[9]. B. Gioli, P. Toscano, E. Lugato, A. Matese, F. Miglietta, A. Zaldei, F. P.Vaccari, Methane and carbon dioxide fluxes and source partitioning in urban areas: The case study of Florence, Italy, *Environmental Pollution*, Vol. 164, 2012, pp. 125-131.

[10]. L. Fei, C. Yan-yan, L. Dong, Simulation of urban gas leakage based on Google Earth, *Procedia Engineering*, Vol. 52, 2013, pp. 220-224.

[11]. A. Merlaud, et al., Small Whiskbroom Imager for atmospherIc compositioN monitorinG (SWING) from an Unmanned Aerial Vehicle (UAV), *Atmospheric Measurement Techniques*, Vol. 11, Issue 1, 2018, pp. 551-567.

[12]. A. R. Al-Ali, I. Zualkernan, F. Aloul, A mobile GPRS-sensors array for air pollution monitoring, *IEEE Sensors Journal*, Vol. 10, Issue 10, 2010, pp. 1666-1671.

[13]. M. Van Damme, L. Clarisse, S. Whitburn, J. Hadji-Lazaro, D. Hurtmans, C. Clerbaux, P.-F. Coheur, Industrial and agricultural ammonia point sources exposed, *Nature*, Vol. 554, 2018, pp. 99-103.

[14]. mapio.net, https://mapio.net/pic/p-1819692

(052)

Real-time Fusion of MEMS Accelerometers/Gyroscopes with Global Navigation Satellite Systems and Road Networks for Enhanced Urban Localization

Hamza Sadruddin [1] **and Mohamed M. Atia** [2]
[1] MASc Student in Department of Electronics, ECE, Carleton University, Canada
[2] Assistant Professor in Department of Systems and Computer Engineering, ECE, Carleton University, Canada
Tel.: 613-520-2600 ex: 5779
E-mails: hamza.sadruddin@carleton.ca, mohamed.atia@carleton.ca

Summary: Fusion of low-cost/low-power MEMS accelerometer/gyroscope sensors with Global Navigation Satellite Systems (GNSSs) is currently used for vehicular localization, internet of things (IoT) tracking and Location-based Services (LBS). However, robust localization in dense urban areas is challenging due to GNSS service interruptions and satellite signal blockage. To overcome this problem, this paper introduces a real-time map-aided MEMS Accelerometer/Gyroscope/GNSS sensor fusion system for enhanced localization in dense urban areas under long GNSS outages. The paper presents an advanced curve-to-curve map-matching algorithms using Hidden-Markov Models (HMM) that fuses, in real-time, an accumulated buffer of Accelerometer/Gyroscope/GNSS sensors measurements using Extended Kalman Filter (EKF). The developed map-aided fusion system was tested on real-road data collected in dense downtown area under long periods of GNSS service interruptions. The results showed robust localization performance under several minutes of GNSS blockage. The developed system is useful for autonomous cars navigation and IoT localization in GNSS-denied areas.

Keywords: GNSS, MEMS Accelerometer/Gyroscope, HMM and EKF

1. Introduction

Global Navigation Satellite Systems (GNSS) are the main localization sensors used in modern electronic systems. GNSS technology have been significantly improved by having more satellites and augmentation systems [1]. However, localization in dense urban areas is still challenging due to GNSS signals multipath or complete outage for extended periods in scenarios like tunnels and/or urban canyons. Fusion of GNSS with dead-reckoning (DR) sensors (e.g. inertial measurement units "IMU") such as low-cost/low-power MEMS accelerometer/gyroscope [1-4] sensors can bridge short GNSS outages. GNSS and accelerometer/gyroscope sensors can be fused using Extended Kalman Filter (EKF). In the fusion process, the localization state transition is triggered by IMU sensors (accelerometer and gyroscopes) while GNSS measurements are used as external observations. The observations in our case is GNSS location and speed measurements. Although the GNSS/IMU fusion provides high-rate robust localization, the localization drifts quickly under GNSS outages leading to significant drifts as illustrated in Fig. 1(a). In addition, under signal reflections and refractions induced by high buildings, GNSS accuracy deteriorate leading to significantly noisy positioning as can be seen in Fig. 1(b). To solve these problems and to provide accurate uninterrupted navigation services under these challenges, the real-time fusion of road map network is proposed. Road maps are currently widely available in modern car systems and high-definition (HD) maps are currently being developed to enable self-driving cars and automated connected vehicles [5]. In contrast to conventional maps that human normally uses to display information, HD maps are different in the level of accuracy, comprehensiveness, resolution and details. HD maps are currently being developed on a large scale as can be seen in [6]. To support automated driving technology where positioning availability in real-time is crucial, this work targets the real-time integration of commercially available maps with GNSS and IMU. While conventional road map network is used in this paper, the same concepts are applicable in HD maps. Hidden Markov Models (HMM) are adopted in this work as it is well suited to perform a curve-to-curve map matching technique. According to state-of-art surveys [7], HMM is known for its accuracy in complex geometry and dense topology road networks.

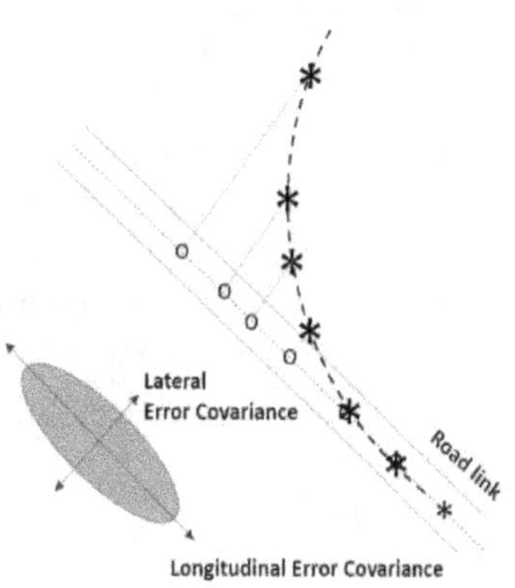

Fig. 1(a). Localization error reset in GNSS outage using Road Networks Maps.

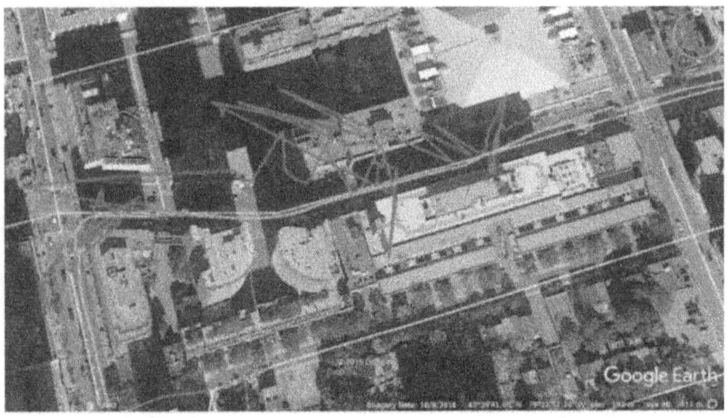

Fig. 1(b). GNSS errors due to multipath in urban areas. Red is GNSS and green is ground-truth solution.

2. INS/GNSS Fusion

Any moving platform can be can be modelled as dynamic system of states vector $x(t)$ that evolves according to the following model:

$$\dot{x}(t) = f(x(t),u(t)) + w(\mathrm{t})\,, \qquad (1)$$

$$y(t) = h(x(t)) + v(\mathrm{t}) \qquad (2)$$

where $f(.)$ is the nonlinear dynamic model, $w(\mathrm{t})$ is the stochastic system noise vector, $u(\mathrm{t})$ is the control signal vector, $y(t)$ is the external measurements vector, $h(.)$ is the nonlinear measurement model and $v(\mathrm{t})$ is a stochastic measurement noise vector. In this work, Kalman filter is adopted due to its real-time efficient performance [1]. As Kalman filter works only under linear systems, the dynamic system above must be linearized. Therefore, first order Taylor series approximation is applied on (1) and (2) to obtain the following linearized system model:

$$\delta\dot{x}(t) = F(t)\delta x(t) + w(t)\,, \qquad (3)$$

$$\delta\dot{y}(t) = H(t)\delta x + \mathrm{v}(t)\,, \qquad (4)$$

$$F(t) = \frac{\partial f(x(t),u(t))}{\partial x}\,, \qquad (5)$$

$$H(t) = \frac{\partial h(x(t))}{\partial x}\,, \qquad (6)$$

The Kalman Filter provides optimal estimation of δx assuming $w(t)$ and $\mathrm{v}(t)$ are zero-mean Gaussian noise vectors with covariance matrices defined by:

$$Q(t) = \left\langle w(t)w(t)^T \right\rangle, \qquad (7)$$

$$R(t) = \left\langle \mathrm{v}(t)\mathrm{v}(t)^T \right\rangle, \qquad (8)$$

and δx is the error vector with zero-mean and a covariance matrix P defined by:

$$P(t) = \left\langle \delta\mathrm{x}(t)\delta\mathrm{x}(t)^T \right\rangle \qquad (9)$$

In discrete form, derivative can be approximated as follows:

$$\delta\dot{x}_k \approx \frac{\delta x_{k+1} - \delta x_k}{T}\,, \qquad (10)$$

where T is the sampling period, equations (3) and (4) can be written in discrete form as follows:

$$\delta x_k = (I + F_k T)\delta x_{k-1} + w_{k-1}\,, \qquad (11)$$

$$\delta y_k = H_k \delta x_k + \mathrm{v}_k \qquad (12)$$

The optimal estimation of the error vector, δx_k given measurements, y_k, is calculated using two steps: prediction

$$x_{k+1} = f(x_k, u_k)\,, \qquad (13)$$

$$P_{k+1} = (I + F_{k+1}T)P_k(I + F_{k+1}T)^T + Q_{k+1}\,, \qquad (14)$$

and update

$$K_{k+1} = P_{k+1}\mathrm{H}_{k+1}^T(\mathrm{H}_{k+1}P_{k+1}\mathrm{H}_{k+1}^T + R_{k+1})^{-1}\,, \qquad (15)$$

$$x_{k+1} = x_{k+1} + K_{k+1}\left[y_{k+1} - h(x_{k+1})\right], \qquad (16)$$

$$P_{k+1} = (I - K_{k+1}H_{k+1})P_{k+1} \qquad (17)$$

3. Enhanced Performance via Map-matching

Localization errors due to GNSS multipath and IMU drifts can be reduced by map-matching and by using the projected position on road map links to reset localization errors as can be seen Fig. 1. This technique is called map-matching (MM)-feedback [2] which greatly enhances the accuracy as seen in Fig. 2. In figure 1, map-matching enhances the error covariance in the lateral direction of the moving object. In Fig. 3, map-matching enhances localization error covariance in both lateral and longitudinal directions of the moving platform. In this work, real-time curve-to-curve map-matching algorithms using Hidden-Markov Models (HMM) is demonstrated.

3. Hidden Markov Models (HMM)

HMMs framework has been widely used to model random processes that are going through a series of hidden states and generating noisy observables. Using state transition probability and state observation (i.e.

emission) probabilities, HMM provides a method to calculate the highest probability sequence of hidden states that interprets the generated observables. In our proposed system, the possible road network map segments were represented as hidden states, whereas the vehicle's location was an observable state. In this case, vehicle's dynamics and road network topology are used to build transition probability matrix while emission probabilities represented the probability that a given state (i.e. road segment) generates an output (vehicle's location). Fig. 3 shows how the problem is modeled as a HMM. Viterbi Algorithm [3] was used to find the sequence of road map segments that most likely to have generated the observed locations. To develop a robust map-matching framework, the vehicle location and orientation history, roads geometry, and map topology constraints were considered in calculating emission and transition probabilities. Fig. 4 shows the system components.

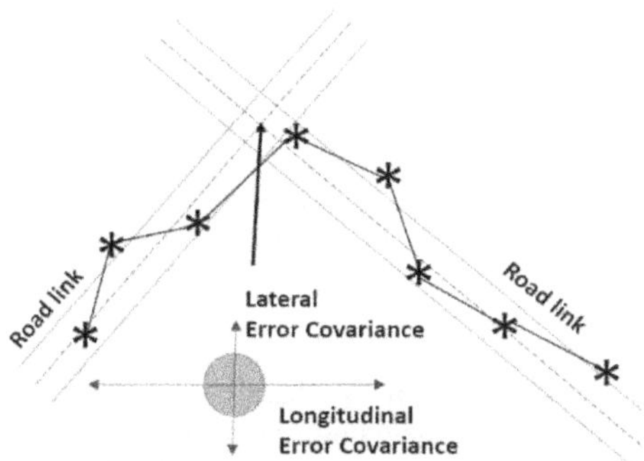

Fig. 2. Map-matching improved accuracy in GNSS outage.

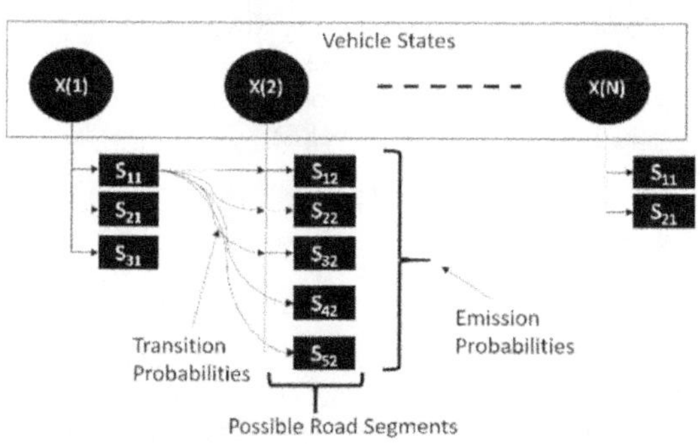

Fig. 3. Map-matching as HMM.

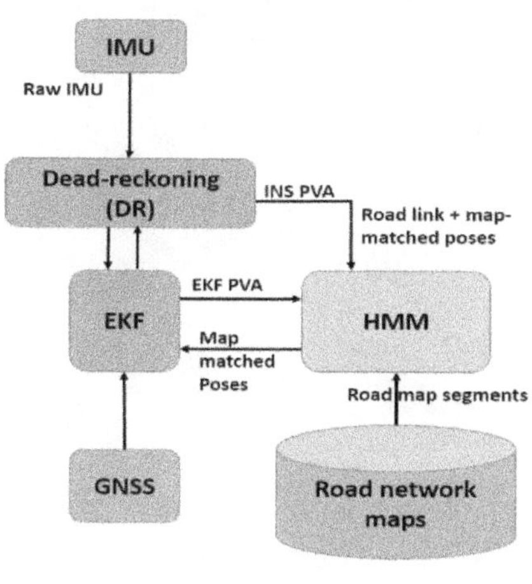

Fig. 4. Overall system block diagram.

4. Experimental Work

The proposed system was realized on the NVIDIA Jetson TX2 embedded system. A complete working logger that can accurately acquire synchronized and time-tagged raw data from both IMU and GNSS has been developed on this platform. Map data access was optimized using the platform storage/retrieval mechanism using cached binary map data. A series of experiments were performed in both GNSS available and denied environments, in order to obtain and validate results. Table 1 shows the specifications of the logger systems used, whereas Fig. 5 shows an experimental car equipped by the developed logger and a ground-truth navigation system ProPak6 from Novatel [8]. All data has been synchronized using the GPS timing.

Table 1. Specifications of the data logger sensors.

Specification	Value
MPU9250 IMU	
Accelerometer technology	MEMS
Accelerometer random noise	0.0294 $(m/s^2/\sqrt{Hz})$
Gyroscope technology	MEMS
Gyroscope noise	0.01 $(°/s/\sqrt{Hz})$
U-blox 8 GNSS Receiver	
Horizontal location accuracy	2.5 m
Velocity accuracy	0.1 m/s

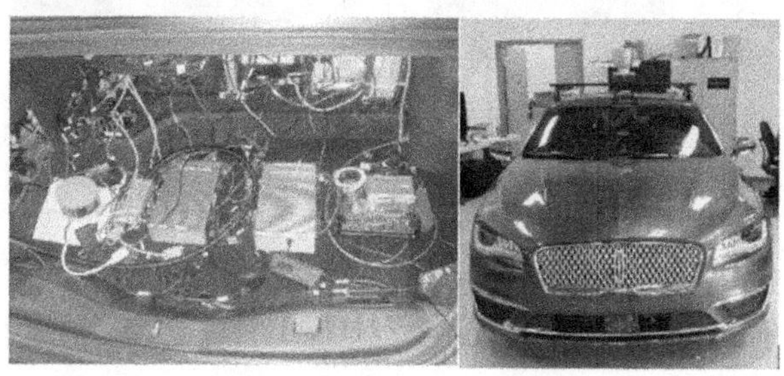

Fig. 5. Experimental car.

5. Results

In this paper, sample of the results in downtown Toronto is demonstrated as can be seen in Fig. 5. Fig. 6 shows the localization error under four twenty-seconds GNSS outages with and without the proposed map-aided fusion system. To test long GNSS outage, a simulated GNSS outage of 20 minutes has been tested. As can be seen from Fig. 7, the system sustained reliable map-matching and enhanced accuracy for the entire 20 minutes with a sub-meter overall localization error.

6. Conclusion

This work introduced an optimized real-time HMM-based map-aided IMU/GNSS sensor fusion system on an embedded platform. HMM has been shown to be a robust and accurate map-matching framework that can support continuous navigation for long periods of GNSS outages. By introducing this proposed real-time implementation of HMM-based map-matching IMU/GNSS fusion, this work will open the door for a new generation of robust and efficient localization systems that will enable several emerging applications such as self-driving cars and internet of things. Although conventional maps have been used in this work, the concept is applicable to emerging HD maps.

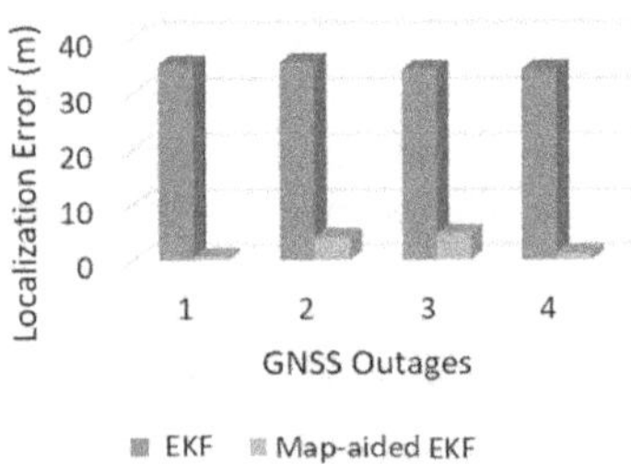

Fig. 6. Map-aided IMU/GNSS fusion performance.

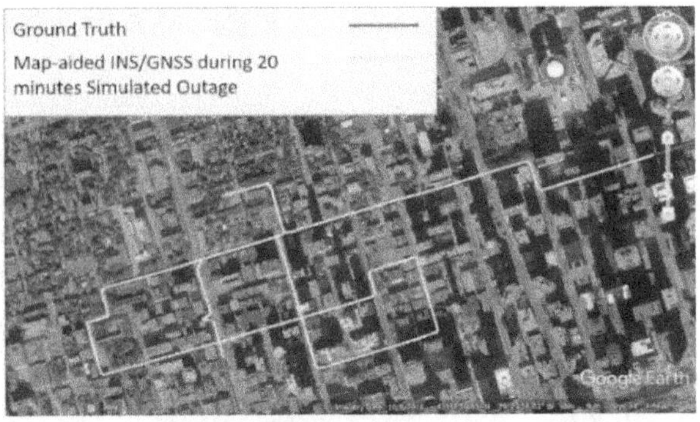

Fig. 7. Improved accuracy in long GNSS outage in Downtown area (20 minutes).

References

[1]. J. Farrell, Aided Navigation GPS with High Rate Sensors, *McGraw-Hill*, New York, 2008.

[2]. M. M. Atia, A. Hilal, C. Stellings, E. Hartwell, J. Toonstra, W. B. Miners, O. Basir, A low-cost lane-determination system using GNSS/IMU fusion and HMM-based multistage map matching, *IEEE Transactions on Intelligent Transportation Systems,* Vol. 18, Issue 11, Nov. 2017, pp. 3027-3037.

[3]. P. Newson, J. Krumm, Hidden Markov map matching through noise and sparseness, in *Proceedings of the 17th ACM SIGSPATIAL International Conference on Advances in Geographic Information Systems (GIS'09)*, Seattle, Washington, 2009, pp. 336-343.

[4]. A. Cismas, I. Matei, V. Ciobanu, G. Casu, Crash detection using IMU sensors, in *Proceedings of the 21st International Conference on Control Systems and Computer Science (CSCS'17)*, Bucharest, Romania, May 2017, pp. 672-676.

[5]. H. G. Seif, X. Hu, Autonomous driving in the iCity – HD maps as a key challenge of the automotive industry, *Engineering*, Vol. 2016, Issue 2, 2016, pp. 159-162.

[6]. S. Niijima, J. Nitta, Y. Sasaki, H. Mizoguchi, Generating 3D fundamental map by large-scale SLAM and graph-based optimization focused on road center line, in *Proceedings of the 26th IEEE International Symposium on Robot and Human Interactive Communication (RO-MAN'17)*, Lisbon, Portugal, 2017, pp. 1188-1193.

[7]. R. Manikandan, R. Latha, A literature survey of existing map matching algorithms for navigation technology, *International Journal of Engineering Sciences & Research Technology*, Vol. 6, Issue 9, September 2017, pp. 326-331.

[8]. NovAtel Inc., Calgary-based, NovAtel ProPak6 Triple-frequency GNSS receiver, https://www.novatel.com/products/gnss-receivers/enclosures/propak6/

(053)

Possibility for Temporal Observation of Thrombus Generated in Extracorporeal Circulator Circuit by Photoacoustic Imaging Using LED

Takahiro Wabe [1], Ryo Suzuki [2], Kazuo Maruyama [2], Yasutaka Uchida [1]
[1] Teikyo University of Science, 2-2-1 Senjyusakuagi, Adati-ku, Tokyo, Japan
[2] Faculty of Pharma-Science, Teikyo University, 2-11-1 kaga,itabasi-ku,Tokyo, Japan
Tel.: + 8169101010, fax: + 8169103800
E-mail: g18na001@st.ntu.ac.jp

Summary: Blood coagulation is measured by using a pressure sensor in a blood circuit, but it is not quick responsive because it is detected by pressure rise caused by coagulation. In this study, we have investigated a method to detect blood clotting at an early stage using photoacoustic imaging, which is thought to be more sensitive. The LED with a wavelength of 850 nm was used as a photoacoustic light source. An ultrasonic wave generated by thermal expansion of mouse blood sealed in a microtube was observed, and also many ripples were observed with time and the coagulation of blood progressed. It was also observed that the waveform considered to correspond to coagulation of blood broadens with time. It was found from the above that there is a possibility that the state of blood clotting can be observed from outside the circuit of the extracorporeal circulation device by using the LED as a light source.

Keywords: Blood coagulation, Photoacoustic imaging, LED, The circuit of the extracorporeal circulation device, Observation with time.

1. Introduction

If the patient undergo extracorporeal blood circulation by methods such as purification therapy, percutaneous cardiopulmonary support, granular leukocyte apheresis, leukapheresis, or auxiliary artificial heart, blood comes into contact with foreign matter of various artificial origins. It is well known that blood contact with foreign matter cause the phenomina such as the distruction of blood cell, coagulation fibrinolysiss system enhancement, or complementary activities, so called thronbus formation(Thrombus occur) [1-3]. Various attempts have been made to prevent blood clotting, and prevention methods using anticoagulants (i.e., heparin) are now mainstream [4, 5]. However, because heparin also acts as a foreign body, it is difficult to completely prevent blood from clotting on the surface of the circuit or in the dialyzer.

In extant extracorporeal circulation circuits, a pressure sensor is used for coagulation detection. This sensor reacts only when blood clotting causes clogging in the circuit or the blood becomes difficult to flow and the pressure. If the response is delayed, the blood will continue to clot and the patient will be at risk. In addition, when blood coagulation was observed, the extracorporeal circulation circuit is changed to a new circuit. However, due to the irregular response, the burden on medical workers is also very large. Furthermore, the clogged circuit has the problem of discarding the patient's blood, which should have been returned to the patient again. If the progress of coagulation in the circuit can be detected in advance, it is possible to carry out an early response to circuit blockage by reverse calculation. Therefore, sensing and details of formation process of blood coagulation in very early stage are required.

Some reports have already been published that measure blood coagulation using photoacoustic imaging technique by using laser. The measurement of the process of thrombus growth by blood coagulation has not been reported yet [6-9].

A solid-state laser is only used as a light source of conventional photoacoustic imaging. The conventional laser device is not suitable for the light source of this study for weight and size, considering that it is necessary to attach to a blood circuit regularly to measure the change of the state of the blood appropriately.

The solid-state laser has a volume ratio of 15000 times, power consumption of 1000 times or more, and a cost ratio of 10 times or more compared to LEDs. As a disadvantage, the output of the LED is inferior compared to the laser.

Then, photoacoustic imaging using LED or LD as a light source is suitable for that purpose.

Considering the LD as a light source, the LD can specify the wavelength and has a high energy density as compared with the LED. However, since a resonator structure is required, it becomes expensive. From these things, we think that LED is suitable for this experiment. In this paper, In order to measure the increase of blood coagulation in the circuit of the extracorporeal blood circulation system, as a preliminary study, photoacoustics are used to measure changes in blood clotting that occur in microtubes.

2. Methods

At the measurement point in the blood purification circuit, the flow velocity is measured at the slowest chamber in the circuit. In photoacoustic imaging, if the flow velocity is too high, it will be difficult to measure, and errors may occur in the measurement results. The chamber has a filter, and the air flow rate is deliberately once relaxed due to the air trap. Fig. 1 shows.

In order to reproduce the environment of the chamber where the flow rate is fairly gentle, this time, blood was put in a microtube and it was sought to visualize the coagulation change process.

Fig. 1 shows the schematic of the experimental equipment used in this study. For the measurement of this experiment, the possibility of measurement of change of blood coagulation is explored using PreXionLED AcousticX (Cyberdyne, Inc.) that meets various conditions. In an extracorporeal circulation device, it is necessary to measure a portion of blood where the flow is slower, such as in the chamber of the dialysis circuit at the blood coagulation measurement site. However, in this experiment, that was not performed. Instead, the microtube was used. After extracting blood from a mouse, it was enclosed in a microtube, and the temporal change of the blood coagulation was observed through transparent plastic. There was approximately 1.5 mL of blood in the microtube, which contained no air. Additionally, the microtube and two LEDs in the light irradiation section were submerged in the water sandwiching the ultrasonic probe of the receiving part. Two LED arrays were attached to both sides of the conventional ultrasonic probe, fixed at approximately 40° with a jig. The distance between the micro tube and the ultrasonic probe with the LED was about 1 cm. The schematic diagram of the photoacoustic experiment is shown in Fig. 2.

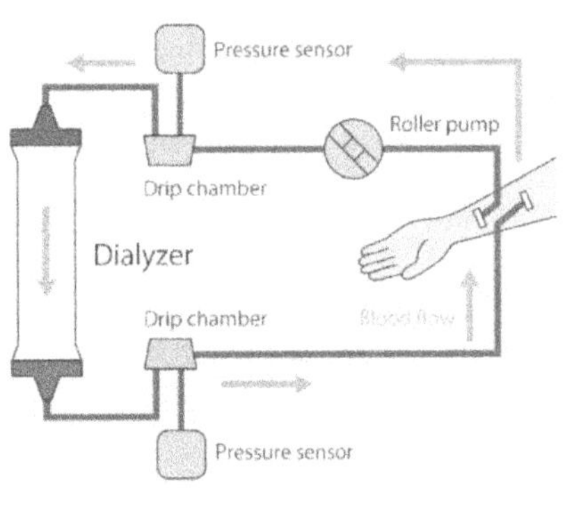

Fig. 1. Flow in the blood purification circuit.

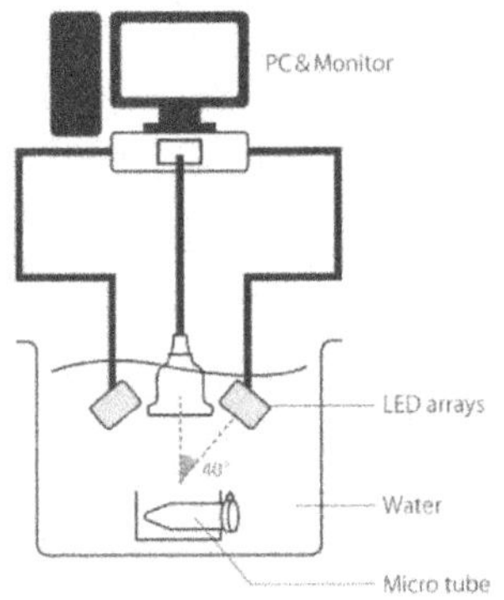

Fig. 2. Schematic of experimental apparatus.

Regarding the light absorption of the liquid and the coagulation blood, we referred to existing data [10]. The light energy was approximately 200 μJ/ pulse. There were two types of ultrasonic probes: 10 MHz and 7 MHz. However, at 10 MHz, the distance resolution improved somewhat, but the depth sensitivity increased 4 to 5 times at 7 MHz. Thus, the 7-MHz ultrasonic probe frequency was chosen.

3. Result

As shown in Fig. 3, we measured the vicinity of the wall surface of the microtube where blood coagulation causes foreign body reaction. The thick black line shown in the imaging unit in Fig. 3 is at the side of the microtube. The lower side is the blood side and the upper side is the underwater side with an arrow showing the boundary.

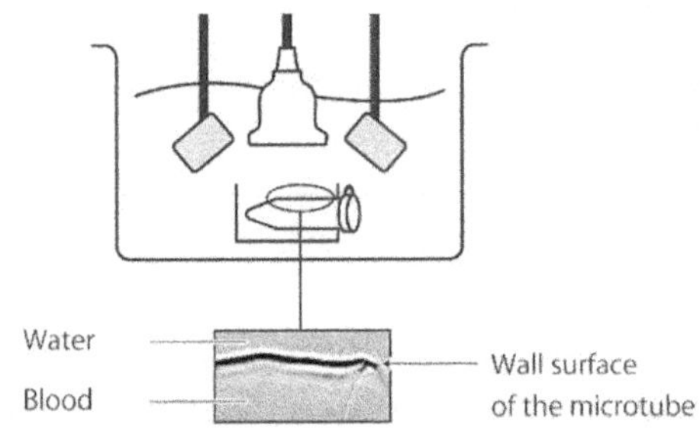

Fig. 3. Correlation of experimental apparatus schematic diagram and imaging chart.

Changes over time in the captured image are shown in Fig. 4. Measurements were performed at four timings immediately after blood injection: blood injection, 2, 5, and 10 min. Figs. 4(a), (b), (c), and (d) correspond to respective pictures. Fig. 4 is rotated 90° from the actual state shown in Fig. 3 for the convenience of subsequent graphing. The microtube shown in Fig. 4 is vertically oriented so that the wall surface is on the left side. As time elapses, the number of white parabolic streaks increased on the blood side of the microtube. The white part (high brightness) of the image represents the elastic wave from the absorber obtaining light energy.

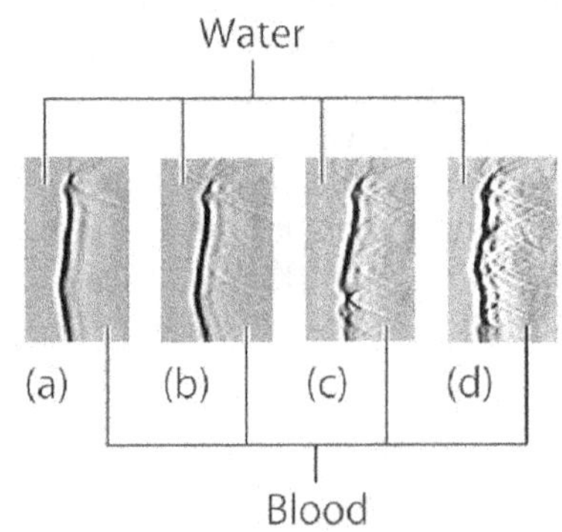

Fig. 4. Photograph by photoacoustic imaging: (a) blood injection; (b) 2 min; (c) 5 min; and (d) 10 min.

The pixel luminance of Fig. 4 is shown in Fig. 5 as a 3-dimensional (3D) gray-scale graph by "imagej" (free software). [11, 12] The x-, y-, and z-axes of the graph are shown as follows:

- x axis: the horizontal distance within the selected range;
- y axis: vertical distance within the selected range;
- z axis: value obtained by calculating the pixel luminance.

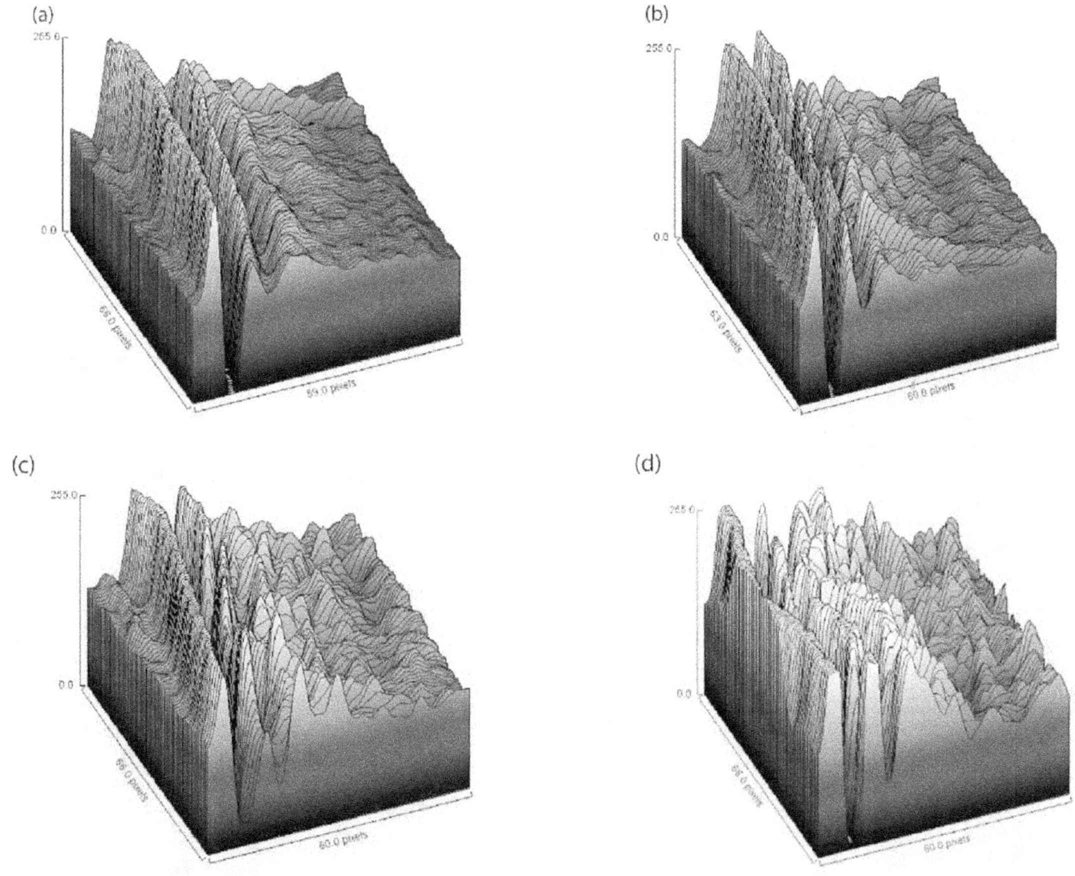

Fig. 5. 3D diagram of luminance part by imagej: (a) blood injection; (b) 2 min; (c) 5 min, and (d) 10 min.

Next, to inhibit coagulation under the same conditions as above, after heparin was mixed, filling the microtube, the image in Fig. 6(a), shows the blood measured photoacoustically, and Fig. 6(b) shows the 3D distribution.

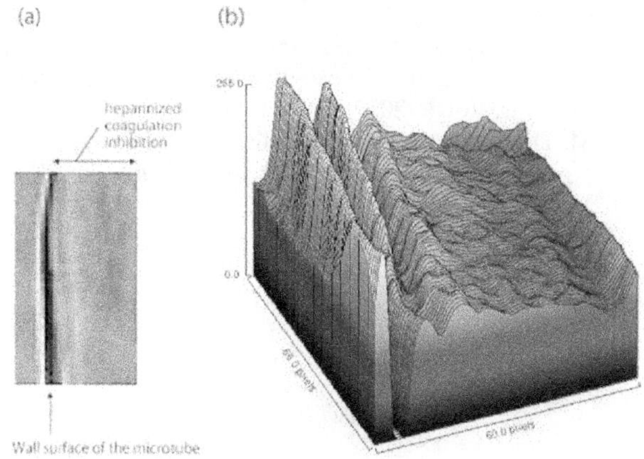

Fig. 6. Photoacoustic imaging of heparinized coagulation inhibition: (a) photograph; (b) imagej 3D diagram of luminance.

Fig. 7 shows an image upon which smoothing was performed (Fig. 5). An instantaneous elastic wave change caused by light energy was not observed. However, the influence of the elastic wave inside the micro tube was. To compare changes over time, immediately after the start of solidification (left figure) and 10 min after solidification start (right drawing) are shown.

4. Discussion

In this experiment, changes in blood coagulation through transparent plastics were measured by photoacoustic tools. However, as shown in Fig. 4, ripples are represented by white streaks, increasing with the passage of time. It can be seen that ripples were generated from the center of the microtube toward the walls. Changes over time occurred in the blood on the right side, but no changes over time occurred in the water outside of microtube. Additionally, in all measurements, because the light energy value irradiated to the microtube did not change, the elastic wave increased, meaning that the state of the blood changed and that the light energy increased over time. The arrow portion in Fig. 5 is the wall surface of the microtube. However, the luminance reaction uniformly increases over time in the blood part of the microtube. As shown in Fig. 6, the elastic wave images caused by coagulation inhibition by heparin are shown, whereas previous figures do not show such a reaction. These changes in photoacoustics over time do not occur in blood where coagulation does not occur. From Fig. 7, it can be seen that the elastic waves are generated inside the microtube as a whole, and that they grow with the time.

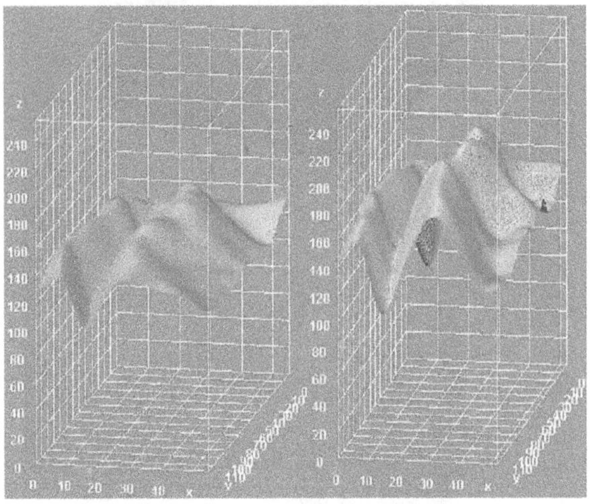

Fig. 7. 3D diagram of luminance part smoothed by imagej.

5. Conclusion

We confirmed time-course changes of coagulated blood using photoacoustic imaging by comparing heparinized blood and coagulated blood with water. The elastic waves measured in this study changed with time based on blood coagulation. Additionally, rather than measuring each individual blood coagulum, coagulation in the microtube was measured using photoacoustic imaging as a change in the entire microtube. Thus, basic examination of blood coagulation, assuming an extracorporeal circulation circuit in the presence of blood flow, is possible. The optimization of wavelength energy and ultrasonic probe frequency in blood coagulation through blood circuits should be studied further in the future.

Acknowledgments

T. Wabe and Y. Uchida would like to thank Mr. N. Sato of Cyberdyne, Inc. Research and Development Dept., for his cooperation in this experiment.

We have no conflicts of interest relationship with any companies or commercial organizations.

References

[1]. J. Utley, Pathophysiology of cardiopulmonary bypass, *Car. Surg.*, Vol. 5, 1990, pp. 177-189.

[2]. Y. Mori, Blood compatible material, *Kobunshi Ronbunshu*, Vol. 42, 1985, pp. 601-615 (in Japanese).

[3]. J. K. Kirklin, A. D. Pacifico, Complement and the damaging effects of cardiopulmonary bypass, *J. Thorac. Cardiovasc. Surg.*, Vol. 86, 1983, 845.

[4]. L. Gott, J. D. Whiffen, R. C. Dutton, Heparin bonding on colloidal graphite surfaces, *Science*, Vol. 142, 1963, 1297.

[5]. Salyer, I. O. at al., New blood compatible polymers for artificial heart applications, in Medical Applications of Plastics (Biomedical Materials Symposium No. I), *Interscience*, New York, 1971.

[6]. H. Zhong, T. Duan, H. Lan, M. Zhou, F. Gao, Review of low-cost photoacoustic sensing and imaging based on laser diode and light-emitting diode, *Sensors*, Vol. 18, 2018, 2264.

[7]. T. J. Allen, P. C. Beard, High power visible light emitting diodes as pulsed excitation sources for biomedical photoacoustics, *Biomedical Optics Express*, Vol. 7, Issue 4, 2016, pp. 1260-1270.

[8]. A. B. Karpiouk, S. R. Aglyamov, S. Mallidi, J. Shah, W. G. Scott, J. M. Rubin, S. Y. Emelianov, Combined ultrasound and photoacoustic imaging to detect and stage deep vein thrombosis: phantom and ex vivo studies, *J. Biomed. Opt.*, Vol. 13, Issue 5, 2008, 054061.

[9]. B. Li, C. Fu, G. Ma, Q. Fan, Y. Yao, Photoacoustic Imaging: A novel tool for detecting carotid artery thrombosis in mice, *J. Vasc. Res.*, Vol. 54, 2017, pp. 217-225.

[10]. R. J. Talbert, S. H. Holan, J. A. Viator, Photoacoustic discrimination of viable and thermally coagulated blood using a two-wavelength method for burn injury monitoring, *Phys. Med. Biol.*, Vol. 52, 2007, pp. 1815-1829.

[11]. C. A. Schneider, W. S. Rasband, K. W. Eliceiri, NIH Image to ImageJ: 25 years of image analysis, *Nature Methods*, Vol. 9, 2012, pp. 671-675.

[12]. M. D. Abramoff, P. J. Magalhaes, S. J. Ram, Image processing with ImageJ, *Biophotonics International*, Vol. 11, Issue 7, 2004, pp. 36-42.

(055)

Direct Sensing of Time-varying Displacement in Nanopositioners by Piezoelectric Ceramics Transducers

A. Bazaei [1,2], M. Boudaoud [2], M. H. Ettefagh [3], Z. Chen [1] and S. Régnier [2]
[1] University of Newcastle, School of Electrical Engineering and Computing, Callaghan, NSW 2308, Australia
[2] Sorbonne University, Institute for Intelligent Systems and Robotics (ISIR), Paris, France
[3] Huazhong University of Science and Technology, School of Artificial Intelligence and Automation, Wuhan, China
Tel.: +61 240238982
E-mail: Ali.Bazaei@newcastle.edu.au

Summary: We address characterization of piezoelectric (PZT) ceramics for direct measurement of stage displacement in piezo-driven nanopositioners. The sensor is compact and cheap, needs no signal conditioning electronics when operating within the sensing bandwidth, and follows time-varying large signal profile of stage displacement accurately. The proposed sensor has a 3σ-resolution of 1.9 nm and its accuracy by a constant scaling factor is less than 27 nm in the sensing bandwidth.

Keywords: Displacement sensor, Piezoelectric sensor, Nanopositioning stage, Time-varying displacement, Scanning.

1. Introduction

High precision measurement of rapidly changing small displacement is required in many micro and nano positioning mechanisms such as nanopositioners for scanning probe microscopy [1-11], microscanner mirrors [12-14], and scanners in selective LASER sintering applications [15-17].

PZT sensors are small in size, low noise, cost-efficient, and fast in response [18]. The voltage induced by external stress in a piezoelectric ceramic is used to measure pressure, acceleration, temperature, strain, and force or to harvest energy [19, 20]. However, it has not been employed for direct measurement of stage displacement in nanopositioners, mainly due to drift and decrease of sensor response at low frequencies from its RC high-pass characteristic. If the displacement varies rapidly such as in scanning tasks, the high-pass decrease issue may not affect the induced voltage. This work reports on large signal characterization of piezoelectric ceramic response for direct measurement of displacement in nanopositioning stages, which have many applications in nanometrology such as scanning probe microscopy.

In [8, 9], stress induced voltage in piezo ceramics were used for vibration damping of nanopositioners, without being used or characterized for direct displacement measurement. In [21], piezo patches were used as a sensor-actuator pair to measure the target displacement. This method, however, requires many additional components, e.g. cantilever, oscillator circuit, and a frequency measurement equipment. In addition, the additional mechanical components and their dynamics can impose a highly negative effect on the displacement sensing bandwidth.

In this paper, we address characterization of a commercial PZT chip as a direct displacement sensor in a nanopositioner. The chip is a multilayer PZT ceramic which is normally used as an actuator.

2. Piezo-Driven Small Nanopositioner with Piezo Cermaic Sensors

Fig. 1 shows a schematic view of the piezo-driven x-y nanopositioner loaded with additional ceramic elements to measure stage displacement along both horizontal axes. Table 1 reports some specifications of the piezo stack actuators and the piezo chip sensors. To eliminate cross-talk between the sensor-actuator pair on each axis, we have separated them by inactive ceramic insulators whose dimensions are the same as the PZT sensor. To calibrate the PZT sensors, the middle stage is extruded 2 mm for direct measurement of displacement by the laser inerferometric vibrometer described in [22]. Voltage amplifiers with gain of 20 are used to drive the piezo stacks.

Table 1. Some specifications of piezo stacks and chips.

Physical Property, (V)	PZT Stack	PZT Chip
Drive Voltage Range	0-150 V	0-150 V
Free Stroke. at 150 V	5.2 μm ± 15 %	2.3 μm ± 15 %
Recommended Preload	100N	100N
Blocking Force at 150 V	250N	250N
Dimensions in mm	2.5×2.5×5.0	2.5×2.5×2.3
Electric Capacitance	100 nF ± 15 %	50 nF ± 15 %
Resonance Frequency	250 kHz	650 kHz
Operating Temperature	-25-130 °C	-25-130 °C
Curie Temperature	230 °C	230 °C

3. Frequency Responses of y-axis

Fig. 2 shows frequency response of stage displacement to the amplifier input, measured by the laser. The dominant resonance mode and 3-dB bandwidth of stage displacement are approximately 17 kHz and 10 kHz, respectively. The measured frequency response from the amplifier input to the PZT sensor voltage exhibits low and high 3-dB corner frequency values at 32 Hz and 6 kHz, respectively, as indicated in Fig. 2.

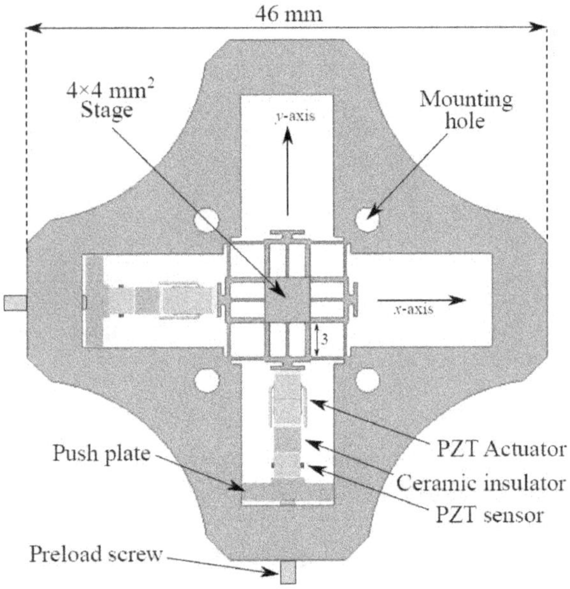

Fig. 1. Schematic view of the nanopositioner.

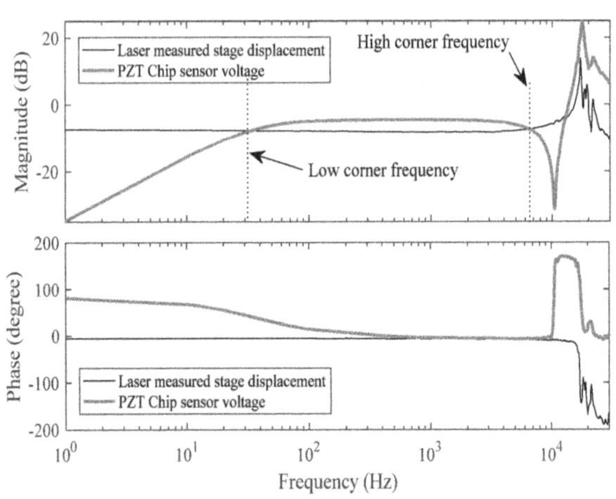

Fig. 2. Frequency responses of y-axis.

The PZT sensor output has a pair of lightly damped complex zeros at 10.5 kHz, which do not appear in the stage displacement response. These zeros create a sharp notch in the frequency response of the PZT chip and set an upper limit on the sensor bandwidth. A modal analysis of the nanopositioner structure reveals that the unloaded nanopositioner has a dominant normal mode along the horizontal axes, as depicted in Fig. 3, at almost the same frequency of the zero pair. At this actuation frequency, the interaction force between the PZT actuator and the stage is considerably reduced as the stage requires nearly zero external force to vibrate at its unloaded normal mode. Since the PZT sensor is primarily responsive to the interaction force, it exhibits such dominant imaginary zeros that does not appear in the actual stage displacement. The existence of such zeros in the interaction force can also be predicted by a mass-spring model as presented in [8].

As the actuation frequency decreases towards zero, the magnitude response of the PZT sensors decrease by 20 dB per decade with a phase lead that tends to 90 degrees. This indicates a single zero at the origin, which is consistent with the response of a first-order high-pass filter made by series combination of a capacitor and a resistor. The total resistance of instruments driven by a PZT sensor during the experiment was measured to be $R = 88\,\text{k}\Omega$. The measured capacitance of the PZT sensors after preloading is around $C = 60$ nF. Hence, the 3-dB corner frequency of the high-pass filter is $(2\pi RC)^{-1} = 30.14\,\text{Hz}$, which is very close to the measured low corner frequencies in Fig. 2 around 32 Hz. The high 3-dB corner frequency of the PZT sensors due to the dominant imaginary zeros is around 6400 Hz. Hence, the frequency range in which the sensor can be used for direct measurement of stage displacement is [32, 6400] Hz. Note that the low corner frequency can further be reduced if the PZT sensor is connected to a load with higher input resistance.

4. Characterization of Piezoelectric Sensor

To calibrate the PZT sensor on the y-axis, we applied a 500 Hz sinusoidal voltage on the PZT stack that swings between 0 and 140 V. Fig. 4(a) shows simultaneously recorded signals from the laser, PZT chip sensor, and the amplifier input. The latter signals are augmented and linearly scaled to have the same extrema as the stage displacement. The PZT sensor voltage closely follows the stage displacement with a sensitivity of 1.377 V/μm. The measurement error shown in Fig. 4(b) indicates minor vibrations around 17 kHz, which is due to excitation of the resonance mode by minor non-smooth profile of sinewave generator, as indicated by the close-up view in Fig. 4(a).

The close overlap of hysteresis curves in Fig. 5 indicates that the PZT sensor precisely follows large deviations of the displacement profile from the almost ideal sinusoidal profile of the actuation input voltage.

Fig. 6 shows results of a similar test at a different frequency but using the scaling factor of 0.726 μm/V for the PZT sensor, which was obtained from the foregoing 500 Hz sinusoidal test. The small measurement error in this validation test reveals high performance of the PZT displacement sensor within the sensing bandwidth.

We further examined the PZT sensor at more frequency values within the sensor bandwidth by large signal sinusoidal excitations. To characterize the sensor accuracy, Table 2 and Fig. 7 report the average absolute error and deviation range of the PZT sensor response, using the foregoing constant scaling factor, respectively. A major contribution to the error is the excitation of resonant mode of the open-loop nanopositioner, which is far beyond the high corner frequency of the sensor. Hence, the actual sensor error within the bandwidth is expected to be much smaller.

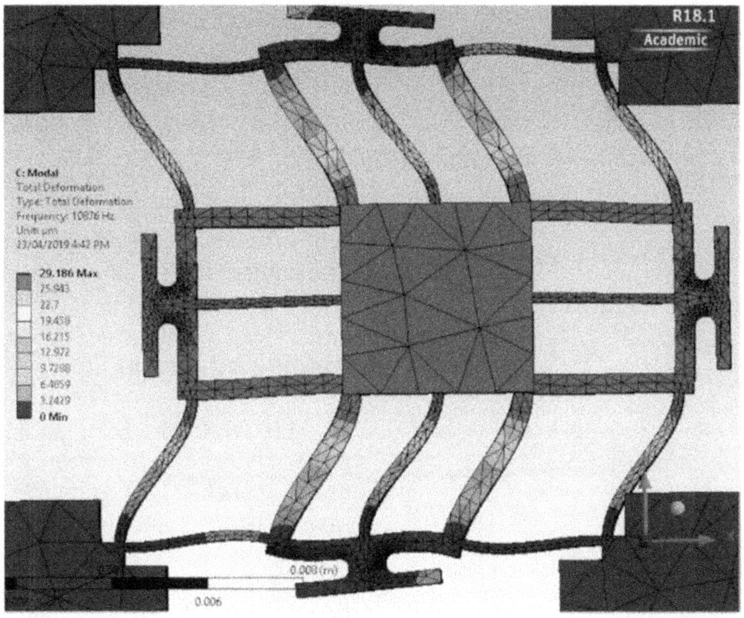

Fig. 3. Magnified mode shape of the monolithic structure, indicating the total deformation pattern along x-axis during the first free vibration mode at natural frequency of 10.88 kHz, whilst the mounting screw holes are fixed and no PZT ceramic is installed in the nanopositioner. The data was predicted by a finite-element modal analysis.

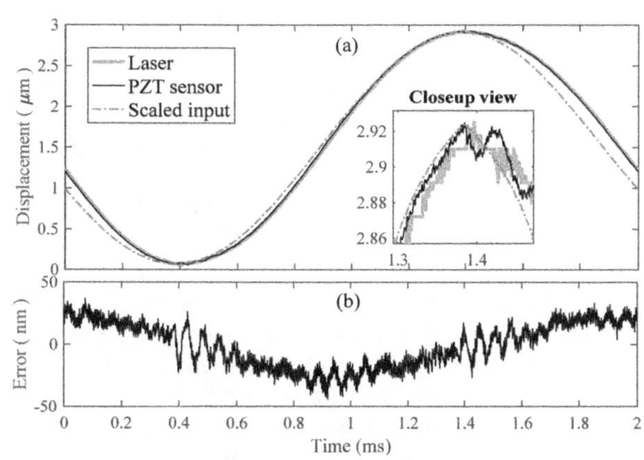

Fig. 4. Calibration test by 500 Hz sinusoidal excitation.

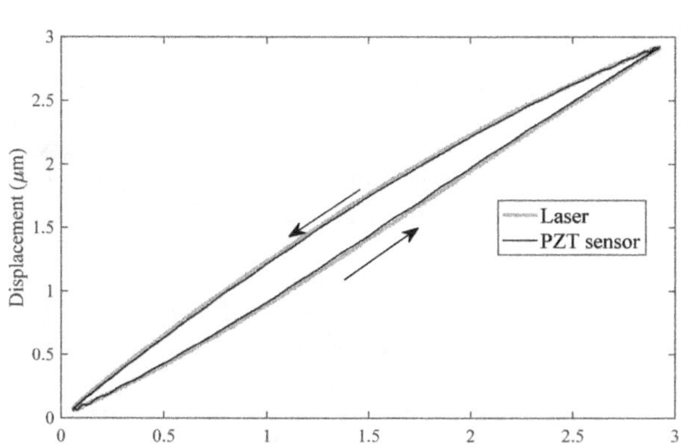

Fig. 5. Hysteresis curves from Laser and PZT sensors.

Table 2. Accuracy of the PZT sensor of y-axis with the constant scaling factor of 0.726 μm/V obtained from the 500 Hz calibration test.

Freq. (Hz)	100	500	786	1000	1583	2178	3204
Error. (nm)	68.1	17.8	19.3	26.9	25.7	48	70

5. Resolution of Piezoelectric Sensor

Resolution of an analog sensor is usually determined from the standard deviation of sensor noise [10, 23]. To characterize the sensor noise, the PZT sensor voltage of y-axis was directly recorded without any filter by a sampling rate of 50 kHz while maintaining the 70 V bias voltage on the y-axis PZT actuator. Using the previously reported scaled factor, the resulting time-domain profile, histogram, and spectral density of the PZT sensor noise are as shown in Fig. 8. The time-domain data indicates that the sensor has a 3σ-resolution of 1.9 nm, which is calculated from the 0.63 nm standard deviation of the noise in a 25 kHz bandwidth. The spectral density of the PZT sensor has an average value of 3.5 pm/√Hz, which is useful to approximate the sensor resolution within a desired bandwidth [18]. Hence, the 3σ-resolution of the PZT sensor with an effective bandwidth of Δf Hz is around $10.5\sqrt{\Delta f}$ pm. The measured value for 3σ-resolution of the PZT sensor noise over a 25 kHz bandwidth is 1.9 nm.

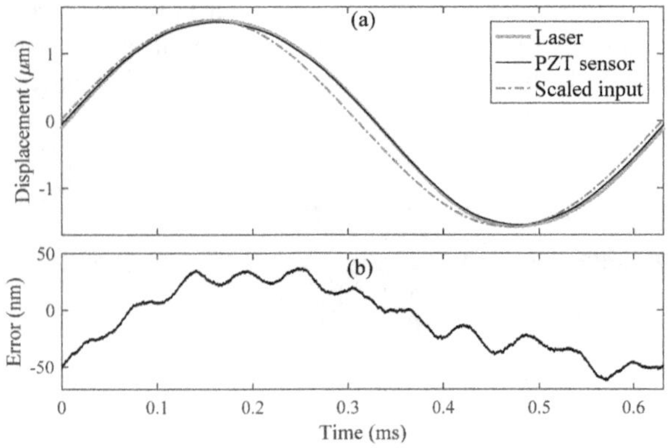

Fig. 6. Validation test at 1583 Hz sinusoidal excitation.

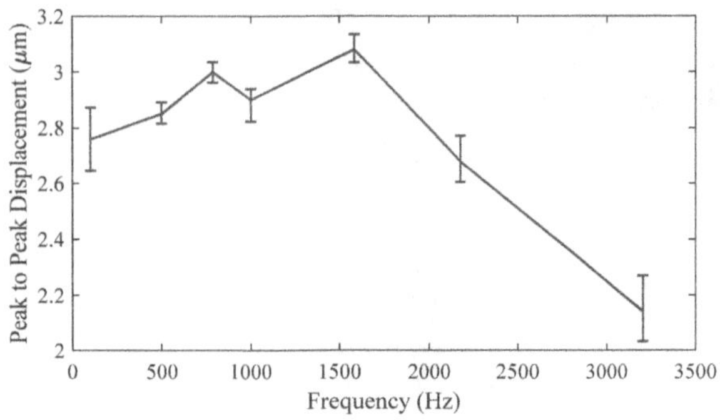

Fig. 7. (a) Evaluation of PZT sensor performance by large signal sinusoidal excitation of y-axis at different frequency values, using a constant calibration coefficient within the sensing bandwidth. The vertical axis represents the peak-to-peak displacement of the stage. The ranges of PZT sensor measurement are indicated by error bars.

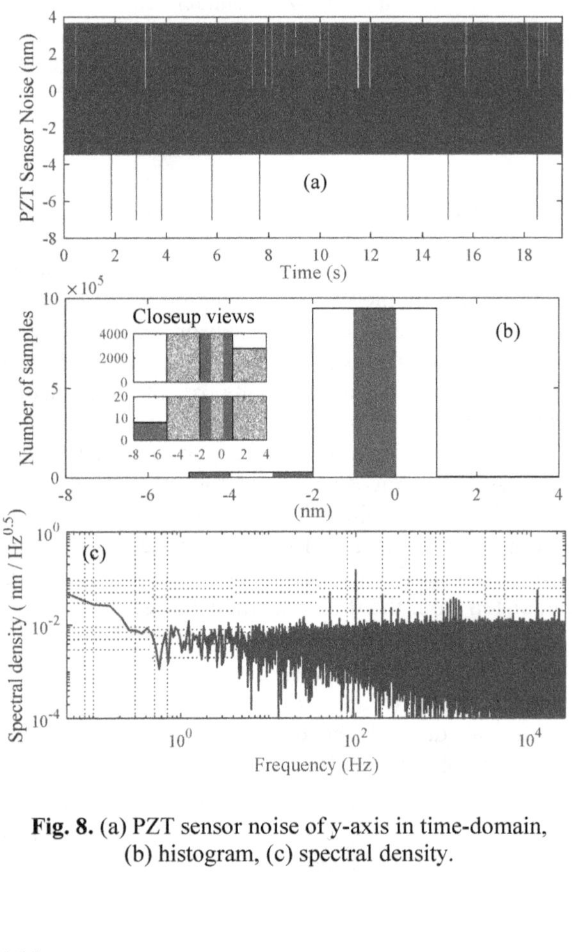

Fig. 8. (a) PZT sensor noise of y-axis in time-domain, (b) histogram, (c) spectral density.

6. Conclusions

Piezoelectric ceramics can be used for real-time measurement of time-varying displacement of nano-positioning stages. The PZT sensor exhibits minimum-phase zeros close to the imaginary axis at the dominant natural frequency of the unloaded nano-positioner. Such zeros do not exist in the response of the stage displacement and limit the high corner frequency of the PZT sensor. However, the sensor has benefits, such as low cost, small noise level, small foot-print, and no required signal conditioning instruments, which makes them an ideal displacement sensor for time-varying applications such as scanning probe microscopy.

Acknowledgement

This work has been partially sponsored by the project PolyREM (Emergence Sorbonne Université), the French government research program Investissements d'avenir through the Robotex Equipment of Excellence (ANR-10-EQPX-44), and the University of Newcastle Australia. We would like to thank Dr. Yuen K. Yong and Prof. Andrew J. Fleming for their valuable advices.

References

[1]. S. S. Aphale, A. J. Fleming, S. O. R. Moheimani, High speed nano-scale positioning using a piezoelectric tube actuator with active shunt control, *IET Micro & Nano Letters*, Vol. 2, 2007, pp. 9-12.

[2]. S. Polit, J. Dong, Development of a highbandwidth XY nanopositioning stage for high-rate Micro/Nanomanufacturing, *IEEE/ASME Transactions on Mechatronics*, Vol. 16, Aug. 2011, pp. 724-733.

[3]. C.-X. Li, G.-Y. Gu, M.-J. Yang, L.-M. Zhu, Design, analysis and testing of a parallel-kinematic high-bandwidth XY nanopositioning stage, *Review of Scientific Instruments*, Vol. 84, Issue 12, 2013, 125111.

[4]. A. San-Millan, V. Feliu-Batlle, S. S. Aphale, Application of a fractional order integral resonant control to increase the achievable bandwidth of a nanopositioner, in *Proceedings of the 20th IFAC World Congress*, Vol. 50, 2017, pp. 14539-14544.

[5]. M. Altaher, S. S. Aphale, High-precision control of a piezodriven nanopositioner using fuzzy logic controllers, *Computers*, Vol. 7, 2018, Issue 1, 10.

[6]. H. Huang, H. Zhao, Z. Yang, Z. Fan, S. Wan, C. Shi, Z. Ma, Design and analysis of a compact precision positioning platform integrating strain gauges and the piezoactuator, *Sensors*, Vol. 12, July 2012, pp. 9697-9710.

[7]. E. Guliyev, B. E. Volland, Y. Sarov, T. Ivanov, M. Klukowski, E. Manske, I. W. Rangelow, Quasi-monolithic integration of silicon-MEMS with piezoelectric actuators for high-speed non-contact atomic force microscopy, *Measurement Science and Technology*, Vol. 23, Issue 7, 2012, 074012.

[8]. A. J. Fleming, K. K. Leang, Integrated strain and force feedback for high performance control of piezoelectric

actuators, *Sensors and Actuators A*, Vol. 161, Issue 1-2, 2010, pp. 256-265.

[9]. Y. K. Yong, A. J. Fleming, High-speed vertical positioning stage with integrated dual-sensor arrangement, *Sensors and Actuators A: Physical*, Vol. 248, Sep. 2016, pp. 184-192.

[10]. C. Lee, S. M. Salapaka, Robust broadband nanopositioning: fundamental trade-offs, analysis, and design in a two degree-of-freedom control framework, *Nanotechnology*, Vol. 20, Dec. 2008, 035501.

[11]. Y. K. Yong, S. O. R. Moheimani, B. J. Kenton, K. K. Leang, Invited review article: High-speed flexure-guided nanopositioning: Mechanical design and control issues, *Review of Scientific Instruments*, Vol. 83, Dec. 2012, 121101.

[12]. K. H. Koh, T. Kobayashi, J. Xie, A. Yu, C. Lee, Novel piezoelectric actuation mechanism for a gimbal-less mirror in 2D raster scanning applications, *Journal of Micromechanics and Microengineering*, Vol. 21, June 2011, 075001.

[13]. T. Aimono, Optical Deflector Including Piezoelectric Sensor Incorporated into Outermost Piezoelectric Cantilever, US Patent 13/846,028, *USA*, Nov. 2013.

[14]. Y. Park, S. Moon, J. Lee, K. Kim, S.-J. Lee, J.-H. Lee, Gimbal-less two-axis electromagnetic microscanner with twist mechanism, *Micromachines*, Vol. 9, May 2018, 219.

[15]. K. An, S. Hong, S. Han, H. Lee, J. Yeo, S. H. Ko, Selective sintering of metal nanoparticle ink for maskless fabrication of an electrode micropattern using a spatially modulated laser beam by a digital micromirror device, *ACS Applied Materials & Interfaces*, Vol. 6, Feb. 2014, pp. 2786-2790.

[16]. K. Bakshi, A. V. Mulay, A review on selective laser sintering: A rapid prototyping technology, *IOSR Journal of Mechanical and Civil Engineering*, Vol. 04, Mar. 2016, pp. 53-57.

[17]. W. Ameen, A. M. Ghaleb, M. Alatefi, H. Alkhalefah, A. Alahmari, An overview of selective laser sintering and melting research using bibliometric indicators, *Virtual and Physical Prototyping*, Vol. 13, July 2018, pp. 282-291.

[18]. A. J. Fleming, A review of nanometer resolution position sensors: operation and performance, *Sensors and Actuators A: Physical*, Vol. 190, Feb. 2013, pp. 106-126.

[19]. G. Gautschi, Piezoelectric Sensorics: Force, Strain, Pressure, Acceleration and Acoustic Emission Sensors, Materials and Amplifiers, *Springer-Verlag*, 2002.

[20]. J. Tichý, et al., Fundamentals of Piezoelectric Sensorics: Mechanical, Dielectric, and Thermodynamics Properties of Piezoelectric Materials, *Springer-Verlag*, 2010.

[21]. M. Umapathy, Piezoelectric based resonance displacement sensor, in *Proceedings of the IEEE Sensors Applications Symposium (SAS'13)*, Galveston, TX, USA, 2013, pp. 193-196.

[22]. M. Boudaoud, Y. Le Gorrec, Y. Haddab, P. Lutz, Gain scheduling control of a nonlinear electrostatic microgripper: design by an eigenstructure assignment with an observer based structure, *IEEE Transactions on Control Systems Technology*, Vol. 23, Issue 4, 2015, pp. 1255-1267.

[23]. I. Cemer, S. Miller, K. Foust, Noise Measurement, *FierceEectronics*, 2011, https://www.fierceelectronics.com/embedded/noise-measurement

(056)

Arrays of Gas Sensing Elements Based on Cerium Oxide-tungsten Oxide Core-shell Single-nanowires

O. Chmela [1], **J. Sadílek** [1], **G. Domènech-Gil** [3,4], **A. Romano-Rodriguez** [3,4], **J. Hubálek** [1] **and S. Vallejos** [1,2]

[1] CEITEC Central European Institute of Technology, Brno University of Technology, Brno, Czech Republic
[2] Instituto de Microelectrónica de Barcelona (IMB-CNM, CSIC), Barcelona, Spain
[3] MIND-Department of Electronics, Universitat de Barcelona (UB), Barcelona, Spain
[4] Institute of Nanoscience and Nanotechnology (IN2UB), Universitat de Barcelona (UB), Barcelona, Spain
Tel.: + 420-5-4114-6153
E-mail: vargas@feec.vutbr.cz or stella.vallejos@imb-cnm.csic.es

Summary: In this work, we present the fabrication and characterization of various parallel-connected single-nanowire gas sensing elements. Each sensing element contains a single cerium oxide-tungsten oxide core-shell nanowire interconnected across a pair of faced nanoelectrodes. The system has been tested in resistive mode towards various concentrations of nitrogen dioxide between 0.25 and 2.5 ppm demonstrating stable and reproducible responses.

Keywords: Gas sensor, Single-nanowire, Tungsten oxide.

1. Introduction

Previous research on single-nanowire sensors and particularly on arrays of single-nanowire sensors demonstrated better efficiency of these systems compared to thin film systems [1]. In addition, gas sensitive nanowires modified superficially with second-phase constituents (e.g., nanoparticle decoration or core-shell structures) have proved generally better sensing properties [2].

In the literature, there are various examples of single-wire based chemical sensors. However, there are only a few reports related to systems based on several individual wires integrated into arrays (for instance, on systems of two electrodes at micro [3] or nano [4] scale) and even less on individual surface modified wires integrated into arrays, despite the enhanced sensing performance reported by these composite structures [2]. In this context, here we present the fabrication of arrays of parallel-connected single-nanowires (based on tungsten oxide modified with cerium oxide) and their gas sensing performance.

2. Fabrication and Processing Steps

The fabrication of arrays of single-wire gas sensors was performed using a combination of standard optical lithographic techniques and electron beam lithography (EBL), as well as dielectrophoresis to align the wires. The system consists of four independent arrays with five or nine faced nanoelectrodes each provided of a third electrode buried to allow the integration of different gas-sensitive nanowires and their operation both in resistive or field-field effect transistor mode (Fig. 1). Tungsten oxide nanowires modified with cerium oxide nanoparticles were achieved using aerosol-assisted chemical vapor deposition process as described previously [5].

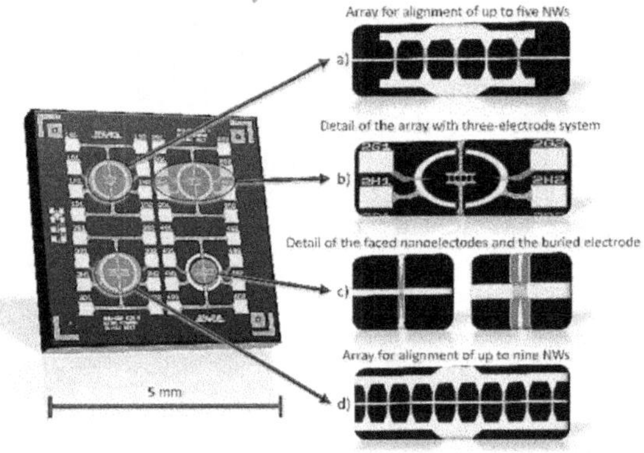

Fig. 1. View of the chip with four independent arrays showing a detailed view of an array for alignment of up to five nanowires (a), the interconnection of the three-electrode system (b, c) and an array for alignment of up to nine nanowires (d).

3. Results

Fig. 2 shows SEM images of the two faced nanoelectrode and the buried third electrode before (Fig. 2a) and after (Fig. 2b) the integration of the tungsten oxide wires modified with cerium oxide via dielectrophoresis process. The fabricated structures allow the integration of only a single-wire in between the faced nanoelectrode circumventing the agglomeration of structures.

Gas sensing characterization of these systems in resistive mode towards various concentration (0.25, 1 and 2.5 ppm) of nitrogen dioxide (NO_2) demonstrated the functionality of the array with the highest responses at an operating temperature of 200 °C (Fig. 3).

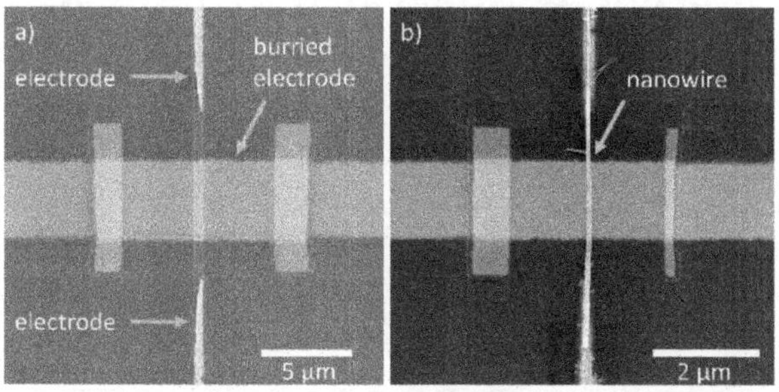

Fig. 2. SEM images of a single system within the array before (a) and after (b) integration of the cerium oxide-tungsten oxide core-shell nanowire.

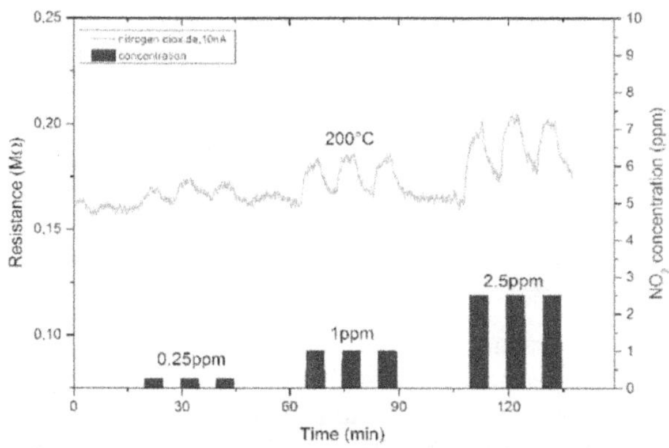

Fig. 3. Sensor response towards various concentrations (0.25, 1, and 2.5 ppm) of nitrogen dioxide at 200 °C.

Further tests of the current dependence of the sensor response proved direct proportionality, registering higher sensor responses for higher current as well as better signal-to-noise ratio (Fig. 4). Generally, the sensor response showed a reproducible behavior with the electrical resistance increasing (Fig. 3) when the sensors were exposed to nitrogen dioxide. This is consistent with other n-type metal oxides reported in the literature including tungsten oxide. These results validate the fabrication process of these systems and pave the way for the test of these structures in field-effect transistor mode.

4. Conclusions

Gas sensing elements based on parallel-connected single-nanowires consisting of tungsten oxide modified with cerium oxide were fabricated. Tests of these elements upon low concentrations of nitrogen dioxide validated the functionality of these systems. Further gas tests and characterization of the elements will be presented at the conference.

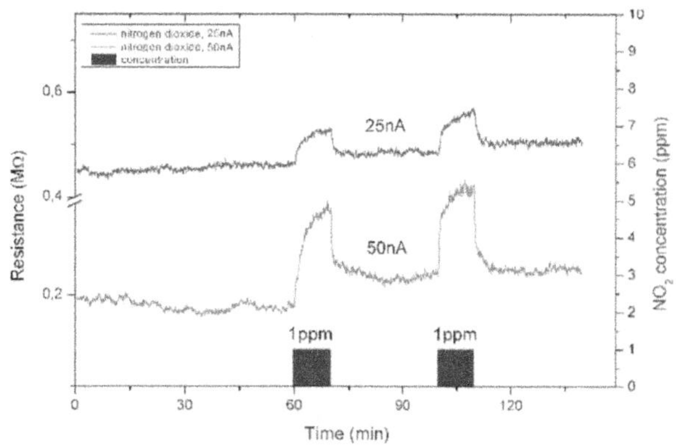

Fig. 4. Sensor response towards 1 ppm of nitrogen dioxide with different currents applied to the system.

Acknowledgements

This work has been supported by the Czech Science Foundation (GAČR) via Grant no. 17-16531S.

References

[1]. E. Brunet, T. Maier, G. C. Mutinati, S. Steinhauer, A. Köck, C. Gspan, W. Grogger, Comparison of the gas sensing performance of SnO_2 thin film and SnO_2 nanowire sensors, *Sensors and Actuators B: Chemical,* Vol. 165, 2012, pp. 110-118.

[2]. D. R. Miller, S. A. Akbar, P. A. Morris, Nanoscale metal oxide-based heterojunctions for gas sensing: A review, *Sensors and Actuators B: Chemical,* Vol. 204, 2014, pp. 250-272.

[3]. J. Guilera, C. Fàbrega, O. Casals, F. Hernández-Ramírez, S. Wang, S. Mathur, F. Udrea, A. De Luca, S. Z. Ali, A. Romano-Rodríguez, J. D. Prades, J. R. Morante, Facile integration of ordered nanowires in functional devices, *Sensors and Actuators B: Chemical*, Vol. 221, 2015, pp. 104-112.

[4]. O. Chmela, J. Sadílek, G. Domènech-Gil, J. Samà, J. Somer, R. Mohan, A. Romano-Rodriguez, J. Hubálek S. Vallejos, Selectively arranged single-wire based nanosensor array systems for gas monitoring, *Nanoscale,* Vol. 10, 2018, pp. 9087-9096.

[5]. M. Tomić, M. Šetka, O. Chmela, I. Gràcia, E. Figueras, C. Cané, S. Vallejos, Cerium oxide-tungsten oxide core-shell nanowire-based microsensors sensitive to acetone, *MDPI Biosensors,* Vol. 8, 2018, 116.

(057)

Extending Bee Hive Health State Monitoring by Integrated Acoustical Sensing and Machine Learning

A. König [1]

[1] TU Kaiserslautern, Institute of Integrated Sensor Systems, Erwin Schrödinger Strasse 12, 67663 Kaiserslautern, Germany

Tel.: + 496312053696, fax: + 496312053889

E-mail: koenig@eit.uni-kl.de

Summary: The importance of insects, and honey bees in particular, for our ecosystem is undisputed. Currently, environmental problems from pesticides to parasites endanger the well being or even the existence of honey bee colonies and insects in general. This imposes an increasing load on skills and activity of hive keepers. Sensors, instrumentation, and machine learning offer solutions to instrument bee hives and, thus, allow to more easily and with less intrusion to monitor bee hives. Like in other industrial disciplines, e.g., Industry 4.0, operation can move from scheduled to event driven activity. The development in MEMS and IoT field in general allows to achieve affordable integrated monitoring solutions. The paper presents, in addition to standard temperature, moisture, and weight monitoring, an approach of acoustical in-hive monitoring with automated decision making and notification implemented in-hive in a SmartComb. Several relevant colony states have been shown to be acoustically detectable by experiments on an instrumented bee hive based on an economically feasible approach. In the next step, the number of discernible states and of involved colonies will be increased.

Keywords: Bee Health monitoring, MEMS microphones, Machine Learning, Industry 4.0.

1. Introduction

The systemic relevance of insects, wild bees, and honey bees in particular, is a well-known fact and found its place in public awareness. Numerous threats from pesticides to parasites, *varroa destructor* in the case of honey bees, threaten these essential contributors to our ecosystem. One consequence is the need for manual pollination, as already exercised in Asia, or the intentions to automate the natural pollination service by bee-like drones, a most recent patent activity to be observed was by Walmart [24]. Activities like the 'More than Honey' documentary of M. Imhoof et al. or the book of Tautz [5] made the role and critical situation of bees and the implications for human life lucid to the public.

Due to these increasing challenges, hive keeping gets increasingly difficult and requires more and more skills and attention from the hive keepers. Traditionally, hive tending works on a scheduled and *post mortem* basis with numerous opening and inspection of the hives, which is invasive, disturbing for the bees and weakening their capabilities to resist the above named increasing challenges. The development in sensor & actuator integration technology, MEMS [7], in particular, Machine Learning, and general automation, e.g., Industry 4.0, which also makes bee drones like [24] possible, allows to introduce capable yet affordable hive instrumentation solutions to closely monitor and manage an increasing number of hives and conduct less invasive event-driven tending of bees with minimized openings at the right time. "Digital" bee keeping moves from scheduled and *post mortem* to predictive and event-driven operation, thus reducing effort, cost, and risk. Major bee research institutions, e.g., in Germany Hohenheim, Veitshöcheim etc. [28-30] have dealt with related issues and goals employing in numerous cases rather costly and lab-size equipment. The ongoing vivid advance in sensor, actuator, electronics, and RF integration technology [7] provides an unprecedented leverage for well performing, yet cheap and disappearing application systems that seem to make Berkeley's Smart-Dust vision getting reality. Relevant examples are IR cameras, which become now available both miniaturized and at dramatically reduced cost, e.g., as an extension to Apple I-Phone or the FLIR Lepton One [7]. MEMS microphones are another perfect example of this development [7]. The surging development of Technical Cognition Systems, comprised of integrated sensors, electronics, RF, energy harvesting, and Machine Learning/AI, offers a leverage and an answer for many issues met in bee keeping. Two major lines of activities can be observed, i.e., *Bee Hive Monitoring* by sensors, embedded systems and custom software, and *Hive Keeper Assistance Systems* by discrete appliances for a group of hives in an apiary. One emerging example is given for varroa screening in an accompanying paper [25].

Here, we will focus on the *Bee Hive Monitoring* with special attention to both affordable and disappearing instrumentation. As in all technical disciplines, the potential market size will determine the feasible technology for the application domain. For instance in [14] bee keeping statistics for the EU are provided, naming 14 M hives in the EU. As professional hive keepers typically have about 100 to 500 hives, price sensitive yet suitable and reliable solutions are required. Market size and cost limitation clearly advocate the use of off-the-shelf components, following up technology driving application fields, e.g., automative, communications (smart phones), and automation/ Industry 4.0.

Numerous activities can be found in the last ten to fifteen years of teams of hive keepers and engineers, complementing the activities of major bee research institutes by practical instrumentation work in the field, e.g., [4]. This vivid spin-off and maker scene mostly focused on hive weighing by scales, temperature and moisture measurement, as well as weather station, tilt, and GPS sensing, employing affordable platforms, like Arduino or Raspberry Pi, in particular the Zero, along with affordable sensors, e.g., load cells for scale assembly for hive weight, e.g., [1, 3, 8-13, 15], or feeder level monitoring. With regard to commercialization numerous bee patents have been filed, e.g., [17-21]. Actuation driven ideas are added here, e.g., for sustaining small colonies or thermal varroa treatment by suitable heaters and application procedures [20, 21].

More recently, acoustical [2, 17, 19] and visual inspection of hives, e.g., for activity/flight hole monitoring [11, 12], are beginning to be added to the instrumentation palette and the resulting data is going to be processed by appropriate methods from machine vision and machine learning [22, 23]. However, in these still few and quire recent approaches, predominantly large scale and rather costly microphone devices are used and camera systems still find restrictions in use to the limited computational power of the affordable computing platforms.

Inspired by intelligent condition monitoring (ICM) and Self-X capabilities added to automation and production systems in IIoT or Industry 4.0 field, the IndusBee 4.0 research activity has been started at ISE [32]. The goal is to achieve miniaturized or integrated, affordable yet reliable sensor systems adapted to the task of bee hive monitoring and process their data by proven to advanced machine learning methods. In this paper, the first step and application of a hive integrated acoustical monitoring system with MEMS microphones and accompanying components will be presented.

2. SmartComb for In-hive Monitoring

For the continued and inobtrusive measurements in a bee colony, a system based on the Raspberry Pi Zero W (OS Jessie) has been set-up in the shape of an instrumented honey comb, denoted as SmartComb. It incorporates two temperature and moisture sensors (DHT22/11) at top and bottom to provide readings at different hive locations. Further it includes HX711 board for hive scale reading. The key feature for this paper is the included I2S MEMS SPH0645LM4H microphone, chosen for low-cost reasons. Additionally, prototypes of high performance Infineon MEMS microphones have been made available for this work, but PDM interface requires additional electronics effort and cost, so that cheap I2S MEMS have been used in first place. Fig. 1 shows the partially assembled SmartComb prototype.

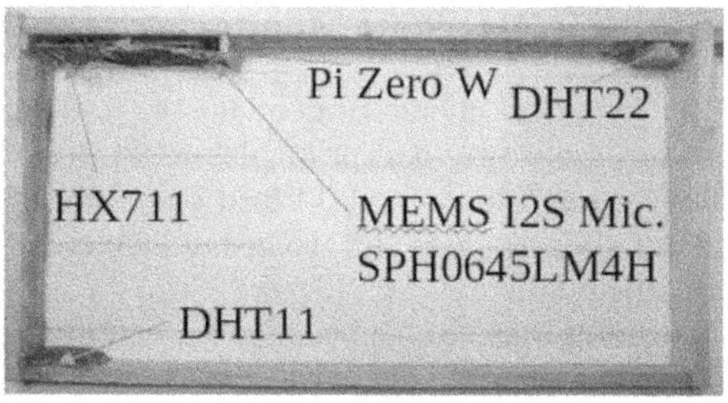

Fig. 1. SmartComb (Zander scale) prototype with acoustical, weight, temperature, and moisture sensing.

After completion, the depicted measurement system was inserted in the colony center of one super and was supplied either from accu or, temporarily, from wired 5 V supply, which limits wiring to just two wire cable from measurement unit to hive exterior. As it is well known, that bee and their byproducts provide a harsh environment, special care on protective packaging of sensors, cables, and embedded system have been expended, e.g., by wrapping sensors in special mesh (s. Fig. 1). The mechanical structure of numerous MEMS sensors can be open to dirt, e.g., wax, propolis, or other hive depris. The MEMS microphone has been mounted in a way avoiding any direct path for contaminating particles, while still being sensitive to in hive sounds, and hopefully, also to vibrations on the comb it is integrated in. The SmartComb of Fig. 1 provides a hive *fingerprint* of two separately located temperature and moisture reading, hive weight reading, and acoustic monitoring data. Additionally, it can be of interest to have separate weight data for each super, e.g., the honey storage super, which in late summer is predominantly used for hosting the feeder module. So, also feeder module weight, proportional to filling level, can be acquired and employed for *event-driven* refilling.

Fig. 2 shows a first prototype of a hive integrated scale for the honey storage/feeder module.

Fig. 2. Separate local scale prototype for honey storage super and feeder module weight monitoring.

The whole setup materials cost of a SmartComb is less than $30 for 'lot size' one, excluding the scale(s) setup. The following experiments have been done with such a SmartComb prototype.

3. Acoustical Hive State Monitoring

The acoustical monitoring of the hive state can follow the same way and principles as in predictive maintenance and Industry 4.0 based on vibration and sound monitoring, e.g., for wear-out detection and failure time prediction, e.g., [2, 6].

Interesting states of the bee hive are enumerated in the following, possibly not exhaustive, list:

1. Ok/Calm or Normal State;
2. Agitated/Disturbed;
3. Knocking/Pecking at hive;
4. Scratching at hive;
5. Swarm mood;
6. Missing Queen;
7. Looting ….. .

The colony states 1-6 are mentioned also in [2, 22, 23] and much more relevant information can be taken from bee generated sounds [17, 19]. Of course, dead silence in case of a perished colony hast to be taken care of in evaluation, too.

From the point of acoustical signal processing and machine learning, two general ways can be pursued after obtaining the raw data from the microphone(s) [6]. On the one hand, the classical approach based on *a priori* knowledge of domain-specific discriminant and invariant features followed by systematic dimensionality reduction and classifier choice and training. This approach is both transparent and amenable to analysis and understanding (*white box*) and will work from small to large amount of labeled data at moderate computational cost. On the other hand, there is the currently very popular field of deep-learning neural networks, that are renowned for learning from scratch and to perform better than classical approaches, yet are not easily amenable to analysis (black *box*), need rich amount of labeled data, and potentially have significant computational requirements in use [6, 22, 23]. It seems worth mentioning, that there is a possible hybrid design approach pursued by ISE that merges the classical knowledge-based approach with learning or fine-tuning of system levels to rapidly achieve a transparent and well performing solution of moderate complexity [25, 26].

Thus, for the processing of the underlying sounds for reasons of transparency and hardware limitations, the classical approach was chosen with first-cut design parameters of, e.g., ROI (700 ms), 16 k samples/s, and feature extraction (MFCC from python_speech_ features), optional dimensionality reduction, and classification (kNN from python sklearn with k = 3). The number of classes was selected to four with 1-4 of the previous list of possibly hive states.

The prototype of was prepared with a foundation for comb completion by bees and inserted into a bee colony, freshly established in spring 2019 (see Fig. 3).

In addition to the possible discernment of various interesting hive states, as enumerated in the previous list, for reasons of data storage and complexity management, a hierarchical approach seems to be favorable and was pursued in the work here. It relates to a common issue known well from, e.g., Industrial Quality Control and Intelligent-Condition-Monitoring (ICM) etc.: It is fairly easy to describe the 'Good' or 'Ok' state and lots of examples are commonly available, but it is extremely hard to *a priori* know all possible deviations from 'Ok' and acquisition of sufficient examples might be hard or even infeasible. This also relates to processing and storage requirements in data acquisition both system design as well as later on-line classification and advocates the use of one-class-classification (OCC) or anomaly or novelty detection in a hierarchical approach. These techniques are widely in use now from network intrusion problems to Industry 4.0 related ICM. The idea will be picked up in the following experimental section along with standard multi-class approach.

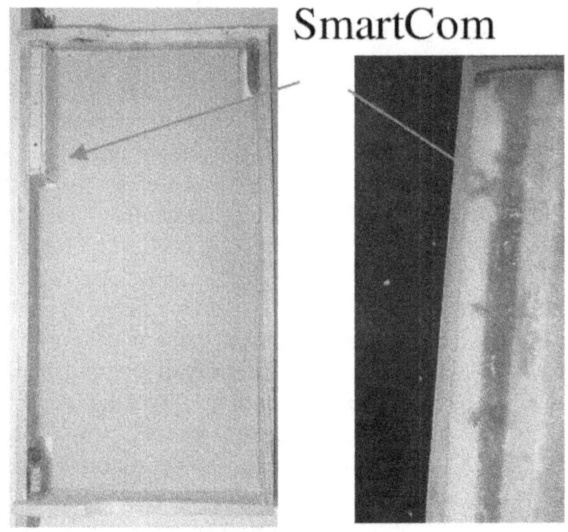

Fig. 3. SmartComb prepared for bee honey comb finishing (left) and inserted in a freshly established colony (right).

4. Experiments and Results

The **SmartComb** of **Fig. 3** was inserted in a living bee population and the hive was stimulated to provoke data of classes 1-4 given in the enumeration of Section 3 while continuously recording with the in-hive MEMS microphone. The obtained recordings were manually windowed, extracted, and labeled according to the meaning of the enclosed acoustical data in each window. Fig. 4 shows a feature space plot based on Sammon's dimensionality reducing mapping of the 13-dimensional MFCC data for the four classes with 25, 20 18, and 14 samples for class 1 to 4, respectively, obtained after data preparation and labeling process.

Dimensionality reduction, e.g., by automated feature selection has been investigated by standard Sequential-Backward-Selection (SBS) with the qsi separability measure [27], but in this particular case, the potential gain for the selected best features *2, 4, 5, 9, and 10* was negligible, so the 13-dimensional data set was used in the following multi-class classification. Training the chosen kNN-classifier with all data and a k = 3 setting returned a resubstitution result of ~97 %

('OK' always 100 %) Following, a hold-out approach was employed where, the data was split up in 53 (15, 15, 12, 11) train and 28 (10, 5, 7, 6) test patterns, where 4 patterns of class 4 were present both in training and testing. Resubstitution gave ~98 % and generalization ~92.86 %.

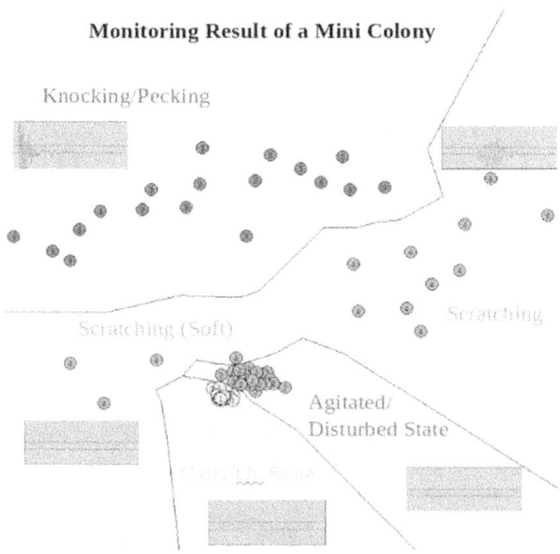

Fig. 4. MFCC feature space plot for MEMS microphone based acoustical hive monitoring data.

For a sound statistical analysis the acquired and labeled data is still very small, but already shows the aptness of the chosen low-cost MEMS microphone, electronics, and assembly in the introduced SmartComb. In a following live test, the trained classifier was activated in the hive and the bee colony in the hive was monitored in response to status and external stimulation/excitation. The data and the classifier decisions were concurrently recorded and evaluated in the following. Fig. 5 shows an excerpt of the 475 conducted live classifications.

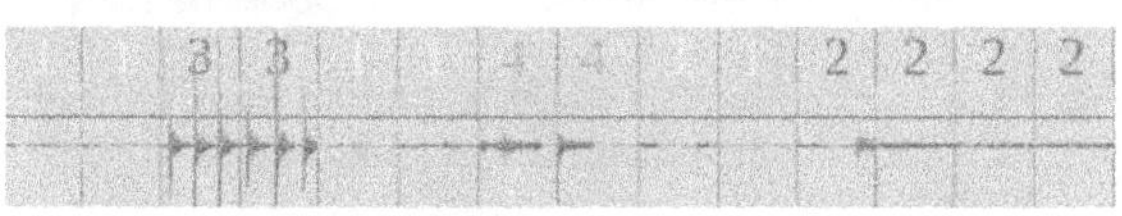

Fig. 5. Live classifications along with acoustical data recorded by MEMS microphone on SmartComb.

The chosen simple approach already worked well in the live test with repeated hive stimulations. Class 1 for 'Ok' state was always correctly discerned from all other sounds, but occasional confusion between class 2 and 4 could be observed, which was to be expected from the lower left of the feature space given in Fig. 4. The common fingerprint of the hive monitoring could now be extended by hive activity state classification as indicated in Fig. 6 for a subset of activated sensors.

The data in Fig. 6 naturally can be graphically displayed and tools and services, e.g., *Thingspeak* [31], are available and frequently employed.

```
Tue Jul 16 12:56:22 2019 0.0Deg 28.0Deg 0.0% 61.0% -1.0g 0.0g [1]
Tue Jul 16 12:56:23 2019 0.0Deg 28.0Deg 0.0% 61.0% -1.0g 0.0g [1]
Tue Jul 16 12:56:25 2019 0.0Deg 28.0Deg 0.0% 61.0% -1.0g 0.0g [2]
Tue Jul 16 12:56:32 2019 0.0Deg 27.0Deg 0.0% 59.0% -1.0g 0.0g [2]
Tue Jul 16 12:56:33 2019 0.0Deg 27.0Deg 0.0% 59.0% -1.0g 0.0g [1]
```

Fig. 6. Fingerprint of hive monitoring by **SmartComb** (First temperature/moisture sensor and global and local scales not in use here) complemented by live acoustical activity state classification.

Sustained departure from the 'Ok' state can be employed to send alerts to the hive keeper and, thus, initiate *event-driven maintenance.* Further, the current 'watch-dog' classifier can also be employed to limit the potentially massive amount of recording data [22] and the following post processing effort for training and classifier refinement, by only recording raw data of not 'Ok' state readings. Fig. 7 tries to elucidate the idea using the data with modified labeling, assuming all not 'Ok' data as being novel or abnormal without trying to distinguish the cause or related class on that level. That corresponds in ICM to perceiving a deviation from the desired state of operation without bothering with the detailed classification of the type of emerging problem yet. This basically allows the training of such a simple 'watch-dog' or OCC classifier just by the presentation of 'Ok' patterns, where a rich amount of examples is commonly available. This can be achieved by various methods and implementations, e.g., by a kNN classifier with a radius limitation, which is also available in the sklearn package. The radius limitation in kNN classification can be set manually as in rejection threshold definition or automatically computed from the available OCC training data. New sensory readings exceeding the set similarity threshold will be rejected, i.e., interpreted as novel or abnormal.

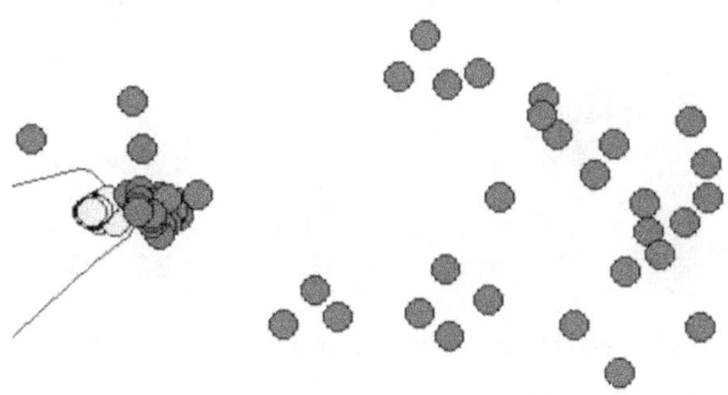

Fig. 7. Feature space plot similar to Fig. 4 with just 'Ok' class on a background of not 'Ok' or abnormal readings.

After this bifurcation, abnormal data might be subject to more detailed determination based on potentially scarce not 'OK' examples and a multi-class approach.

The value of the regarded approach with integrated and low-cost devices also depends on long term stability. The deployed SmartComb has been in use in the hive for several month by now and has been completed by the bees even as part of the brood nest (see Fig. 8). The microphone is still working fine, but the classification accuracy for new live tests is degrading. Fig. 9 shows the growing colony, which changes the acoustical properties of the hive and even in actually calm state it is generating a stronger signal, which leads to occasional confusion with the agitated/disturbed state of the previous colony size. Further considerations on microphone(s) placement, external sounds, colony size and evolution etc. have to be included in the design activities for a long term stable and invariant hive state determination. This also includes the issue of exporting a found state classifying system solution from one hive to one or several other hives, similar to the issues and approach in [26].

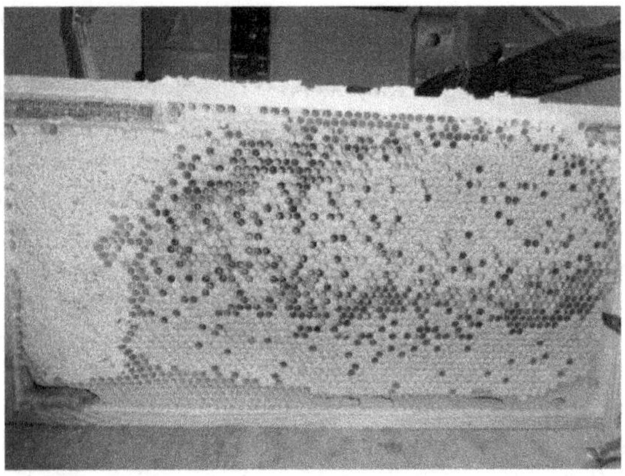

Fig. 8. SmartComb temporarily extracted from hive after several months of insertion in the bee colony.

Fig. 9. SmartComb in the bee hive with growing colony.

5. Conclusions

The paper discussed recent issues and needs in bee keeping and transfer options of knowledge and methods from automation, (I) IoT, ICM, and Industry 4.0 domains to provide solutions. An effective, disappearing, and low-cost hive-integrated approach, benefiting from the ongoing advance in micro-/nano-technologies, for the field of bee hive monitoring and colony activity and state assessment was introduced and demonstrated. In future work in our IndusBee 4.0 project [32], extension to a more comprehensive list of states, advanced evaluation and adaptation strategies, deployment to a larger number of colonies, establishing of an apiary AI (see Fig. 10), and addition of sensors, e.g., for flight hole activity monitoring (see Fig. 11) with payload assessment will be pursued.

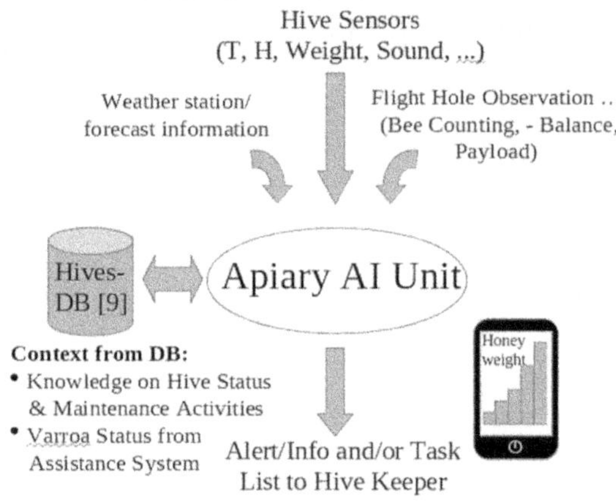

Fig. 10. Apiary AI proposed for aggregating and managing both hive monitoring and hive keeper assistance systems data of multi-colonies with sophisticated alerting, information, and scheduling for hive keepers.

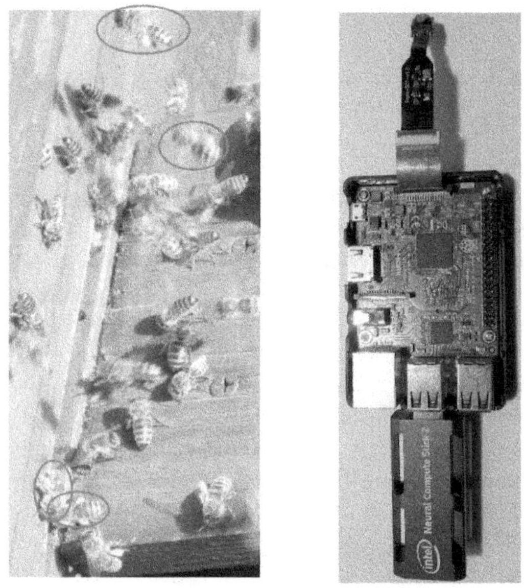

Fig. 11. Idea of embedded flight hole monitoring with Intel NCS stick as one potential accelerator (Mock-up).

References

[1]. Beekeeping with Arduino, https://www.hackster.io/pvalyk/beekeeping-with-arduino-4216bb

[2]. Modular Sensor Systems for Industry 4.0, https://moses-pro.de/

[3]. OSBeehives, https://www.osbeehives.com/

[4]. P. Wallich, Beehackers – Cheap widgets are like honey to hive keepers, *IEEE Spectrum*, May 2011, pp. 20-21.

[5]. J. Tautz, Phänomen Honigbiene, *Spektrum Akademischer Verlag (Elsevier)*, 2007

[6]. W. Zhang, G. Peng, C. Li, Y. Chen, Z. Zhang, A new deep learning model for fault diagnosis with good

anti-noise and domain adaptation ability on raw vibration signals, *Sensors*, Vol. 17, 2017, 425.
[7]. Yole Developpement, Technology Reports and Roadmaps, www.yole.fr
[8]. Ken Meyer, https://hackaday.io/project/beemon
[9]. BeeManager, https://www.beemanager.de/
[10]. Honey Bee Counter using Arduino, https://duino4projects.com/honey-bee-counter-using-arduino/
[11]. IoBEE, http://users.telenet.be/iobee/index_en.html
[12]. BeeCounter, http://duino4projects.com/honey-bee-counter-using-arduino/
[13]. BeeMonitor, https://beemonitor.org/
[14]. M.-P. Chauzat, L. Cauquil, L. Roy, S. Franco, P. Hendrikx, M. Ribiere-Chabert, Demographics of the European Apicultural Industry, https://www.ncbi.nlm.nih.gov/pmc/articles/PMC3827320/pdf/pone.0079018.pdf
[15]. HoneyPi, https://www.honey-pi.de/
[16]. V. Suta, et al, Apiary Monitoring System, Patent Application WO 2015/048308 A1, *Solutionbee LLC*, 25.09.2014.
[17]. J. Bromenschenk, C. Henderson, R. Seccomb, S. Rice, R. Etter, Honey Bee Acoustic Recording and Analysis System for Monitoring Hive Health, US Patent 7549907B2, *USA*, 2007.
[18]. E. C. Cook, Compositions and Methods for the Treatment and Management of Bee Hives and the Bees Therein, U.S. Patent 7,922,559 B2, *USA*, 2011.
[19]. T. J. Brundage, Acoustic Sensor for Beehive Monitoring, US Patent 8,152,590 B2, *USA*, 2012.
[20]. G. Sinanis, G. Lambropoulos, Beehive Frame and Comb Foundation for Controlling Varroa Miites, US Patent 8,272,921 B2, *USA*, 2012.
[21]. R. Linhart, Method of a Thermal Treatment of Bee Colonies and a Device for Pursuance of This Method, US Patent 9,363,984 B2, *USA*, 2016.
[22]. V. Kulyukin, S. Mukherjee, P. Amlathe, Toward audio beehive monitoring: deep learning vs. standard machine learning in classifying beehive audio samples, *Appl. Sci.*, Vol. 8, 2018, 1573.
[23]. I. Nolasco, A. Terenzi, S. Cecchi, S. Orcioni, H. L. Bear, E. Benetos, Audio-based identification of beehive states, *arXiv:1811.06330v2*, Feb. 2019.
[24]. C. Gohd, Autonomous Robot Bees Are Being Patented by… Walmart?, https://futurism.com/robot-bees-drones-walmart
[25]. A. König, VarroaCounter – Towards Automating the Varroa Screening for Alleviated Bee Hive Treatment, in *Proceedings of the 5th International Conference on Sensors Engineering and Electronics Instrumentation Advances (SEIA'2019)*, 2019, pp. 244-247.
[26]. M. Eberhardt, S. Roth, A. König, Industrial application of machine-in-the-loop-learning for a medical robot vision system – Concept and comprehensive field study, *J. Computers and Electrical Engineering (CEE)*, Vol. 34, March, 2008, pp. 111-126.
[27]. K. Thongpull, A. König, Advance and case studies of the DAICOX framework for automated design of multi-sensor intelligent measurement systems, *Technisches Messen*, Vol. 83, Issue 4, 2016, pp. 234-243.
[28]. Deutscher Imkerbund, https://deutscherimkerbund.de/152-Adressen_Forschungsinstitute
[29]. IBRA, https://www.ibrabee.org.uk/
[30]. JKI, https://www.julius-kuehn.de/en/bee-protection/
[31]. IoT Analytics Thingspeak, https://thingspeak.com/
[32]. IndusBee 4.0, https://www.eit.uni-kl.de/koenig/gemeinsame_seiten/projects/IndusBee_40.html

(058)

Effect of Thin Film Interconnect Inelasticity on MEMS Pressure Sensor Hysteresis

Y. Hamid [1,2], D.A. Hutt [1], D.C. Whalley [1] and R. Craddock [2]
[1] Wolfson School of Mechanical, Electrical and Manufacturing Eng., Loughborough University, UK
[2] Druck Ltd a division of BHGE, Leicester, LE6 0FH, UK
Tel.: + 44 (0)1162317893
E-mail: y.hamid@lboro.ac.uk / youssef.hamid@bhge.com

Summary: Thermal hysteresis dominates precision errors in absolute low pressure piezoresistive Silicon-based MEMS pressure sensors. A population of 10 experimental unpackaged "floating" bare devices were subjected to pressure and temperature cycles resulting in a mean hysteresis error of initially 0.062 % of full scale (FS) that reduced to 0.036 %FS after the third cycle. Within the limited 5 to 65 °C thermal cycles used in this study, this hysteresis is not predicted by the elastic recovery regime and is attributed to low temperature creep phenomena caused by stresses, that develop as a result of the coefficient of thermal expansion (CTE) mismatch between the Al-1 %Si bond pads and the Si substrate.

Keywords: Thermal hysteresis, piezoresistive pressure sensor, Al-1 %Si metalization pads.

1. Introduction

Al-1 %Si thin film metalizations are a classical packaging element for silicon MEMS devices that make use of wire bonds as electrical interconnects. The bond pads are typically sputtered onto MEMS wafers, sintered at 450 °C then cooled to room temperature [1]. The 20.7×10^{-6} K^{-1} CTE mismatch between the Al-1 %Si bonds pads and Si substrate result in the thin film being in tension at room temperature, with stresses that may be beyond the elastic regime. When subsequently thermally cycled from 25 to 450 °C, 1 µm thick Al-Si films start at 25 °C in the elastic return region before recrystallization at temperatures varying from 75 to 175 °C depending on the load history [2]. Several studies have analyzed hysteresis effects in piezoresistive pressure MEMS devices e.g. [3-5], where the cyclic temperature range is from -55 to +125 °C and the devices are structurally bonded to transistor outline (TO) headers. Having extra packaging elements, one cannot segregate thin film effects from other packaging effects. In this study the piezoresistive chips were therefore tested in an unmounted "floating" bare package, ensuring any effect observed is likely to be due principally to the thin film. This, to the authors' knowledge, has not been previously accomplished. In this study, even for a significantly reduced thermal range compared with these earlier studies, hysteresis errors limit the performance of the MEMS device.

2. Packaging Method

This study used an experimental MEMS absolute pressure device as shown in Fig. 1 and having dimensions of $1.65\times1.65\times1.37$ mm^3. To obtain floating devices, the MEMS chip was temporarily bonded to a stainless-steel TO enclosure using photoresist to retain the die in place during wire bonding. Acetone was then used to remove the photoresist, leaving the device free. Visual inspection of 10 devices was carried out to make sure they were clean and only supported by the wire bonds in a floating configuration. After an initial 5 to 65 °C conditioning cycle, the devices were then tested over 3 cycles of 50 to 200 kPa absolute pressure and from 5 to 65 °C as shown in Fig. 2.

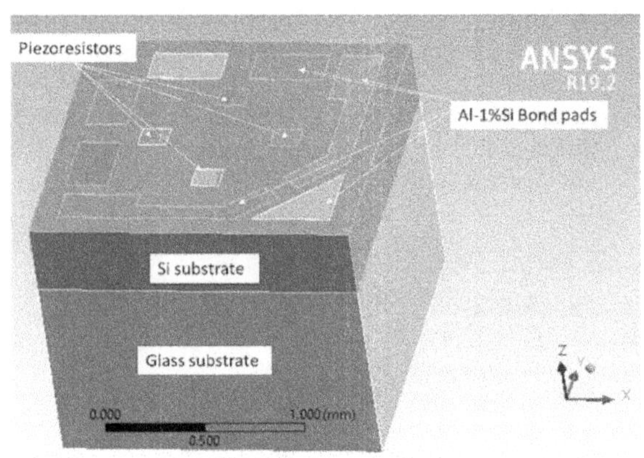

Fig. 1. Isometric view of pressure MEMS device with Al-1 %Si bond pads.

3. Figs. and Tables

An ANSYS® R19.2 FEA analysis was carried out that estimated the Von Mises (VM) stress in the metalization to be 280 MPa at 25 °C after cooling from 450 °C with a 100 kPa applied pressure as a boundary condition. This value was obtained using a bilinear hardening model for Al taken from the Granta material property library. The VM stress value exceeded the 150 MPa 0.2 %-offset yield limit for a thin film predicted by Gardner and Flinn [1] and would therefore be expected to result in significant plastic strain. During the limited thermal cycle range carried out here, the metalization would therefore be expected to remain in the elastic return regime, thus yielding a linear behaviour.

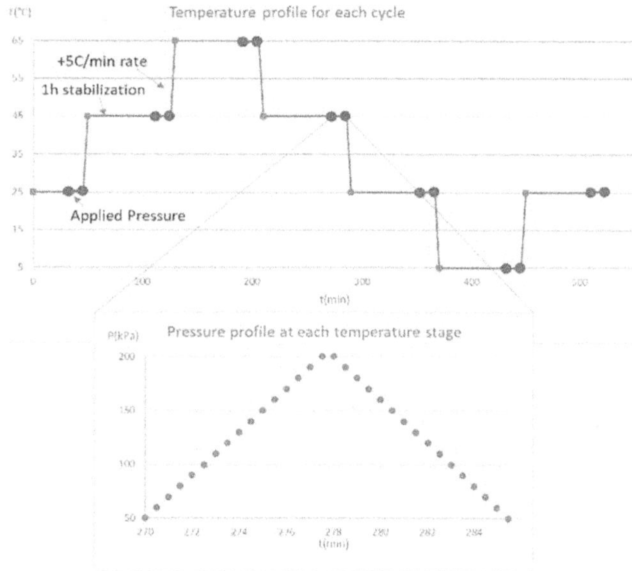

Fig. 2. Temperature and pressure profile for the first cycle. Each temperature has a stabilization period of 1 h at the end of which the pressures are applied.

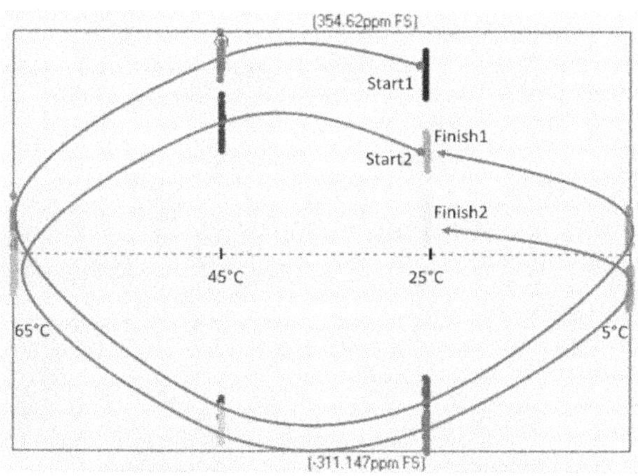

Fig. 3. Hysteresis error plot for an individual sensor over two temperature cycles.

Fig. 4. Hysteresis drop over a 5 to 65 °C temperature cycle range for a population of 10 units. The average values decrease from 0.062 %FS to 0.036 %FS.

The sensor's Wheatstone bridge voltage output and diode voltage (a measure of temperature) were fitted with a third order polynomial function to calculate the "fitted pressure", which would be the pressure returned by a commercial pressure sensor. Hysteresis errors were then calculated as the error between the actual applied pressures and fitted pressures, which can be seen in Fig. 3. This shows a hysteresis error plot for an individual sensor over the first two temperature cycles, where each data group represents a 50 to 200 to 50 kPa pressure cycle. Data groups vertically aligned are for pressure cycles obtained at the same temperature, while the vertical axis is the residual error which ranges from -0.031 to +0.035 % of full scale (FS) pressure. It can be seen that the error is mainly dominated by thermal hysteresis. Fig. 4 shows a box plot of hysteresis values after each thermal cycle. The average hysteresis for the first cycle error was 0.062 % full scale (FS) peak-to-peak, which reduced to 0.036 %FS at the third cycle. Although only three cycles are used here, the results indicate that the thermal hysteresis is improved by cycling the units.

4. Conclusions

Within the limited 5 to 65 °C temperature range, the reduction in thermal hysteresis with the number of cycles cannot be explained by the thermal expansion mismatch directly, as the stresses are predicted to be in the elastic recovery region. Therefore, stress relaxation and creep phenomena at low temperature are believed to be the source of the observed hysteresis and its stabilization over the cycles. Further low temperature creep properties of the Al-1 %Si are therefore needed for use within a more detailed FEA model to validate this creep-induced hysteresis hypothesis.

AcknowWith best wishes,
Sincerely,

Prof. Sergey Y. Yurish,
Editorledgements

This work was partly funded by the EPSRC Centre for Doctoral Training in Embedded Intelligence grant number EP/L014998/1.

References

[1]. D. S. Gardner, P. A. Flinn, Mechanical stress as a function of temperature in aluminum films, *IEEE Trans. Electron Devices*, Vol. 35, Issue 12, 1988, pp. 2160-2169.

[2]. W. D. Nix, Mechanical properties of thin film, *Metall. Trans. A*, Vol. 20A, 1989, pp. 2217-2245.

[3]. J. A. Chiou, S. Chen, Thermal hysteresis analysis of MEMS pressure sensors, *J. Microelectromechanical Syst.*, Vol. 14, Issue 4, Aug. 2005, pp. 782-787.

[4]. E. Eiper, R. Resel, Thermally-induced stresses in thin aluminum layers grown on silicon, *Powder Diffr.*, Vol. 19, Issue 01, 2004, pp. 74-76.

[5]. Å. Sandvand, E. Halvorsen, H. Jakobsen, In situ observation of metal properties in a piezoresistive pressure sensor, *J. Microelectromechanical Syst.*, Vol. 26, Issue 6, 2017, pp. 1381-1388.

(059)

Correlation between Weather Conditions and Respiratory Pathologies in Gran Canaria by the Use of Remote Sensing

I. Caballero-Leiva [1], N. Marrero Betancort [2], J. J. Rodriguez Betancor [3], D. Rodriguez-Esparragón[4], J. Marcello [5]

[1,2,4,5] Institute of Oceanography and Climate Change, IOCAG. University of Las Palmas de Gran Canaria, Campus of Tafira. 35017, Las Palmas, Spain
Tel.: +34 928457357
E-mails: isabel.caballero@ulpgc.es, nerea.marrero102@alu.ulpgc.es, jrodbeth@gobiernodecanarias.org, dionisio.rodriguez@ulpgc.es, javier.marcello@ulpgc.es
[3], Gran Canaria Dr. Negrín University Hospital, 35010, Las Palmas, Spain
Tel.:+34 928450000

Summary: The Sahara dust is well known to have an adverse impact on human health. The Canary Islands, due to their vicinity to the Sahara desert, are frequently affected by the Sahara aerosols carried by the winds. This study analyses the correlation between climatic (Sea Surface Temperature, Aerosol Optical Thickness, Wind Intensity and Wind Direction) and clinical (emergency admissions due to respiratory pathologies) variables in the island of Gran Canaria and for a period of 14 years. The bivariate lineal correlation (Pearson) shows statistical significance between accumulated monthly averaged values of Sea Surface Temperature and the emergency admissions. There is also a significant correlation between the climatic variables Wind Intensity and Aerosol Optical Thickness.

Keywords: Remote sensing, Health risks, Climate change, SST, AOT.

1. Introduction

Canary Islands are frequently influenced by the Sahara dust that crosses the Atlantic Ocean sweeping the Spanish archipelago. This meteorological event is commonly known as "CALIMA" (Cloud, Aerosols and Ice Measurements in the Sahara Air Layer). This paper is part of a wider research project which studies the relationship between the CALIMA and the weather conditions and their influence on respiratory, psychic and allergic conditions.

The influence of Sahara dust in human health is well studied not only in several European countries [1-6] but also across the Atlantic ocean, all the way in USA [7]. The emergency services of the Canary Islands are very familiar with the effect that this event has in the population´s health, noticing an increase in the attendance of patients during the days when the event is present (Doctor Negrín University Hospital, admissions team personal comment).

Sahara dust is transported along several paths around the globe; over the North Atlantic Ocean to North America and South America, across the Mediterranean to Southern Europe and sometimes as far north as Scandinavia or east as Middle East [8]. Considering the proximity of the Macaronesian Region to the desert´s coast and that the dust sweeps along the Canary Island when the wind blows towards the Atlantic, the relevance of studding the impact of these aerosols in the Canarian population seems evident. Furthermore, there are studies looking into the possibility of the Trend Winds rotation to the east [9] which would increase the periods of time in which the Canarian archipelago is exposed to the Sahara dust. Therefore, it is crucial for the medical services of the Canary Islands to understand what the consequences of such change in the wind climate could mean in the health of their residents in order to plan the adaptation of their services.

The project counts with the Admission Department of one of the main hospital in the island of Gran Canaria (Doctor Negrín University Hospital). They provide not only the clinical data but also their medical expertise to the study.

The section of the study presented in this paper shows results of the early stages of the project and focuses on the analysis of climatic and clinical parameters, their correlation and the influence on respiratory conditions in the Gran Canarian population.

2. Methods

2.1. Study Area

The Canary Islands are located in the west coast of the African Continent and are part of the Macaronesian Region. Its climate is characterized by the predominance of the Trade Winds, which reach the islands with a north-northeasterly direction [2]. The wind climate of the islands is driven by the Azores high pressure system which dominate below approx. 1500 m a.s.l. [10]. When the dominant wind direction changes to easterly and soutly the islands receive Sahara air, which is characterized by warm temperatures, high relative humidity and the presence of suspended Sahara dust (CALIMA) [1]. This happens most frequently during the months of summer, July to August [3].

2.2. Data

The data presented in this study covers the period of January 2003 to December 2017, 14 years in total.

2.2.1 Climatic Data

The climatic data presented in this work are Wind Direction (WD), Wind Intensity (WI), Sea Surface Temperature (SST) and Aerosol Optical Thickness (AOT), all expressed in monthly averages.

Wind data was obtained from the National Centre of Atmospheric Prediction and the National Centre for Atmospheric Research of the United States of America (NCEP/NCAR). They consist on a re-analysis of monthly averages of the u and v components of the wind measured at 10 meters above sea surface, both near shore and offshore. The data is provided in a matrix of 192 pixels of longitude and 94 pixels of latitude (Gaussian grid), with a processing level 4, and a geographical resolution of 0.25°×0.25° (approx. 25 km × 25 km). It is provided in HDF5 format [11]. From the Gaussian grid, 8 time series were extracted around the archipelago. After comparison of WD and WI for all points and corroboration that there was no significant special variation of the wind data in the study area, an average was performed and so one single (averaged) time series was used for the analysis performed in the present study. Fig. 1 shows the location of the 8 NCEP extracted data points.

AOT [12] and SST [13] are a product of MODIS (Moderate Resolution Imaging Spectroradiometer), aboard the NASA satellite Aqua and presented in monthly averages. The data is provided in maps, the 4 km resolution was selected for the tasks performed in this study. Six data points (time series) were extracted from the mentioned map. An average of the 6 data points was performed and so single averaged time series of AOT and SST are presented in this work. The AOT used is AOT_869. Fig. 2 shows the location of the 6 AOT and SST extracted data points.

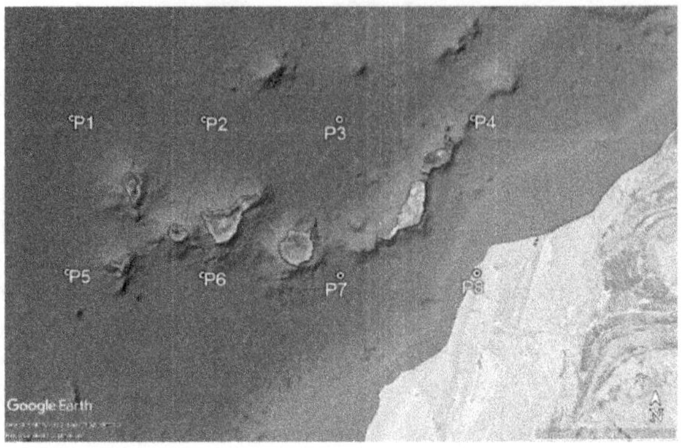

Fig. 1. NCEP data points locations. P1 to P8. (Google Earth Pro. Image Landsat/Copernicus).

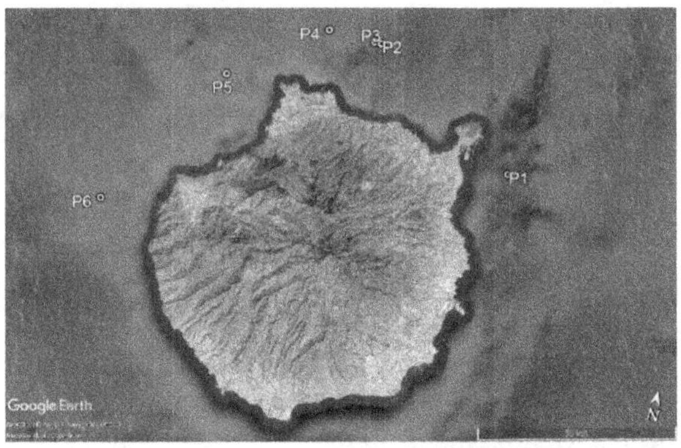

Fig. 2. MODIS-Aqua data points location. P1-P6. (Google Earth Pro. Image Landsat/Copernicus).

Fig. 3 to Fig. 5 display the monthly time series of climate data for the entire study-period.

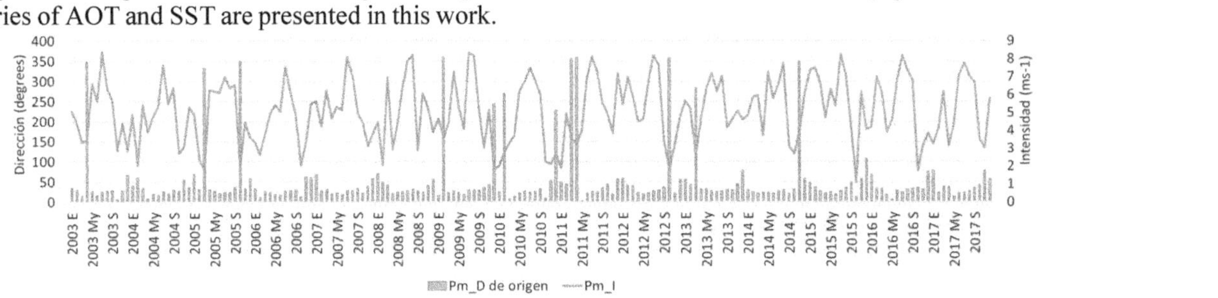

Fig. 3. Wind Intensity (Pm_I, orange line) and Wind Direction (Pm_D de origen, orange line) averaged time series of 8 data points. Monthly average. Period 2003-2017.

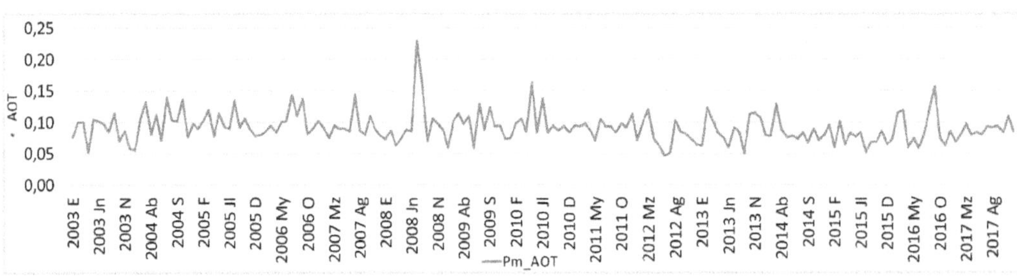

Fig. 4. Aerosol Optical Tickness (Pm_AOT) averaged time series of 6 data points. Monthly average. Period 2003-2017.

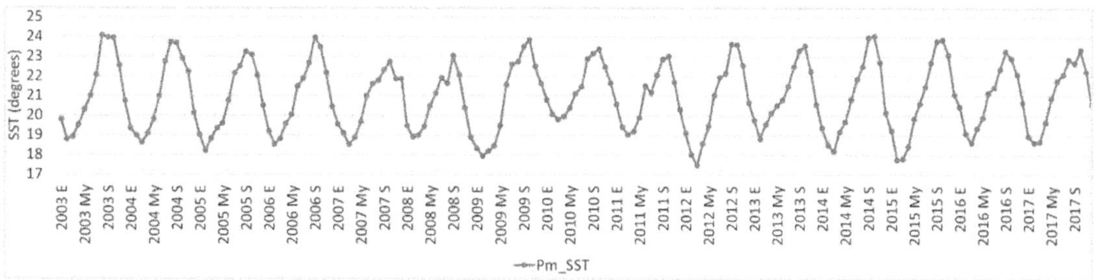

Fig. 5. Sea Surface Temperature (Pm_SST) averaged time series of 6 data points. Monthly average. Period 2003-2017.

2.2.2 Clinical Data

The clinical data was provided by the admissions team of the Gran Canarian Doctor Negrín University Hospital. The data consists on the number of daily emergency admissions due to breathing pathologies (IUPR). Since the climatic data was obtained in monthly averages, the daily clinical data was accumulated into the same time step.

The data was converted to 1000 %, for a population of 350000 citizens that were assigned to the Hospital when the study took place (14).

Fig. 6 displays the monthly clinical data in rates per thousand and for the entire study-period.

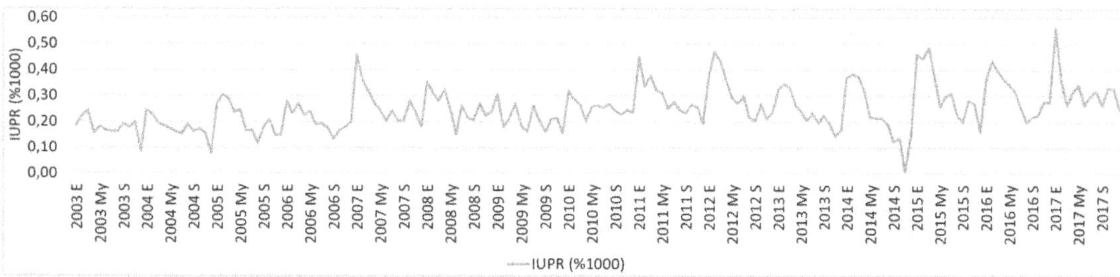

Fig. 6. Emergency admissions due to breathing pathologies (IUPR). Monthly accumulations. Period 2003-2017.

3. Results

The data analysis presented in this article consists on bivariate linear correlations and were applied to all pairs of variables. The Pearson coefficient was calculated for all pairs of climatic variables as well as between the climatic and clinical variables. Other than correlation analysis applied to the whole data series as one, seasonal analysis were also performed, as well as annual trends.

The following figures show some of the results obtained.

Fig. 7 shows the accumulated monthly averaged values of SST vs IUPR for the 14-year period, that is, the annual trends. The trend lines show an inverse relationship between these two variables. For the first trimester of the year, both variables see a progressive decrease, but while IUPR keeps dropping until the end of the year (with excuse of small raises in June and September), the Sea Surface Temperature increases from March to September, to drop again from then to December. The temperature ranges between a maximum of 25° seen in September and 20° in March.

As for the IUPR, data varies between 0.36 ‰ cases in January and a minimum of 0.18 ‰ in December.

The Pearson correlation coefficient between the monthly accumulated values of SST and IUPR (% 1000) for the 2003-2017 period is -0.661 ($p < 0.001$).

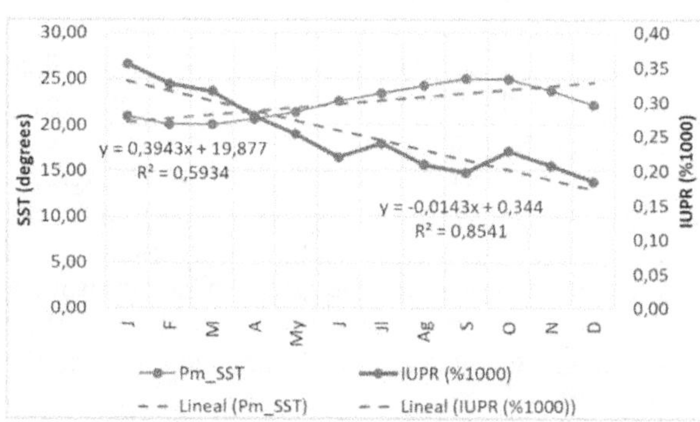

Fig. 7. Accumulated monthly averages of SST (C, purple line) vs IUPR (%1000, green line) and corresponding trendlines. Period 2003-2017.

Fig. 8 shows two of the climatic variables plot against each other, Wind Intensity (WI) vs the Aerosol Optical Thickness (AOT) both in monthly averages and for the full study-period. The correlation coefficient of this pair is the only one that, analised in (non accumulated) monthly averages gives a significant statistical correlation, with Pearsons coefficient being 0.256 ($p < 0.001$).

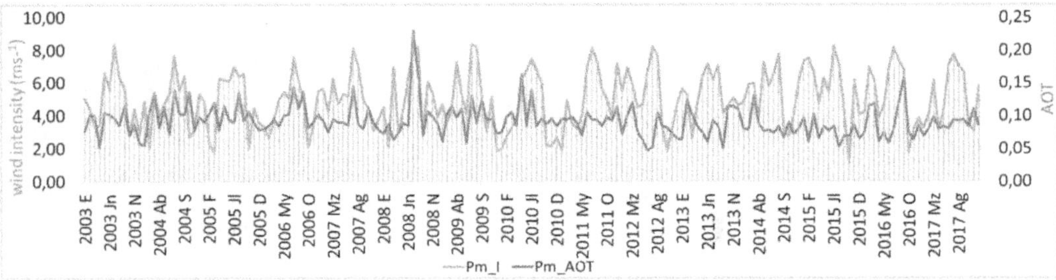

Fig. 8. Wind Intensity (ms-^{1}, orange line) vs AOT (pink line). Monthly average. Period 2003-2017.

Fig. 9 shows the time series of AOT vs IUPR for monthly averages of the trimester March-April-May, for every year from 2003 to 2017. Visually, it is easy to identify that for most years, the AOT and IUPR present the same trends, in some cases (2006 and 2013) it is almost parallel. It is noticeable the year 2009 for the opposite event; the graphs are mirrored.

However, even though there is visually a clear relationship between the amount of aerosol in the atmosphere and the number of recorded respiratory medical admissions, the correlation coefficient is not statistically significant (0.083 (sig. (bilateral) = 0.262)). Ongoing investigations are taking place in this respect.

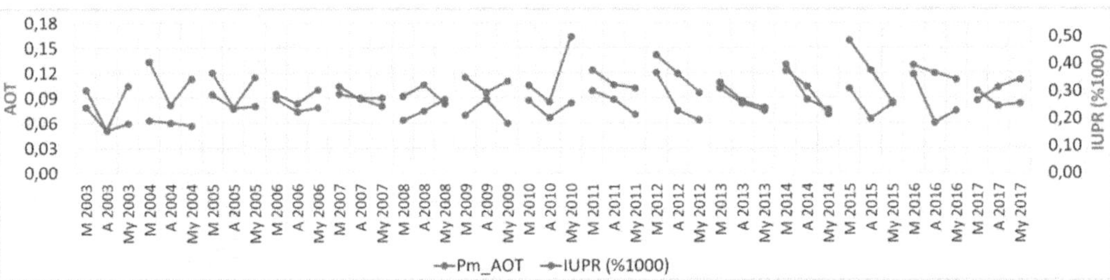

Fig. 9. AOT (adimentional, pink line) vs IUPR (%1000, green line). Monthly averages for March, April and May. Period 2003-2017.

4. Conclusions

The results presented in this study show significant correlations between some climatic variables and the respiratory health status of the grancanarians. When looking into the correlation among the predicting variables (climatic variables), there is significant correlation between Wind Intensity and the Aerosol thickness. This is expected considering that the stronger the wind blows over the Sahara, the concentration of dust resuspended into the atmosphere increases as well as the stronger the wind, the further it is transported across the ocean and through the archipelago.

Analyzing the data in monthly averages for the 14 years period, the aerosols do no show significant correlation with any other climatic variable. In future studies the plan is to use wind data in a shorter temporal resolution (daily). Considering that the CALIMA events lasts in average 3 days at the time, using monthly averages might mask the data preventing, for instance, the expected correlation between wind direction and aerosol concentrations, to be significant. Same applies to the correlation between Wind Direction and clinical data. It was expected to find a clear correlation between these two; aerosols reach the archipelago when the wind blows westwards or northwards over the Sahara. No correlation is shown in our study so the idea is to repeat the correlation analysis using daily data. It is also important to consider that the AOT data provided by MODIS-Aqua represents all solid and liquid particle in the full atmospheric column. Another option to depurate the data for correlation analysis would be to apply an algorithm to calculate the particles of a size between 2.5 μm and 10 μm, being this range the usually analised in human health related studies. However, even though correlation analysis applied to the full length of the period do not show significance, when looking into the plot for shorter periods (i.e. trimester March-April-May) there seems to be a visual relationship between AOT and IUPR, showing a possible seasonal influence in the relationship of climatic and clinical variables. Further studies are carried out at the time this article is released.

When accumulating the data sets monthly for the full14-year period, the variables SST and IUPR present significant correlation, however, it is important to note that this correlation does not necessary imply causality. The correlation between these two variables is slightly negative, which is expected considering that respiratory conditions tend to increase during winter, when the temperature drops, and decrease during the summer time.

For future studies, in order to get a better understanding of the influence of climatic factors like Wind Direction, it is advisable to use shorter temporal resolution, preferably daily time steps. In the case of Aerosols, considering working with specific particle sizes in advisable.

The final aim of this project is to help the Canary Islands health system be prepared for the changes that climate change might force in the number and nature of emergency admissions in the island of Gran Canaria. Considering the correlations found in this study, the well-known adverse impact of CALIMA in human health and the potential increase of CALIMA episodes due to climate change in the Macaronesian Region, it is essential to keep investigating the relationship between climatic and clinical parameters in the Canary Islands.

Acknowledgements

This work has been supported by the PASTOR (ProID2017010072) project, funded by the Department of Economy, Industry, Commerce and Knowledge of the Canary Islands Government, and the European Fondo Europeo de Desarrollo Regional (FEDER).

This study would not have been possible without the collaboration of the Admissions Team of the Gran Canarian Doctor Negrín University Hospital.

References

[1]. S. Mallone, M. Stafoggia, A. Faustini, G. Paolo Gobbi, A. Marconi, F. Forastiere, Saharan dust and associations between particulate matter and daily mortality in Rome, Italy, *Environ. Health Perspect.*, Vol. 119, Issue 10, 2011, pp. 1409-1414.

[2]. S. Z. Sajani, R. Miglio, P. Bonasoni, P. Cristofanelli, A. Marinoni, C. Sartini, et al., Saharan dust and daily mortality in Emilia-Romagna (Italy), *Occup. Environ. Med.*, Vol. 68, Issue 6, 2011, pp. 446-451.

[3]. I. Menéndez, E. Derbyshire, T. Carrillo, E. Caballero, J. P. Engelbrecht, L. E. Romero, et al., Saharan dust and the impact on adult and elderly allergic patients: The effect of threshold values in the northern sector of Gran Canaria, Spain, *Int. J. Environ. Health Res.*, Vol. 27, Issue 2, 2017, pp. 144-160.

[4]. M. Sanuy, E. Duo, W. S. Jäger, P. Ciavola, J. A. Jiménez, Linking source with consequences of coastal storm impacts for climate change and risk reduction scenarios for Mediterranean sandy beaches, *Nat. Hazards Earth Syst. Sci.*, Vol. 18, Issue 7, 2018, pp. 1825-1847.

[5]. P. Kassomenos, A. Analitis, K. Katsouyanni, E. Kougea, E. Samoli, Does the presence of desert dust modify the effect of PM10 on mortality in Athens, Greece?, *Sci Total. Environ.*, Vol. 409, Issue 11, 2011, pp. 2049-2054.

[6]. L. Perez, A. Tobias, X. Querol, N. Künzli, J. Pey, M. Viana, et al., Coarse particles from saharan dust and daily mortality, *Epidemiology*, Vol. 19, Issue 6, 2019, pp. 800-807.

[7]. J. M. Prospero, O. L. Mayol-Bracero, Understanding the transport and impact of african dust on the Caribbean Basin, *Bul. Am. Meteorol. Soc.*, Vol. 94, Issue 9, 2013, pp. 29-37.

[8]. N. J. Middleton, A. S. Goudie, Saharan dust storms: nature and consequences, *Earth-Science Rev.*, Vol. 56, 2001, pp. 179-204.

[9.] Grupo de Procesado de Imágenes y Teledetección ULPGC, Servicios de estudios de la evolución climática de las caracteristicas del viento en la región Canarias-Souss Masaa Dràa, con imagen satélite en el marco del projecto CLIMATIQUE, 2014, https://studylib.es/doc/8262032/resumen-ejecutivo

[10]. J. Carrillo, J. C. Guerra, E. Cuevas, J. Barrancos, Characterization of the marine boundary layer and the trade-wind inversion over the sub-tropical North Atlantic, Boundary-Layer Meteorol., Vol. 158, Issue 2, 2016, pp. 311-330.

[11]. Kalnay, The NCEP/NCAR 40-year reanalysis project, *Bull. Amer. Meteor. Soc.*, Vol. 77, 1996, pp. 430-470.

[12]. NASA Goddard Space Flight Centre. Moderate Resolution Imaging Spectroradiometer (MODIS) Aqua {aot_869} Data, https://oceandata.sci.gsfc.nasa.gov/MODIS-Aqua/Mapped/Monthly/4km/aot_869/

[13]. MODIS-Aqua Level-3 Mapped Sea Surface Temperature Data, https://oceancolor.gsfc.nasa.gov/data/10.5067/AQUA/MODIS/L3M/SST/2014/

[14]. Gobierno de Canarias S de S, https://www3.gobiernodecanarias.org/sanidad/

(060)

Mechanoluminescent Pulse Pressure Sensors

K. V. Tatmyshevskiy
Vladimir State University, 87 Gorky Str., 600000, Vladimir, Russia
Tel.: + 79051456012, fax: + 7 (4922)3 35242
E-mail: tatmysh@mail.ru

Summary: Mechanoluminescence (ML) is a visible or infrared radiation of solid bodies that occurs under the mechanical influence. The brightest luminescence is observed under the influence of pulse pressures. Mechanoluminescent sensors (MLS) are based on the principle of direct conversion of the input mechanical pressure to the output optical signal. Such sensors can be used in conjunction with fiber-optic data transmission channels, which provides high noise immunity to electromagnetic interference. MLS in the form of a flexible film of a large surface area allows to realize the function of visualizing fields of impulse pressures or tactility. This article describes the physical mechanism of the MLS, made from ZnS: Mn. ML occurs under the excitation of centers of luminescence (CL) by moving charged dislocations that accompany plastic deformation of the sensor. A mathematical model of the sensor and the results of the calculation of the main characteristics are given. The dependences of the parameters of the output optical signal from the parameters of the input pressure pulses are shown.

Keywords: Mechanoluminescence, Mechanoluminescent pulse pressure sensor, Dislocation kinetics.

1. Introduction

Mechanoluminescent sensors (MLS) are based on the principle of direct conversion of the mechanical energy of an elastoplastic deformation into the energy of optical radiation, which is recorded by a photodetector. Structurally, MLS is a thin-layer element, which includes fine crystalline phosphor powder (ZnS:Mn) and a transparent binding material. Mn atoms play the role of luminescence centers (CL) in these sensors. The optimum thickness of the MLS does not exceed 20...100 μm, which is determined by the particle size distribution of the powder and the characteristics of light propagation in fine powder layers. Film MLS in conjunction with a matrix photodetector allows you to visualize and record pulse pressure fields [1].

The advantages of MLS are high noise immunity to electromagnetic interference, the ability to transmit signals over fiber lines, the absence of moving parts, hardness and planarity [2]. Such sensors can be used, for example, to control the operation of car airbags.

2. Mathematical Model

2.1. Mechanoluminescence Mechanism

It has been shown that mechanoluminescence (ML) in crystalline compounds is a consequence of the processes of motion of dislocations that accompany the plastic deformation in crystals [3]. Dislocations in A_2B_6 semiconductors, and in particular ZnS, have a strong electrical charge. In the process of plastic deformation, CL interacts with the electric field of moving charged dislocations, which leads to the excitation of CL with their subsequent radiative transitions.

The physical model of MLS is based on the phenomenon of tunneling of electrons in impurity CL in an electric field of moving dislocations arising from plastic deformation of a crystal. During plastic deformation, the dislocations begin to move within the volume of the crystal. While the plastic deformation continues, excitation processes of CL prevail. After its termination, an exponential decrease in luminescence is observed due to the transition of excited CL to the stationary state.

2.2. Radiation Kinetics Equation

The specific flux of optical radiation in MLS is determined by the expression

$$\mathrm{R(t)} = \frac{\eta}{\tau}\exp\left(-\frac{t}{\tau}\right)\int_0^{t_\sigma} N_{CL}\, 2r_{in}(t)\, \tilde{N}_{mD}(t)\, \tilde{V}_D(t)\,\mathrm{dt}, \qquad (1)$$

where η is the energy of a light quantum; τ is the time constant corresponding to the exponential attenuation of the intracenter ML characteristic of Mn atoms; t_σ is the pressure pulse duration time; N_{CL} is the total number of CL in the sensor, on the area of 1 mm^2; $r_{in}(t)$ is the interaction radius of a moving dislocation with CL; $\tilde{N}_{mD}(t)$ is the average density of mobile dislocations; $\tilde{V}_D(t)$ is the average dislocation velocity. Here, the integrand describes the kinetics of excitation of CL during the action of the pressure pulse t_σ, and the expression before the integral describes the kinetics of luminescence attenuation.

2.3. Sensor Deformation Equation

The kinetics of excitation depends on plastic strain ε_p, the speed of which is determined by the Orowan expression [4]

$$\dot{\varepsilon}_p = |\vec{b}|\widetilde{N}_{mD}(t)\,\widetilde{V}_D(t), \qquad (2)$$

where $|\vec{b}|$ is the modulus of the Burgers vector characterizing the lattice distortion by the dislocation. The stress-strain state of MLS for the case of uniaxial application of pressure is determined by the expression

$$\sigma_1 = E\left(\varepsilon_1 - \frac{4}{3}\varepsilon_p\right) = \sigma(t) - E\varepsilon_{1p}, \qquad (3)$$

where σ_1 is the principal stress applied perpendicular to the MLS plane, E is the Young's modulus, ε_1 is the total strain in the form of the sum of the elastic ε_{1e} and plastic ε_{1p} components, $\sigma(t)$ is the pressure pulse. Expression (3) shows that stresses increase with an increase in total strain ε_1, and decrease due to plastic ε_{1p}. Here the case of quasistatic deformation, i.e. observance of uniformity of plastic deformations across the thickness of the MLS was considered. This condition is fulfilled either at low strain rates or at small ratios of the thickness of the sensor to its diameter.

Dislocation parameters are found by Gilman expressions [5]

$$\widetilde{N}_{mD} = \widetilde{N}_{tD}\exp\left(-\widetilde{N}_{tD}/\widetilde{N}_{cr}\right), \qquad (4)$$

$$\widetilde{N}_{tD} = \widetilde{N}_{tD0} + M\varepsilon_p, \qquad (5)$$

where $\widetilde{N}_{tD}$ is the average total dislocation density; $\widetilde{N}_{cr}$ is the critical value of the total dislocation density corresponding to the end of the yield area in the $\sigma(\varepsilon)$ diagram, $\widetilde{N}_{tD0}$ is the initial value of the total density of dislocations; M is the dislocation multiplication factor.

The average dislocation velocity can be determined by the expression

$$\widetilde{V}_D(t) = k_s c_{tr}\exp\left(D_f/\sigma_1\right), \qquad (6)$$

where k_s is the correction factor, zeroing the speed of dislocations at pressures below the yield strength; c_{tr} is the shear wave velocity in the crystal; D_f is the effective stress of internal friction.

2.4. Calculation Results

Examples of calculations are shown in Fig. 1 and Fig. 2.

The system of equations (1)-(6) is a mathematical model of MLS. The joint solution of these equations, taking into account the coefficient of radiation output from the MLS surface, made it possible to calculate the output optical signals depending on the parameters of the input pressure pulses of half-sinusoidal form. In the calculations it was assumed that manganese CL radiate at a wavelength of 580 nm. The time constant of ML attenuation was determined experimentally [6] and has the value of $\tau = 150$ µs.

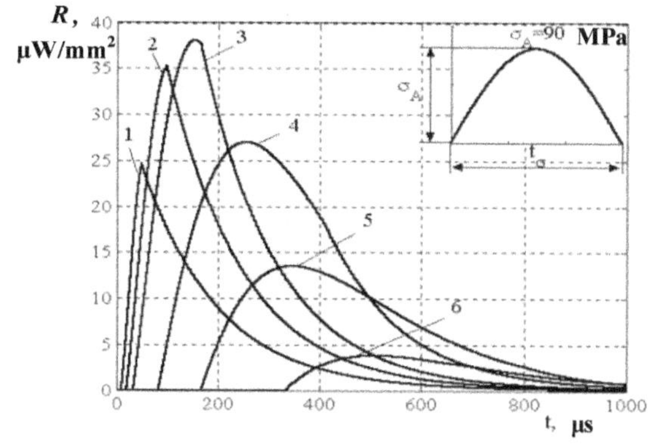

Fig. 1. Estimated dependences of the energy luminosity $R(t)$ under the action of pressure pulses of the same amplitude $\sigma_A = 90$ MPa and different duration t_σ: 1 – $t_\sigma = 60$ µs; 2 – $t_\sigma = 120$ µs; 3 – $t_\sigma = 200$ µs; 4 – $t_\sigma = 500$ µs; 5 – $t_\sigma = 1000$ µs; 6 – $t_\sigma = 2000$ µs.

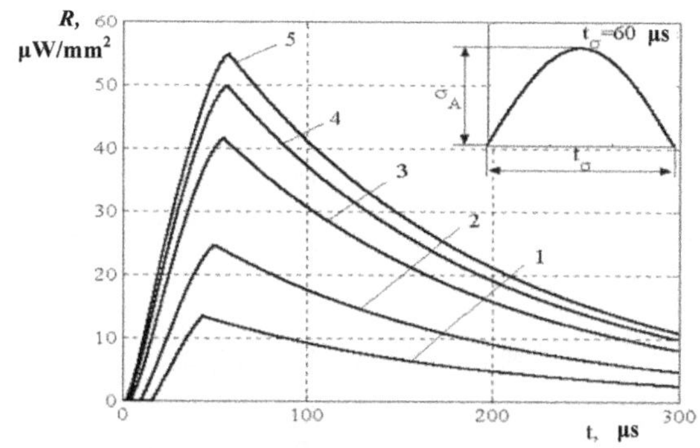

Fig. 2. Estimated dependences of an energy luminosity $R(t)$ of MLS under an influence of pressure impulses of identical duration $t_\sigma = 60$ µs and varying amplitude σ_A: 1 – $\sigma_A = 60$ MPa; 2 – $\sigma_A = 90$ MPa; 3 – $\sigma_A = 180$ MPa; 4 – $\sigma_A = 270$ MPa; 5 – $\sigma_A = 360$ MPa.

3. Conclusions

The developed mathematical model allows determining the MLS output signal in absolute units of energy luminosity. The model takes into account the features of the MLS design. The MLS transformation function has, firstly, a sensitivity threshold, and secondly, a substantially non-linear character. The intensity of ML is sufficient for being registered by modern photodiodes. The calculation results are confirmed by experimental data [6].

References

[1]. A. Feng, S. Michels, A. Lamberti, P. F. Smet, Mechanoluminescent materials: A new way to analyze stress by light, in *Proceedings of the 18th International Conference on Experimental Mechanics (ICEM'18)*, Vol. 2, Brussels, Belgium, 1-5 July 2018, 492.

[2]. I. Sage, G. Bourhill, Triboluminescent materials for structural damage monitoring, *Journal of Materials Chemistry*, Issue 11, 2001, pp. 231-245.

[3]. Yu. A. Ossipian, et al., Electronic Properties of Dislocations in Semiconductors, *Editorial URSS*, 2000.

[4]. J. W. Taylor, Dislocation dynamics and dynamic yielding, *Journal of Applied Physics*, Vol. 36, Issue 10, 1965, pp. 3146-3150.

[5]. J. J. Gilman, Dislocation dynamics and response of materials to impact, *Applied Mechanics Reviews*, Vol. 21, Issue 8, 1968, pp. 767-783.

[6]. N. Yu. Makarova, K. V. Tatmyshevskiy, A table for experimental study of mechanoluminescent pulse pressure transducers, *Instruments and Experimental Techniques*, Vol. 49, Issue 1, 2006, pp. 135-140.

(061)

Noise Analysis of PCB-based, Double-core Planar Fluxgate Sensor

M. Ortner [1], **A. Roshanghias**[1], **M. Lenzhofer**[1] **and T. Becher** [2]
[1] Silicon Austria Labs, Sensor Systems, 9500 Villach, Austria
[2] Federal Institute of Santa Catarina, SC 88020-300 Florianopolis, Brazil
Tel.: +43 4242 5600300 200, fax: + +43 4242 5600300 400
E-mail: michael.ortner@ctr.at

Summary: In previous studies it was shown that planar fluxgate structures can be improved by applying an asymmetric second core to emulate a closed core system. In this case, the sensors power consumption is massively reduced while sensitivity improves at the same time. It is unclear, however, whether the double core structure functions as simple signal amplification and also amplifies the noise or if it truly enhances the signal to noise ratio of the device. In this work an experimental study is presented to answer this question.

Keywords: Magnetic sensors, Planar fluxgate, PCB based sensor technology.

1. Introduction

Fluxgate sensors are vector magnetometers that feature a wide range of resolutions spanning from picotesla to millitesla [1]. Due to their simplicity and robustness they have several applications for geomagnetic surveys, space explorations, handheld probes as well as medical applications [2].

The major short come of the fluxgate, however, is that its sensitivity scales with its size making them bulky with typical sensor head dimensions in the centimeter range. Miniature sensors were realized by microfabrication [3] and integrated circuit technology [4]. In addition, planar fluxgate structures were developed at the turn of the century [5], based on the idea that coils and core are fabricated as thin, stacked layers. When combined with PCB technology they have become a cheap possibility to realize a magnetic sensor and functionalize the PCB e.g. for direct on-PCB current sensing [6, 7].

2. Double Core Structure and Proposal

The fluxgate sensor response is given by the peak height of the second harmonic of the periodic output voltage which depends directly on the external magnetic field to be measured [6].

It was proposed in previous works to use a double-core structure to enhance the sensitivity of such devices [6-8]. The idea is that the double-core structure emulates a closed core system so that core saturation is achieved at lower excitation current. For a PCB based-device with a size of 18×6 mm² sensitivities of up to 10 mV/(μT*Hz) were achieved for 0.2 Watt power supply [7]. However, in these previous studies sensitivities were measured without regard of the noise. While the double core structure easily enhances the sensitivity by a factor of 2 [8], noise levels were not indicated which means that this structure might simply function like a signal amplifier.

In this work the critical detectivity (noise over sensor sensitivity) is studied for a single- and a double-core system.

2. Experimental Layout

A four-layer PCB with trace width and spacing of 75 µm, thickness of 35 µm, resulting in a total thickness of 500 µm and three adjacent coils of 60 windings respectively in a total structure size of 18×6 mm² is used. The planar core structures are fabricated of a cobalt-based film of Metglas 2714A with 15 µm thickness, initial relative permeability of $\mu_r \approx 10^6$ and saturation magnetization of B_s = 0.57 T and a ferromagnetic alloy VITROVAC 6155 U55 with lower initial permeability of $\mu_r \approx 1330$, saturation magnetization of B_s = 0.99 T and foil thickness of 23 µm. A photo of the sensor can be found in Fig. 1.

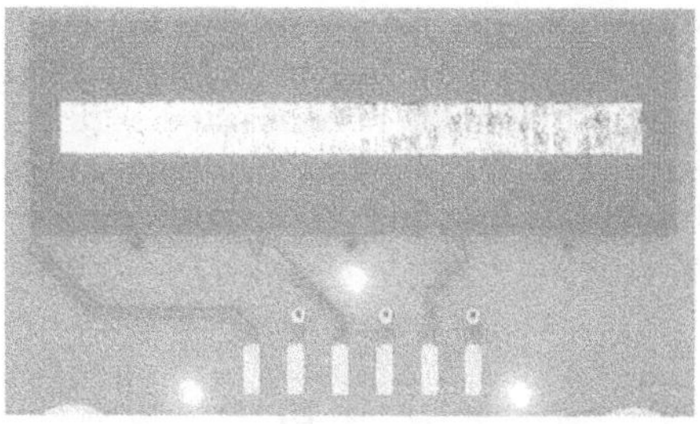

Fig. 1. Photo of the sensor device.

For precise fabrication the core was cut from the foil by pico-second laser cutting to conserve the magnetic properties and was then adhesively bonded to the PCB using a Micro Assembly Station (FINEPLACER® Lambda) with expected precision of ±1 µm.

Finally a printed 3D Helmholtz coil system was developed to fully compensate the geomagnetic field avoiding influences from other components and to apply an external field in a controlled way. The automatic mechanism compensates the geomagnetic field using a 3D GMR sensor with a precision below 1 μ T including offset drift and temperature compensation. The sensor also functions as a reference sensor for the applied external field. A photo of the Helmholtz coil is shown in Fig. 2.

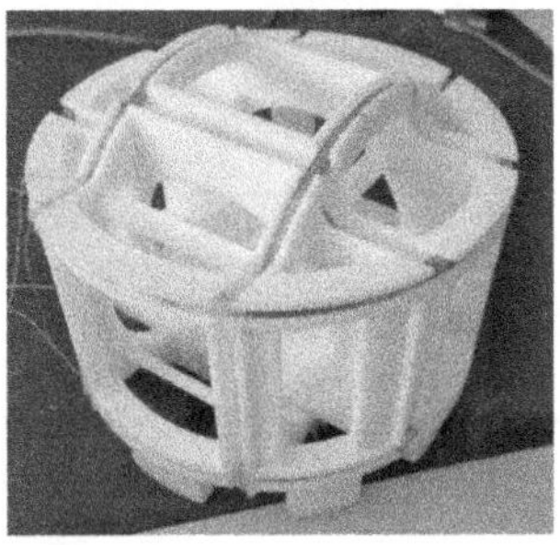

Fig. 2. Photo of the test setup. The fluxgate sensor is tested in the HC coil together with a reference GMR sensor.

4. Experimental Results

The sensor response of the single- and the double-core samples are displayed in Figs. 3a and 3b for different input voltages as function of the external magnetic field. It can be seen that the sensor exhibits a saturation behavior: Initially the sensor response (at zero-field) is low, increases with increasing voltage and reaches a limit which is around 6 V for the single-core and 3.5 V for the double-core. This confirms the original motivation, to make the system more energy efficient.

In Fig. 3c the 0-point sensitivity is shown as a function of the frequency for all voltages and both sensor samples. The sensitivity should increase linearly with the frequency until eddy currents start to suppress the sensor response. This behavior can be observed, however, the jump at 130-160 kHz, which can be observed for all voltages and samples is of unknown origin. Fig. 3c shows that the double-core sensitivity is generally higher as of the single-core, which was also found in previous studies [8].

Finally, Fig. 3d shows the 0-point noise. It is critical to notice that this noise shows the very same behavior as the sensitivity in Fig. 3c. This means that with increasing sensitivity the noise increases as well which strongly supports our theory-based conjecture that the double core structure simply functions as a passive signal amplifier.

5. Conclusion

Detailed analysis of the detectivity of planar single- and a double-core PCB based fluxgate structures was performed. It was shown that double-core structure functions as a passive signal amplifier and does not enhance detectivity. However, the double core structure is energetically more efficient as the full sensitivity is reached at lower input voltages.

References

[1]. S. Tumanski, Modern magnetic field sensors – A review, *Organ Stowarzyszenia Elektryków Polskich*, Vol. 10, 2013, p. 13.

[2]. P. Ripka, Chapter 3, in Handbook of Magnetic Materials, Vol. 18, *Elsevier*, 2009, pp. 347-420.

[3]. H.-S. Park, Development of micro-fluxgate sensors with electroplated magnetic cores for electronic compass, *Sensors and Actuators A*, Vol. 114, Issue 2-3, 2004, pp. 224-229.

[4]. W.-S. Huang, A novel 3D CMOS micro-fluxgate magnetic sensor for low magnetic field detection, in *Proceedings of the IEEE SENSORS Conference*, 2010, pp. 1791-1794.

[5]. I. Vincueria, Flux-gate sensor based on planar technology, *IEEE Transactions on Magnetics*, Vol. 30, Issue 6, 1994, pp. 5042-5045.

[6]. M. Lenzhofer, A new approach for a planar miniaturized PCB based high sensitivity fluxgate sensor design, in *Proceedings of the AMA Conference SENSOR*, 2017, pp. 41-45.

[7]. N. Navaei, Analysis of single- and double core planar fluxgate structures, in *Proceedings of the MDPI EUROSENSORS 2018 Conference*, 2018, pp. 134-136.

[8]. P. Ripka, Symmetrical core improves micro-fluxgate sensors, *Sensors and Actuators A*, Vol. 92, Issue 1-3, 2001, pp. 30-36.

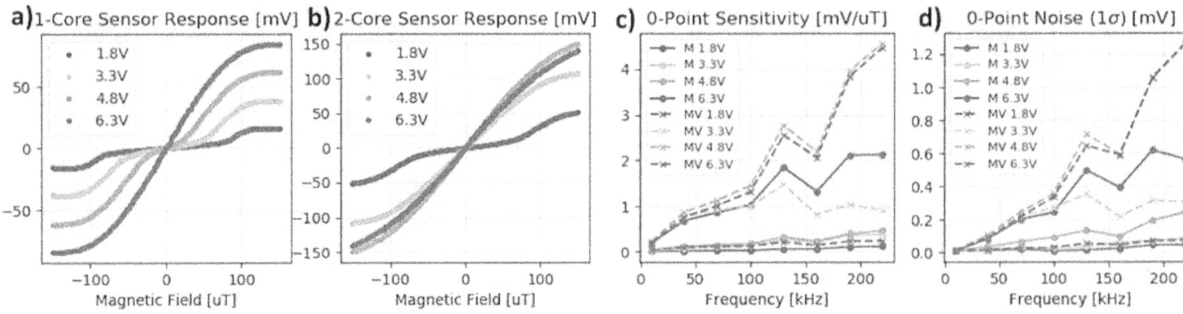

Fig. 3. Experimental results: a) and b) show the sensor responses, c) the sensitivities and d) the noise behavior.

(062)

New Approach to Sensing Amphiphilic Electrochemically Active Compounds

T. Navrátil [1], **A. Rajcová** [2], **R. Jerga** [2], **V. Müllerová** [2], **S. Zacharov** [3] **and J. Skopalová** [2]
[1] J. Heyrovský Institute of Physical Chemistry of the Czech Academy of Sciences, Dolejškova 3, 182 23 Prague 8, Czech Republic
[2] Department of Analytical Chemistry, Palacký University Olomouc, Faculty of Science, 17. listopadu 12, 771 46 Olomouc, Czech Republic
[3] Department of Occupational Medicine, First Faculty of Medicine, Charles University and General University Hospital in Prague, Na Bojišti 1, 128 00 Prague 2, Czech Republic
Tel.: + 420266051111
E-mail: navratil@jh-inst.cas.cz

Summary: Determination of amphiphilic molecules in various matrices belongs to frequently solved analytical tasks. For this purpose, we have developed an electrochemical sensor based on accumulation of the analyte in a phospholipid (asolectin) membrane formed on the surface of a glassy carbon electrode. The determination has been carried out applying direct current or differential pulse voltammetry (using standard addition method and calibration line method). As an example, determination of capsaicin and similar compounds (generally capsaicinoids (CPDs)) in different Chilli peppers is described. Two different methods have been tested: direct method on unmodified electrode and transfer method on asolectin modified electrode. The results achieved by differential pulse voltammetry have been statistically compared with those achieved using HPLC/UV.

Keywords: Voltammetry, Electrochemistry, Modified surface, Electrode, Sensor, Amphiphilic molecule, Capsaicinoids.

1. Introduction

Development of new sensors for analysis of biologically active compounds (BACs) and their medically or environmentally important metabolites is one of the most important trends in modern analytical chemistry. For that reason, the electrochemical methods (ECMs) can be successfully applied [1]. Their main advantages consist in low investment and running costs, simplicity, and portability. Great attention has been focused on surface modifications of bare electrodes with chemical, biochemical, and/or biorecognition elements. They can serve as smart (tailor-made) BAC sensors (e.g., for selective detection of DNA damaging compounds, etc.).

Phospholipid membrane (PLM) modified electrodes belong to this group. Biological membrane exists as a surface, at which the hydrophobic parts of phospholipids (PLs) are protected from water, while hydrophilic ones are in contact with the aqueous medium or with a polar surface [2]. Due to presence of the lipophilic part, the amphiphilic molecules (e.g., dopamine) are capable to be accumulated in the supported PLMs. Increased amount of such electrochemically active molecule can be voltammetrically determined [2]. Various materials can support PLMs, e.g. amalgams, glassy carbon. Similarly, different PLs (pure or mixtures – e.g. asolectin [3]) can be used to form PLMs.

In this contribution, we describe a simple way of preparation of phospholipid modified glassy carbon electrodes (PLMGCEs) and their applications on the example of determination of capsaicinoids (CPDs) in model solutions and in chilli peppers.

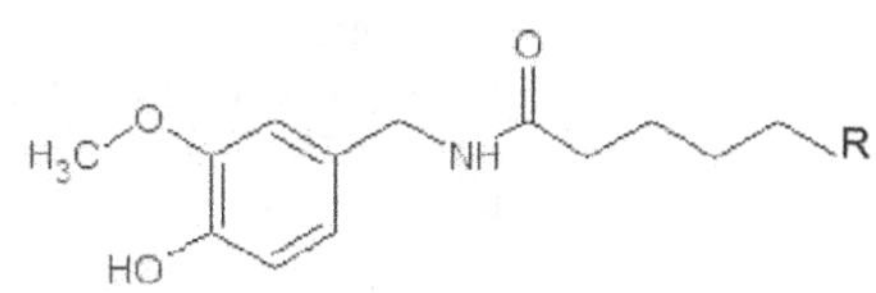

Fig. 1. Structure formula of common part of capsaicinoids (CPDs) (e.g. for capsaicin R is CH = CH-CH(CH_3)$_2$).

2. Experimental

2.1. Chemicals and Devices

Capsaicin (≥ 95 %) was obtained from Sigma-Aldrich. Asolectin from soybean (lecithin ca 25 %) used for modification of glassy carbon electrode (GCE) was from Fluka (Switzerland). Britton-Robinson buffer (BRB) solution (pH 2) was used as the supporting electrolyte (SE).

CPDs in some chilli peppers were analyzed (Goat Horn Pepper, Habanero, Naga Jolokia, and Carolina Reaper). All voltammetric measurements were performed using Autolab PGSTAT128N (Metrohm, The Netherlands) in three-electrode system with the working GCE/PLMGCE, reference Ag|AgCl|3M-KCl electrode, and auxiliary Pt electrode.

2.2. Preparation of PLM on GCE Surface

After mechanical polishing on alumina slurry, GCE was electrochemically activated in NaCl solution, (E_{ac} = 1.5 V, t_{ac} = 3 min). Then, the electrode was rinsed and dried. Asolectin solution in hexane was pipetted on the GCE surface and immersed into NaCl solution, in which the bilayer PLM was formed spontaneously. The PLM quality was tested by cyclic voltammetry in ferrocyanide solution [4].

2.3. Voltammetric Methods of Capsaicinoids Determination

CPDs were extracted from the weighted chilli peppers samples using maceration (15 min), sonication (30 min) in ethanol and filtration.

Two differential pulse voltammetric (DPV) methods were applied for the determination of CPDs in the extracts [4]:

DPV-A (direct method on unmodified GCE): An aliquot of the extract was added to SE (BRB, pH 2) and DPV measurement was performed directly with GCE (E_{in} = +0.3 V, E_{fin} = +1.0 V, scan rate 20 mV s^{-1}, pulse amplitude 50 mV, pulse width 80 ms). Before each measurement, the GCE was cleaned using alumina slurry.

DPV-B (transfer method on the PLMGCE): The PLMGCE was immersed into the ethanolic extract diluted with water (1:4, *v*/*v*) for 1 min. Then, it was rinsed by distilled water, and finally placed into the vessel with clean SE. DPV measurements were performed at the same parameters as in DPV-A. PLMGCE was cleaned from the accumulated CPDs by immersion in distilled water for 10 min.

Quantification of CPDs was performed by the method of calibration curve and by the standard addition method. Capsaicin was used as a standard to which all results were referred.

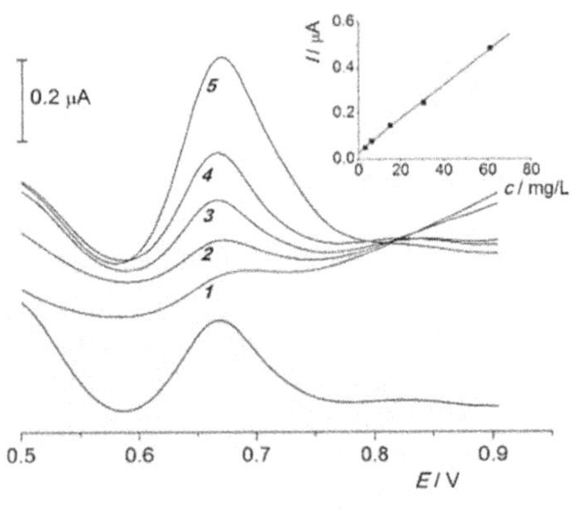

Fig. 2. DPV-B curves of capsaicin on PLMGCE at concentrations 3, 6, 15, 30, 60 mg/L (***1-5***) and of Naga Jolokia extract (red). Respective calibration curve (inset).

3. Results and Discussion

CPDs as lipophilic substances accumulate effectively in the phospholipid layer. This feature was used for their isolation from a complex plant extracts in which less lipophilic components are present that can interfere at direct voltammetric analysis. The response of the PLMGCE sensor in DPV-B method increases linearly with capsaicin concentration in the range of 3-60 mg/L (Fig. 2). Table 1 summarizes some results of CPD determination by both tested DPV methods and a comparative HPLC method. The voltammetrically achieved results involving all electrochemically active CPDs somewhat differ from those of HPLC (capsaicin only). The results of the DPV-B method with PLMGCE correlate much better with the HPLC results (correlation coefficient r = 0.9) than the direct method DPV-A results with the unmodified electrode (r = 0.6). Repeatability of the DPV-B method (expressed as RSD of 5 repeated measurements at capsaicin conc. level 15 mg/L) is better than 15 %.

Table 1. Content of capsaicin (g/kg) determined using tested DPV methods and HPLC (standard addition method).

	DPV-A	DPV-B	HPLC/UV
Naga Jolokia (extract)	2.8	2.7	3.5
Habanero (extract)	8.2	0.7	1.0
Habanero (powder)	7.6	7.7	11.9
Goat horn (powder)	1.9	3.0	4.5
Carolina reaper (powder)	31.3	18.1	12.8

4. Conclusions

The developed sensor for analysis of amphiphilic electrochemically active compounds, based on the GCE modified with asolectin PLM in transfer mode, proved to be suitable for the voltammetric determination of CPDs in various chilli peppers.

Acknowledgements

The authors thank Czech Science Foundation for financial support (T. N. the project GAČR No. 17-03868S and J. S. the project GAČR No. 17-05387S).

References

[1]. J. Barek, J. Fischer, T. Navratil, K. Peckova, Yosypchuk B. J. Zima, Nontraditional electrode materials in environmental analysis of biologically active organic compounds, *Electroanalysis,* Vol. 19, Issue 19-20, 2007, pp. 2003-2014.

[2]. H. Tomkova, R. Sokolova, T. Opletal, P. Kucerova, L. Kucera, J. Souckova, J. Skopalova, P. Bartak, Electrochemical sensor based on phospholipid modified glassy carbon electrode – Determination of paraquat, *Journal of Electroanalytical Chemistry,* Vol. 821, pp. 33-39.

[3]. T. Navratil, I. Sestakova, V. Marecek, Supported phospholipid membranes formation at a gel electrode and transport of divalent cations across them, *International Journal of Electrochemical Science,* Vol. *6*, Issue 12, 2011, pp. 6032-6046.

[4]. A. Rajcova, Study of electrochemical behaviour of capsaicin on carbon electrodes and its analytical application, MS Thesis, *Palacký University Olomouc,* Olomouc, Czech Republic, 2018.

(064)

Development and Mathematical Description of a Bistable Hydrogel-based Sensor Switch for Monitoring of the Relative Humidity

N. Gulnizkij [1] and G. Gerlach [1]

[1] University Dresden, Solid-State Electronics Laboratory, Mommsenstrasse 15, 01062 Dresden, Germany
Tel.: +49 351 463 43794, fax: +49 351 463 32320
E-mail: nikolai.gulnizkij@tu-dresden.de

Summary: This contribution presents a hydrogel-based bistable sensor switch with a switching hysteresis for the measurement of the relative humidity in ambient air. The sensor enables the switching at targeted humidity threshold values, for example in buildings and containers. The sensor does not require any internal electrical energy supply for the monitoring of the relative humidity and the operation, since the required energy is taken from the environment. To provide reliable switching without arcing the sensor characteristics has to show a hysteresis. This contribution deals with the principle concept of a sensor for the implementation of such a switching hysteresis based on a mono-/bimorph beam partly covered with swellable hydrogel and provides a simple mathematical model to describe the switching hysteresis in dependence on the geometry parameters, the material properties, the pre-deflection due to a corresponding axial force and the buckling due to the hydrogel swelling. To prove the switching hysteresis and to validate the mathematical model, a sensor switch with switching hysteresis was developed. Based on the obtained results a miniaturized silicon-based sensor switch is proposed.

Keywords: Hydrogel-based bistable sensor switch, Switching hysteresis, Bimorph effect, Relative humidity, Beam theory.

1. Introduction

Autonomous sensors that receive their energy from an energy harvester or directly from the environment have the potential to save energy for applications in numerous sectors [1, 2]. For humidity sensing, the swelling behavior of a structured water vapour-sensitive hydrogel is used. This swelling provides the mechanical energy to deflect the silicon bending plate by using the bimorph effect (Fig. 1).

The switching will be triggered to close the microcontact by passing a threshold humidity value. As long as this defined threshold is not reached, the electrical microcontact remains open. The swelling behavior of the hydrogel shows a high sensitivity with regard to slight changes in relative humidity resulting in a corresponding opening and closing of the microcontact [2-4]. To avoid voltage peaks and arcs during the switching process that could arise and destroy the microcontact, a bistable characteristic with hysteresis is needed.

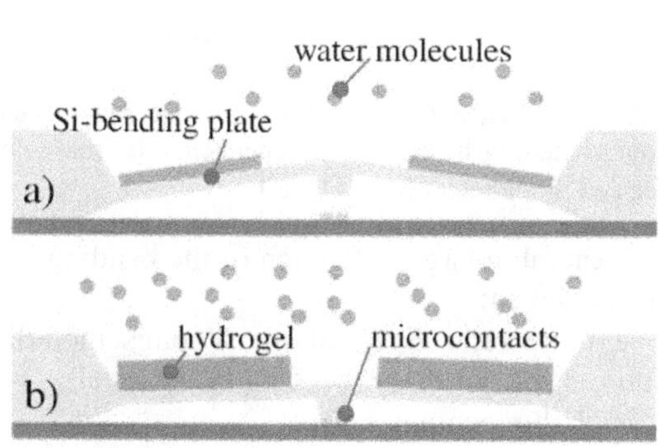

Fig. 1. Humidity threshold sensor switch, consisting of a bimorph bending plate with patterned hydrogel layer and microcontact, a) open, b) closed [3, 4].

2. Switching Hysteresis

Fast switching can be achieved by an S-shaped curve for the deflection-versus-humidity characteristics (Fig. 2). This can be reached by implementing a pre-deflection due to a corresponding axial force. The bending of the bimorph in the direction of the contacts leads to a sudden buckling and the associated closure of the contacts.

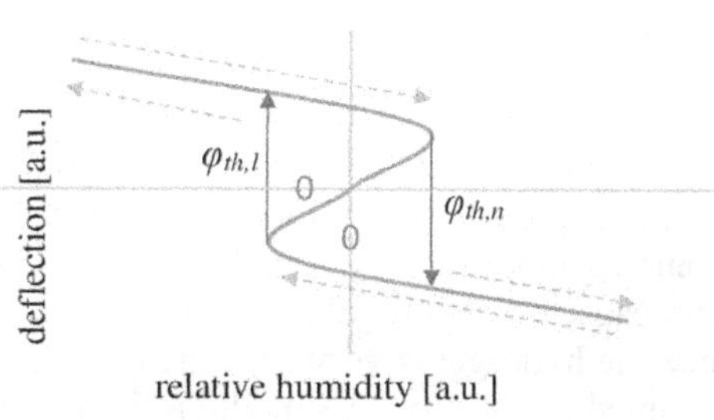

Fig. 2. S-shaped deflection-versus-humidity curve. The switching points are marked with vertical arrows at the two φ_{th} points.

3. Sensor Model without Axial Compression

To calculate the deflection of the bending plate from Fig. 1 in dependence both on the coating degree and, the ratio of the hydrogel-coated part of the plate to the uncoated part, a simple mechanical model for monomorphic and bimorpic beams was used that is based on the beam theory [5]. Fig. 3 shows the left half of the sensor switch from Fig. 1.

Here the beam is divided in two rigid beam parts plus the localized bending compliance in between, both for l_B and l_P, i.e. the hydrogel-coated and the uncoated part. The length l_B characterizes the hydrogel-coated part with the torsional or bending compliance n_B and l_p the length of the uncoated part with the compliance n_P. $l = 2(l_P+l_B)$ yields the total length of the bending beam b is the beam width; h_1, h_2 are the thicknesses and E_1, E_2 the Young´s moduli of the beam and the hydrogel, respectively. If the hydrogel swells then the beam will be deflected by the movement s due to the bimorph effect (Fig. 3c). This is also assumed for the hydrogel-coated beam part with length l_B and compliance n_B.

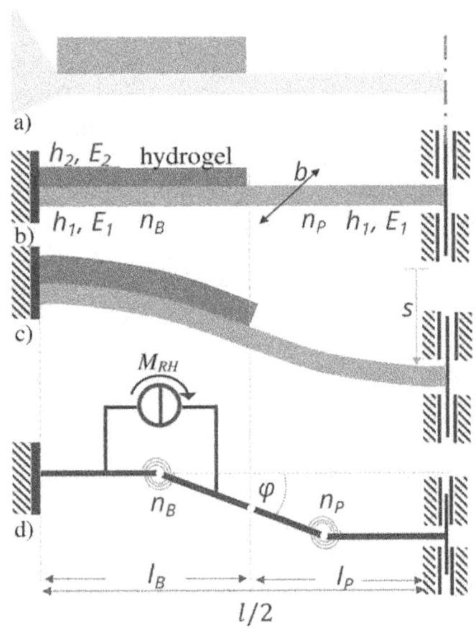

Fig. 3. Half bending beam of the sensor switch as shown in Fig. 1, a) setup, b) model in undeflected and c) deflected state, d) model with lumped components. s deflection, φ deflection angle, M_{RH} moisture-induced moment source. Indices: 1 beam, 2 hydrogel, P uncoated, B hydrogel or hydrogel-coated.

Fig. 3d shows the mechanical model of a beam-like sensors switch. The length-dependent compliance of the beam is expressed by the torsion spring n_B and the inner bending moment M_{RH} from the bimorpic part of the beam caused by hydrogel swelling.

Since the hydrogel is very soft, the compliance n_B is determined by the Young´s modulus E_2 of the sheet metal as is for n_P:

$$n_B = 12\frac{l_B}{Ebh^3}, (1)\ n_P = 12\frac{l_P}{Ebh^3} \quad (1)$$

The total compliance n the humidity moment M_{RH} is acting on amounts to

$$n = \frac{n_P \cdot n_B}{n_P + n_B} \quad (2)$$

$$M_{RH} = -\frac{b}{2}\cdot\frac{E_1E_2h_1h_2(h_1+h_2)}{E_1h_1+E_2h_2}\cdot\varepsilon_{RH} \quad (3)$$

Assuming that the thickness of the hydrogel layer and the bending beam are equal ($h_1 \approx h_2$) it yields [5]:

$$M_{RH} \approx -E_2h_2^2b\cdot\varepsilon_{RH}, \quad (4)$$

where ε_{RH} is the moisture-dependent strain of the hydrogel. The deflection of the bending beam can be described for this simple model by:

$$s = \frac{l}{2}\varphi = \frac{l}{2}nM_{RH} \quad (5)$$

From equations (2), (4) and (5) it follows for the deflection of the bending beam:

$$s = -6\frac{E_2l^2}{E_1h}\frac{l_B}{l}\left(1-\frac{l_B}{l}\right)\cdot\varepsilon_{RH} \quad (6)$$

As can be seen, the deflection s depends significantly on the ratio E_2/E_1 of the Young´s moduli of the hydrogel and the sheet metal, on the geometric dimensions of the hydrogel layer, on the moisture-dependent swelling strain ε_{RH} and on the coating degree l_B/l. A maximum deflection s_{max} is obtained for this simplified model if the ratio l_B/l is 0.5. Fig. 4 shows the calculated curve progression of the beam deflection in dependence of the material parameter from Table 1 and by variating of the ratio l_B/l.

Table 1. Material properties and geometry parameters of beam sheet metal and hydrogel.

Parameter	Beam n_B | n_P	Hydrogel
Length [mm]	0 to 50 | 0 to 50	0 to 50
Width [mm]	12 | 12	12
Thickness [mm]	0.03 | 0.03	0.03
Youngs Modulus [N/mm²]	210000	300 at 98 %RH
Strain (ε_{RH}) at 98 %RH	0	0.06

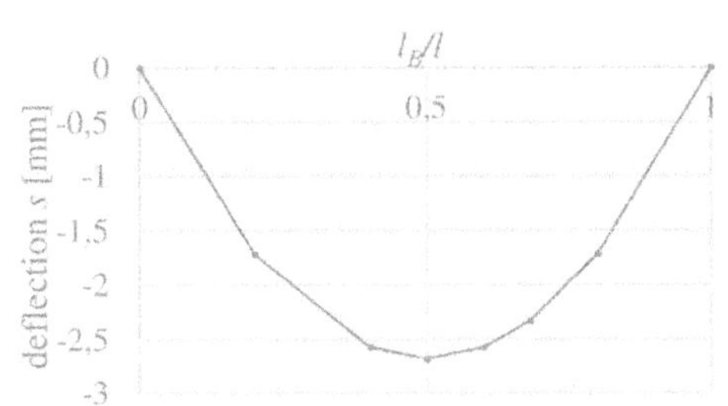

Fig. 4. Calculated deflection of a beam-like sensor switch without switching hysteresis in dependence on the coating degree l_B/l. Parameters from Table 1.

In general, a large deflection of the bending beam can be achieved:

- By a large ratio E_2/E_1 of the Youngs moduli of hydrogel and substrate (bending beam: steel);
- Through thin bending beams;
- By a large swelling strain ε_{RH} or expansion coefficient α_{RH} as a result of a change in relative humidity.

4. Sensor Model with Axial Compression

As mentioned in Section 2, a switching hysteresis, i.e. an S-shaped deflection-versus-humidity curve can be achieved by implementing a pre-deflection due to an axial force [8]. This can be considered in the model from Fig. 3 by adding an axial translational spring n_R, the pre-deflection angle φ_0 and an axial force F_A and with the assumptions $l_B = l_P = l/4$ and $n = n_B || n_P$ (Fig. 5).

To achieve a sudden buckling the pre-deflection of the bending beam is deflected in the opposite direction compared to the deflection direction resulting from the humidity change via M_{RH}. When swelling, the hydrogel-covered part again acts as a bimorph providing a bending moment source depending on the changing humidity.

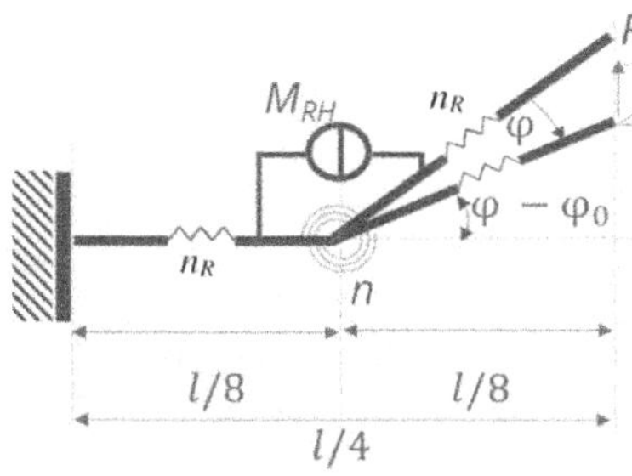

Fig. 5. Mechanical model of the left half of the setup of Fig. 3 with a pre-deflected bending beam with length $l/4$, compliance n, humidity-induced bending moment M_{RH} and axial pre-deflection angle φ_0 due to axial compression. Pre-deflection in opposite direction with respect to the humidity-caused deflection according to Fig. 3.

From the model in Fig. 5, the equilibrium of the moments yields [6]:

$$M_{RH} - F_A \frac{l}{4}(\varphi - \varphi_0) + F_A(\varphi - \varphi_0)^3 = 0 \quad (7)$$

Eq. (7) shows that M_{RH} depends not only linearly on φ but also cubically. This causes the S-shaped curve of Fig. 2 with the two threshold values for switching when humidity is increasing and decreasing, respectively.

5. Experimental

To prove the concept, a macroscopic sensor switch was manufactured from a beam with uniform thickness and width (Table 2, Fig. 6). For sufficient adhesion of the hydrogel to the metal surface, the metal surface was pre-treated. The bending beam was cleaned in a solution made of distilled H_2O, 25 % ammonia and 25 % peroxide (2:1:1) at 85 °C for 15 minutes. Afterwards, the sheet metal surface was treated with oxygen plasma via a plasma etching system (VTD: DREVA 450) at a pressure of 15 Pa at 400 W for 5 min. Then a 3 ml solution of the adhesion-promoter agent made of 0.9 ml (3-Aminopropyl)triethoxysilane (APTES) and 99.1 ml H_2O was applied to the plasma-treated surface and heated in an oven at 100 °C for 10 min [7]. The hydrogel blend was made from 15 wt% poly(vinyl alcohol) (PVA) and 7.5 wt% poly(acryl acid) (PAA) in a mass ratio of 4:1 [8, 9]. Both polymer powders were purchased from Sigma-Aldrich. The powders were solved in deionized water at 70 °C and mixed until a homogenous solution was achieved. To form the hydrogel blend both polymer solutions were mixed at 60 °C for 2 h. With respect to the coating degree of 0.5, the beam was partly covered with the hydrogel solution. For this, 5 ml of the polymer solution was partly deposited to the prepared surface by a pipette and dried at 22 °C for 6 h. For thermally cross-linking of the structured PAA/PVA, the hydrogel-covered beam was baked in an oven at 120 °C for 20 min.

As shown in Fig. 6a, the bending beam was clamped in a fixation and loaded with an axial force by an adjusting screw.

The deflections of the sensor switch was determined in a humidity chamber and measured by means of a camera. The relative humidity was varied with a bubble system and monitored by means of a humidity sensor (HYTELOG USB, B+B sensors). Fig. 6b depicts the measured deflection of the macroscopic sensor switch in dependence on the relative humidity and the calculated deflection from Eq. (7) with the parameters from Table 2.

Table 2. Material properties and geometry parameters of beam sheet metal and hydrogel.

Parameter	Sheet n_B | n_P	Hydrogel
Length [mm]	25 | 25	25
Width [mm]	12 | 12	12
Thickness [mm]	0.03 | 0.03	0.03
Youngs Modulus [N/mm²]	210000	3500 at 5 %RH 300 at 90 %RH

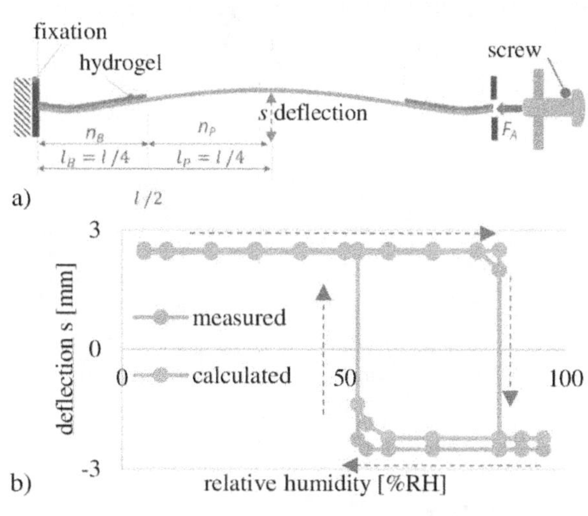

Fig. 6. Beam-like sensor switch with switching hysteresis: a) sensor setup with axial clamping, b) measured and calculated deflection s in dependence on the relative humidity. Values taken from Table 2.

From the results of Fig. 6b, two switching points at 58 %RH and 80 %RH were ascertained. In comparison to the measured results, the calculated deflection values show a deviation of up to 8 % (Fig. 6b). In compression to the measured hysteresis the calculated hysteresis from Eq. (7) shows a good agreement for the switching points at 58 %RH and 80 %RH.

6. Silicon-based Miniaturized Sensor Switch

To make sensors small and cost-efficient silicon-based miniaturized sensors are targeted. As shown in Fig. 1 flexure plates can be used as known from commercial silicon-based pressure sensors. Parts of the plate are covered with hydrogel to provide the needed humidity-dependent bending moment. The pre-deflection can be implemented by covering the inner part of the silicon plate with silicon dioxide (Fig. 7). SiO_2 has a thermal expansion coefficient ($0.5 \cdot 10^{-7}$ K^{-1}) that is smaller than that of the silicon plate ($2.3 \cdot 10^{-6}$ K^{-1}) [10].

Deposition was done via a PE-CVD (plasma-enhanced chemical vapour deposition) process using a shadow mask.

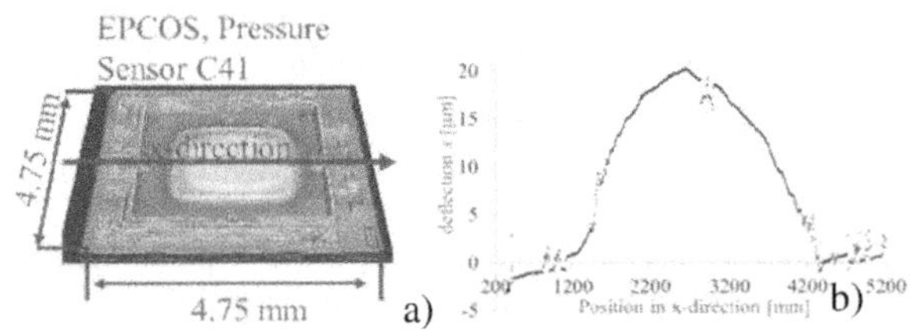

Fig. 7. a) Pressure sensor with an 18.8 μm thick silicon bending plate and a 2.5 μm thick SiO_2 layer deposited in the middle part of the plate, b) pre-deflection along the red arrow.

Fig. 8 shows the deflection *s*, when the humidity content was changed gradually between 25 %RH and 98 %RH. From the results, two switching points at 59 %RH and 85 %RH were observed as expected.

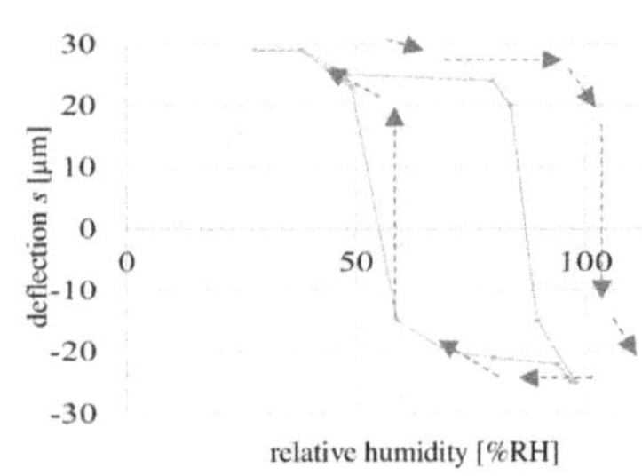

Fig. 8. Deflection *s* of a silicon-based sensor switch with both patterned SiO_2 layer (inner part of the plate) and structured hydrogel layer (outer part) in dependence on relative humidity.

7. Conclusion

To calculate the humidity-based beam deflection depending on the hydrogel pattern, based on the beam theory a mechanical model was developed. A maximum of the deflection s_{max} is reached when the degree of coverage l_B/l amounts to 0.5. The mechanical model was complemented with an axial force F_A to describe the switching hysteresis for safe and fast opening and closing of the switch. To demonstrate the switching hysteresis a beam-like sensor switch was manufactured from a metal sheet with uniform width and thickness partly covered with a hydrogel layer (coating degree 0.5). It was axially compressed with a force F_A leading to the targeted switching hysteresis. The width of the hysteresis can be varied by adjusting the force F_A. To create miniaturized MEMS-based sensor switches commercial silicon pressure sensors have been used, where the pre-deflection and the corresponding axial compressive force was evoked by a SiO_2 layer structured in the middle part of the Si-bending plate.

References

[1]. L. T. Chen, C. Y. Lee, W. H. Cheng, MEMS-based humidity sensor with integrated temperature compensation mechanism, *Sensors and Actuators A*, Vol. 147, 2008, pp. 522-528.

[2]. T. Frank, A. Steinke, G. Gerlach, Binary ZeroPower Sensors: An alternative solution for power-free energy-autonomous sensor systems, *Microsystem Technologies*, Vol. 18, Issue 7-8, 2012, pp. 1225-1231.

[3]. C. Bellmann, R. Sarwar, A. Steinke, T. Frank, H. F. Schlaak, G. Gerlach, Development of a humidity micro switch based on humidity sensitive polymers, in *Proceedings of the 23rd Micromechanics and Microsystems Europe Workshop (MME'12)*, Ilmenau, Germany, 2012.

[4]. C. Bellmann, Leistungslose Feuchteschwellwertschalter, PhD Thesis, *TU Dresden*, 2016.

[5]. U. Marschner, G. Gerlach, E. Starke, A. Lenk, Equivalent circuit models of two-layer flexure beams with excitation by temperature, humidity, pressure, piezoelectric or piezomagnetic interactions, *Journal of Sensors and Sensor Systems*, Vol. 3, 2014, pp. 187-211.

[6]. N. Gulnizkij, G. Gerlach, Autonomous threshold sensor for humidity measurements, *Journal of Sensors and Sensor Systems*, July 2019, submitted.

[7]. N. Gartmann, C. Schütze, H. Ritter, D. Brühwiler, The effect of water on the functionalization of mesoporous silica with 3-aminopropyltriethoxysilane, *Journal of Physical Chemistry Letters*, Vol. 1, 2010, pp. 379-382.

[8]. J. Jose, F. Shehzad, M. A. Al-Harthi, Preparation method and physical, mechanical, thermal characterization of poly(vinyl alcohol)/poly(acrylic acid) blends, *Polymer Bulletin*, Vol. 71, Issue 11, 2014, pp. 2787-2802.

[9]. K. F. Arndt, A. Richter, S. Ludwig, J. Zimmermann, J. Kessler, D. Kuckling, H. J. Adler, Poly (vinyl alcohol)/poly (acrylic acid) hydrogels, FT-IR spectroscopic characterization of crosslinking reaction and work at transition point, *Acta Polymerica*, Vol. 50, Issue 11-12, 1999, pp. 383-390.

[10]. G. Gerlach, W. Dötzel, Introduction to Microsystem Technology, *John Wiley & Sons*, 2008.

(072)

Carbon Fiber Brush Electrode for Detection of Bioactive Compounds Transported Across Biomimetic Membranes

J. Skopalová, D. Riman, R. Jerga and P. Barták
Palacký University Olomouc, Faculty of Science, Department of Analytical Chemistry, 17. Listopadu 12, 771 46 Olomouc, Czech Republic
Tel.: +420-585634442, fax: +420-585634433
E-mail: jana.skopalova@upol.cz

Summary: Carbon fiber brush electrode was designed as a flexible sensor for amperometric detection of bioactive compounds diffusing across the supported phospholipid membrane in a miniaturized two-compartment electrochemical cell. Asolectin, a mixture of phospholipids from soybean, was used to form a model phospholipid membrane simulating a real cell membrane. Applicability of the designed sensor was demonstrated on the monitoring of dopamine diffusion across the asolectin covered polycarbonate membrane whereat the equilibrium state was achieved within 20 hours.

Keywords: Carbon fiber brush electrode, Asolectin, Phospholipid membrane, Dopamine, Amperometry.

1. Introduction

Model phospholipid membranes (PLMs) simulating natural cell membranes can be very useful in studying the penetration of biologically important compounds (BICs) into the intracellular space, regardless of whether these species are characteristic of living organisms or are xenobiotics such as drugs and pesticides. Many of these compounds are electroactive and can therefore be detected by electroanalytical techniques.

Recently, a glass two compartment cell has been constructed to study the transport of metal ions complexes with low molecular organic acids through artificial PLMs, where the samples were taken from the cell and analyzed in an external voltammetric vessel [1]. In our work, we present the miniaturized design of the two-compartment cell adjusted for on-line monitoring of the transport of electroactive BICs across the supported PLM (sPLM). For this purpose, carbon fiber (CF) brush electrode (CFBE) was designed as a flexible electrochemical sensor.

CFBEs have found wide spectrum of applications. Advantages of carbon fibers such as high electrical conductivity, large specific surface, flexibility and ability to be designed and tailored for specific applications are often exploited in electrochemical reactors for effective generation of desired reactants or electricity in microbial fuel cells [2, 3]. Recently they have been used as substrates for atmospheric solids analysis probe mass spectrometry in combination with electrochemical generation of oxidation products [4] and as an electrode substrate for a phospholipid modified electrochemical sensor for the detection of dopamine and its metabolite [5].

2. Experimental

2.1. Design of Carbon Fiber Brush Electrode

The working CFBE was made of carbon fibers extracted from carbon cloth (KTC-03, Karbotechnik, Czech Republic). A bundle of heat-treated carbon fibers (500 °C, 30 min) was tied with 0.25 mm copper wire, serving as electrical contact, pulled in a glass capillary (1.3 mm i.d.) and sealed with epoxide, leaving CFs protruded outside the capillary [5]. Before the use CFs protruding from the tube were ultra-sonicated in hexane, methanol and distilled water. Electrochemical cleaning and activation of carbon fibers were performed in sulfuric acid solution (0.1 mol/dm^3) by potential cycling between -1.0 V and +1.5 V (50 cycles, scan rate 0.5 V/s). Finally the CFBE was washed in distilled water.

2.2. Design of Two-compartment Electrochemical Cell with Supported Phospholipid Membrane

The cell (Fig. 1) consists of two compartments made of 2 ml glass vials with drilled holes for the working and auxiliary electrodes, for electrolyte and sample filling, and with a coaxial tubes for the reference electrodes. Polycarbonate membrane separating both compartments serves as a substrate for sPLM.

The polycarbonate membrane (PCM) was covered with a phospholipid layer using asolectin solution (20 g/dm^3) in hexane dropped on the circular polycarbonate membrane filter (12 mm diameter). After hexane evaporation, the modified membrane was inserted into two-compartment cell (between the two vials). The formation and self-assembling of the phospholipid layer in the pores of the filter was made in solution of NaCl (9 g/dm^3) and monitored using electrochemical impedance spectroscopy (EIS) in four-electrode configuration. A compact phospholipid layer was formed in 2 hours and was stable for at least 24 hours.

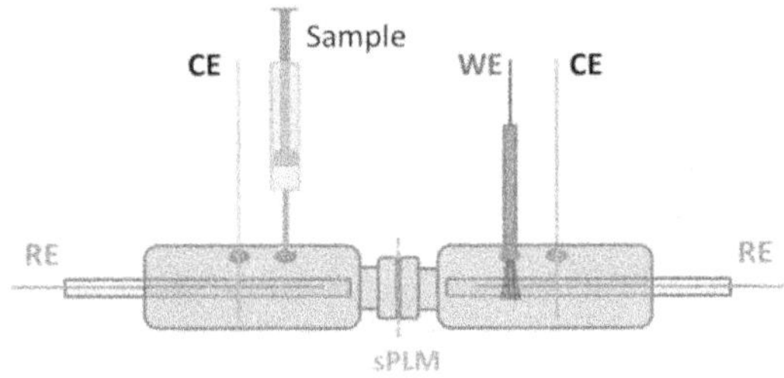

Fig. 1. Scheme of two-compartment electrochemical cell for monitoring of transport of compounds across the supported phospholipid membrane (sPLM). WE – working carbon fiber brush electrode, RE – reference Ag/AgCl/3M KCl electrode, CE – auxiliary Pt electrode.

All electrochemical measurements were carried out on Autolab PGSTAT128N (Metrohm Autolab, The Netherlands).

3. Results and Discussion

The diffusion of an electroactive compound across the asolectin sPLM was tested using amperometry in three-electrode system with working CFBE. Based on the previous results and experiences [5], dopamine was chosen as the model compound.

First, the amperometric monitoring of dopamine was tested in the cell with unmodified PCM. One compartment of the cell with CFBE sensor was filled with phosphate buffer solution of pH 6.5 as the supporting electrolyte. Solution of dopamine in the phosphate buffer solution of pH 6.5 was injected into the second compartment. Amperometric measurement was carried out on CFBE at potential +0.3 V which corresponds to the limiting current of dopamine oxidation. Current signal was recorded every 5 min (Fig. 2). Equilibrium concentration of dopamine in both compartments was reached approximately within 17 hours.

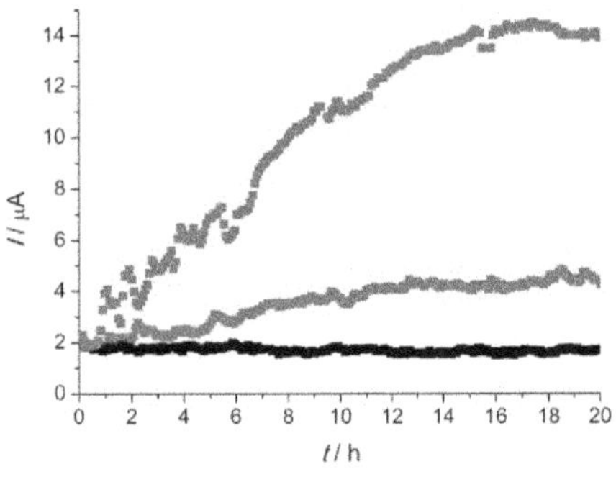

Fig. 2. Amperometric monitoring of dopamine diffusion across polycarbonate membrane without (green) and with asolectin layer (red). Supporting electrolyte – phosphate buffer solution of pH 6.5 (black).

Much slower diffusion of dopamine across the membrane was observed in the experiment performed with sPLM (Fig. 2). The control experiment was performed with asolectin sPLM in absence of dopamine indicating stability of the amperometric signal in the measuring system.

Three times lower current signal of dopamine achieved with sPLM compared to uncovered PCM results from strong binding of dopamine to asolectin PLM [5]. Similar experiment with $K_4[Fe(CN)_6]$ instead of dopamine provided constant current signal within 24 h and thus proved the sPLM prevents diffusion of hydrophilic ionic compounds and is selectively permeable for lipophilic substances.

4. Conclusions

CFBE was proved to be a suitable sensor for amperometric monitoring of lipophilic electroactive BICs diffusing across the sPLM. Similar design of the cell with CFBE sensor is currently developed for the monitoring of partition of lipophilic compounds in the phospholipid liposome/water system.

Acknowledgements

This work was supported by the Grant projects of Czech Science Foundation (17–05387S) and Palacký University Olomouc (IGA_PrF_2019_028).

References

[1]. M. Parisova, T. Navratil, I. Sestakova, J. Jaklova Dytrtova, V. Marecek, Influence of low molecular weight organic acids on transport of cadmium and copper ions across model phospholipid membranes, *International Journal of Electrochemical Science,* Vol. 8, Issue 1, 2013, pp. 27-44.

[2]. G. S. Xia, Y. H. Lu, H. B. Xu, Electrogeneration of hydrogen peroxide for electro-Fenton via oxygen reduction using polyacrylonitrile-based carbon fiber brush cathode, *Electrochimica Acta,* Vol. 158, Issue 2015, pp. 390-396.

[3]. A. J. Hutchinson, J. C. Tokash, B. E. Logan, Analysis of carbon fiber brush loading in anodes on startup and performance of microbial fuel cells, *Journal of Power Sources*, Vol. 196, Issue 22, 2011, pp. 9213-9219.

[4]. J. Skopalová, P. Barták, P. Bednář, H. Tomková, T. Ingr, I. Lorencová, P. Kučerová, R. Papoušek, L. Borovcová, K. Lemr, Carbon fiber brush electrode as a novel substrate for atmospheric solids analysis probe (ASAP) mass spectrometry: Electrochemical oxidation of brominated phenols, *Analytica Chimica Acta,* Vol. 999, pp. 60-68.

[5]. R. Jerga, V. Müllerová, J. Štěpánková, P. Barták, H. Tomková, J. Rozsypal, J. Skopalová, Phospholipid-modified carbon fiber brush electrode for the detection of dopamine and 3,4-dihydroxyphenylacetic acid, *Monatshefte Fur Chemie,* Vol. 150, Issue 3, 2019, pp. 359-400.

(073)

An Overview of Magnetometer Sensors Performance for the Purposes of Traffic Flow Monitoring

M. Hodoň [1], O. Karpiš, [1] P. Ševčík, [1] J.Miček and A.Kociánová [2]
[1] University of Žilina, Department of Technical Cybernetics, Univerzitná 8215/1, 01026, Slovakia
[2] University of Žilina, Department of Highway Engineering, Univerzitná 8215/1, 01026, Slovakia
E-mail: michal.hodon@fri.uniza.sk

Summary: The paper deals with the problematic of vehicle detection in scenarios, where it is difficult to use standard methods. The measurement scenario is hereby described with a car route located in remote mountains. The route is the main interconnection between Slovakia and Czech Republic. Due to its importance, it is necessary to monitor the traffic flow. For this purpose, the special sensor is being developed, which is based on the implementatiom of anisotropic magneto-resistive (AMR) sensory. In this paper we focused on the performance of actually available AMR sensor magnetometers, and their comparison at specific conditions.

Keywords: Magnetometer, Sampling, Performance, Power consumption, Vehicle detection.

1. Introduction

The world of information technology is now more than ever characterized by interconnecting of individual systems. According to IoT Analytics [1] more than 17 billion devices are currently connected, of which about 7 billion are Internet of Things (IoT). It should be noted that smart phones, tablets and laptops are not included in this amount. The number of active IoT devices is expected to rise to 10 billion by 2020 and to 22 billion by 2025.

The largest number of devices are currently connected via short-range technologies (e.g. Bluetooth, ZigBee, Z-wave).

In summary, these technologies are referred to as Wireless Personal Networks (WPAN). It is expected that a large part of the future growth of IoT devices will be provided by the development of low-power Wide Area Networks (LPWANs). The main standards in this area (Sigfox, Lora, NB-IoT) currently account for more than 25 million connected devices, up to 2 billion by 2025. Very low power consumption (i.e. long battery life, possibility to use green power supply) and a very large range that can be more than 20 km promises increase to the large number of applications. All these technologies use mainly free frequency bands for communication. Obviously, the only way to achieve the aforementioned growth with the limited resources such as the frequency band, is to use it as efficiently as possible [2].

Most of the IoT data goes to public or private could storage where are saved, processed and visualized. This trend is unlikely to change in the future. But what has to be changed is the qualitative content of data transmitted to cloud servers. Due to limited capacity of wireless technologies, it is impossible to transfer all data in raw, unprocessed state. The further, the greater is the importance is to have the data preprocessing directly at the place of the capturing. Moving data processing closer to its place of origin (of moving applications and functionalities closer to the user) is currently a much discussed topic and is generally referred to as “Edge computing”. The importance of this trend in the IoT field is also evident from the activities that the largest cloud companies such as Microsoft, Google and Amazon are developing in this area. As an example, Google Edge TPU is a processor that is focused on using machine learning and artificial intelligence “on the edge” [3].

Thanks to the continuous development of technologies in all areas of electronics (from sensors, through microcontrollers to radio modules), the number of IoT applications is growing rapidly. Lower power consumption, as well as higher computing power for even the simplest processors, also allow the use of sensor types that could not be used in the past due to the large amount of data they produce. This applies mainly to image and sound recording but also to other signals with higher data volume (vibration measurement, magnetic field changes, etc.).

The relatively high computational power available at the place of signal acquisition allows the data volume to be reduced to an acceptable level with respect to the communication channel used.

2. Problem Description

Vehicle detection technologies was massively developed in the last couple decades. Their developments went hand in hand with the requirements, which are currently put on the measurement strategies, where the accuracy is viewed as a necessity. According to the resources available, different measurement methods can be applied, with different results reached. For example with camera and image recognition techniques, different statistics can be derived, such as number of vehicle of different types, speed averages, problematic drivers, etc. Problems only occur, where the measurement scenario provide only a limited number of resources but the measurement requirements are still high. For this purpose it is necessary to build the special sensory devices.

Considering the fact that most of all street vehicles have noteworthy measures of steel material in their body, AMR sensor magnetometers represent a good technical mean for identifying vehicles. As mentioned authors in [4], application of such sensors for the purposes of traffic monitoring provides the cost-effective and easy-to-install solution. After the application of Discrete Fourier Transform and Principal Component Analysis (PCA) for extracting features from magnetic signatures, the detected vehicles can be easily classified into three different categories. Beside not so demanding application as is the detection of static vehicle at, for example, parking lot [5], most application ask for the real sensor performance, so just the features with time-stamp can be extracted from the measured magnetic field. It is therefore evident, that signal pre-processing at the sensor side, or "sensor edge", is very important.

Different approaches can be applied at the sensor side, with different performance requirements. For example authors in [6] proposed and implemented a simple algorithm based on a nonlinear transformation of the measurements as correlation-based driving direction classification. In [7], simple five-state machine process algorithm was developed and implemented for real-time vehicle detection and counting. Speed estimation there relied on precise time-stamping of vehicle arrival and departure instants, which depended on T-sync algorithm accuracy. However, sometimes it is necessary to add some other supporting sensors to increase the measurement accuracy. In this case particle filtering can play the role. In [8], an accelerometer sensor was added into the system. The paper showed that this approach can serve a kind of feasible solution. For the sensor fusion, multirate particle filter was used.

Examples above proved that AMR sensors can be considered as reliable, low-power and cost-effective technology, which is suitable for vehicle detection. An integration of AMR sensors into the traffic infrastructure opens the space for novel applications, aimed at the topic of intelligent transportation systems (ITS). These sensors can for example measure the traffic flow, predict congestion, or adaptively control traffic routes [9].

3. ITS Sensor Structure

As it was mentioned within examples above, the utilization of AMR sensory would be very tough without signal preprocessing at the sensor side, or eventually, signal post-processing at the side of final application. Therefore, though the AMR sensor can be considered as a very important part of the final application, there are also other parts (Fig. 1):

- Control unit – this is crucial part of any ITS sensor, because it controls all algorithmization processes. Parameters such floating point operations per second (FLOPS) as well as power requirements are the key features to be considered when selecting appropriate controller.
- Power – in most application scenarios, the power cords are missing, therefore requirements on power efficiency are very important. Different energy harvesting techniques are obviously used as supporting technologies for the sensor lifetime extension.
- Memory – in applications where the data are stored at the sensor side, it is necessary to integrate power efficient solution for the data storage (SD card 50 mA @ 3.3 V, SPI Flash (25 mA @ 3,3 V).
- Communication subsystem – communication with user or with parent application should be primary used for the data download, secondary for the settings adjustments. Communication can be event-based (button, shake,..) or synchronous waking and sending the data in defined time window. Wireless (Bluetooth, Xbee, ISM proprietary protocols ...) or wired (USB, RS485, Ethernet ...) means can be used for these purposes.
- Sensor subsystem – beside magnetometers, which are going to be described in another section, other supporting sensors could take place according to the application specifics. For example GNSS sensors for timing purposes, sound sensors for the classification of voices, or gas sensors for the measurements of environmental conditions.

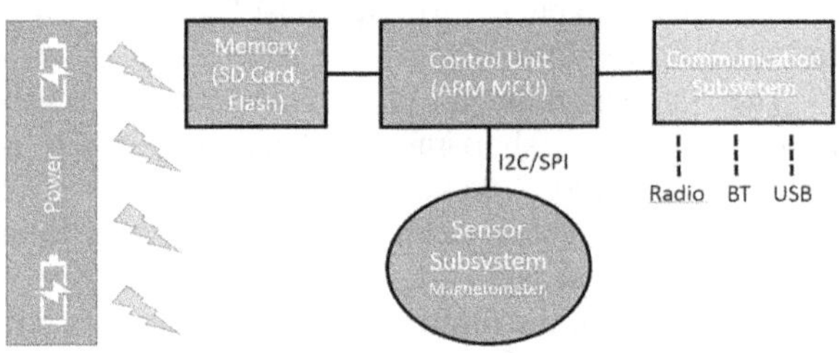

Fig. 1. ITS sensor structure.

4. Magnetometer Selection

There are several digital AMR magnetometers as SoC applicable for ITS applications available at the market with different performance and quality measures. To select the right one, first the application scenario has to be described.

The desired application scenario comes out from the project "Innovative monitoring and analysis of traffic on the cross-border road network", which is aimed at the development of special low-cost, low-power, wireless, portable magnetic traffic counter for the analysis of the traffic density at the cross-border area between Slovakia and Czech Republic. This area can be described with mountains and forests, where the electricity is provided rarely. The proper sensor should be able to detect the disturbances of magnetic field relied to its proximity, and thus to detect the vehicle size and direction (Fig. 2).

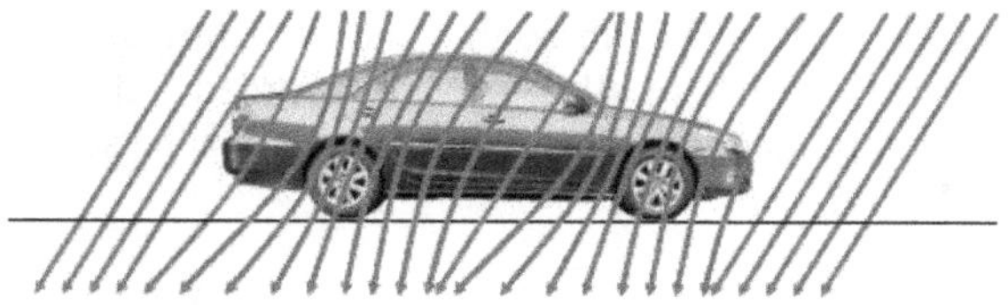

Fig. 2. Earth's magnetic field through vehicle [20].

For this purpose a market analysis was performed, resulting in the choice of 10 magnetometers for the further evaluation. The list of chosen magnetometers is provided in the Table 1, together with the basic parameters which were extracted from [10-19]. We were focused on chips, which are without any problems available at the market either from re-sellers or from specified suppliers in amount of single pieces. From the reference sources is also obvious, that we were focused on the magnetometers with digital output for minimization of power consumption.

From the table is evident, that there are some parameters, which can be left out from the comparison, since they are almost the same among all of the sensors. What is important in terms of the aimed measurements is following:

- Output noise – lowest (around 0.3 uT and less) – to enable the sensor to measure accurately and invulnerable to any noise source;
- Measurement sensitivity – highest (around 0.1 uT/LSB and higher) – to enable the sensor to sense any change of magnetic field properly;
- Measurement range – low – to provide the best measurement quality to sense with highest sensitivity in relatively low measurement range;
- Output data rate – highest – to sample the data at highest possible speed (not to loose any information);
- Power consumption – lowest – power consumption of whole device is very important.

What is however also important, and what could be easily investigated in laboratory conditions is the measurement repeatability, resp. the spread of measurement data of each magnetometer axis under the same laboratory conditions. In terms of measuring the variability of data spread, the standard deviation S is the preferred and most used measure as a square root of the variance (1):

$$S_K = \sqrt{\frac{1}{N-1}\sum_{i=1}^{N}|M_{Ki} - \mu_K|^2}, \tag{1}$$

where μ is the mean of measured value in all three axes X, Y, Z of magnetometer of N measures, where $K \in \{X, Y, Z\}$ (2):

$$\mu_K = \frac{1}{N}\sum_{i=1}^{N} M_{Ki} \tag{2}$$

Interpretation of the coefficient of variation (CV) for each magnetometer can be then used to interpret the variability of a sample relative to its mean by dividing the standard deviation by the mean (3):

$$CV = \frac{S}{|\mu|} \tag{3}$$

Table 1. List of the tested magnetometers [10-19].

	Magnetometer Type	Output Noise	Sensitivity [uT/LSB]	Measurement Range +- [uT]	Output Data Rate [Hz]	Current Consumption [uA]	Idle [uA]
1	**LSM303AGR**	0.3 uT RMS	0.15	5000	100/150	950/250 @ 100 Hz	2
2	**LIS2MDL**	0.3 uT RMS	0.15	5000	100/150	950/250 @ 100 Hz	1.5
3	**LSM303AH**	0.3 uT RMS	0.15	5000	100/150	950/250 @ 100 Hz	2.5
4	**LSM303C**	0.35 uT RMS	0.058	1600	80	270/40 @ 20 Hz	6
5	**LSM9DS1**	0.35 uT RMS	0.014 - 0.058	1600	80	600 @ 20 Hz	6
6	**MPU9250**	N/A	0.15	4900	100	280 @ 8 Hz	3
7	**FXOS8700CQ**	<100 nT/Hz @ 100 Hz	0.1	1200	800	575 @ 400 Hz	2
8	**MLX90393**	Varies acc. to settings	0.16 - 5.8	4800	700	100 @ 10 Hz	2.4
9	**RM3100**	30-15 nT	0.05 - 0.013	800	1600/n	70-260 @ 8 Hz	1
10	**HMC 5983**	N/A	0.07 - 0.40	100-800	220	100 @ 7.5 Hz	2

To properly measure the data, as to interpret the inputs for (1), (2) and (3), the special measurement set was developed and installed as can be seen in the block schematic in Fig. 3.

The measurement scenario consisted of the rotating arm which was fixed on the precisely driven stepper motor which was controlled through the AtMega8 MCU and two H-bridges. On the arm was placed strong neodymium magnet which was moved across the defined track. The track consisted of 42 measurement points, on which the magnet stops for exactly 1200ms. Every tested magnetometer from the group available in Table 1, was place exactly on the same position fixed in a certain distance of 1cm from the magnet installed at the rotating arm. So during the measurement of each magnetometer, same measuring conditions were achieved:

- Same distance between the moving magnet and the statically placed measuring unit;
- Same movement of rotating arm, characterized with the same speed and the same track;
- Same position of magnetometer relative to the rotating arm.

The magnetometers were set to the maximum output data rates with no filter implemented, thus the real raw data could be measured. All magnetometers were interconnected through I2C interface with Raspberry Pi v 3.0, through which all settings were adjusted and where all data were recorded. Influence of rotating magnet to the magnetometer can be seen in the Fig. 4.

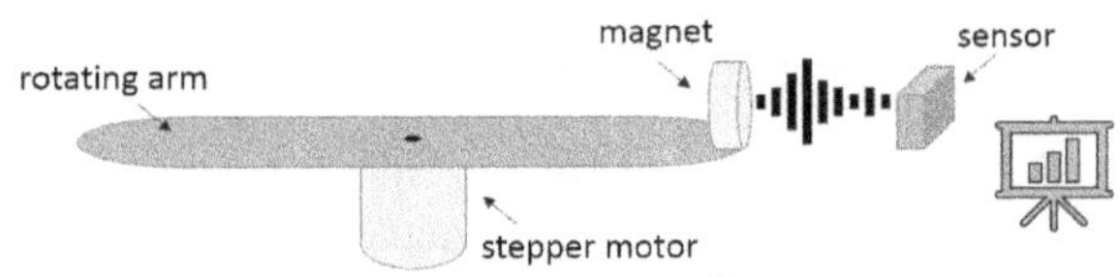

Fig. 3. Block schematic of the measurement set.

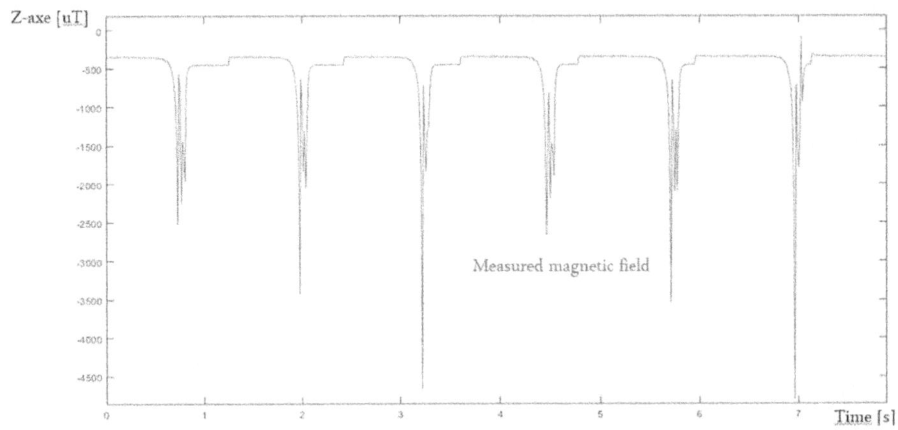

Fig. 4. Influence of rotating magnet to the magnetometer.

Whole test scenario in real can be seen in the Fig. 5.

The measurement results can be seen in the Table 2 as the interpretation of the coefficient of variation for X, Y, and Z axes of every tested magnetometer.

From the results is evident, that magnetometers LSM303AGR, LIS2MDL and LSM303AH has probably the same AMR core. From the results is also evident that LSM303C, LSM9DS1, FXOS8700CQ and RM3100 provides the best results. They will be further investigated in the scenario of real traffic.

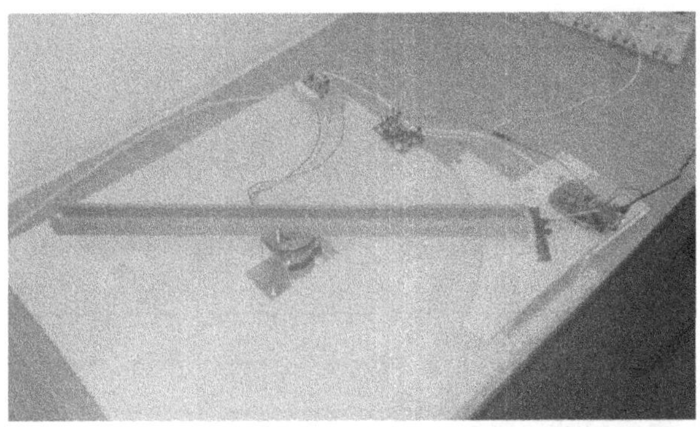

Fig. 5. Test scenario.

Table 2. Measured results.

	CV(X)	CV(Y)	CV(Z)
LSM303AGR	0.0750	0.0075	0.0332
LIS2MDL	0.0748	0.0073	0.0331
LSM303AH	0.0748	0.0073	0.0332
LSM303C	0.0037	0.0209	0.0100
LSM9DS1	0.0207	0.0038	0.0149
MPU9250	0.0062	0.0237	0.1391
FXOS8700CQ	0.1010	0.0441	0.0221
MLX90393	0.2560	0.2837	0.0583
HMC 5983	0.0273	0.0086	0.0283
RM3100	0.0031	0.0042	0.0106

5. Conclusion

In this paper the basic selection of market-available magnetometers was made. According to measurement results, the smaller groups of magnetometers for further evaluation was selected. If we consider the output data rates of single magnetometer as the important parameter, which serves the additional information about the measured signal, we can select the best two sensors FXOS8700CQ and RM3100, on which the final ITS application could be built.

Acknowledgement

This contribution is the result of the project Innovative monitoring and analysis of traffic on the cross-border road network, ITMS 304011P777 supported by the Interreg V-A SK-CZ Cross-border Cooperation Program.

References

[1]. State of the IoT 2018: Number of IoT Devices Now at 7B – Market Accelerating, https://iot-analytics.com/state-of-the-iot-update-q1-q2-2018-number-of-iot-devices-now-7b/

[2]. K. Mekki, et al., A comparative study of LPWAN Technologies for large-scale IoT deployment, *ICT Express*, Vol. 5, Issue 1, 2019, pp. 1-7.

[3]. Edge TPU, https://cloud.google.com/edge-tpu/

[4]. H. Tafish, W. Balid, H. H. Refai, Cost effective vehicle classification using a single wireless magnetometer, in *Proceedings of the IEEE International Wireless Communications and Mobile Computing Conference (IWCMC'16)*, Paphos, Cyprus, 5-9 Sept. 2016, pp. 194-199.

[5]. A. Gadžović, N. Lekić, I. Radusinović, Automation of parking services using UHF RFID technology and magnetometers, in *Proceedings of the 21st Telecommunications Forum (TELFOR'13)*, Belgrade, Serbia, 2013, pp. 597-600.

[6]. N. Wahlström, R. Hostettler, F. Gustafsson, W. Birk, Classification of driving direction in traffic surveillance using magnetometers, *IEEE Transactions on Intelligent Transportation Systems*, Vol. 15, Issue 4, Aug. 2014, pp. 1405-1418

[7]. W. Balid, H. Tafish, H. H. RefaiIntelligent, Vehicle counting and classification sensor for real-time traffic surveillance, *IEEE Transactions on Intelligent Transportation Systems*, Vol. 19, Issue 6, June 2018, pp. 1784-1794.

[8]. R. Hostettler, P. M. Djurić, Vehicle tracking based on fusion of magnetometer and accelerometer sensor measurements with particle filtering, *IEEE Transactions on Vehicular Technology*, Vol. 64, Issue 11, Nov. 2015, pp. 4917-4928.

[9]. K. Zaatouri, Mo. H. Jeridi, T. Ezzedine, Adaptive traffic light control system based on WSN: Algorithm optimization and hardware design, in *Proceedings of the 26th International Conference on Software, Telecommunications and Computer Networks*, Split, Croatia, 13-15 Sept. 2018, pp. 1-6.

[10]. LSM303AGR Datasheet, DocID027765 Rev 10, *STMicroelectronics*, 2018.

[11]. LIS2MDL Datasheet, DocID030621 Rev 5, *STMicroelectronics*, 2018.

[12]. LSM303AH Datasheet, DocID027766 Rev 7, *STMicroelectronics*, 2018.

[13]. LSM303C Datasheet, DocID024975 Rev 2, *STMicroelectronics*, 2014.

[14]. LSM9DS1 Datasheet, DocID025715 Rev 3, *STMicroelectronics*, 2015.

[15]. MPU-9250 Product Specification Revision 1.1, Document Number: PS-MPU-9250A-01, *InvenSense Inc.*, 06/20/2016,

[16]. FXOS8700CQ Technical Datasheet, *NXP Semiconductors*, April 2017,

[17]. MLX90393 Triaxis® Magnetic Node Datasheet, Revision 003, *Melexis*, 2017.

[18]. RM3100 & RM2100 Sensor Suite User Manual, Doc 1017252 R07, *PNI Sensor Corporation*, 2015.

[19]. HMC5983 Datasheet, Form # 900425, *Honeywell International Inc.*, 2011.

[20]. Honeywell Application Note – AN218 Vehicle Detection Using AMR Sensors, *Honeywell International Inc.*, 2005.

(074)

Adaptable Power Consumption Profiles for Wearable Localization Devices

Manuel Faustino [1], **Jorge Calado** [1], **João Sarraipa** [1] **and Ricardo Jardim-Gonçalves** [1]
[1] CTS, UNINOVA, DEE, Faculdade de Ciências e Tecnologia, Universidade NOVA de Lisboa, 2829-516 Caparica, Portugal
E-mails: m.faustino@campus.fct.unl.pt, jsc@uninova.pt, jfss@uninova.pt, rg@uninova.pt

Summary: The technology has progressed in recent years as well as the average life expectancy, which has led to a visible increase in the number of elderly people. As people get older, the health conditions usually degrade. As consequence, the help needed in daily routines changes considerably. Thus, a lot of technology solutions is being developed with the aim to try to mitigate this problem. There are a number of diseases that cause a progressive decline in the person's physical and cognitive capabilities, which are intrinsically associated with the advancing age. Alzheimer's is one of the most common diseases in this specific area and one of those that needs more solutions. Since those suffering from this disease can suffer from cases of wandering that in turn can lead to traumatic experiences. This paper analyzes the use of dynamic power consumption profiles that deal with the activation and deactivation of the various functions offered by wearable localization devices, increasing their active stage duration in support of people with dementia.

Keywords: IoT, Wireless sensors, Dynamic power profiles, Geographical location, Healthcare, Alzheimer, Dementia.

1. Introduction

Over the past few years, there has been enormous technological growth at all levels and in a wide range of areas of our society [1].

Similarly, in recent years we have witnessed a constant increase in life expectancy, leading to a marked increase in the number of elderly people. A report from United Nations [2] indicates that at the time (2013) the number of elderly people in the world is about 841 million, which is four times higher than the 202 million that lived in 1950. Besides that, they also predict that in 2050 there will be 2 billion of elderly people in the world. As we can observe in literature [3], as people get older, the health conditions usually degrade. Consequently, the help needed for people in these advanced ages to do their daily routines changes considerably.

Dementia [4] is the term used to describe symptoms of a group of diseases that cause a progressive decline in the person´s physical and cognitive capabilities, and is mainly associated with the advancing age. This particular problem needs specific solutions, mainly regarding the localization devices. Alzheimer is the most common cause of dementia, which being a progressive and degenerative disease affecting cognitive capabilities, can lead to cases of wandering. Due to this specific health situation, simply forgetting to recharge any device would greatly reduce their effectiveness.

Since wireless sensors can operate independently and with low power consumption, they can play an important role in helping to mitigate this mentioned problem. Wearables are a particular case of these kind of wireless sensors. Since they can be attached to any piece of clothing makes them adequate to be used in this problem resolution scenario, even more when associated to modern communication solutions and effective control over the power consumption, enabling the devices to run for several years without having any recharge or battery replaced.

Having this in mind the authors conceived an application scenario that uses a wearable device to handle wandering situations of people who suffer from any dementia condition. Additionally, the proposed wearable device will transmit its current localization to a central system which processes all the information and tries to predict any case of wandering, avoiding traumatic experiences to the person with dementia (PwD). To achieve these predictions the system needs to have pre-defined safe and unsafe zones for each specific user, which the central system will rely on to later communicate to the device making the device dynamically adapt its uptime working duration to save energy.

In conclusion with this work the authors intend not only to develop a wearable localization device, but mainly to configure it to be self-adaptable in relation to the level of potential dangerous for a wandering situation of its user. This solution uses a pre-defined set of communications protocols and behaviors establishing a set of power consumption profiles with the objective to improve the device working uptime and effectiveness.

This paper is organized as follow. In Section 2, related solutions are analyzed. Section 3 describes the system architecture. Experimental setup and test methodology are presented in Section 4. The results are presented in Section 5 and the conclusions are given in Section 6.

2. Related Work

To design a device with the mentioned type of requirements, the authors took into consideration the integration of different technologies and methods of communication, such GSM [5, 6], NB-IoT [7], LoRa

[8], among others, in parallel to geolocation modules like GPS [9].

Numerous wearable solutions, capable of localization similar to the one proposed by authors, were identified and some of them are presented in the following.

In [10], the authors proposed a system platform that tracks the location and type of motion of elderly people inside their homes in order to reduce the worries of the respective caretaker.

In terms of technology, the system consists in a wearable device composed by a microcontroller board (Raspberry Py) and a motion sensor that uses BLE [11] technology. Both combined with a set of BLE scanners, send data to a MQTT [12] server and the server publishes that received data to a subscribing computer, which then processes that information in order to track the user position and type of motion.

A different solution is presented in [13] that enables the monitoring of the elderly health status continuously, through the integration of some sensors and also guarantees indoor and outdoor localization. This solution also has a remote reasoning system that receives all the data produced and does all the processing that allows generating alerts in cases where they are justified. In terms of technology, the proposed system is composed by a GPS module which obtains the outdoor position of the user and sends it to a GPRS module which is responsible for send that user position to the system central storage. To perform the indoor localization the system uses BLE technology as seen in the previous example. The authors of this solution also propose a method to ensure a longer battery life which is through the use of the BLE module only when the user is located indoor and the GPS module in case it is located outdoor, being switched off otherwise.

However, none of these solutions proposed a device that works with dynamic energy efficiency solution based in multi situations, establishing a set of power energy usage profiles. Thus, in the authors' opinion, it represents a pertinent downside to what concerns to wireless devices, once, as we know these devices work without access to an "inexhaustible" source of energy.

Thus, this paper tries to present a solution that allows to respond to this necessity through the development of a wearable device that integrates the aforementioned technologies, that will work together to obtain the geographical location of the person who is using the device and, at the same time, the device has the ability to transmit and receive information with the central data storage (cloud). This exchange of information between the device and the cloud will allow the detection of potential wandering events by PwD, sending an alert to the respective caregivers in such cases.

3. System Architecture

Based in this scenario, and bearing in mind the aforementioned goal of having a device that is energy efficient, the proposed solution incorporates a central system of data storage (cloud), which will do the part that concerns to the processing of all data generated by our device, since this function requires a high power consumption that could not be supported by our device without drastically reducing its lifetime.

The communication between the device and the cloud platform is established recurring to a machine to machine communication protocol, MQTT, which stands for Message Queue Telemetry Transport. MQTT is a publish/subscribe messaging protocol built using Transmission Control Protocol (TCP) and is a protocol for resource-constrained devices [14].

Using this publish/subscribe protocol it is easy to establish the connection between the device and the cloud. The device subscribes to some topics to receive the messages published by the platform and vice-versa.

One of the reasons why the authors decided to use MQTT for communications between the device and the cloud platform is because this protocol uses several levels of QoS (Quality of Service). This allows the possibility of having the confirmation that the message was delivered through the four-way handshake method ensuring that the message is successfully transmitted only once, thus saving resources at the energy level.

In order to use MQTT as a message transport protocol between the device and the cloud platform, it is necessary to use some wireless technology that supports Internet Protocol (IP) use over TCP or UDP connections [15]. Taking into account the requirements of this project, the low-power wide-area technologies and especially the cellular network connections made with Narrowband IoT (NB-IoT) technology are the best solution. NB-IoT is a Low Power Wide Area Network (LPWAN) radio technology standard and is designed to provide, in a low power spectrum, a long distance network connectivity. NB-IoT focuses mainly on providing long device battery life, low device complexity, support for a large number of devices and high network coverage, to enable the reach of devices in a wide range of locations [16].

In conjunction with the NB-IoT, the proposed device also includes a GPS module and antenna, so that it is possible to obtain the location of the person who is using the device to later send this information to the cloud.

Furthermore, the device will work with efficient power management profiles that will control which modules of the device will be connected at any given time. The device subscribes to a specific MQTT topic related to these profiles and the management of these profiles will be sent by the cloud platform to the device, and the device receiving this information will then adapt to the scenario in question.

4. Methodology

Firstly, the focus of this work was to develop the software needed for the device to work by integrating the various technologies already discussed in this article, enabling the device to dynamically adapt its configurations relating to each of these technologies. Secondly, in order to analyze the extent of autonomy and efficiency, by adapting a dynamic power consumption profile to the device, a methodology was developed to perform the tests to validate the proposed theory.

4.1. Experimental Setup

To prove the proposed theory, the work presented followed a methodology that was based on precise experimental measurements, for the power consumptions of all the components that can integrate a device that has the purpose of being autonomous and energetically efficient.

Since it is known that each of the communication modules present in the device can operate with various energy levels of consumption, corresponding to each mode of operation, the measurements made aim to identify these modes, calculate their duration in time and measure the corresponding power consumption.

Additionally, since these devices operate at a constant supply voltage, power consumption measures can be carried out by measuring and monitoring the absorbed current by the device. In this process it was used the shunt resistor method to perform such energy measurements due to its low cost, small size, and relative simplicity, while granting reasonable accuracy [17].

In this method, the sense resistor (which is called shunt) is connected in series between the power supply and the device, the voltage drop across the shunt resistor is measured at the resistor terminals using a Voltmeter connected to them. Through this method and recurring to Ohm´s Law it is possible to calculate the current consumed by the device at a given time, and corresponding power draw.

The experimental setup for the measurements included a digital multimeter, a 10 Ohm shunt resistor, a Lipo 3.7 V 800 mAh rechargeable battery and the Sodaq Sara SFF board [18] with two antennas attached, one for the NB-IoT and one for the GPS, as shown in Fig. 1.

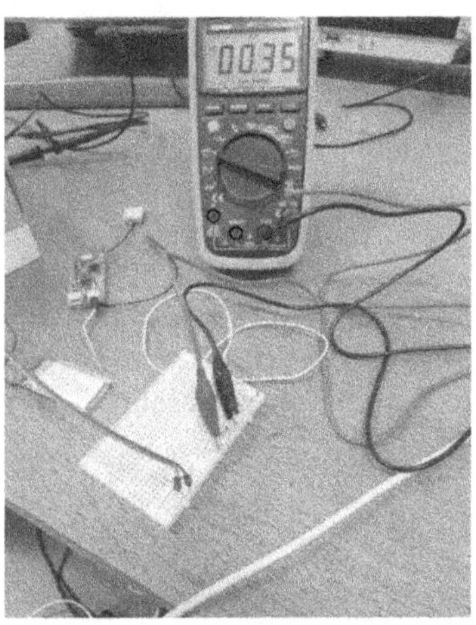

Fig. 1. Measurement setup.

4.2. Test Methodology

To decide on the time length, T, for the tests which should ensure an observable linear behavior of battery drop, preliminary tests were run to measure how long it takes for NB-IoT and GPS to completely drain the LiPo 3.7 V 800 mAh battery. The preliminary tests were run for 6 hours long, with transmission or location query intervals set to 5 s. In the case of NB-IoT, the payload size is set to 1 byte and in the case of GPS the message payload is 10 bytes.

Fig. 2 describes the steps that the software in the device contemplates: it starts by initializing the device, after device initialization the device goes into sleep mode for a few seconds, waking up after that time and switching on the NB-IoT communication module to try to connect the device to the network.

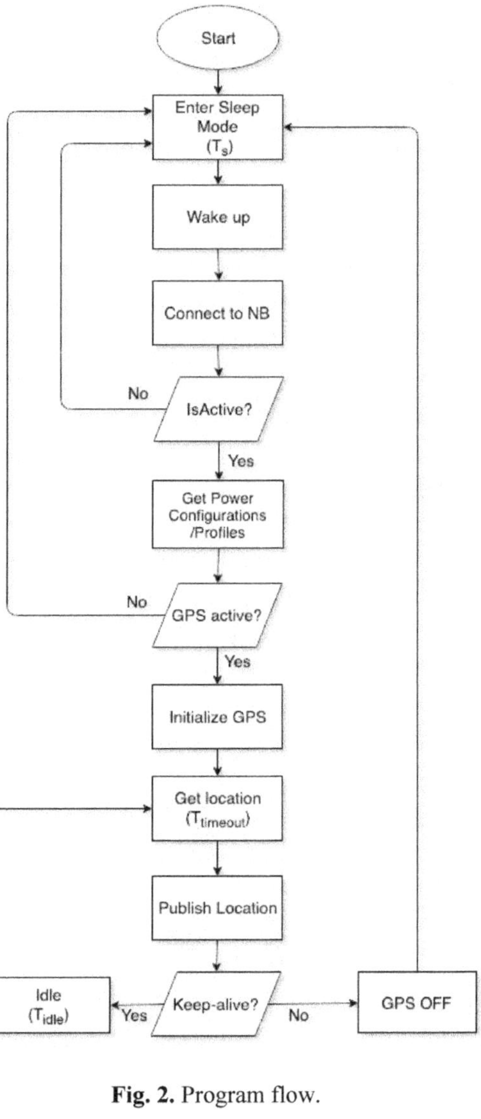

Fig. 2. Program flow.

After establishing this connection, the device tries to connect to the MQTT broker to be able to exchange messages with a certain payload with the platform. The device then goes to sleep mode (for a pre-defined time) after which it wakes up and waits for a message indicating whether the device should be activated or not. If no activation command is received the device re-enters sleep mode. Otherwise, the device enters the "normal operation mode".

First it starts by obtaining the power profiles configurations, which are sent by the platform, indicating whether the GPS should remain active after the cycle is complete, and also the interval of time for the communications made with NB-IoT. Then the device initiates the process of obtaining the current location publishing the data to the platform, completing the cycle.

5. Results

The tests were conducted with the purpose of validating if the program flow described above, is the one that best suits to the objective of improving the device uptime and effectiveness, taking into account the consumptions of the different components that make up the device in the different phases/states present in our program.

5.1. Experimental Results

The individually measurements of the consumptions were performed in each one of the states presented in the program flow, using the method mentioned in 4.1, to understand to what extent each of the program main features influence the device's instantaneous battery consumption.

In the diagram, there are 8 different states that were measured in the tests:

- ***Sleep mode***: represents the deep sleep mode of the Sodaq SFF board, which means that neither the NB-IoT nor GPS modules are connected. By default, the sleep time (T_s) is set to 10 seconds but it is dynamically configurable by the program.
- ***Wake up***: this state is when the device exits sleep mode and is able to reattach to the NB-IoT.
- ***Connect to NB***: this state is after the wake-up state and is where the device reattaches to the NB-IoT network in order to reestablish the connection with the cloud and waits for the message indicating whether the device should be active or not.
- ***Get configurations***: in this state the device is waiting for the platform to publish a message to the topic subscribed by the device that indicates the configurations concerning all modules of the device. Receiving this message, the device dynamically changes his configurations accordingly. If there are no new configuration messages from the platform, the device exits this state after 10 s and keeps his normal behavior keeping the current configurations.
- ***Initialize GPS***: this state occurs if the configurations message indicates that the GPS module must be switched on and if it is switched off when the message is received. If this module is already on, the device will only change the sample rate for the location acquisition if the value is different from the current one.
- ***Find fix***: in this state the device tries to get the coordinates of the current position. Whether this operation is successful or not the device then posts a message to the subscribed topic by the platform indicating the coordinates obtained on success, or a default error message on failure. The timeout ($T_{timeout}$) for obtaining the location is set to 30 seconds but is dynamically configurable through the settings message, after this time if the GPS was unable to get an exact location the error message is sent.
- ***Publish location***: This state represents the phase at which the device publishes the message generated in the previous state to the platform. After this state, if the current device configurations indicate that the GPS should be kept on then the delay (T_d) configured as GPS sample rate is applied, that is, this time represents how many seconds will the device wait before trying to get the GPS position again. Otherwise, the GPS is turned off at the end of the acquisition of the GPS coordinates and the device returns to sleep during time T_s.
- ***Idle***: if the current settings indicate that the GPS should be turned on the device stays in idle mode for the T_{idle} time indicated in the configurations message.

For each of these different states, 10 tests were made, and the results shown in Table 1 were obtained by calculating the average of these 10 results.

Table 1. Sodaq power consumption in each state.

State	Voltage (mV)	Current (mA)	Time (s)	Power (mW)
Sleep	0.0015	0.00015	T_s =10	$2.3e^{-4}$
Wake up	0.142	0.0142	5	0.002
Connect to NB	0.893	0.0893	15	0.08
Idle	0.219	0.0219	T_{idle}=10	0.005
Get configurations	0.272	0.0272	10	0.074
Initialize GPS	0.321	0.0321	10	0.01
Find fix	0.600	0.0600	30	0.036
Publish Location	1.089	0.1089	10	0.12

The results obtained in this experiment show that the current consumption is higher when the device connects to NB, when is publishing the message to the cloud and when the GPS is trying to find fix. And as expected, the states that consume the least are the sleep state, the wake-up state, idle, and GPS module initialization.

Taking these results as a starting point, the next objective was to verify whether or not they validate the program flow, in order to understand if there are energy savings as a result of turning off the GPS after each location acquisition and also if there is significant savings if the NB-IoT is disconnected after some time, sending the device to deep sleep mode.

The previous results were obtained when the GPS was turned off between each of the location obtaining cycles. Therefore, to try to understand the impact of keeping GPS turned on continuously, the same tests were performed under the same conditions, but this turn, keeping the GPS on. The results of these tests can be found in Table 2.

Table 2. Sodaq power consumption with GPS on.

State	Voltage (mV)	Current (mA)	Time (s)	Power (mW)
Get configurations	0.772	0.0772	10	0.06
Find fix	0.580	0.0580	30	0.034
Publish Location	2.390	0.2390	10	0.571

As can be seen, keeping GPS turned on has a significant impact on the device power consumption, which in these tests represents an increase in both states, the one of getting the configurations and the other one of publishing the location for the platform, increasing its respective instantaneous power consumption to more than double in both cases compared to the previous tests.

To test for significant savings by disconnecting the NB-IoT module from the network between any message exchange between the device and the platform, the authors performed similar tests but this time turning off the NB-IoT module after each of its related operation and reconnecting it before each of the operations where it is indispensable. The results of these tests can be found in Table 3.

Table 3. Sodaq power consumption detaching NB.

State	Voltage (mV)	Current (mA)	Time (s)	Power (mW)
Get configurations	0.994	0.0994	10	0.099
Find fix	0.282	0.0282	30	0.008
Publish Location	1.690	0.1690	10	0.286

With these results it was possible to verify that by turning off the NB-IoT module we were able to reduce the energy consumption of the state that takes longer in our program which is the "Find fix" state, compared to the results presented in Table 1. This is due to the fact that we turned off the NB-IoT module which has a significant associated energy cost. In relation to the other two states presented in this table, it is possible to denote a slight increase in the energy consumption of each of these states, which is due to the need to reconnect the NB-IoT module to the network before performing each of the operations associated with these states, that inevitably implies an associated energy cost. However, this cost is compensated as this network reconnect operation through NB-IoT does not take many seconds to be successfully performed.

Therefore, the authors defined that the full cycle of NB-IoT communication in case GPS is turned off is composed of "Sleep", "Wake up", "Connect to NB" and "Get configurations". In case GPS is turned on a complete NB-IoT communication cycle followed by acquiring GPS current position is composed of "Sleep", "Wake up", "Connect to NB", "Get configurations", "Initialize GPS", "Find fix", "Publish Location" and "Idle".

These two complete cycles are the ones that will be taken into account when calculating the consumptions of the device as a whole.

5.2. Power Consumption Profiles

Finally, another factor to take into account in profiling, and perhaps the most important in this study, is that the device will be used by people with dementia, which adds some variables when profiling power consumption. Such as the fact that the person is at home or outside, which influences the interval when we must wake up the device so that it is communicating with the platform to prevent the person from wandering. Another variable is also the location of the person when they are outside, as they may be in a location known to the person, i.e. as usual location, or in an unknown or unusual place, which also has to be taken into account when setting the time between communications.

With all this in mind the authors propose two different power profiles to test how different configurations influence the battery life of the device under the same usage profile. These profiles are shown in Table 4 and represent a more relaxed stance concerning the frequency of sleep time (profile 1) and a more intense stance (profile 2).

The time assigned to variable T_s in this table reflects the time in which the device will stay in sleep mode, in other words, the time set in T_s indicates which is the time interval in which each the device should wake up in order to perform the full cycle, either for NB-IoT or GPS. The time defined in T_{idle} indicates the time the device should remain in idle but without turning off the GPS and NB-IoT modules.

Table 4. Power profiles for person with dementia.

Profiles	Home	Outside (regular location)	Outside (irregular location)
Profile 1	$T_s = 30\,min$ $T_{idle} = 0\,min$	$T_s = 10\,min$ $T_{idle} = 5\,min$	$T_s = 1\,min$ $T_{idle} = 30\,s$
Profile 2	$T_s = 15\,min$ $T_{idle} = 15\,min$	$T_s = 5\,min$ $T_{idle} = 1\,min$	$T_s = 30\,s$ $T_{idle} = 30\,s$

To understand to what extent these defined profiles meet the final objective, the authors have defined a generalized usage scenario for a PwD. This scenario assumes that the person is most of the time at home (≈80 %), leaving home during 20 % of the time. This scenario also assumes that during the time the PwD is outside he/she walks 10 % of the time in proper known (regular or familiar) areas and the other 10 % in unknown (irregular or uncommon) areas.

As such, to determine the average current consumption of the device, it is necessary to calculate the average consumption of one full cycle recurring to the equation:

$$I_{AVG} = \frac{\sum_s t_s \times I_s \times n_s}{\sum_s t_s}, \quad (1)$$

where "s" represents each state of the full cycle, t_s the time spent in each state, I_s the current consumption in that state and n_s is the number of times that full cycle occurs in time interval t_s.

For profile 1, when the user is at home, the device is in sleep mode for 30 minutes and wakes up to perform a full cycle of GPS and NB-IoT, which takes 80 seconds. As the current consumption for the sleep state is 0.0015 mA as shown in Table 1 and recurring to formula 1 the current consumption for one full cycle is 6.22 mA.

As such the average current consumption associated to *Home* location is given by:

$$I_{Home} = \frac{0.0015 \times 1800 \times 2 + 6.22 \times 80 \times 2}{3600}, \quad (2)$$

which gives $I_{Home} = 0.273$ mA for the time interval of 1 hour. Recurring to the same method the current consumption for the user location Outside (regular location) is $I_{Outside\ reg} = 0.829$ mA and when the user is outside but in an irregular location the device current consumption is $I_{Outside\ irreg} = 8.29$ mA.

Using the usage scenario described above, as the user spend 80 % of the time in home, 10 % outside in a regular location and other 10 % outside but in an irregular location, it is possible to calculate the weighted average of current consumption associated with profile 1 configurations, which is given by:

$$I_1 = 0.8 \times 0.273 + 0.1 \times 0.829 + 0.1 \times 8.29, \quad (3)$$

which results in $I_1 = 1.13$ mA as weighted average consumption.

Therefore, using these results, it is possible to estimate the battery life of our device, which according to the DigiKey battery life calculator [19] is given by:

$$800\ \text{mAh}/1.13\ \text{mA} \times 0.7 = 495.6\ \text{h} \approx 21\ \text{days}, \quad (4)$$

800 mAh represents the battery capacity, 1.13 mA is the current draw by the device under test, and 0.7 is the battery discharge factor that takes into account external factors that may influence battery life.

Performing the same method for profile 2, gives $I_{Home} = 0.554$ mA, $I_{Outside\ reg} = 1.66$ mA and $I_{Outside\ irreg} = 16.6$ mA which results in a weighted average consumption of $I_2 = 2.27$ mA. In this case, the battery could stand for 10 days without recharging.

Using profile 1 configurations enables the device to last for three weeks on a single battery charge, which is a near-ideal scenario believing that the PwD will have visits by his family or by one caregiver, at least once a month. In this way, recharging of the device can be guaranteed, thus not compromising the proper functioning of the device. Profile 2 is indicated for a PwD that has more regular visits, for example once a week.

6. Conclusions

In this paper the authors presented the work to design and test a prototype of a wearable device that can be used by people who suffers from dementia. Due to the primary user's inconstancy to remember to charge it daily, such devices need to have an extended energy autonomy. In that sense, a set of adaptable energy profiles are proposed, enabling the optimization of power consumption.

By leveraging the device components configuration dynamically, the proposed energy profiles manage to extend the battery life by more than a week and up to almost three weeks, depending on the usage scenario.

One of the possible implementations of this wearable device can be the integration into a sole of the shoe, as it does not cause disturbance to the person wearing it and it is less likely that the person will forget to use it.

This work will continue to be further improved by testing more profiles and refining those already defined, as well as by dynamically switch the profile according to the remaining battery level, by extending the same principles to additional low power communication technologies, and by testing the scenarios in real life trials.

Acknowledgements

The authors acknowledge the European Commission for its support and partial funding and the partners of the research project CARELINK, AAL-CALL-2016- 049 funded by AAL JP, and co-funded by the European Commission and National Funding Authorities of Ireland, Belgium, Portugal and Switzerland.

References

[1]. J. Gubbi, R. Buyya, S. Marusic, M. Palaniswami, Internet of Things (IoT): A vision, architectural

elements, and future directions, *Futur. Gener. Comput. Syst.*, Vol. 29, Issue 7, 2013, pp. 1645-1660.

[2]. U. Nations, UN, World Population Ageing, http://www.un.org/esa/population/publications/worldageing19502050/, 2013, pp. 8-10.

[3]. K. Bhargava, G. Mcmanus, S. Ivanov, Fog-centric Localization for Ambient Assisted Living, in *Proceedings of the International Conference on Engineering, Technology and Innovation (ICE/ITMC'17)*, 2017, pp. 1424-1430.

[4]. M. Prince, R. Bryce, E. Albanese, A. Wimo, W. Ribeiro, C. P. Ferri, The global prevalence of dementia: a systematic review and metaanalysis., *Alzheimers. Dement.*, Vol. 9, Issue 1, Jan. 2013, pp. 63-75.

[5]. ETSI – 2G, https://www.etsi.org/technologies/mobile/2g

[6]. Cellular Standards for the Third Generation, https://www.itu.int/osg/spu/ni/3g/technology/#Cellular Standards for the Third Generation

[7]. Release 13, https://www.3gpp.org/release-13

[8]. LoRa Alliance, https://lora-alliance.org/

[9]. R. Bowie, Global Positioning system/wide area augmentation system, *Aviat. Saf. Lett.*, October 2010.

[10]. N. E. Tabbakha, W. Tan, C. Ooi, Indoor location and motion tracking system for elderly assisted living home, in *Proceedings of the International Conference on Robotics, Automation and Sciences (ICORAS'17)*, 2017, pp. 1-4.

[11]. Bluetooth Technology, https://www.bluetooth.com/bluetooth-technology/

[12]. A. Banks, R. Gupta, MQTT Version 3.1.1, OASIS Committee Specification Draft 02 / Public Review Draft 02, https://docs.oasis-open.org/mqtt/mqtt/v3.1.1/csprd02/mqtt-v3.1.1-csprd02.html

[13]. L. Mainetti, L. Patrono, A. Secco, and I. Sergi, An IoT-aware AAL system for elderly people, in *Proceedings of the International Multidisciplinary Conference on Computer and Energy Science (SpliTech'16)*, 2016, pp. 1-6.

[14]. M. Collina, M. Bartolucci, A. Vanelli-Coralli, G. E. Corazza, Internet of Things application layer protocol analysis over error and delay prone links, in *Proceedings of the 7th Adv. Satell. Multimed. Syst. Conference (ASMS'14)*, 2014, pp. 398-404.

[15]. A. Larmo, A. Ratilainen, J. Saarinen, Impact of CoAP and MQTT on NB-IoT system performance, *Sensors*, Vol. 19, Issue 1, 2019, 7.

[16]. A. Höglund, *et al.*, Overview of 3GPP Release 14 enhanced NB-IoT, *IEEE Network*, Vol. 31, Issue 6, 2017, pp. 16-22.

[17]. S. Ziegler, R. C. Woodward, H. H. Iu, S. Member, L. J. Borle, Current sensing techniques: A review, *IEEE Sensors Journal*, Vol. 9, Issue 4, pp. 354-376, 2009.

[18]. Sodaq Sara SFF Board, https://shop.sodaq.com/sodaq-sara-sff-r412m.html

[19]. Digikey Battery Life Calculator, https://www.digikey.com/en/resources/conversion-calculators/conversion-calculator-battery-life

(076)

Detection of the Glial-fibrillary-acidic-protein Biomarker with an Ultra-high-frequency Surface-acoustic-wave Lab-on-a-Chip

M. Agostini, G. Greco, M. L. Vieri, F. Amato and M. Cecchini
NEST, Istituto Nanoscienze-CNR and Scuola Normale Superiore, Piazza San Silvestro 12, 56127 Pisa (Italy)
Tel.: +39 050 509502
E-mail: matteo.agostini@sns.it

Summary: Glial-fibrillary-acidic-protein (GFAP) has been recently discovered as a biomarker for brain-related pathologies. Severe and mild traumatic injuries, intracerebral hemorrhage and glioblastoma multiforme cancer are all addressable with this circulating biomarker. The aim of this work is to detect GFAP with the best achievable limit of detection by means of an ultra-sensitive surface-acoustic-wave (SAW) based lab-on-a-chip (L oC), which we recently demonstrated. First, we develop several detection strategies by means of gold quartz-crystal-microbalance (QCM) sensors. A polyethylene glycol (PEG) chain is used for antifouling, protein-G for antibody-orientation, and antibody-splitting protocol for high-surface coverage. Then, we exploit the best-performing functionalization for the detection of GFAP with our ultra-high-frequency (UHF) SAW-based L oC. This SAW-L oC can be further developed and engineered with the aim of realizing a point-of-care biosensing platform for the detection of multiple brain-pathology biomarkers.

Keywords: Surface-acoustic-wave (SAW), Lab-on-a-chip, Brain biomarker, Glial-fibrillary-acidic-protein (GFAP), Microfluidics, Biosensor.

1. Introduction

GFAP is a promising biomarker for severe brain-related pathologies in humans [1, 2]. The interest on GFAP as a biomarker is related to two main aspects. First, the importance of the pathologies that can be addressed: severe traumatic injuries, intracerebral hemorrhage and glioblastoma multiforme cancer. Second, the possibility of detecting this biomarker in circulating blood. Lab-on-a-chips (L oCs) have recently drawn significant attention as ideal technologies for the detection of biomarkers. This interest stems from the possibility of miniaturizing and speeding up the assays, and contemporary decrease costs, increase portability, facilitate integration with other technologies. In the last years an increasing number of works have demonstrated biosensors able to detect this protein, exploiting various types of surface functionalizations. The sensitivity and selectivity are the main features that an ideal L oC is expected to exhibit during the detection. This is of paramount importance in detecting biomarkers at low concentrations in biological fluids, as in the case of GFAP in blood.

In this context, Rayleigh surface acoustic waves (SAWs) are an effective mean for realizing L oCs. SAWs interact with fluids and activate a number of microfluidic phenomena, such as mixing, droplet actuation, microparticle manipulation, nebulization and micropumping, to name but a few. In a previous work, we showed ultra-high-frequency (UHF) SAW acoustic resonators fabricated and proposed for biosensing applications. Their limit of detection was better than that of standard commercial gravimetric sensors [3, 4].

Here, we design and test three functionalization strategies with a quartz-crystal-microbalance (QCM). The aim is to maximize the sensitivity and selectivity of the GFAP detection with the SAW-L oC, which contains several UHF SAW resonator biosensors. The first strategy exploits a PEG layer for antifouling, the second the protein-G for antibody-orientation and the third the antibody-splitting protocol for an enhanced surface-coverage. We then exploit the best-performing functionalization for the GFAP detection with our UHF SAW-based L oC.

2. Biosensing Experiments

2.1. Functionalization Strategies Design and Testing

We compare three different functionalization strategies (Fig. 1) for GFAP detection in clean buffers and in fetal bovine serum (FBS). The experiments are performed with gold-plated QCM sensors. The first strategy exploits a polyethylene glycole (PEG) layer for antifouling (F1) and two different antibodies against two different epitopes (F1a and F1b). The second strategy is based on the protein-G capability of binding and orienting the anti-GFAP antibody (F2). In the third, the anti-GFAP is split in two parts with a reducing agent and the gold surface is directly covered with the reduced antibody (F3). The GFAP detection capability and selectivity are addressed. Experiments have been carried out with increasing concentrations of GFAP injected on the functionalized sensors. The resulting QCM shifts are reported in Fig. 2, where a clean phosphate buffered saline (PBS) is used. The experiments showed that F1a is the most sensitive functionalization for the GFAP detection. This results was further confirmed with the FBS buffer, evaluating also the selectivity of the detection.

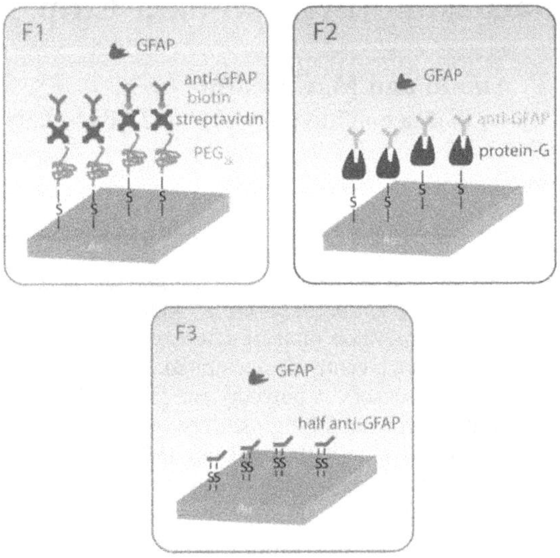

Fig. 1. Functionalization strategies. F1 exploits the PEG antifouling effect, F2 the protein-G binding and orientation of the antibody and F3 the high surface-coverage of the reduced antibody.

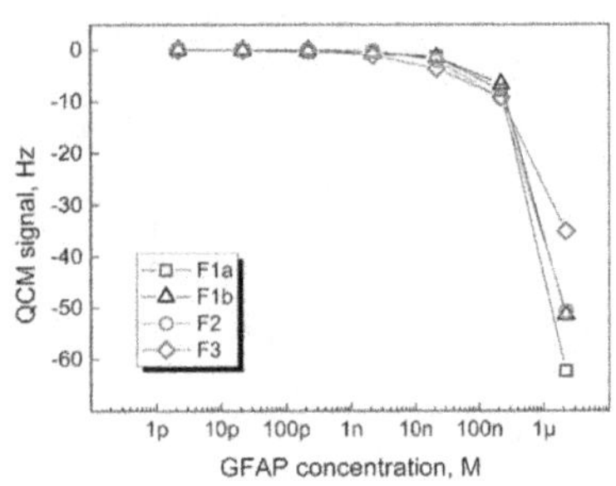

Fig. 2. QCM calibration with GFAP. The QCM signal goes from 0 (no protein bound to the QCM sensor) to negative frequencies upon molecules immobilization.

2.2. Glial-fibrillary-acidic-protein Detection with the Surface-acoustic-wave Based Lab-on-a-chip

The combination of SAW-biosensing and SAW-microfluidics on the same chip allowed the effective detection of GFAP. Fig. 3 shows different resonance frequency shifts of the resonator biosensors contained in the SAW-L oC developed in this work upon exposure to different GFAP concentrations. The two distributions become significantly separated for GFAP concentrations higher than 22.8 pM. The UHF-SAW resonators multiplexing and the SAW-induced recirculation of the fluid inside the microchambers were fundamental for the detection of GFAP with antibodies immobilized on the biosensor surface (F1a functionalization).

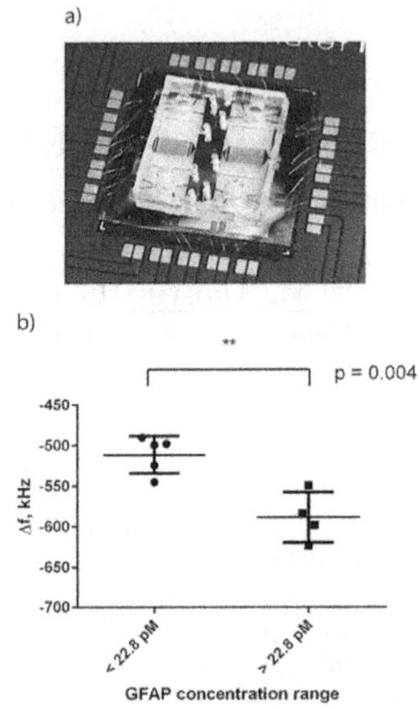

Fig. 3. GFAP detection with the UHF-SAW L oC. a) Photo of a device developed in this work. b) Resonance frequency shifts of the SAW resonator biosensors upon exposure to different GFAP concentrations.

3. Conclusions

In this work, we designed and tested by QCM different functionalization strategies for the GFAP biomarker detection. We then used the best-performing functionalization with our ultra-sensitive UHF SAW-based L oC. GFAP was significantly detected at clinically-relevant concentrations (higher than 22.8 pM). This SAW-L oC can be further developed and engineered with the aim of realizing a point-of-care biosensing platform for the detection of multiple brain-pathology biomarkers.

Acknowledgements

This work has been funded by the FAS projects GLIOMICS and SENSOR of Regione Toscana.

References

[1]. J. Tichy, S. Spechtmeyer, M. Mittelbronn, E. Hattingen, J. Rieger, C. Senft, C. Foerch, Prospective evaluation of serum glial fibrillary acidic protein (GFAP) as a diagnostic marker for glioblastoma, *Journal of Neuro-Oncology*, Vol. 126, 2016, pp. 361-369.

[2]. J. Lei, G. Gao, J. Feng, Y. Jin, C. Wang, Q. Mao, J. Jiang, Glial fibrillary acidic proteina s a biomarker in severe traumatic brain injury patients: A prospective cohort study, *Critical Care*, Vol. 19, 2015, pp. 1-12.

[3]. M. Agostini, G. Greco, M. Cecchini, A Rayleigh surface acoustic wave (R-SAW) resonator biosensor based on positive and negative reflectors with sub-nanomolar limit of detection, *Sensors and Actuators B: Chemical*, Vol. 254, 2018, pp. 1-7.

[4]. A. Mujahid, F. Dickert, Surface acoustic wave (SAW) for chemical sensing applications of recognition layers, *Sensors*, Vol. 17, 2017, pp. 1-26.

(077)

High-resolution Data Acquisition and Control System for Performance Monitoring of Domestic Solar Water Heaters

E. L. Meyer, J. C. Nwodo and O. K. Overen
University of Fort Hare, Institute of Technology, 1 King Williams Town Road., 5700 Alice, South Africa
Tel.: +27 (0) 40 602 2086, fax: + 27 (0) 40653 0665
E-mail: jnwodo@ufh.ac.za

Summary: The design of a data acquisition and control (DAC) system for on-site and remote performance monitoring of the solar water heater (SWH) systems at the Fort Hare Institute of Technology (FHIT) is presented. This system incorporates a number of sensors which measure meteorological data as well as the temperature of the water in the collector of the SWH system. At 04h00, the water in the collector is drained into a 500-litre tank, which serves as a heat sink. Subsequently, the collector is refilled with fresh water from the main water line. The entire system is autonomous and this paper presents the DAC system as well as the dynamics between the different inlet and outlet temperatures of a high-pressure and low-pressure SWH systems.

Keywords: Solar water heater, Data acquisition and control, Meteorological data, Performance monitoring, Apparent heat gain.

1. Introduction

Various energy demand projection, be it present or future has shown that the current practice of using non-renewables source in the generation of power [1] for heating systems is not sustainable. Over the past decade, renewable and sustainable ways of water heating have been on the increase through the utilization of solar energy [2]. Despite the readily availability of solar heating equipment [2, 3], they are widely used in homes in the Eastern Cape and other provinces in South Africa [4].

In order to monitor the performance of typical household solar water heater (SWH) systems, a data acquisition system (DAS) is necessary. Such a system according to Ahmed [5] comprises a storage device, analog, and digital processing modules, computer, computer interface, and software.

An optimized solar water heater system designed by Li and Liu [6] claimed to have achieved high-performance by applying the predictive power of machine learning. This approach also provided an accurate and quick method for measuring desired parameters. These parameters range from ambient temperature, relative humidity, water temperature at various points of the system, irradiance, volume of water used up, and wind speed and direction. The available solar irradiance is key to the optimum performance of solar water heater systems.

Several researchers have worked towards improving the performance of domestic solar water heating systems by studying and analyzing off the shelf solar water heater systems. Studies by Kulkarni *et al.* [7] attempted to improve the performance of solar water heater system by passing fluids in several stages, maintaining peak variable flow rate within the collector, maintaining a reduced constant flow rate within the collector, as well as providing water to the inlet of the collector from mains and from the storage tank.

In the study, the DAS was designed to measure the thermal state of the fluid of the SWH systems, meteorological data as well as ensure that the high and low pressure SWH systems drain at 04h00. The DAC drain the systems into 500 litres retaining tank by actuating the solenoid valves retrofitted on both systems. Once the drain cycle is complete, the DAC refills the cistern with cold water from the municipality for the high and low pressure SWH systems. The results obtained, therefore, provides the basis and reference for the adaptability and application of the system.

The purpose of this study is to design a data acquisition system capable of monitoring the performance of household solar water heater systems with high data resolution, enabling a detailed analysis of the dynamic changes in water temperatures at various points in the closed systems.

1.1. Theoretical Concept

The performance of these solar water heaters is governed by the temperature stratification of the water tanks and of course the thermosyphon effect under no load conditions. The efficiency η of such system is given by:

$$\eta = mC_{HP} (T_{HP}out\text{-}T_{HP}in)/GA_{HP}, \qquad (1)$$

where m is the mass of water, C_{HP} the specific heat capacity of water, $T_{HP}in$ and $T_{HP}out$ is the temperature of water flowing in and out of the collector of the high-pressure SWH respectively, G is the plane-of-array (POA) irradiance and A is the aperture area of the solar collector.

2. Site Description and Methodology

2.1. SolarWatt Park Alice

SolarWatt Park at the University of Fort Hare is located at latitude 32.8° south and longitude 26.8° east at an altitude of 540 m in Alice, Eastern Cape South Africa.

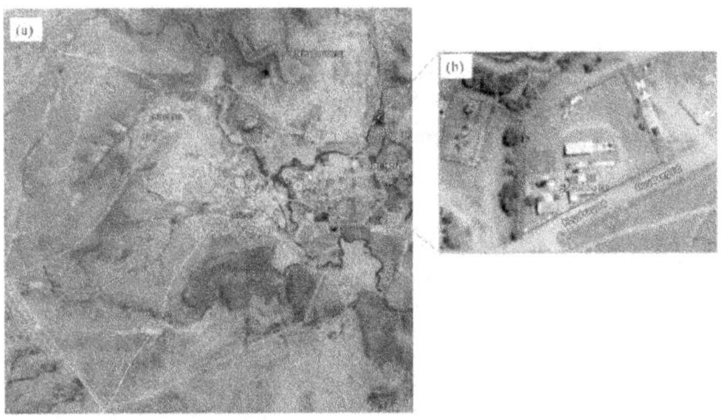

Fig. 1. (a) Google Earth Map of Alice showing; (b) SolarWatt Park.

The climate in Alice in Fig. 1(b) falls under the temperate interior (Zone 2) climate of South Africa [8]. The climate of Alice consists of mild winter and hot summer with an average dry bulb temperature of 15 °C and 29 °C, respectively.

A westerly wind prevails in winter, while in summer, an easterly wind is predominant. According to [9,10], in a year throughout Alice, the average wind speed is 2.5 m/s while the average solar radiation in summer and winter is 606.06 W/m^2 and 346.17 W/m^2 respectively.

Fig. 2 shows the outdoor test rig with the *high-pressure* (HP) and *low-pressure* (LP) solar water heaters (SWHs). The photovoltaic thermal systems shown are not part of this paper.

Fig. 2. Outdoor test rig indicating the HP and LP solar water heaters.

During this study, the SWHs were initially filled at 03h30 before sunset, exposed to solar energy throughout the day, and hot water drained again at 04h00. All the while various temperatures and irradiances (global POA, downward Infrared irradiance, and total irradiation) were measured at 5-second intervals. Other parameters included the main inlet temperature (T_{in}), collector inlet temperature (T_{in_col}), collector outlet temperature (T_{out_col}), geyser outlet temperature (T_{out_g}) as well as ambient temperature (T_a) and humidity (R_H). Fig. 3 shows the high resolution data acquisition and control (DAC) system used to fill and drain the SWHs and continuously measuring the aforementioned parameters.

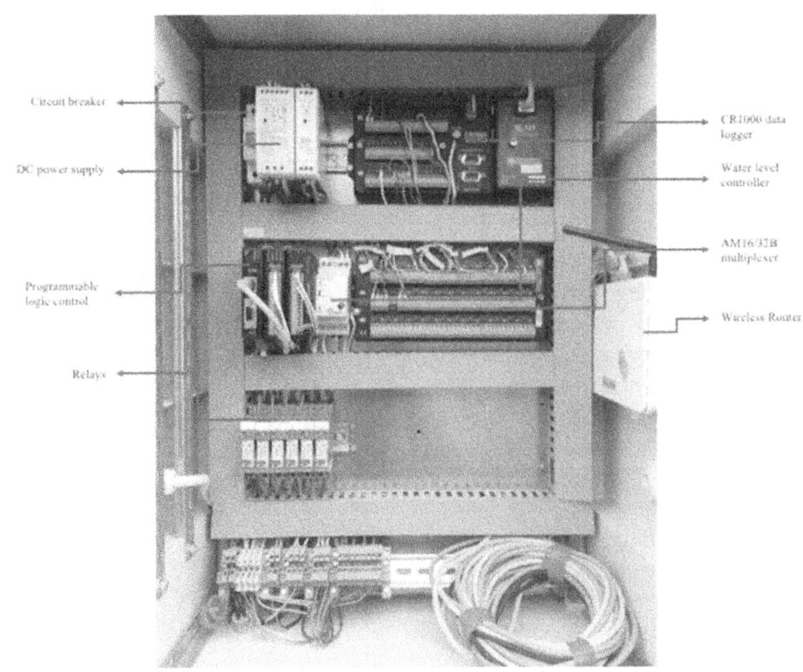

Fig. 3. Data Acquisition and Control system used to measure and control various sensors.

All sensors used in monitoring the SHW and the meteorological data were carefully selected to achieve the desired outcome. These sensors connected to the multiplexer and CR1000 data logger. The DAC in Fig. 3 is remotely monitored through the wireless router as indicated in the figure. The most important parameters considered are irradiance, water temperature, ambient temperature, and relative humidity.

2.2. Solar Radiation Sensor

The Kipp & Zonen CMP10 radiometer used is integrated with a thermopile which is a passive thermal sensing element for detecting irradiance. The thermopile has a thermocouple junction pairs that are connected in series. The measurement (hot) junction of one of the thermocouple takes in thermal radiation, which increases its temperature. The difference between the measurement and a fixed temperature reference (cold) junction produces a voltage directly proportional to the differential temperature created. This is referred to as the thermoelectric effect [11].

The principle of operation of thermopiles of one radiometer to another is the same, their construction, however, might differ from one model to another. Also, the sensitivity of a given radiometer thermopile depends on its physical characteristics [12]. The specification of the pyranometer (radiometer) used in the study as obtained from [12] has a spectral range of 0.3-2.8 μm, sensitivity of 9.24 μV/W/m², and response time of < 5 s. The response time for the pyranometer refers to the delay to respond to incident radiation. Furthermore, the speculated response time is the time taken by the pyranometer to deliver 95 % for any measurement taken after a step-change in irradiance.

2.3. Temperature and Humidity Sensors

The ambient air temperature and corresponding relative humidity (RH) were measured by the sun shielded HMP 60 temperature relative humidity probe while the water temperature at various locations on the SWH system was measured using the KM07 immersion probe Type K thermocouple. The temperature and RH probe measures air temperature via its platinum resistance temperature (PRT) detector [13, 14]. The air temperature measurement specifications indicated in [15] for the HMP 60 probe shows that its accuracy is ± 0.6 %, measurement range of 0 to 100 % RH (-40° to +60 °C).

The KM07 immersion probe Type K thermocouple is made from a waterproof 316 (Food Grade) stainless steel with a domed end sensor. The hermetically sealed thermocouple allows the probe to be suspended within a water storage tank or at any point within the pipes of a system and should be in contact with the working fluid. The specification of the Type K thermocouple used in this study obtained from [16] is presented in Table 1.

Table 1. KM07 immersion probe Type K thermocouple.

Parameters	Measurement range	Response time (s)	Accuracy (±)
Temperature (°C)	-50 to 250	0.1	0.25 %

A naturally aspirated radiation shield with six plates accommodates the outdoor air temperature measuring probe. The radiation shield is white in colour, which allows it to reflect solar radiation. Also, the louvre allows the free flow of natural air through the shield. This enables the probe to be in close contact with the ambient air temperature (solar effect excluded) and water vapour [17]. The KM07 immersion probe Type K thermocouple probe to wire junction is sealed with a waterproof adhesive-lined heat-shrink. This allows the thermocouple to be used for permanent measurement of sentinel points for Legionella in combination with a thermocouple wall port (TCWALLPORT).

3. Results and Discussion

Fig. 4 shows the temperatures measured at the warm outlet of the SWH geysers at a 5-second interval.

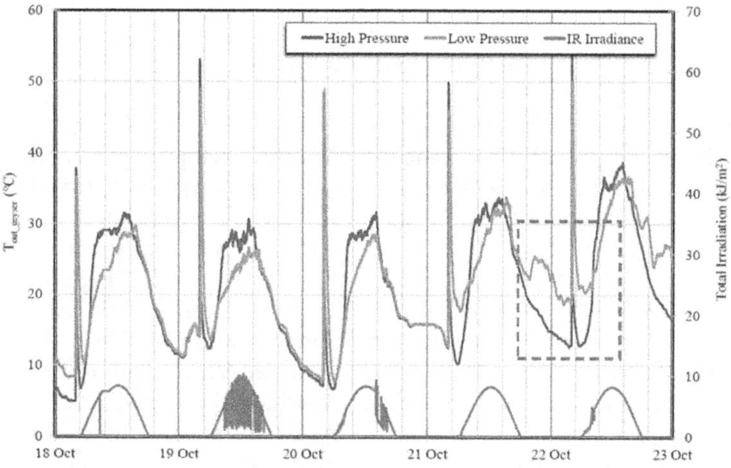

Fig. 4. High resolution temperature and irradiance measurements over a 5 day period.

Both high- (HP) and low-pressure (LP) geysers were drained at 04h00, hence the daily spike highlighted around 04h00. Interesting to note is that the LP geyser maintains the outlet after sunset (~18h30) as indicated in the red square of Fig. 4. This may be due to the fact that the LP is an indirect system, however, it may also result in higher standing losses for the LP.

Also noteworthy is that the LP outlet reaches its maximum daily temperature roughly 2 hours later than the HP outlet.

Fig. 5 shows the difference between outlet temperatures and main cold inlet temperature. When this difference is greater than zero, conventional wisdom implies that there is a net heat gain.

However, for the LP SWH with its evacuated tubes, the natural convection phenomenon brings about the effect of a constant heat flux causing the apparent temperature at the outlet to remain high during the night.

Fig. 6 shows the hot water temperature produced by the system on a cloudy day.

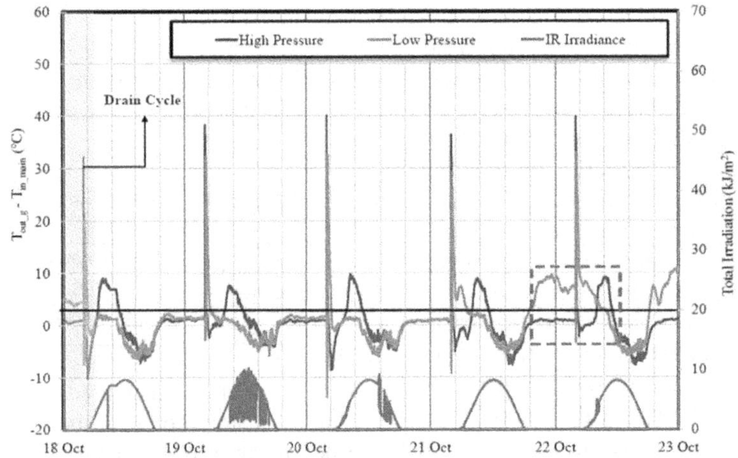

Fig. 5. Geyser outlet temperature minus main cold water inlet.

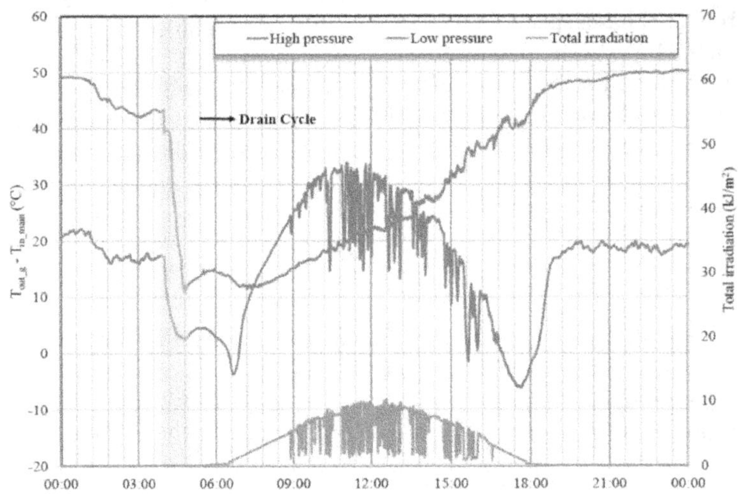

Fig. 6. Partly cloudy day system output.

The hot water generated on a cold day (partly cloudy day) as seen in Fig. 6, suggests that irradiance to a large extent influences the performance of SWH systems. During the drain cycle for the HP and LP highlighted in Fig. 6, a gradual drop in water temperature is noticed. Since hot water is less dense than cold water, the former would flow through first followed by cold water. The water temperature fluctuates for the high-pressure system with a maximum of about 32 °C due to the drastic change in irradiance, while the temperature for the low-pressure system (indirect system) increases gradually with a maximum of about 27 °C at 13h00.

To further understand the impact of irradiance on the water temperature of the SWH systems, Fig. 7 shows the geyser outlet water temperature on a clear sky day.

Since irradiance in any location varies throughout the day as seen in Fig. 7, the water being heated follows a similar trend as the irradiance, although a slight shift is noticed as it takes some time for water to be heated. At the end of the drain cycle, cold water is introduced into the drained cisterns. The heat trapped by the collectors is transferred to the water hence the slight increase in temperature in the region marked A.

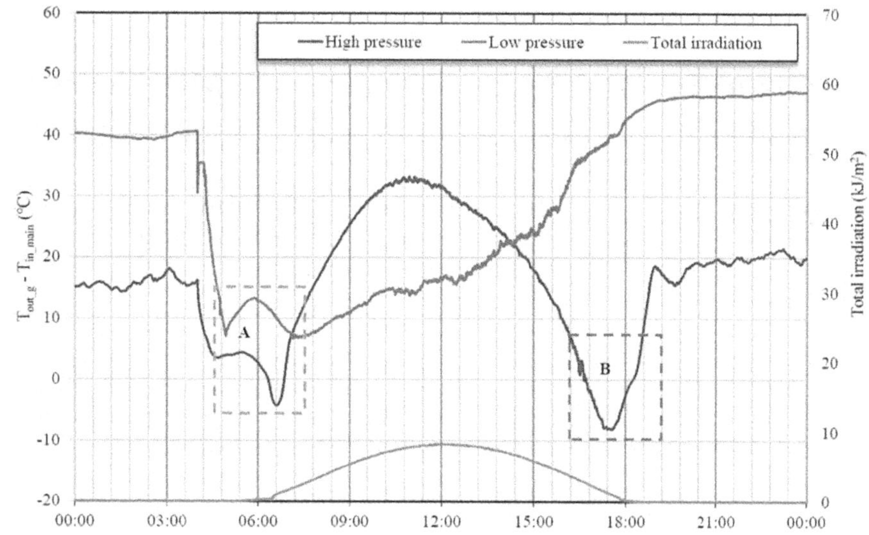

Fig. 7. Clear sky day system output.

In order to determine the heat gain/loss in the area marked B, it was assumed that the heat in the system is directly proportional to change in water temperature (Q α ΔT).

A peak water temperature of 48°C for the LP was observed for the entire day, while the HP's maximum water temperature of 32°C was obtained.

4. Conclusions

The high resolution DAC system provides valuable insights into the dynamics of both *high-pressure* and *low-pressure* SWH. It is evident that the method of determining system efficiency is of the utmost importance. Also, the available irradiance plays a vital role in ensuring the productivity of SWH systems.

Acknowledgements

The authors acknowledge the financial support of the South African National Research Foundation (NRF), Department of Science and Technology (DST), Eskom Tertiary Education Support (TESP) programme, and Govan Mbeki Research & Development Centre (GMRDC) at the University of Fort Hare.

We also wish to thank Messrs. Melvin Adonis and Vincent Schoeman for their technical support.

References

[1]. A. M. Krishna, K. P. Rao, M. B. Prakash, N. Ramchander, Data Acquisition system for performance monitoring of solar photovoltaic (PV) power generation, *Int. J. of Eng. Res. & Tech.*, Vol. 1, Issue 7, 2012, pp. 1-6.

[2]. A. Shafieian, M. Khiadani, A. Nosrati, A review of latest developments, progress, and applications of heat pipe solar collectors, *Ren. & Sust. En. Rev.*, Vol. 95, 2018, pp. 273-304.

[3]. Y. Deng, Z. Quan, Y. Zhao, L. Wang, Z. Liu, *Energy Conversion and Management*, Vol. 106, 2015, pp.1039-1047.

[4]. Department of Minerals and Energy Solar Water Heating, http://www.energy.gov.za/files/swh_frame.html

[5]. S. N. Ahmed, Physics and Engineering of Radiation Detection (2nd Ed.), *Elsevier*, 2015.

[6]. H. Li, Z. Liu, Performance prediction and optimization of solar water heater via a knowledge-based machine learning method, in Handbook of Research on Power and Energy System Optimization, *IGI Global*, 2018, pp. 55-74.

[7]. G. N. Kulkarni, S. B. Kedare, S. Bandyopadhyay, Optimization of solar water heating systems through water replenishment, *Energy Conversion and Management*, Vol. 50, Issue 3, 2009, pp. 837-846.

[8]. SABS 2011. SANS 204:2011: Energy Efficiency in Buildings, *South Africa*, 2011.

[9]. D. C. U. Conradie, South Africa's climatic zones: Today, tomorrow, in *Proceedings of the International Green Building Conference and Exhibition*, 2012, pp. 1-9.

[10]. O. K. Overen, et al., A commercial building lighting demand-side management through passive solar design, *IOP Conf. Ser.: Earth Environ. Sci.*, Vol. 290, 2019, 012148.

[11]. W. Tang, K. Yang, J. He, J. Qin, 2010. Quality control and estimation of global solar radiation in China, *Solar Energy*, Vol. 84, Issue 3, 2010, pp. 466-475.

[12]. Kipp & Zonen 2012, Solar Monitoring Stations, http://www.kippzonen.com/Download

[13]. N. Yamazoe, Y. Shimizu, Humidity sensors: Principles and applications, *Sensors and Actuators*, Vol. 10, Issue 3-4, 1986, pp. 379-398.

[14]. Y. Moser, M. A. M. Gijs, Miniaturized flexible temperature sensor, *Journal of Microelectromechanical Systems*, Vol. 16, Issue 6, 2007, pp. 1349–1354.

[15]. HMP60 Temperature and Relative Humidity Probe *Campbell Scientific Inc*., Utah, 2010.

[16]. RS Components Ltd. RS PRO Immersion Temperature Probe, Type K Sensor, https://za.rs-online.com/web/p/temperature-probes/7887144/

[17]. Solar Radiation Shields: For Temperature/Relative Humidity Sensors, http://campbellsci.com/solar-radiation-shields.

(079)

Effect of the Channel Length in the Response of a MIS Transistor Sensor with Optical Gain for Nano-watts Light Signal

J. Hernández-Betanzos [1], A. A. Gonzalez-Fernandez [1], J. Pedraza [1] and M. Aceves-Mijares [1]
[1] Electronics Department, INAOE, Apdo. 51, Puebla, Pue. 72000, México
E-mail: maceves@inaoep.mx

Summary: In this work, we study the optical response of MOS-like transistors with a Si_3N_4 integrated waveguide as the dielectric on the gate, and P-type substrate with 1×10^{12} cm^{-3} acceptor concentration and different channel length. Simulation results show the possibility to integrate this kind of MIS transistor as detectors in an electrophotonic circuit compatible with CMOS fabrication process and obtaining electrical gain for low power light signals (below 400 nW).

Keywords: Electrophotonics, Integrated photonics, Planar waveguide, Si photonics.

1. Introduction

Electrophotonics is an area whose goal is to integrate operations involving electrons and photons in a monolithic chip of silicon [1]. Lab-on-a chip for chemical and biological sensing applications could take advantage of the totally integrated silicon electrophotonic chips. One of the principal limitations for these circuits is the lack of an integrated silicon light sources. A viable recent alternative to overcome this is the integration of light emitting devices using nanostructured materials based on silicon [2]. However, such devices have relatively low emitting power, and light detectors with higher sensitivities are needed. Silicon eletrophotonic chips have been fabricated and light detection has been reported using a nitride waveguide and p-n junction as a photodetector [3, 4]. Light propagation and evanescent field absorption in a silicon detector-waveguide system are described in [2, 4].

In this paper, a MOS-like transistor with an integrated waveguide is proposed as a highly sensitive photodetector for electrophotonic applications. The device is studied through computer simulation. The dielectric material in the transistor is replaced by silicon nitride, which is simultaneously used as waveguide, having effectively an electrophotonic metal-insulator-semiconductor (MIS) structure. The goal is to identify the channel length effects on the electrical gain of optical signals with power below 400 nW injected directly from the gate area in order to move forward in.

2. Device Description

The structure was simulated using the Silvaco platform [5, 6]. The substrate doping concentration was 1×10^{12} cm^{-3}, nitride thickness was 140 nm and the studied channel lengths (L) were 40 µm, 20 µm and 10 µm. Fig. 1 shows the simulated structure.

Electric and optical simulations were realized using Atlas [6]. The drain voltage was fixed to 5 V, and different gate voltage (V_{GS}) values were studied. A light beam was applied from the gate dielectric in normal direction towards the substrate, with 600 nm wavelength (λ) and optical powers (P_{in}) varying from 0 nW to 400 nW.

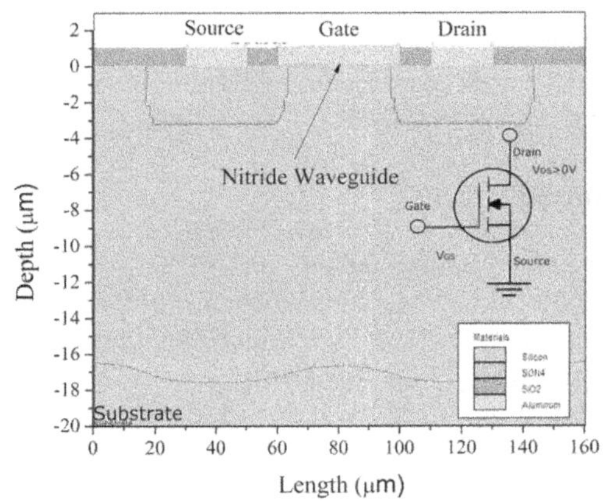

Fig. 1. MIS Transistor Structure Proposed.

3. Simulation Results

3.1. Electrical Simulations

Fig. 2 shows the drain current at dark condition (I_{Ddark}) as a function of V_{GS} for three transistors with different channel lengths. The drain voltage was 5 V.

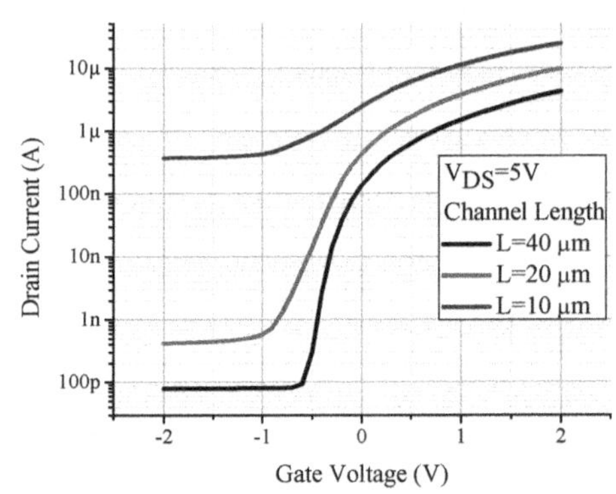

Fig. 2. Drain Current of MIS transistor at dark condition. I_{Ddark} tends to increase when the channel length is reduced.

3.2. Optical/electrical Simulations

Fig. 3 shows the drain current under the different illumination conditions for the transistor with 40 μm channel length.

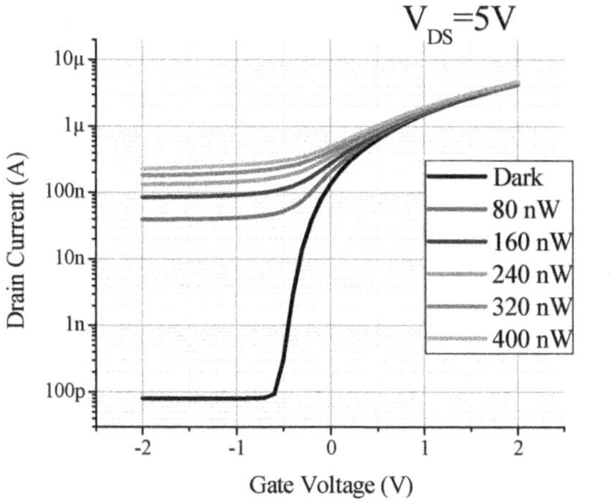

Fig. 3. Drain current under illumination. L is 40 μm. λ is 600 nm and drain voltage is 5 V. The presence of light increases the drain current values.

In order to show only the drain current contribution due to the optical power, the dark current was subtracted from the total drain current I_D. Optical drain current is:

$$I_{Dopt} = I_D - I_{Ddark} \quad (1)$$

This current was compared with the maximum available photocurrent provided by the light emitter I_{ph}, which can be regarded as a measure of the incident photons expressed as a current, and is defined as [6]:

$$I_{ph} = q\frac{P_{in}\lambda}{hc}, \quad (2)$$

where q is the electron charge, h is the Planck Constant, c the light speed. For 400 nW light power, and 600 nm wavelength. I_{ph} is 194 nA. Fig. 4 shows the I_{Dopt} under such illumination conditions for different channel lengths and different gate voltages.

The photogenerated drain current is greater than the I_{ph} value, and is increased when the channel length is decreased. This means the existence of an optical photocurrent gain.

4. Conclusions

A MIS transistor with an integrated nitride waveguide was studied through computer simulation. The transistor was biased with different gate voltages and was illuminated directly from the gate considering visible light and power varying from 0 nW to 400 nW. The results show that the drain current increases as the light intensity increases. An optical detector with current gain that is a function of the gate length is observed. Such gain increases as the channel length decreases. This means that the device can be used to improve the detection of light intensities in the nW range in visible wavelengths in electrophotonic systems.

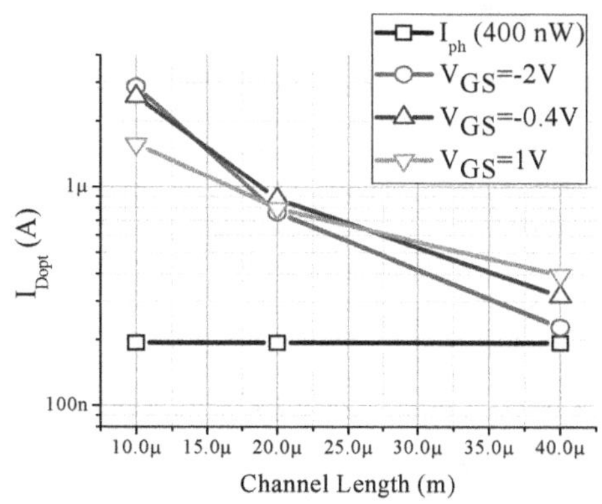

Fig. 4. Optical drain current for different channel lengths and gate voltage. Drain voltage is 5 V.

References

[1]. J. Juan-Colás, A. Parkin, K. E. Dunn, M. G. Scullion, T. F. Krauss, S. D. Johnson, The electrophotonic silicon biosensor, *Nat. Commun.*, Vol. 7, 2016, 12769.

[2]. J. Alarcón-Salazar, G. V. Vázquez, A. A. González-Fernández, I. E. Zaldívar-Huerta, J. Pedraza-Chávez, M. Aceves-Mijares, Waveguide-detector system on silicon for sensor application, *Adv. Mater. Lett.*, Vol. 9, Issue 2, 2018, pp. 116-122.

[3]. A. A. González-Fernández, J. Juvert, M. Aceves-Mijares, C. Domínguez, Monolithic integration of a silicon-based photonic transceiver in a CMOS process, *IEEE Photonics J.*, Vol. 8, Issue 1, 2016, 7900213.

[4]. K. Misiakos, et al., All-silicon spectrally resolved interferometric circuit for multiplexed diagnostics: A monolithic Lab-on-a-Chip integrating all active and passive components, *ACS Photonics*, Vol. 6, Issue 7, 2019, pp. 1694-1705.

[5] Athena User's Manual, *Silvaco Inc.*, 2013.

[6] Atlas User' s Manual, *Silvaco Inc.*, 2015.

(082)

Secure PUF: Physically Unclonable Function Based on Arbiter with Enhanced Resistance Against Machine Learning (ML) Attacks

Mohammad El-Hajj [1,2], **Ahmad Fadlallah** [1], **Maroun Chamoun** [2], **Ahmed Serhrouchni** [3]
[1] University of Sciences and Arts in Lebanon
E-mails: {m.elhajj, a.fadlallah}@usal.edu.lb
[2] Saint Joseph University, Lebanon
E-mail: maroun.chamoun@usj.edu.lb
[3] Telecom ParisTech, France
E-mail: ahmed.serhrouchni@telecom-paristech.fr

Summary: Latest cryptographic algorithms are based on the storage of secret information (stored in RAM), which is prone to a number of attacks like reverse-engineering, cold-boot, side channel, device tampering, etc. Therefore new methods should be developed to enable secure transmission of data and authentication especially in IoT sectors. In the past few years promising hardware security primitive is used Physical Unclonable Functions (PUFs), which is a lightweight one-way function to extract a unique identity to each end device based on physical factors introduced during manufacturing. In this paper, we introduced a new scheme of Arbiter PUF to create an identity for each device. We have provided a scheme that is prone to machine learning attacks. Our scheme avoids the shortcoming of several previously proposed related PUF based authentication protocols that is summarized in a benchmark done in this paper.

Keywords: Internet of Things, IoT, Security, Authentication, PUF, Arbiter, Machine Learning (ML).

1. Introduction

The Internet of Things (IoT) will consist of billions of physical objects which are embedded with electronics, software and sensors. Such devices will be connected to the Internet infrastructure to collect, exchange, analyze, and even being involved in decision making without any human intervention [1]. IoT will be deployed in a wide spectrum domains such as public health, smart cities, smart homes, environmental monitory, intelligent transportation system, waste management, chemicals, military, agriculture, etc.[1, 2].This wide-spectrum domains mean the generation of huge amount of security-based information [3].However, many challenges have been emerged: the variety of connectivity protocols and communication methods [4],the power consumption where most of the end devices are battery powered [5] and deployed in environments that are hard to reach for battery recharging or maintenance; like in internet of Medical things (IoMT) [6] and Wireless Sensor Networks(WSNs) [7].

One of the most security concerns in the use of connecting objects in everyday people life is the authentication [8], and to address such concern many cryptographic algorithms have been introduced but the majority of devices that will constitute IoT are Constrained devices1 [1] which make majority of traditional cryptographic algorithms infeasible [9].

The traditional tackle was based on storing the cryptographic keys in a Non-Volatile Memories (NVMs) [10] like FRAM, EEPROM [11], etc. Nevertheless, such tackle is at risk due to physical attacks such as side channel [12], invasive and semi-invasive attacks [13], and reverse engineering.

Physically Unclonable Functions (PUFs) could be the solution for many security breaches mentioned above [14].In this paper we will explain PUF and how it can be deployed as a hardware security solution for some IoT security challenges, and design and implement a new PUF-based architecture to be used for authentication of IoT devices.

The rest of this paper is organized as follow: Section 2 presents the classification of PUF based on its data sets, then we discussed the applications that benefits from the use of PUF. While Section 2 discusses the usage of non-linear combination generators to overcome modelling attacks and its security analysis, Section 4 describes the new class of PUF circuit. Finally, Section 5 concludes the paper and mentions the future work.

2. Classification of PUF and Its Applications

There are number of classifications of PUF, we will focus on the last one due to its direct impact on the applications based on PUF. The overall classification of PUFs is shown in Fig. 1:

1) Classification based on the physical construction proper- ties: Intrinsic PUFs [15] vs. Non-Intrinsic PUFs [16].

[1] Limited CPU, memory, and power resources, thus they don't have enough resources to handle extra functionalities and protocols.

2) Classification based on the implementation technology used: Non-Electrical vs. Electrical V.S. Silicon based PUFs [17].
3) Classification based on the number of Challenge Response Pairs (CRPs): Weak vs. Strong [15], and the properties of this classification are discussed below.

Strong PUF [18]: are PUFs that support a large number (exponential with respect to the physical components in a circuit) of CRPs with a specification that it is impossible to read out all the CRPs within a feasible time, so that an adversary can't derive unknown CRPs from a prior knowledge of a limited number of CRPs.

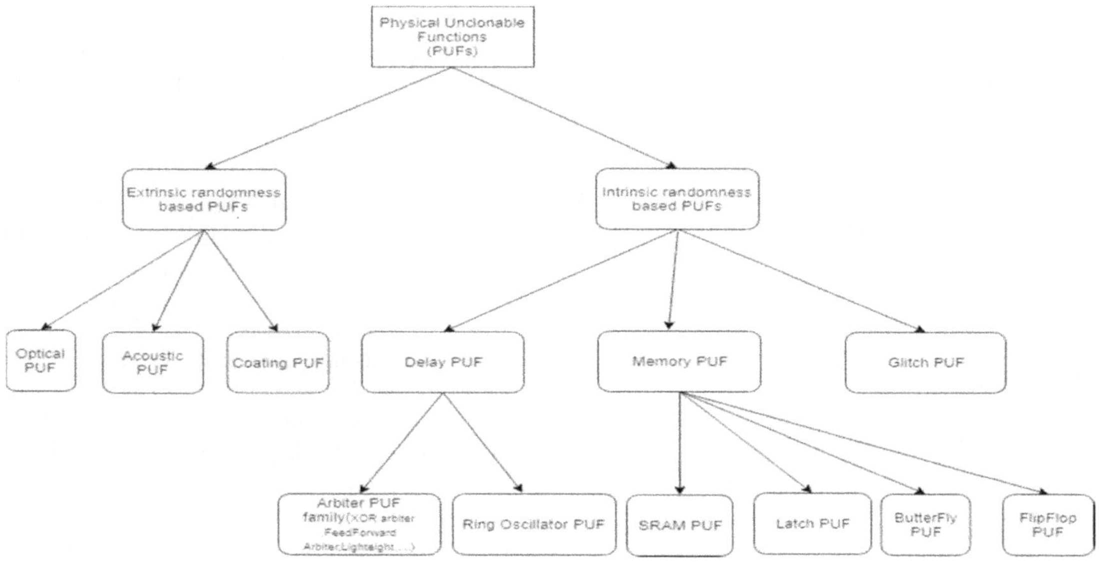

Fig. 1. PUF Classification.

Specifically Strong PUFs should not allow ML modeling reverse engineering which could benefit from the knowledge of subset CRPs then using them to train a ML algorithm to model the circuit [15].Once a model for a PUF has been derived, two main security features of PUF are no longer hold: unpredictable and unclonability. This type of PUF is ideal for circuit identification [19] and secret key generation [20].

Weak PUFs [21]: are PUFs that support a limited number of CRPs or even one CRP. They are mainly used for the generation of cryptographic secret key that is needed for cryptographically based security use-cases.This type is not suitable for circuit identification due to the limited number of CRPs, otherwise it will be prone to replay and side channel attacks. The literature shows that this type of PUF is less prone to modelling attacks [22].

Controlled PUFs [23]: are PUFs that use Strong PUFs as the main component, with the addition of a control logic(algorithm) surrounding the PUF.The main idea behind the control logic is the encapsulation of the input and output from the outer environment. Such architecture has the ability to block modelling attacks, unless the output of the String PUFs are realized then the Controlled PUF could be broken down. There are some cryptographic applications (smart card authentication, private storage in computers, etc.) that require certain CRPs to be unknown to prevent the cloning of the PUF circuit.

Regarding the applications, PUFs can be used to realize two security objectives: Key generation and Authentication [14].

(A) Key generation

Security services like confidentiality, data authentication, and data integrity are achieved by using either encryption or message authentication which need a secret key. Weak PUFs circuits are used to avoid the need to store keys on the chip, the output of the PUFs are used as secret key. The main challenge is that, there is no commitment that the same outputs will be produced from the same set of inputs due to the noise and environmental variations. To overcome this limitation, an error correction process is used like Bose-Chaudhuri- Hocquenghem codes (BCH) codes [24], Binary Symmetric Channel (BSC) [25], Syndrome Construction [26], Index-Based Syndrome Coding [27], Complementary index-Based Syndrome Coding [28], and Differential Sequence Coding [29]. Thus weak PUFs can produce secret key that can be used for security applications that needs a cryptographic key, the process is made up of two main phases: Initialization phase and the Reconstruction phase:

- *Initialization phase:* The PUF circuit generates a re sponse (output), the error correction code for this output is saved for later use.
- *Re-construction phase:* In this phase, the same output should be generated. The PUF is feed by an input to produce an output, the output generated will be corrected using the error correction code from the initialization phase to produce the same output. For security services that need a key to proceed, the output generated is hashed down to a specific length

to be used as a cryptographic key (AES, RSA, etc.). The whole process is shown in Fig. 2.

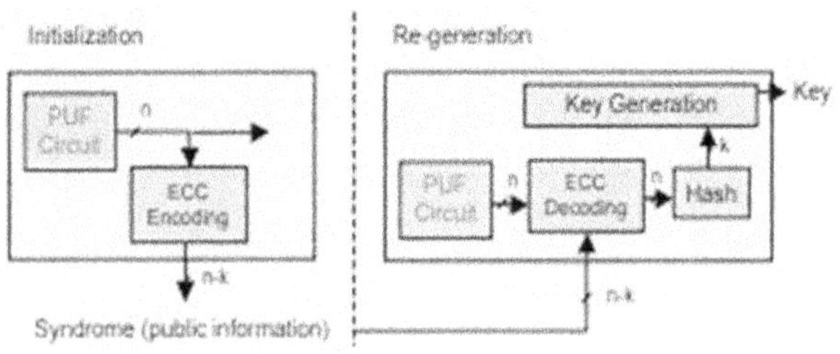

Fig. 2. Cryptographic key generation.

(B) Authentication

Machine-to-Machine (M2M) communication in IoT use case is an essential task, thus the identification of devices should be verified to build the trust. PUF can be used to authenticate devices without the need of highly cost cryptographic modern protocols. PUF-based authentication can even deployed to extremely resource constraint devices (e.g., Radio-Frequency Identification (RFID) and Near Field Communication (NFC) devices).

The authentication of the devices is achieved via the use of Challenge/Response protocol. PUF provides devices with set of bit-strings responses to challenges that uniquely identify each device, the identification is achieved by using strong PUFs to harden the work done by an adversary to build a clone of the device.

Authentication process consists of two main phases: Initialization phase and the Verification phase.

Initialization phase:

This phase is done before the utilization of the device, a set of CRPs from each PUF instance is stored in secure database.

Verification phase:

Upon request of authenticity of the device, a set of stored CRPs are chosen randomly and applied to the device. The resulted response is compared to the one stored inside the database, if there is a match then the device is authenticated, otherwise it is considered a fake device. It is important that a large set of CRPs are stored (Strong PUF) to eliminate the need of reusing CRPs and being vulnerable to man-in-the-middle attacks [30].The whole process is shown in Fig. 3.

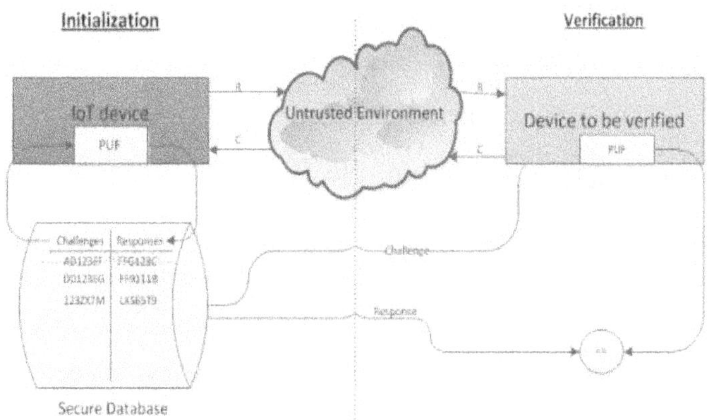

Fig. 3. PUF-based authentication process.

As a conclusion for this section, an overview is done with the experimental results for different PUFs proposed in the literature, and how they are affected by machine learning attacks to model them is shown in Table 1.

3. Related Work

We will focus in the related work based on adding nonlinearity to combine multiple PUFs structure leading to secure PUF or possibly harden modelling attacks. The only function used to combine multiple PUFs according to the state-of-art was XOR function, which is considered insecure against modelling attacks. In the current literature there are attempt to use alternative and arbitrary combining functions for multiple PUFs such as Linear Feedback Shift Registers (LFSR). To generate a pseudo-random key stream, LFSR is used as a cryptographic abstractions in stream ciphers.

Authors in [40] proposed a Ring Oscillator PUF structure based on LFSR in which they built a True Random Number Generator (TRNG) that could be used in key generation. In [41] authors proposed a PUF scheme that benefits from LFSR as a random generator, and the random generated is used as a challenge for the RO PUF. The work done is like introducing a controlled PUF by pre-processing the challenge before being used as an input to the PUFs, and the output produced by the PUFs is post-processed by a re-configurable component before being used as an authentication key. No security analysis is done for the proposed solution. While authors in [42] proposed an implementation for LFSR in silicon, they used 128 circuit block and 3 XOR gates among them, to enable fine uniqueness of the response produced, they introduced the LFSR in which the output of the 1st circuit block is fed back to the inputs of 99th, 101th, and 126th circuit blocks with the XOR gate. The output of the 1st circuit block is the output of circuit block. Authors didn't provide a security analysis for the above architecture, although they have shown by experiments the uniqueness and reproducibility of their proposed solution. Authors in [43] introduced a new Strong PUF scheme that is based on the use of a Weak PUF and the response will be an input to multiple LFSR, the output produced is then combined by a Boolean function *f* to produce random and secure response. The experimental results have shown that the PUF design is unique and reliable, security analysis is doe and the authors have shown that there model is not resistance to modelling attacks. Authors in [44] proposed a new PUF scheme based on the use of configurable LFSR to increase the response length, they used the LFSR output as the control signal inputted to the multiplexers. The experimental results have shown that the PUF design is unique and reliable, but no security analysis is done. In [45] authors proposed a new RO PUF scheme that is based also on the use of LFSR that generates a random signature that is derived from the random characteristics of D-Latch to generate a random clock. The experimental results show that the proposed design can be used to generate considerably unique and reliable responses even with environmental variations.

Table 1. Experimental results of PUFs.

PUF Type	Challenge	Response	Uniqueness	Reliability	Resistance to Modeling attacks
Optical[31], [32]	laser positioning	Speckle Pattern vector	49%	75%	Yes
Coating[33]	Sensors	Capacity measurement	50%	95%	Yes
Arbiter[34]	Delay line	Arbiter	49%	95%	No
Ring Oscillator[35]	Ring	Counter	48%	98%	Yes*
Feed-Forward Arbiter[34]	Delay line	Arbiter	38%	90.2%	No
XOR Arbiter[35]	Delay line	Arbiter	45%	91%	No
Light-Weight secure PUF[36]	Delay line	Arbiter	?	?	No
SRAM[37]	Memory Address	power-up state	49%	88%	Yes
Latch[38]	Latch	State of NOR	49%	97%	Yes
ButerFly[39]	Butterfly Cell	State of Cell	50%	94%	Yes

* It is vulnerable to modelling attacks if used as Strong PUF.
? Not presented in the literature.

4. Proposed Solution

As Table 1 shown above, current PUF implementations are prone to modelling attacks using machine learning techniques such as SVM, LG, and ES.A new scheme of PUFs (based on Arbiter) will be introduced in this paper in which the challenge of one PUF is derived from the challenge of another PUF. In such circuits, the output response is dependent on the delay process as traditional arbiter PUF and the input challenge to one of PUFs is also dependent on the delay process too. The process variations of one of PUFs will fed a bent function before being considered as a challenge of the other PUF. The overall process is shown in Fig. 4.

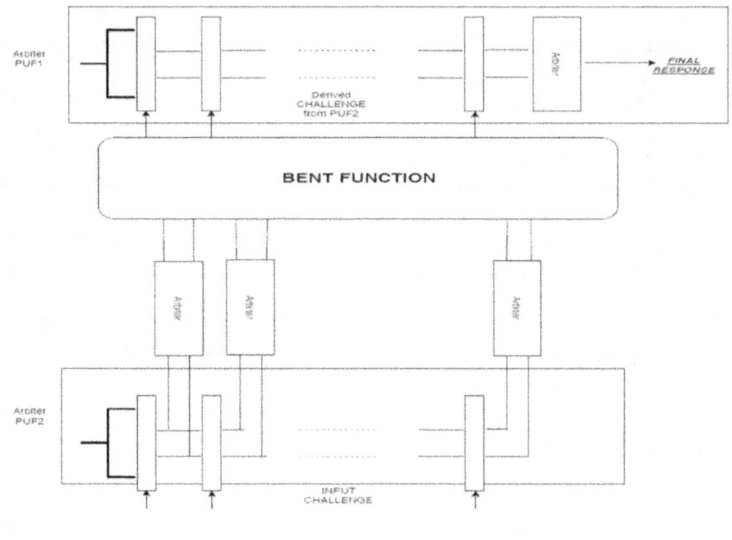

Fig. 4. Secure PUF with Bent schematic.

5. Conclusion

This papers proposed a novel design with the theme that the challenge of one PUFs is derived from another one. To increase its resistance against modelling attacks using machine learning Bent functions are used between the multiple PUFs. Security analysis of the proposed solution will be done in future work to prove the resistance of such solution to modelling attacks by explaining Bent functions and its characteristic.

References

[1]. M. El-hajj, M. Chamoun, A. Fadlallah, A. Serhrouchni, Taxonomy of authentication techniques in internet of things (IoT), in *Proceedings of the* IEEE *15th Student Conference on. Research and Development (SCOReD'17)*, 2017, pp. 67-71

[2]. M. El-hajj, M. Chamoun, A. Fadlallah, A. Serhrouchni, Analysis of authentication techniques in internet of things (IoT), in *Proceedings of the 1st IEEE Cyber Security in Networking Conference (CSNet'17)*, 2017, pp. 1-3.

[3]. L. Atzori, A. Iera, G. Morabito, The Internet of Things: A survey, *Computer Networks*, Vol. 54, Issue 15, 2010, pp. 2787-2805.

[4]. S. Al-Sarawi, M. Anbar, K. Alieyan, M. Alzubaidi, Internet of things (IoT) communication protocols, in *Proceedings of the 8th IEEE International Conference on Information Technology (ICIT'17)*, 2017, pp. 685-690.

[5]. M. El-Haii, M. Chamoun, A. Fadlallah, A. Serhrouchni, Analysis of cryptographic algorithms on IoT hardware platforms, in *Proceedings of the IEEE 2nd Cyber Security in Networking Conference (CSNet'18)*, 2018, pp. 1-5.

[6]. H. Suo, J. Wan, C. Zou, J. Liu, Security in the Internet of Things: A review, in *Proceedings of the International Conference on Computer Science and Electronics Engineering (ICCSEE'12)*, Vol. 3, 2012, pp. 648-651.

[7]. G. Han, L. Liu, J. Jiang, L. Shu, G. Hancke, Analysis of energy- efficient connected target coverage algorithms for industrial wireless sensor networks, *IEEE Transactions on Industrial Informatics*, Vol. 13, Issue 1, 2017, pp. 135-143.

[8]. M. El-hajj, A. Fadlallah, M. Chamoun, A. Serhrouchni, A survey of internet of things (IoT) authentication schemes, *Sensors*, Vol. 19, Issue 5, 2019, 1141.

[9]. W. Trappe, R. Howard, R. S. Moore, Low-energy security: Limits and opportunities in the Internet of Things, *IEEE Security & Privacy*, Vol. 13, Issue 1, 2015, pp. 14-21.

[10]. M. T. Ghoneim, M. M. Hussain, Review on physically flexible nonvolatile memory for internet of everything electronics, *Electronics*, Vol. 4, Issue 3, 2015, pp. 424-479.

[11]. Performance and Energy Comparison of F-Ram and EEPROM devices, Report, Colorado, USA, 2012, https://www.digikey.at/Web%20Export/Supplier%20 Content/ramtron-1140/pdf/ramtron-tech-fram-eeprom-comparison.pdf

[12]. K. Zhao, L. Ge, A survey on the internet of things security, in *Proceedings of the Ninth International Conference on Computational Intelligence and Security (CIS'13)*, Washington, DC, USA, 2013, pp. 663-667.

[13]. S. P. Skorobogatov, Semi-invasive attacks: A new approach to hardware security analysis, PhD Thesis, *Citeseer*, 2005.

[14]. C. Herder, M.-D. Yu, F. Koushanfar, S. Devadas, Physical unclonable functions and applications: A tutorial, *Proceedings of the IEEE*, Vol. 102, Issue 8, 2014, pp. 1126-1141.

[15]. R. Maes, I. Verbauwhede, Physically unclonable functions: A study on the state of the art and future research directions, in Towards Hardware-Intrinsic Security, *Springer*, 2010, pp. 3-37.

[16]. M. Roel, Physically unclonable functions: Constructions, properties and applications, *Katholieke Universiteit Leuven*, Belgium, 2012.

[17]. C. Wachsmann, A.-R. Sadeghi, Physically unclonable functions (PUFS): Applications, models, and future directions, *Synthesis Lectures on Information Security, Privacy & Trust*, Vol. 5, Issue 3, 2014, pp. 1-91.

[18]. U. Ruhrmair, H. Busch, S. Katzenbeisser, Strong PUFS: Models, constructions, and security proofs, in Towards Hardware-Intrinsic Security, *Springer*, 2010, pp. 79-96.

[19]. K. Lofstrom, W. R. Daasch, D. Taylor, IC identification circuit using device mismatch, in *Proceedings of the IEEE International Solid-State Circuits Conference (ISSCC'2000)*, 2000, pp. 372-373.

[20]. G. E. Suh, S. Devadas, Physical unclonable functions for device authentication and secret key generation, in *Proceedings of the* 44th *ACM/IEEE Design Automation Conference (DAC'07)*, 2007, pp. 9-14.

[21]. U. Ruhrmair, F. Sehnke, J. Solter, G. Dror, S. Devadas, J. Schmidhuber, Modeling attacks on physical unclonable functions, in *Proceedings of the 17th ACM Conference on Computer and Communications Security*, 2010, pp. 237-249.

[22]. U. Ruhrmair, J. Solter, PUF modeling attacks: An introduction and overview, in *Proceedings of the Europe Conference & Exhibition on Design, Automation & Test (DATE'14)*, 2014, pp. 1-6.

[23]. B. Gassend, D. Clarke, M. Van Dijk, S. Devadas, Controlled physical random functions, in *Proceedings of the 18th Annual Computer Security Applications Conference*, 2002, pp. 149-160.

[24]. N. Alimohammadi, S. B. Shokouhi, Secure hardware key based on physically unclonable functions and artificial neural network, in *Proceedings of the 8th IEEE International Symposium on Telecommunications (IST'16)*, 2016, pp. 756-760.

[25]. B. Colombier, L. Bossuet, V. Fischer, D. Hely, Key reconciliation protocols for error correction of silicon PUF responses, *IEEE Transactions on Information Forensics and Security*, Vol. 12, Issue 8, 2017, pp. 1988-2002.

[26]. Y. Dodis, L. Reyzin, A. Smith, Fuzzy extractors: How to generate strong keys from biometrics and other noisy data, in *Proceedings of the International Conference on the Theory and Applications of Cryptographic Techniques (EUROCRYPT'04)*, 2004, pp. 523-540.

[27]. M.-D. Yu, S. Devadas, Secure and robust error correction for physical unclonable functions, *IEEE Design & Test of Computers*, Vol. 27, Issue 1, 2010, pp. 48-65.

[28]. M. Hiller, D. Merli, F. Stumpf, G. Sigl, Complementary IBS: Application specific error correction for PUFS, in *Proceedings of the IEEE International Symposium on*

Hardware-Oriented Security and Trust (HOST'12), 2012, pp. 1-6.

[29]. M. Hiller, M. Weiner, L. Rodrigues Lima, M. Birkner, G. Sigl, Breaking through fixed puf block limitations with differential sequence coding and convolutional codes, in *Proceedings of the 3rd International Workshop on Trustworthy Embedded Devices*, 2013, pp. 43-54.

[30]. U. Ruhrmair, On the security of puf protocols under bad PUFS and PUFS-inside-PUFS attacks, *IACR Cryptology ePrint Archive*, Vol. 2016, 2016, 322.

[31]. R. Pappu, B. Recht, J. Taylor, N. Gershenfeld, Physical one-way functions, *Science*, 2002, Vol. 297, Issue 5589, pp. 2026-2030.

[32]. U. Ruhrmair, C. Hilgers, S. Urban, A. Weiershauser, E. Dinter, B. Forster, C. Jirauschek, Revisiting optical physical unclonable functions, *IACR Cryptology ePrint Archive*, Vol. 2013, 2013, 215.

[33]. P. Tuyls, G.-J. Schrijen, B. Skoric, J. Van Geloven, N. Verhaegh, R. Wolters, Read-proof hardware from protective coatings, in *Proceedings of the International Workshop on Cryptographic Hardware and Embedded Systems (CHES'06)*, 2006, pp. 369-383.

[34]. J. W. Lee, D. Lim, B. Gassend, G. E. Suh, M. Van Dijk, S. Devadas, A technique to build a secret key in integrated circuits for identification and authentication applications, in *Proceedings of the IEEE Symposium on VLSI Circuits*, 2004, pp. 176-179.

[35]. B. Gassend, D. Clarke, M. Van Dijk, S. Devadas, Silicon physical random functions, in *Proceedings of the 9th ACM Conference on Computer and Communications Security*, 2002, pp. 148-160.

[36]. M. Majzoobi, F. Koushanfar, M. Potkonjak, Lightweight secure PUFS, in *Proceedings of the IEEE/ACM International Conference on Computer-Aided Design (ICCAD'08)*, 2008, pp. 670-673.

[37]. J. Guajardo, S. S. Kumar, G.-J. Schrijen, P. Tuyls, FPGA intrinsic PUFS and their use for IP protection, in *Proceedings of the International Workshop on Cryptographic Hardware and Embedded Systems (CHES'07)*, 2007, pp. 63-80.

[38]. Y. Su, J. Holleman, B. Otis, A 1.6 pj/bit 96 % stable chip-id generating circuit using process variations, in *Proceedings of the IEEE International Solid-State Circuits Conference (ISSCC'07)*, 2007, pp. 406-611.

[39]. S. S. Kumar, J. Guajardo, R. Maes, G.-J. Schrijen, P. Tuyls, The butterfly PUF protecting IP on every FPGA, in *Proceedings of the IEEE International Workshop on Hardware-Oriented Security and Trust (HOST'08)*, 2008, pp. 67-70.

[40]. N. Arasavalli, A. Pawar, et al., Design and implementation of linear feedback shift register based physical unclonable function, *IJARCCE*, Vol. 6, 2017, pp. 388-392.

[41]. B. R. Rani, S. C. Sreedha, Design and implementation of physical unclonable functions for device authentication and secret key generation using ASIC, IJRAET, Vol. 4, 2015, pp. 34-43.

[42]. Y. Ogasahara, Y. Hori, H. Koike, Implementation of pseudo linear feedback shift register physical unclonable function on silicon, in *Proceedings of the IEEE International Symposium on Circuits and Systems (ISCAS'16)*, 2016, pp. 758-761.

[43]. S. Hou, Y. Guo, S. Li, A lightweight LFSR-based strong physical unclonable function design on FPGA, *IEEE Access*, Vol. 7, 2019, pp. 64778-64787.

[44]. B. Srinivasu, P. Vikramkumar, A. Chattopadhyay, K.-Y. Lam, COLPUF: A novel configurable LFSR-based PUF, in *Proceedings of the IEEE Asia Pacific Conference on Circuits and Systems (APCCAS'18)*, 2018, pp. 358-361.

[45]. F. Amsaad, A. Sherif, A. Dawoud, M. Niamat, S. Kose, A novel FPGA-based LFSR PUF design for IoT and smart applications, in *Proceedings of the IEEE National Aerospace and Electronics Conference (NAECO'18)*, 2018, pp. 99-104.

(085)

Oxygen Gas Sensing Technologies New Features in Combustion Process

P. Shuk and C. McGuire
Rosemount Measurement and Analytical, Emerson Automation Solution,
6021 Innovation Blvd, Shakopee, MN 55379-4676, USA
Tel.: +1 952 204 6030, fax: + 1 952 204 2054
E-mail: Pavel.Shuk@Emerson.com

Summary: Zirconia oxygen gas sensing technology is well known for reliable and interference free oxygen measurements in combustion environment. Less known new zirconia oxygen sensor features like humidity or stoichiometric are already implemented on the market and new carbon monoxide (CO) breakthrough measurement are discussed in details. These new features are extending zirconia oxygen analyzer applications and making combustion control more reliable and efficient.

Keywords: Combustion analysis, Flue gas, Potentiometric oxygen gas sensing, New features for gas sensing.

---

1. Introduction

The oxygen is measured to optimize the combustion process efficiency [1-2], to control the product formation and quality and to minimize runaway combustion leading to the explosion. Oxygen excess in the combustion process is measured in the flue gas to control and optimize the minimum of oxygen concentration in flue gas for safe operation and process high efficiency (Fig. 1).

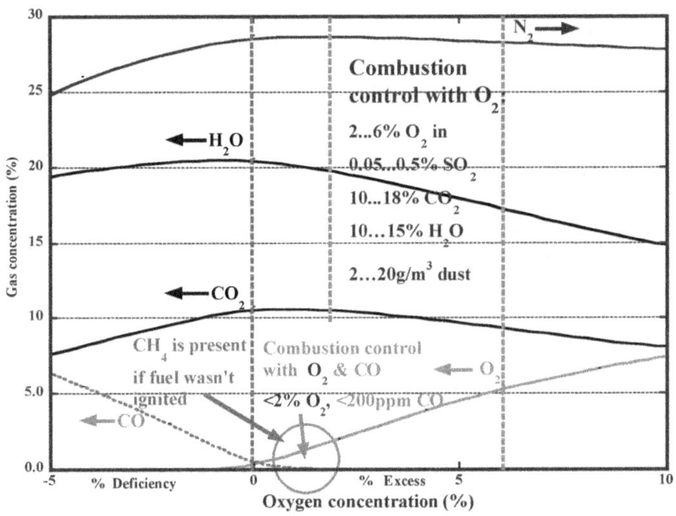

Fig. 1. Combustion flue gas diagram with methane control needed for safety [3].

Zirconia potentiometric oxygen gas sensing technology widely used nowadays in O_2 analyzers for different industrial combustion applications and in automotive Lambda sensors was invented in 1961 by Möbius and Peters [4] and Weissbart and Ruka [5] and was rapidly accepted in the power industry. All industrial zirconia oxygen sensors are based on an electrochemical cell with oxygen ion conducting yttria stabilized zirconia (YSZ) solid electrolyte and two platinum-based cermet electrodes printed and sintered on the opposite sides of the zirconia ceramics exposed to the process and reference gases.

Differential oxygen chemical potentials on the oxygen sensor electrodes would develop electromotive force (EMF), E, according to the Nernstian equation:

$$E = \frac{RT}{4F} \ln \frac{p(O_2)''_{\text{Process}}}{p(O_2)'_{\text{Ref}}}, \qquad (1)$$

with R universal gas constant, T the process temperature in Kelvin (K) and F the Faraday number.

By using fixed oxygen partial pressure on the reference electrode, e.g., air with $p(O_2) = 2.1*10^4$ Pa, the sensor signal of the thermally balanced oxygen sensor will be only depending on the sensor temperature. Zirconia potentiometric oxygen gas analysis was established by H.-H. Möbius [6] and is permitting measurements of free or chemically bonded oxygen even in reducing conditions with very low equilibrium concentration of oxygen (<1 ppb) released at high temperature (Fig. 2).

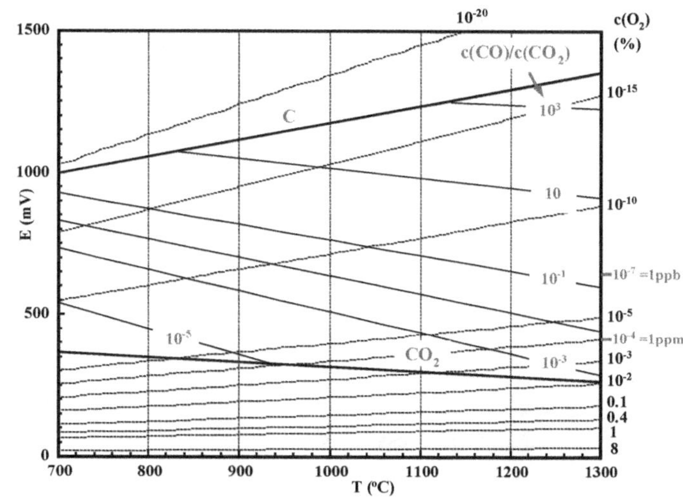

Fig. 2. Oxygen potentiometric temperature diagram.

Zirconia O_2-probe was applied with some very good success in the combustion control of power generation boilers, hot stoves for steelmaking, heating and combustion exhaust gas control of coke ovens for steelmaking, lime and cement kilns combustion control, incinerator combustion control, combustion control of heating furnaces for oil refinery and petrochemical industry and many other applications.

2. New Features for Oxygen Gas Sensing Technology

Moisture feature implemented in some oxygen analyzers on the market is based on wet in-situ and dry extractive (after removing moisture) oxygen concentration measurements:

$$c(O_2)_{dry} = c(O_2)_{wet} * 100\ \%/[100\ \% - c(H_2O)], \quad (2)$$

considering oxygen concentration is unchanged and/or stable in the process. Two zirconia oxygen analyzers might be serially installed with the first probe measuring oxygen in wet environment and the 2nd probe after condensing the water [7]. As an option, moisture concentration might be calculated from in-situ oxygen concentration measured by zirconia probe and extractive oxygen measurement using portable oxygen analyzer after removing moisture (dry).

In the regular mode oxygen analyzer is measuring the remaining oxygen excess in the combustion process. CO is formed when the combustion of carbon-based fuel is mixed with not sufficient amount of oxygen to complete the combustion:

$$C_xH_y + O_2 \rightarrow CO_2 + H_2O + CO, \quad (3)$$

Carbon monoxide, CO is a product of the imperfect combustion of the hydrocarbon fuels and is almost always formed to some degree when something is burning. Platinum electrode in oxygen sensor is well known as a catalyst and will burn most of the combustibles, e.g., carbon monoxide (CO) reducing the oxygen concentration on zirconia sensor electrodes. This oxygen sensor signal reduction may be considered as an indication of CO presence especially in the CO breakthrough combustion scenario [8] and might be used for CO concentration detection (Fig. 3), especially by employing derivative of a raw signal, dE/dt (Fig. 4).

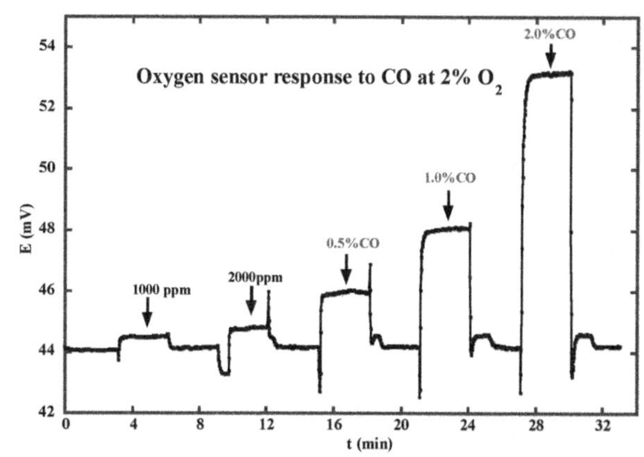

Fig. 3. CO concentration effect on zirconia oxygen sensor signal [8].

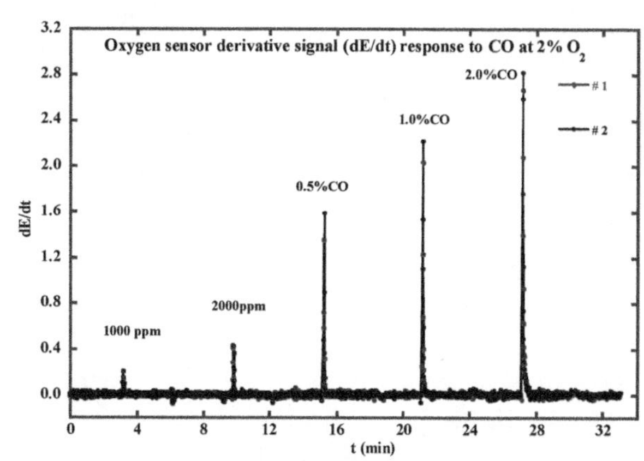

Fig. 4. O_2-sensor derivative signal (dE/dt) response to CO at 2 % O_2 [8].

Stoichiometric feature used for the indication of fuel rich condition defined as oxygen deficiency is based on zirconia oxygen sensor signal dependence on reducing gas concentration, e.g., CO, and assumption of complete oxidation of CO on oxygen sensor, consuming oxygen ($CO + 1/2O_2 = CO_2$). Calibrated oxygen sensor will indicate oxygen deficiency, concentration needed to bring combustion process back to the stoichiometric point.

3. Conclusions

New zirconia oxygen sensor features like moisture, CO breakthrough measurements and stoichiometric with additional information on combustion process would make combustion control more efficient providing new application opportunities.

References

[1]. P. Shuk, Process Zirconia oxygen analyzer – State of art, *Technisches Messen*, Vol. 77, Issue 1, 2010, pp. 19-23.

[2]. P. Shuk, Oxygen gas sensing technologies application: a comprehensive review, in Sensors for Everyday Life, Smart Sensors, Measurement and Instrumentation (S. C. Mukhopadhyay, et al., Eds.), *Springer*, Heidelberg, New York, 2017, pp. 81-107.

[3]. P. Shuk, C. McGuire, E. L. Brosha, Methane gas sensing technologies in combustion: a comprehensive review, *Sensor & Transducers*, Vol. 229, Issue 1, 2019, pp. 1-10.

[4]. H. Peters, H.-H. Möbius, Procedure for the gas analysis at elevated temperatures using galvanic solid electrolyte elements, DD-Patent 21673, *Germany*, 1961.

[5]. J. Weissbart, R. Ruka, Oxygen gauge, *Rev. Sci. Instrum.*, Vol. 32, 1961, pp. 593-595.

[6]. H.- H. Möbius, Basics of oxygen gas potentiomitric analysis, *Z. Phys. Chem.*, Vol. 230, Issue 5-6, 1965, pp. 396-416 (in Deutsch).

[7]. AQUATEC® 2000, https://www.enotec.com/en/AQUATEC-2000-O2-H2O-Analyzer

[8]. P. Shuk, R. Jantz, J. Kramer, Oxygen Sensor for CO Breakthrough Measurements, US Patent Application 2017/0003246 A1, *USA*, 1/5/2017.

(087)

Ultrasonic Flowmeter for Leakage Detection in Water Mains

F. Huber [1], **M. Schwarz** [1] **and B. Zagar** [1]
[1] Johannes Kepler University Linz, Institute for Measurement Technology, Altenberger Straße 69, 4040 Linz, Austria
Tel.: + 43 732 2468 5920
E-mail: franz.huber@jku.at

Summary: This contribution deals with the measurement of very low flow velocities for the use in household water mains to detect small leaks. For this purpose, the ultrasonic time-of-flight principle is exploited. This method is based on different propagation velocities and thus propagation times of the ultrasonic waves with and against the flow direction. This inherently requires the measurement of very small differences between the propagation times in opposing directions. A method to reduce the requirement of resolution of the time measurement or to increase the accuracy of the measurement of the flow velocity for a given resolution is the so-called sing-around technique. In this method, the length of the propagation path is virtually lengthened by sending a new ultrasonic pulse in the same direction immediately after receiving one at the end of the propagation path. This is done N times for both propagation directions and the flow velocity is calculated from the difference of the aggregate propagation times. We present the mechanical setup as well as the necessary electronics for the implementation of the sing-around technique for the measurement of very low flow velocities.

Keywords: Ultrasonic flow sensor, Sing-around technique, Speed of sound, Velocity measurement.

1. Introduction

In this paper we present an ultrasonic flowmeter that should be able to resolve flow velocities in water down to a few µm/s which allows for the continuous monitoring of household water mains for the detection of very small leaks that would otherwise go unnoticed over a potentially long period of time.

In order to measure to a detection limit of 5 drops per minute from a faucet for a typical water pipe diameter of ¾ inch it is necessary to reliably measure temperature and pressure compensated velocities over extended periods of time including those assumed to exhibit zero intended flow.

In a previous work [1] we were able to demonstrate that ultrasonic flowmeters are able to resolve flow velocities of less than 1 mm/s. To increase the sensitivity to measure flow velocities lowered by the factor of 20 we concentrate on examining the so-called sing-around technique of processing the ultrasonic returns [2, 3].

The paper is structured as follows: first we present the sing-around technique including a range of limiting influences that have to be taken into account. The experimental setup is discussed in detail in the next section. The paper is concluded by the interpretation and discussion of measurement data and a brief conclusion.

2. Ultrasonic Sing-around Technique

To measure the velocity of a flowing medium ultrasonically it is necessary to determine the propagation time of sound pulses over a very well-known distance. For this purpose, an ultrasonic transducer generates short duration sound bursts that propagate through the medium to the receiver. The propagation times are measured in two directions. This can be seen in Fig. 1.

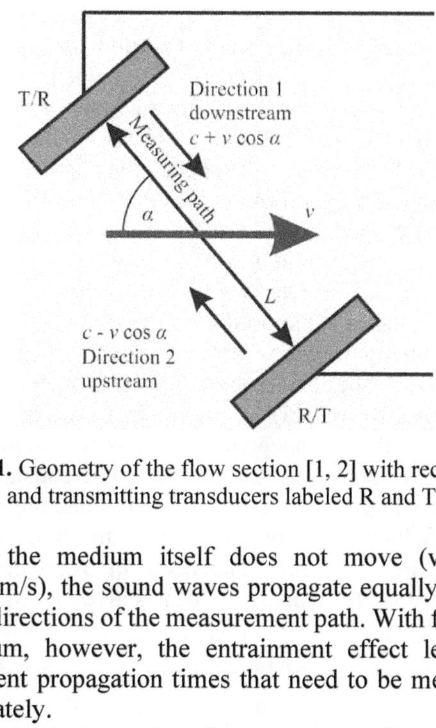

Fig. 1. Geometry of the flow section [1, 2] with receiving and transmitting transducers labeled R and T.

If the medium itself does not move (velocity $v = 0$ m/s), the sound waves propagate equally fast in both directions of the measurement path. With flowing medium, however, the entrainment effect leads to different propagation times that need to be measured accurately.

To calculate the flow velocity from these propagation times, it is necessary that the speed of sound c is constant to a very high degree over the duration of the two-way measurement. This assumption might not be valid if the temperature or the pressure of the medium changes gradually over time, because the speed of sound depends on both of them: $c = f(p, T)$ [4].

In the realization of the sensor, we assume that the medium is flowing at a velocity v under an angle α with respect to the measuring path. Fig. 1 shows two opposing ultrasonic transducers both able to act as transmitter and receiver, since they are exploiting the reciprocal piezoelectric effect [5]. c is the speed of sound in the medium, which in our case is water. The length of the measuring path is L.

The sound propagation time decreases in the direction of the flow (t_d) because the sound waves are sped up and increases against the flow (t_u) because the sound waves are slowed down. The results of the sound propagation times are given in the relationships in Eqs. (1) and (2). Solving these equations for the unknown velocity v leads to Eqs. (3) and (4).

$$t_d = \frac{L}{c + v \cdot \cos(\alpha)}, \quad (1)$$

$$t_u = \frac{L}{c - v \cdot \cos(\alpha)}, \quad (2)$$

$$v = \left(\frac{L}{t_d} - c\right) \cdot \frac{1}{\cos(\alpha)}, \quad (3)$$

$$v = \left(c - \frac{L}{t_u}\right) \cdot \frac{1}{\cos(\alpha)} \quad (4)$$

However, it is necessary to know the speed of sound c and the fraction L/t_u or L/t_d, respectively, to a high degree of accuracy. This on the other hand means that both have to be known at least to the desired resolution for v. Hence, the speed of sound needs to be known to the sought flow velocity resolution of µm/s, which is very difficult to attain.

Equations (3) and (4) show that only knowing either c or L/t_u or L/t_d, respectively, with the needed accuracy makes it impossible to calculate the velocity of the flowing medium. To work around this stringent requirement, the difference of the propagation times according to Eq. (5) is used. This results in an expression that is independent of this requirement, if the speed of sound can be assumed to be constant to a very high degree within the duration of one measuring cycle, which can be estimated to last a few milliseconds at a time. The result of using the time difference for the calculation of v is given in Eq. (6).

$$\Delta t = t_u - t_d, \quad (5)$$

$$v = \frac{\Delta t}{t_u \cdot t_d} \frac{L}{2 \cdot \cos(\alpha)} \quad (6)$$

In order to demonstrate the required accuracy of the propagation time measurements a short evaluation with the known parameters distance L = 120 mm, angle α = 23° and speed of sound c = 1484 m/s in water is given: The propagation time for the stationary medium is 80.9 µs and the propagation time difference for an assumed flow velocity of v = 100 µm/s is only 10 ps. This is a very small time difference that is very difficult to measure since there is few hardware available to be able to even resolve in the ps range.

Therefore, it is expedient to strive for an effective measuring distance extension. This is achieved by the sing-around principle [2], which is shown in Fig. 2. Ideally after receiving the sound pulse at the receiving transducer, the control electronics immediately triggers the generation of the next pulse in the same direction. The triggered control electronics then causes the transmitter to send out a new sound pulse. Overall this happens up to a predetermined upper limit of N passes, whereby the length of the measuring path is virtually extended. This results in an accordingly increased total propagation time. As a result this also leads to a time difference increased by the factor N. E.g., if the sing-around circuit has run 100 times with the values of the previously mentioned example the necessary time difference resolution increases from 10 ps to 1 ns.

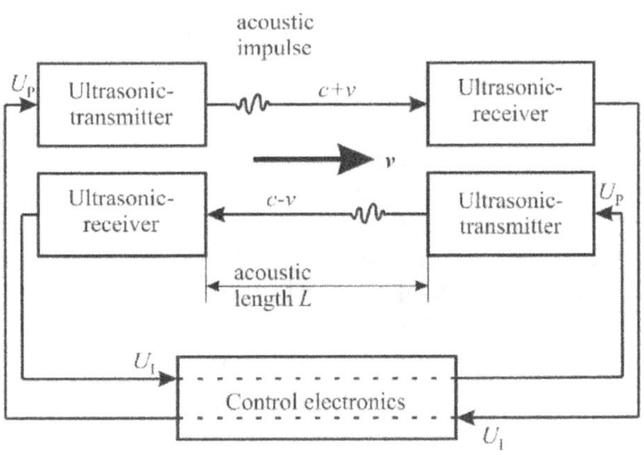

Fig. 2. Sing-around technique [2]. The control electronics is used for the signal detection, to emit the next pulse, to control the number of runs N and to measure the propagation times and thus for determining the flow rate.

Since the ultrasonic transducers, the electrical paths and the control electronics have nonzero group delays there will always be an additive time delay. This dead time must be constant to a very high degree. Thus, the selected electronics should not jitter in a larger range than 10 ps in order to obtain the necessary proper numerical correction by the control electronics. This corresponds to a frequency of 100 GHz, which is higher by the factor of 100.000 than the nominal frequency of the ultrasound transducers used.

A very important part of the sing-around system is the control electronics. It detects the received signal of the receiving ultrasonic transducer. When the detection has taken place, the transmission of the next sound pulse with the ultrasonic transmitter is initiated. The control electronics is also required for triggering of consecutive pulses and the measurement of the total runtime of the N passes and thus for determining the flow rate. A detailed description of the control electronics is given in Section 3.

3. Mechanical Setup and Electronics of the Sing-around System

There are several options for setting up the measuring path: Clamp-on, which means that the transducers are attached to the outside of the pipe wall, or invasively, meaning that the transducers are fixed in a bore in the measuring section's wall. Furthermore, time-serial or time-parallel operation is feasible.

These different possibilities all have their own advantages and disadvantages. One advantage of the clamp-on technique compared to the invasive measurement is that it does not influence the flow of the medium in the pipe. Furthermore, it is easy to change the measurement location. Other advantages are that there are no disturbances of the flow of the medium or pressure losses.

Due to the additional transitions of the sound waves through the pipe walls, the damping is higher for the clamp-on case. Furthermore, there is a greater measurement uncertainty due to the not exactly known geometry as well as thickness and the speed of sound of the pipe wall material [2]. Therefore, due to the loss of amplitude caused by the reflections off the interfaces between the clamped-on transducer and the pipe as well as the pipe and the water we chose an invasive setup. The attenuation in the water depends on the frequency, the temperature and the pressure.

Time-serial operation means that the propagation times for the two directions are measured one after another, for N runs each at a time. In a time-parallel case, both measurements are carried out simultaneously, which is possible since the sound waves propagating in different directions do not influence and interfere with each other.

A big difference between the time-serial and time-parallel measuring systems is the different electrical and physical complexity. For the time-parallel option the entire electronics is needed twice whereas for the time-serial case it is necessary to switch between the transmitting and the receiving function of the transducers after measuring in one direction.

On the physical side it is necessary that both ultrasonic transducers are stimulated at the same instance in time, if propagation times allow to do so. If this is not possible a second measuring path and four ultrasonic transducers instead of two are needed.

The advantage of the time-parallel variant is that if the speed of sound c changes as a result of changes in temperature or pressure, this does not affect the measurement results due to the natural averaging of the sing-around method. The same applies to changing flow velocities [2]. The disadvantage of this method is that the geometric conditions are difficult to match for both flow paths.

Fig. 3 shows the realized measuring section of the flowmeter. Therein the ultrasonic transducers are visible in black glued into a transparent pipe section. They are fixed into the flow channel with a combination of silicone and hot glue. The silicone is necessary for the construction to be waterproof. The hot glue is required for the transducers to be firmly mounted, so that they cannot move out of their position.

The flow channel was manufactured by 3D printing using the principle of stereolithography. To connect the pipe with the 3D printed flow measuring section a two component adhesive is used. The construction includes two possible measuring paths (1-4 and 2-3) to be able to measure either time-serial or time-parallel.

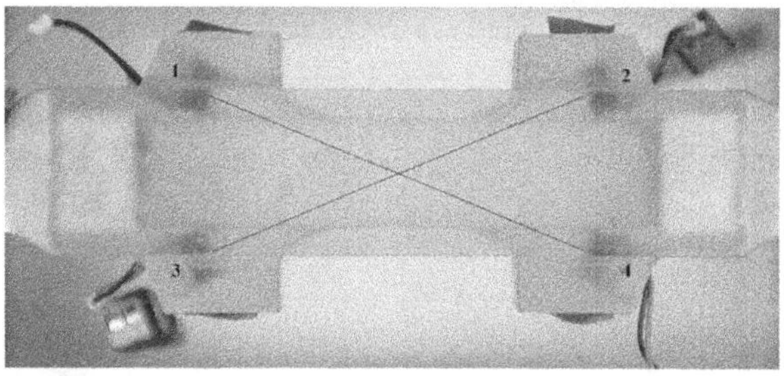

Fig. 3. Measuring section with ultrasonic transducers in black (1, 2, 3, 4) and the two measuring paths (1-4, 2-3).

To excite the transducers it is necessary to pay attention to the following aspects. The ideal signal to stimulate the transmitting ultrasonic transducer would be a Dirac impulse because it is infinitely short. Hence, all frequencies are effectively stimulated, thus also the resonance frequency of the transducer. Because only high-end pulser-receiver systems are getting close to the ideal excitation signal, an excitation signal specifically geared towards the resonance frequency has to be used for the low-cost case.

The amplitude of the excitation signal is chosen so that the amplitude of the received signal is maximized in the allowed range of the electronics to simplify the detection. Another criterion is that the received signal should fade fast enough, within one or two cycles at best. This is necessary to simplify the detection and to not affect the subsequent received signal.

Since we want the transducers to oscillate with the highest possible amplitude the signal has a frequency close to their nominal frequency [2]. Furthermore, the resonance must be strongly damped, so that the signal decays quickly, which is caused by the water.

To generate sound pulses fulfilling the given requirements, the STEVAL-IME011V1 evaluation board based on the STHV748 ultrasound pulser from STMicroelectronics is used. With the chosen pulser circuit it is only possible to generate rectangular pulses of discrete pulse periods with multiples of 6 ns and with amplitudes of up to ±95 V to obtain signals with arbitrary pulse durations. The signal of the ultrasonic pulser is sent to the transmitting ultrasonic transducer. Its triggering will be described later.

Fig. 4 shows the output signal U_P of the ultrasonic pulser used for the excitation. The ringing of the rectangular pulses is due to the finite rate of increase of the ultrasonic pulser. It can be seen that the period is 996 ns, which is matched to the resonance frequency of 1 MHz ± 0.5 % of the ultrasonic transducers. The signal U_{P2} shown in Fig. 4 consists of two rectangular pulses to excite the transducer to produce a signal with a high amplitude. If the excitation will be done with this signal the receiving signal would decay exponentially which can be seen in the upper part of Fig. 5.

In order to detect the received signal, it is necessary to recognize its peak. A detailed description follows in the description of the control electronics. To simplify the recognition of the received signal a faster decay and larger signal differences are desirable, thus a third pulse is applied. This third rectangular pulse is phase shifted by 180° with respect to the first two rectangular pulses which interferes destructively. This can be seen in the signal U_P in Fig. 4.

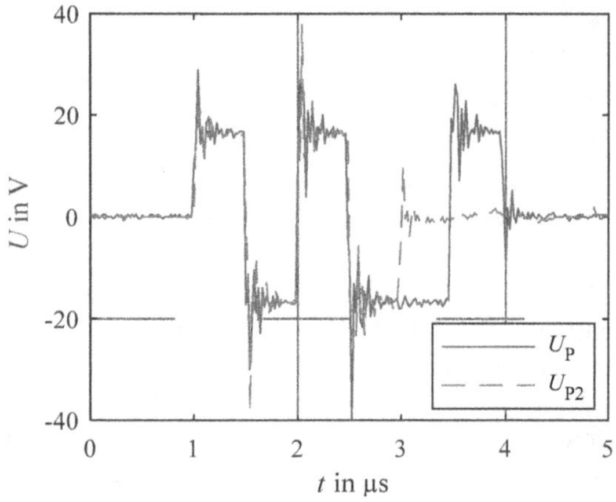

Fig. 4. Excitation signals U_P and U_{P2} for the transmitting ultrasonic transducers. The added rectangular pulse in U_P is phase shifted by 180° to simplify the detection of the received signal due to destructive interference.

As a result, the received signal decreases faster in amplitude due to the induced destructive interference [6]. The received signal with this extinction is shown in the bottom of Fig. 5. The negative effect of this is that the received signal has a smaller amplitude.

For accurate time measurement, the most negative peak must be detected. The simplification achieved by the additional phase shifted pulse is that the fast decrease of the received signal increases the difference of the amplitudes of the two most negative peaks. This can be seen in Fig. 5.

The voltage difference of the two most negative peaks is greater due to the destructive interference in the lower graph. As a result, the detection of the receiving signal is simplified because fluctuations in amplitude due to contamination of water or deposits on the transducers or the tube walls have less influence.

Fig. 5 shows that the received signal of the ultrasound transducer changes its shape due to the destructive interference caused by the excitation signal. The reason for the slight sawtooth shape of the signal U_I is a group delay distortion.

In order to measure both propagation times accurately, i.e. with the desired uncertainty of less than 10 ps, it is necessary to always trigger the clock at the same relative phase at least within 62.8 µrad within the pulses' complex envelopes. One way to guarantee this condition can be met is to accurately detect all signal's zero crossings to identify the zero crossing closest to the envelope's maximum elongation. This task is performed by the control electronics' fast finite state machine. Its schematic is shown in Fig. 9.

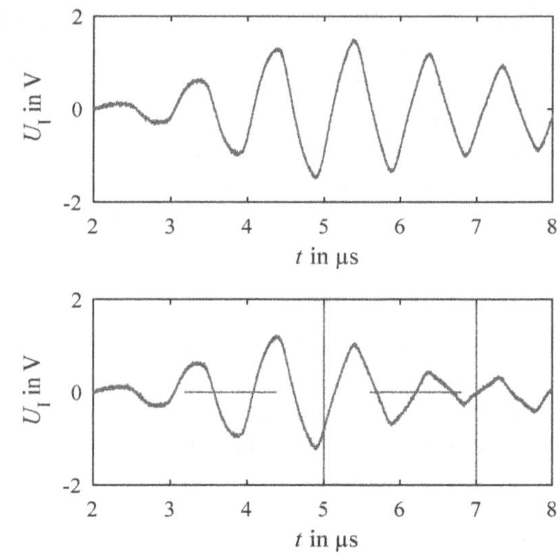

Fig. 5. Received signal U_I without (top) and with (bottom) an additional phase shifted pulse.

The input of the high-speed comparators C1 and C2 is the voltage signal U_I which is the output signal of the receiving ultrasonic transducer. Comparator C2 compares the input voltage with a customized reference voltage level U_{ref} to detect the most negative peak of the received signal.

The most negative peak instead of the most positive peak is chosen since it is the more pronounced peak in that particular setup, which simplifies its detection. The output voltage of the comparator U_{C2}, which is an analog electronics component, is level-shifted to obtain the proper voltage levels for the following RS flip-flop. If the input voltage U_I passes C2's threshold voltage from below the RS flip-flop is set and subsequently reset at the immediately following zero-crossing detected by C1. This event triggers the emission of the next ultrasonic pulse via the subsequent high speed logic circuit.

Fig. 6 shows an example of the signals which are used for the described detection. U_T is the inverted output voltage of the RS flip-flop that is used instead of the non-inverted output because the ultrasonic pulser requires a rising edge to be triggered. In this case the voltage will be set to a logic low when the most negative peak was detected and it will be reset to a logic high after the next zero-crossing. With this rising edge the required condition for accurate propagation time measurement is fulfilled. It is used for time measurement and for triggering the ultrasonic pulser.

However, there is always a certain dead-time between the triggering and the actual beginning of the ultrasonic pulse. This must be constant within very stringent limits to be corrected. The challenge in this regard especially are the clocked parts of the sing-around system.

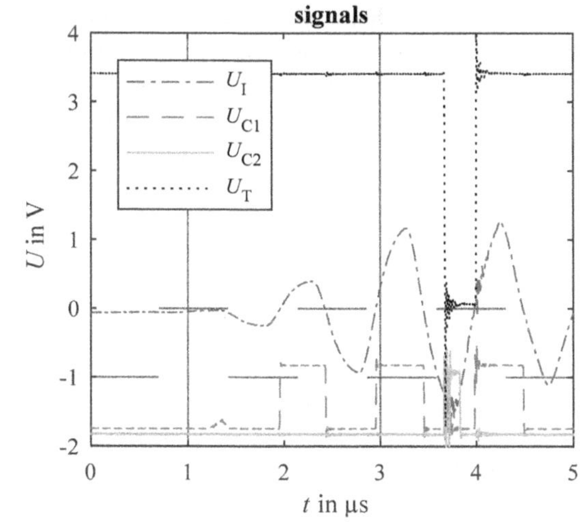

Fig. 6. Signals of the control electronics, input voltage U_I, comparator voltages U_{C1} and U_{C2}, trigger voltage U_T.

The microcontroller (µC in Fig. 9) is used for the initialization and communication with the time-to-digital converter (TDC) via an SPI interface. The microcontroller is also needed to count the predefined number of sing-around runs. This leads to propagation times extended by a factor of N, which are measured with the time-to-digital converter.

A very important part for the time measurement is the logic block between the microcontroller and the TDC. This block prevents an additional dead-time caused by the microcontrollers clock. The reason why this could occur is that the time difference between the triggering and the emitting of the signal by the microcontroller can fluctuate. The solution for this difficulty is the logical interconnection of the microcontroller and the output of the RS flip-flop. This logic block triggers the time measurement of the TDC with the edge of the RS flip-flop instead of the edge of the microcontroller's output signal.

4. Characterization and Measurement Results

The challenge in characterizing the different parts of the built sing-around system is that it is more accurate than standard test tools.

First, the uncertainty of the analog parts between the input U_I of the comparators and the output U_T, used to trigger the ultrasonic pulser for the next pulse, was investigated. For generating the input signal U_I, a high performance signal generator was used. The time measurement was performed with an oscilloscope. The resulting uncertainty of the measurement is in the range of the uncertainty already induced by the signal generator.

This proved that the crucial components of the high accuracy time measurement are the clocked parts of the system. The problem with the clocked elements is that they do not react immediately when the corresponding edge is applied. They start with the next edge of their internal clock. Because of this, there is an additional variable dead time. This problem is bypassed by the logic block for the used microcontroller.

As it turned out, the major challenge is the clocked part for pulse generation. Every time the signal U_T triggers the pulser an additional dead time is added. It was determined by measurements that this dead time shows strong fluctuations. For this purpose, a corresponding pulse was applied with a signal generator periodically. This pulse starts the ultrasonic pulser. We measured the time required by the ultrasonic pulser from triggering the start pulse until the emission. The result of this measurement is shown in the histogram in Fig. 7. It can be seen that the dead time fluctuates within a range of approximately 0.4 µs.

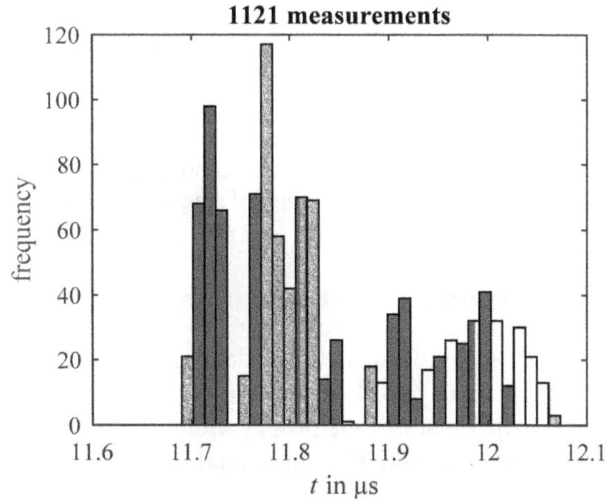

Fig. 7. Histogram of the measured dead time of the ultrasonic pulser.

Various measurements were carried out for further characterization. In order to be able to make statements regarding the number of sing-around runs and the number of measurements both values were varied. The influence of the number of sing-around runs is shown in Fig. 8. In each case 1000 measurements were carried out for 10 and 100 runs. The histograms were calculated from the median and the mean triple absolute deviation. Outliers outside these boundaries were removed. As shown in Fig. 8, the fluctuation range increases with the number of passes. The same applies to the number of outliers.

To avoid this difficulty, the medians were considered. It turned out, that these also fluctuate but with a smaller fluctuation range. For example, two series of measurements with 1000 runs each were compared for 10 sing-around runs. The difference of the two medians was 3 ns. With this fluctuation in the time measurement, it is not possible to achieve the desired resolution for flow measurement in household water mains. For the given example it is necessary to measure time differences of 100 ps over 10 runs.

As has been shown, the challenge is the clocked component of pulse generation because its dead time varies greatly. That is why the next step is to find a more suitable kind of pulse generation with a very constant dead time. In the best case the pulse generation should work analogue, thus without relying on an internal clock.

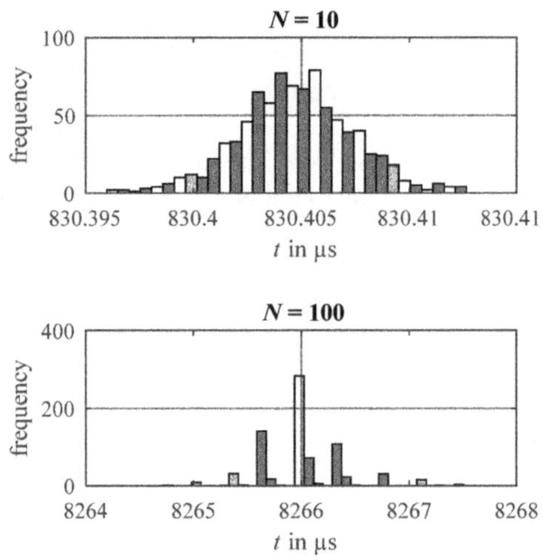

Fig. 8. Histograms of the measured propagation time without flow. N is the number of sing-around runs.

5. Conclusion and Outlook

An ultrasonic flow meter for leakage detection was introduced. It is based on the sing-around technique using the difference of the propagation times caused by the flowing medium. The functional principle and the electrical and mechanical setup of the measuring system were presented. Since the desired resolution of the measurement of the flow velocity is not yet achieved further steps will be taken. For this purpose, the pulser that is used to generate the excitation will be replaced by a more suitable component.

References

[1]. M. Schwarz, B. Uhl, B. G. Zagar, M. Stur, Ultrasound-based measuring system for the flow velocity of liquids and gases in drainage pipes in tunnels, *TM – Technisches Messen*, Vol. 86, Issue 2, 2019, pp. 93-103.

[2]. J. Gätke, Akustische Strömungs- und Durchfluß-messung, *Akademie-Verlag Berlin*, 1991.

[3]. J. Delsing. The zero flow performance of a sing-around type flow meter, in *Proceedings of the IEEE Symposium on Ultrasonics (IUS'90)*, 1990, pp. 1541-1544.

[4]. V. A. Belogol'skii, S. S. Sekoyan, L. M. Samorukova, S. R. Stefanov, V. I. Levtsov, Pressure dependence of the sound velocity in distilled water, *Measurement Techniques*, Vol. 42, Issue 4, 1999, pp. 406-413.

[5]. J. A. Jensen, Estimation of Blood Velocities Using Ultrasound, *Cambridge University Press*, 1996.

[6]. A. B. Raine, N. Aslam, C. P. Underwood, S. Danaher, Development of an ultrasonic airfow measurement device for ducted air, *Sensors*, Vol. 15, Issue 5, 2015, pp. 10705-10722.

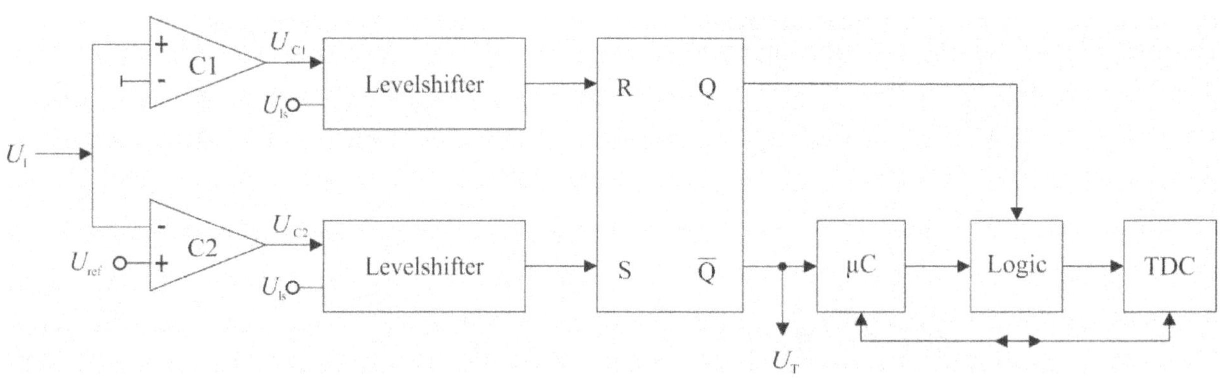

Fig. 9. Schematic of the control electronics. U_I is the output signal of the receiving transducer. Comparator C1 performs the zero-crossing detection of U_I and C2 the detection of the necessary peak. U_T is used to trigger the ultrasonic pulser and for the microcontroller (μC) to control the time measurement which is done by the TDC. The logic block is used to prevent an additional dead time caused by the microcontroller.

(089)

New Method for $NO/NO_2/NH_3$ Measurements and differentiation with a Single Sensor

B. Ya. Farber [1], **G. Graves** [1], **Y. C. Li** [2], **T. Rash** [2] **and N. W. Currier** [2]
[1] Zircoa Inc., 31501 Solon Rd., 44139 Solon, OH, USA
[2] Cummins Inc., 1900 McKinley Avenue, 47201, Columbus, IN, USA
Tel.: + 14403497236, fax: + 14403497209
E-mail: bfarber@zircoa.com

Summary: A new approach (Pulse Discharge Technique, PDT) for measurements and differentiation between different exhaust gas components critical for diesel emission control is presented. Electrodes of a conventional lambda sensor are activated by short high amplitude voltage charging pulses of equal amplitude and opposite polarity and kinetics of self-discharge is measured during the pauses separating the charging pulses. PDT demonstrates high sensitivity (in the ppm range) to NO, NO_2 and NH_3 and low cross-interference to O_2, CO, CO_2, HC and H_2O. Differentiation between different gas components is accomplished by parameter analysis of self-discharge kinetics and charge transfer characteristics during the charging pulses. Models explaining gas sensitivity under the conditions of PDT are presented and discussed.

Keywords: NO_x sensor, NH_3 sensor, Lambda probe, Pt/YSZ electrodes, Polarization, Charge transfer.

1. Introduction

Currently, the technology of choice for the abatement of NO_x emissions from heavy-duty diesel engines and coal fired power plants is the Selective Catalytic Reduction (SCR) using urea as a reducing agent [1].

Efficiency of SCR based NO_x abatment depends on complex chemical reactions taking place within the NH_3-NO/NO_2 reacting system in the presence of O_2 and H_2O [2].

For optimization of the SCR control algorithm, independent measurements and differentiation between individual exhaust components is highly desirable in a single sensor. However, available commercial sensors for NO_x [3] and NH_3 [4] measurements do not provide NO/NO_2 differentiation and there is no commercial sensor for separate NO_2 measurements.

Simple and robust oxygen (lambda) sensors are dominating the exhaust sensor field since 1976 [5].

Recently, it was shown that these sensors can be used for multi-gas measurements due to development of Pulse Discharge Technique (PDT) [6].

2. PDT Method

The PDT is based on charging the sensor electrodes with a series of high amplitude (2.5 V) voltage pulses (~ 100 ms) of equal duration and opposite sign. Pulses are separated by pauses, when sensor self-discharge is monitored. Kinetics of the sensor discharge provides information on the concentration and identification of the exhaust components.

Fig. 1 shows a schematic of the PDT method.

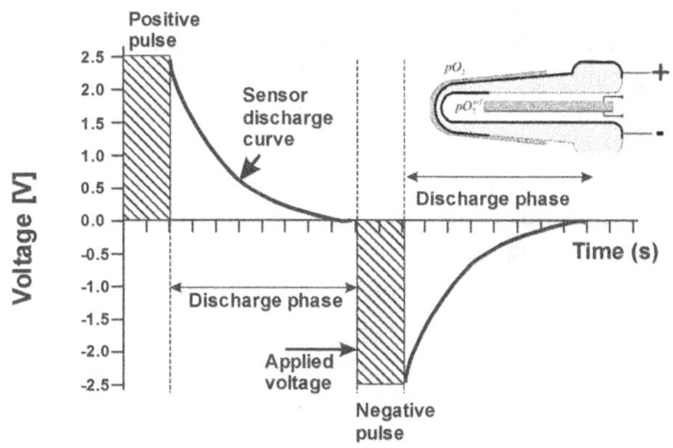

Fig. 1. Schematic of the PDT method.

Sensor output under conditions of PDT is recorded either by measuring voltage at a fixed time during the pause or by linearization of the discharge curve in Voltage versus Log(t) coordinates and extrapolation to a fixed pause time of 10 sec.

3. NO, NO_2, NH_3 differentiation

Fig. 2 shows oxygen sensor response under conditions of PDT to variable concentrations of NO, NO_2 and NH_3.

Sensor response is qualitatively different for different ranges of excess NO, NO_2 and NH_3 concentrations.

In the range of excess concentration of NO_x (NO+ NO_2) relative to NH_3, amplitude of the sensor signal is increasing by absolute value proportionally to the logarithm of NO_x concentrations following both negative and positive voltage pulses (see (a) and (b) in Fig. 2). However, excess NH_3 relative to NO_x concentration results in increase of the sensor output following negative pulses but decrease by absolute values following positive pulses.

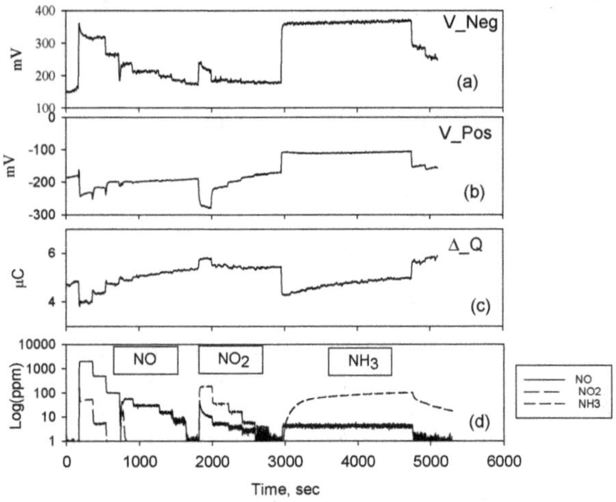

Fig. 2. Sensor output following negative voltage pulses (a), positive voltage pulses (b), differential charge transfer (c), applied gas concentrations (d).

Under conditions of PDT sensor demonstrates stronger response following negative voltage pulses in the range of excess NO but shows weaker response in the range of excess NO_2. Differential charge transfer between positive and negative voltage pulses is decreasing with increasing NO and NH_3 concentrations but increasing with increased NO_2 concentration (see (c) in Fig. 2).

Fig. 3 demonstrates close agreement between PDT estimated and applied NO concentrations.

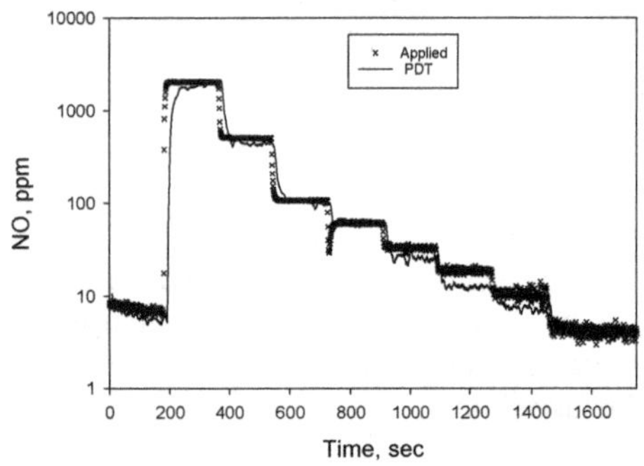

Fig. 3. Applied versus measured NO values.

Specific features of the O_2 sensor response under conditions of PDT can be explained by considering formation and decomposition of Pt oxide under oxidizing and reducing conditions following negative and positive voltage pulses. During negative voltage pulses oxygen ions are pumped from the reference side of the O_2 sensor to the measurement electrode resulting in formation of transient platinum oxide (PtO), which is capable of charge storage at the electrode-electrolyte interface [7]. Reaction between formed PtO and NO/NO_2 results in accelerated charge dissipation proportional to the concertation of gaseous species. Positive voltage pulses result in complete reduction of formed PtO and partial reduction of ZrO_2 in the vicinity of the electrode/electrolyte interface leading to F-center formation. Reaction between F-centers and NO_2 may be a contributing factor for increased NO_2 sensitivity following positive voltage pulses.

Competing adsorption/desorption mechanisms on the fully reduced electrode/electrolyte surface between NO/NO_2 and NH_3 maybe a contributing factor explaining different responses for NO_x and NH_3 following positive voltage pulses.

2. Conclusions

Combination of the Oxygen sensor and PDT provides a unique opportunity for measurements and differentiation between different exhaust components critical for SCR control optimization with a single sensor.

References

[1]. O. Krocher, Selective catalytic reduction of NO_x, *Catalysts*, Vol. 8, 2018, pp. 459-463.

[2]. I. Nova, C. Ciardelli, E. Tronconi, D. Chatterjee, M. Weibel, NH_3-NO/NO_2 SCR for diesel exhausts after treatment: Mechanism and modelling of a catalytic converter, *Topics in Catalysis*, Vol. 42, Issue 1, 2007, pp. 43-46.

[3]. N. Kato, Y. Hamada, H. Kurachi, Performance of thick film NOx sensor on diesel and gasoline engines, *Journal of Engines*, Vol. 106, 1997, pp. 1246-1253.

[4]. D. Y. Wang, M. Shost, J.-H. Yoo, D. Cabush, D. Racine, Ammonia sensor for closed-loop SCR control, *Cars - Electron. Electr. Syst.*, Vol. 1, Issue 1, 2009, pp. 323-333.

[5]. J. Riegel, H. Neumann, H. M. Wiedenmann, Exhaust gas sensors for automotive emission control, *Solid State Ionics*, Vol. 152, 2002, pp.783-800.

[6]. B. Farber, Method of Sensor Conditioning for Improving Signal Output Stability for Mixed Gas Measurements, *US Patents:* US 7,585,402; US 9,304,101; US 10,156,545; US 10,180,408; EP 1891422, USA, 2009.

[7]. H. Popke, E. Mutoro, C. Raiss, B. Luerssen, M. Amati, M. K. Abyaneh, L. Gregoratti, J. Janek, The role of platinum oxide in the electrode system Pt(O-2)/yttria-stabilized zirconia, *Electrochimica Acta,* Vol. 56, Issue 28, 20111, pp. 0668-10675.

(090)

A Study on Low Latency Serial Power Line Tunnels for Sensor Data Transmission

K. Kellner [1], D. Hochwarter [2], M. Brandl [1]
[1] Danube University Krems, Department for Integrated Sensor Systems, Dr. Karl Dorrek Str. 30, 3500 Krems
[2] Innovation Management HELLA Austria GmbH Vienna, Austria
Tel.: 0043 2732 893 2621
E-mail: karl-heinz.kellner@donau-uni.ac.at

Summary: In systems where sensor data has to be forwarded to a host system, power line communication can be an alternative to wireless or wired bus solutions. Furthermore, a serial communication unit is present in almost every microcontroller. Therefore, there is a natural demand for cost efficient serial tunnels over power line, still marginally supported by industry. Especially when low latency is needed as for example in local interconnect networks or controller area networks. In this paper, several key points on serial tunneling over power line are discussed. A commercial available modem using quadrature phase shift keying, a direct sequence spread spectrum based modem and a straightforward, cost efficient, open source modem are compared. A brief description of the working principle is given. Practical use cases on two automotive vehicles are studied.

Keywords: Sensor data transmission, Power line communication, Local interconnect network, Controller area network, Modulation.

1. Introduction

In power line communication (PLC), state of the art modems utilizing orthogonal frequency-division multiplexing (OFDM) provide high-speed data communication with up to several hundred Megabits per second [1]. But when transmitting sensor data, speed is not always on top of the requirement list. Especially, tunnelling of an (existing) serial bus like local interconnect network (LIN) or controller area network (CAN) over PLC is not a straightforward task for most commercial modems. This is not only because of costs per node, but also by limitations caused by latency. Low latency necessary for LIN over PLC is supported only by a few devices e.g. SIG60 (YAMAR Ltd., Israel). It is utilizing quadrature phase shift keying (QPSK). A modem for this purpose based on direct-sequence spread spectrum (DSSS) is presented in [2], but it is still under development.

For CAN over PLC there is, up to now (May 2019), no commercial real time tunnel available. Although the DCAN500 (YAMAR Ltd. Israel) was announced some years ago, it is also still under development. As a further alternative, a simple, cost efficient, easy-to-rebuild modem is presented here and compared to these systems in a practical use case.

2. Key Considerations

In replacing the conventional LIN or CAN physical layer by PLC, latency is a key parameter. In a LIN network there is a master node which requests replies from several slave nodes. Request and response has to be performed within a narrow time window given by a scheduler [3]. In CAN, things are even worse due to the unique arbitration scheme on the physical layer, which forces the latency to be below one bit time [4].

Moreover, all methods containing a serial to parallel conversion and vice versa produce unwanted latency. Let's consider a serial transmission of one single byte with an OFDM based tunnel. Suppose the tunnel is capable of transmitting one byte at once, on 8 different frequencies. But unfortunately, the communication can only start after clocking in the last bit of the serial stream. At the receiver, clocking out the first bit starts soonest after receiving the last bit at the transmitter. Even by neglecting acquisition (preamble detection, PLL locking, stabilization time of AGC etc.) and transmission time itself, there is a minimum latency of 8 bit times. The same problem prevents efficient use of forward error correction methods. This is also addressed by [5] in detail observing latency of a QCA7000 (Qualcomm Inc., USA) based home plug system.

Next to consider is synchronisation (acquisition), necessary for synchronised demodulation which usually takes several bit times and produces unwanted latency.

A fast acquisition time (<4 bit times, SIG60) is a solution, but has the drawback of misinterpreting noise as the start of a data frame. Another method is keeping all nodes synchronized by e.g. a master node, which is experimentally performed by [2]. An additional solution is asynchrony demodulation e.g. on-off keying (OOK) with the drawback of lower interferer robustness but lower complexity.

3. Methods, Test Setup and Results

A brief overview of the open source modem is given now. It utilizes a conventional OOK scheme modulating marks (on) and spaces (off) on a sinusoidal carrier. In Fig. 1, a block diagram is shown. The receiver consists of a ceramic narrow band filter, a voltage variable attenuator (VVA), an amplifier, an envelope detector, a discriminator, and an adaptive threshold correction (ATC). The ATC keeps the threshold level of the discriminator near the optimum at half of the signal level. It is, together with the filter, a key element for achieving optimal performance.

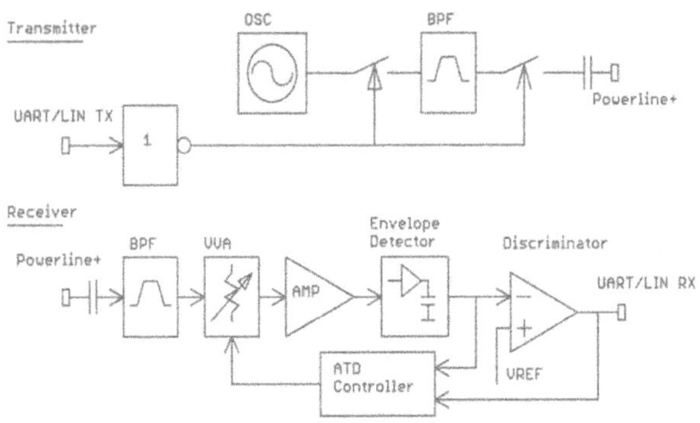

Fig. 1. Block diagram of the proposed OOK modem.

The minimum threshold is 5 mVpp. The latency is below one bit time.

The practical usability of all three systems was tested in a Volkswagen Touran 1.6 BMT TDI (YOM: 2011) and a Peugeot 308 SW TDI (YOM: 2010) with LIN protocol at 19200 Baud. One node acts as master, a second node as slave. Master requests and slave responses with 8 byte LIN frame, 50 packets/s. A minimum of 1000 LIN packets are sent and the packet loss rate was calculated. Packet loss is recognized by LIN framing error or LIN checksum error. Devices under test are SIG60 micro modules (YAMAR), DSSS Modems from [2] and the proposed OOK Modem. Motor, window lifters (periodically activated), radio, cabin ventilation and air condition (AC), headlights, break lights, blinker, wiper were running during tests. Transmission level for all modems was set to 500 mVrms. The SIG60 carrier runs at 6.5 MHz, DSSS, and OOK both at 8 MHz, all modems use capacitive coupling to the PL.

Table 1 shows measured results under different setups. Receive levels are measured with ATO1104 oscilloscope (Shenzhen Micsig Instruments Co., Ltd. China).

Table 1. Measurement results.

Vehicle	Device	Master Port	Slave Port	M->S A[dB]	S->M A[dB]	Packet Loss
Peugot 308	SIG60	Front 12V cigarette lighter socket	BT	>50	>50	100 %
	DSSS			>50	>50	90 %
	OOK			>50	>50	100 %
	SIG60		LS	15	24	0 %
	DSSS			24	29	0 %
	OOK			24	29	1 %
VW Touran	SIG60		BT	50	38	100 %
	DSSS			26	27	0 %
	OOK			26	27	1 %
	SIG60		LS	1	5	0 %
	DSSS			12	11	0 %
	OOK			12	11	1 %
BT	Plus at battery, minus at generator housing					
LS	Luggage cabin 12 V socket					
A	Attenuation = 20 log (500 mVrms / RX level)					

4. Conclusions

Cost efficient low latency serial tunnels that are able to substitute the physical layer of UART, LIN and CAN are rarely supported by industry.

We tested a commercial QPSK modem, a DSSS-based research system, and a low cost free hardware OOK based solution under practical conditions in two different automotive vehicles utilizing LIN protocol. As expected, reliability correlates with the attenuation of the signal. The DSSS Modem shows the lowest packet loss, but this might be due to the stronger attenuation of the SIG60's carrier frequency. Under moderate attenuation, the OOK modem packet loss is around 1 %. This is sufficient for sensor readout and can be considered as a practicable low cost alternative. For choosing the optimal operating frequency, a pretest of the transfer function is suggested.

Acknowledgements

This work is co-founded by the government of Lower Austria and the European Regional Development Fund (ERDF). Project ID: WST3-F-5030664/004-2017.

References

[1]. C. Cano, et al., State-of-the-art in power line communications: From the applications to the medium, https://arxiv.org/pdf/1602.09019.pdf

[2]. M. Brandl, K. Kellner, T. Posnicek, D. Hochwarter, Performance evaluation of a DQPSK and a DSSS PLC-modem for vehicular applications, in *Proceedings of the IEEE International Symposium on Power Line Communications and its Applications (ISPLC'19)*, Prague, Czech Republic, 3-5 April 2019, pp. 1-5.

[3]. LIN Specification ISO 17987, https://microchipdeveloper.com/lin:specification

[4]. C. Watterson, Optimizing CAN node bit timing analog devices technical article MS-2678, https://www.analog.com

[5]. A. Ogunleye, Design and Development of an Embedded HomePlug Green*PHYTM* Web Based System, http://hdl.handle.net/10019.1/98543.

(091)

Design and Implementation of Environment Monitoring System Based on ZigBee

Fuzheng Zhang [1], **Weile Jiang** [2], **Qijing Lin** [1,3,4,5] **and Hao Wu** [1]
[1] State Key Laboratory for Manufacturing Systems Engineering, Xi'an Jiaotong University, 710049, Xi'an, China
[2] Institute of Heritage Sites & Historical Architecture Conservation, Xi'an Jiaotong University, Xi'an, 710049, China
[3] Collaborative Innovation Center of High-End Manufacturing Equipment, Xi'an Jiaotong University, Xi'an, 710054, China
[4] Xi'an Jiaotong University Suzhou Institute, Suzhou, 215123, China
[5] State Key Laboratory of Mechanical System and Vibration, Shanghai Jiaotong University, Shanghai, 200240, China
Tel.: + 862982668616
E-mail: jiangwl@xjtu.edu.cn

Summary: ZigBee technology with the characteristics of low complexity, ad-hoc network, low cost, low power consumption and high security, is widely used for wireless sensing. Wireless sensor network is an interdisciplinary application between multi-sensor technology and wireless network, and has the broad application prospects in many fields such as intelligent agriculture and smart home. This paper designs and realizes an indoor environment monitoring system based on ZigBee and wireless sensor network. The system uses end-nodes to collect temperature, humidity, PM2.5 and other environmental parameters. The collected data is transmitted to the coordinator through wireless sensor network and finally to the monitoring system software by serial port protocol. The online monitoring platform with functions of real-time data display, storage and over-limit warning is developed by using the LabVIEW software. The test results show that the system is stable, easy to generate a network system, speedy and reliable to transmit data, and flexible to place nodes.

Keywords: ZigBee, Wireless sensor network, Environment monitoring, LabVIEW, Monitoring platform.

1. Introduction

Based on the characteristics of high information integration and multi-disciplinary intersection, Wireless Sensor Network (WSN) is concerned by various countries. SN involves embedded computer technology, communication technology, sensor technology and distributed information processing technology. It can collect various environmental parameters in the area covered by the wireless network, and process the information accordingly, and finally transmit them to the users who need the information [1]. At present, the network protocols applied to WSN mainly include Wireless Fidelity (WiFi), Bluetooth, General Packet Radio Service (GPRS), ZigBee, etc. Among them, ZigBee technology has the advantages of low complexity, self-organizing network, low cost, low power consumption and high security. It is widely used in smart agriculture [2-6], smart home [7-9], chemical detection [10-13] and medical rehabilitation [14-17].

In recent years, ZigBee technology, because of its superiority, has been used for environmental monitoring [2-17]. Zhang et al. [5] designed an intelligent control system for the environmental monitoring in greenhouses. This system realizes the monitoring of environmental parameters in greenhouses, e.g., temperature and humidity. However, the type of sensors used is single and the number of nodes is too large; Gong et al. [9] designed an environment monitoring system in smart home based on ZigBee. This system has the advantage of the real-time monitoring. However, the parameters monitored by the system are few and the software lacks the functions of data storage and over-limit warning. In addition, the package of nodes has not been designed; Ni et al. [14] studied a telemedicine system based on ZigBee, EDGE, Ethernet and embedded LINUX system. This system realizes the functions of home self-care, remote diagnosis, etc., but the system is too complicated. At present, most researches on ZigBee technology are only applied to specific occasions, i.e., it is not universal.

This study aims to realize indoor environment monitoring with the characteristics of low complexity, ad-hoc network, low cost, low power consumption and high security. In order to avoid the problems of complicated wiring, poor mobility, high cost and energy loss caused by traditional optical fiber, coaxial cable and other wired monitoring, this study designed and built an environmental monitoring system based on ZigBee technology. The system can effectively reduce the number of nodes and improve the utilization of nodes. This paper developed a new type of terminal node package box to achieve waterproof and dustproof protection for node modules. The online monitoring platform for this system was developed by LabVIEW software to realize data display, storage and over-limit warning of the collected indoor environmental data.

2. Monitoring System Design

2.1. System Design

As shown in Fig. 1, the monitoring system is mainly composed of three parts: a terminal node, a coordinator and an online monitoring platform. All the sensors are installed on the terminal node. The acquired environmental data by the terminal node is wirelessly transmitted to the coordinator module and finally sent to the monitoring platform. If the distance between the terminal node and the coordinator is too far to achieve effective transmission, power amplifiers or routers can be used to extend the transmission distance. The system layout is shown in Fig. 2.

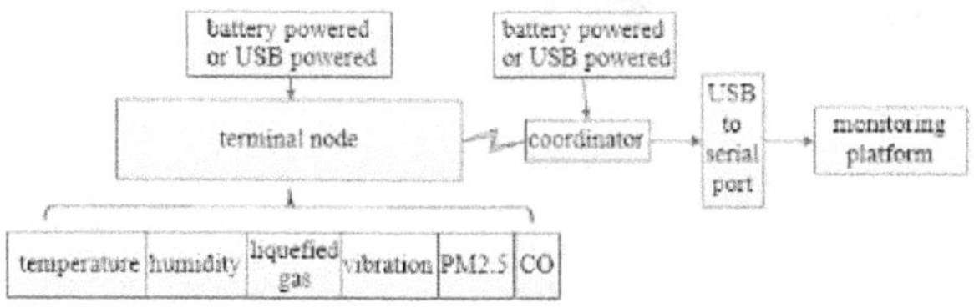

Fig. 1. System framework.

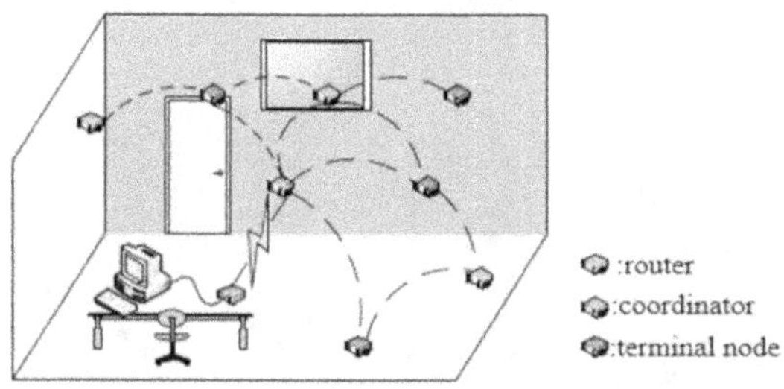

Fig. 2. System layout.

The coordinator is connected to the PC by a USB cable. In addition, the terminal node, router, and coordinator are identical in hardware. They are only different in functions by burning different codes.

2.2. Sensor Parameters

The environmental parameters monitored by the system are temperature, humidity, fine particles (PM2.5), carbon monoxide, liquefied gas and vibration, the parameters of each sensor used in this system are listed in Table 1. Among them, the liquefied gas sensor and the vibration sensor can only realize the function of over-limit warning.

Table 1. The list of sensor parameters.

Name	Scope	Accuracy	Type
temperature &humidity	0~50 °C	±2 °C	DHT11
	20~95 %	±5 %	
PM2.5	0~8000 pcs	±1 um	GP2Y1014AU0F
carbon monoxide	10-1000 ppm	±5 ppm	MQ-7
liquefied gas	0/1		MQ-2
vibration	0/1		SW-1801P

2.3. Package Design

The node module package box of the system is made of acrylic sheet and is assembled by organic adhesive (acrylic glue), which has the advantages of simple design, simple assembly and low cost. By using the transparent acrylic sheet, the user can directly observe the condition of each sensor from the outside. The layered design inside the package box can isolate the node module from each sensor module, which can protect the node module from dust and water, and also make it easy to replace the sensor module directly. The node package box is shown in Fig. 3.

Fig. 3. Node package box.

2.4. Monitoring Platform

The user monitoring platform of this system is written by LabVIEW software. LabVIEW is a program-development environment developed by National Instruments (NI) and its program is written by graphical G language [18-20]. The login system of the system is shown in Fig. 4, the users enter the monitoring platform by the username and password set up by themselves. The user monitoring platform designed in the system is shown in Fig. 5.

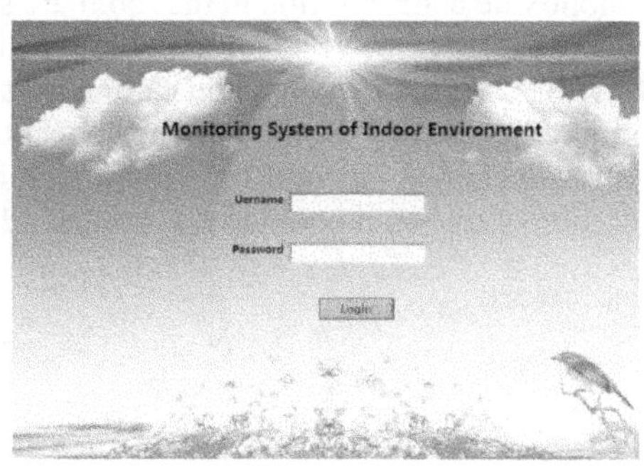

Fig. 4. Login system.

It has the functions of real-time display of indoor environmental parameters, e.g., temperature, humidity, PM2.5 and concentration of carbon monoxide. For indoor parameters of liquefied gas and vibration, the monitoring platform has only the function of over-limit warning because of sensor's performance limitations. According to the specific conditions of the monitored indoor environment, the upper limit and lower limit of the vibration and concentration of liquefied gas can be set as artificial values. In addition, the monitoring platform also has the function of data storage for each indoor environmental parameter, which can automatically generate a historical record of monitoring data in the background to facilitate data analysis and prediction.

3. Experimental Results and Analyses

The environmental parameters monitored by the terminal node module were temperature, humidity, PM2.5 concentration, carbon monoxide concentration, liquefied gas concentration and vibration. Among them, the monitoring form of liquefied gas and vibration is over-limit warning, while other environmental parameters can be detected to specific values. The environmental data is collected by the system for a duration of 24 hours in December in an indoor room in Xi'an, and the environmental changes are observed and analyzed from 6 am on the first day to 6 am on the next day.

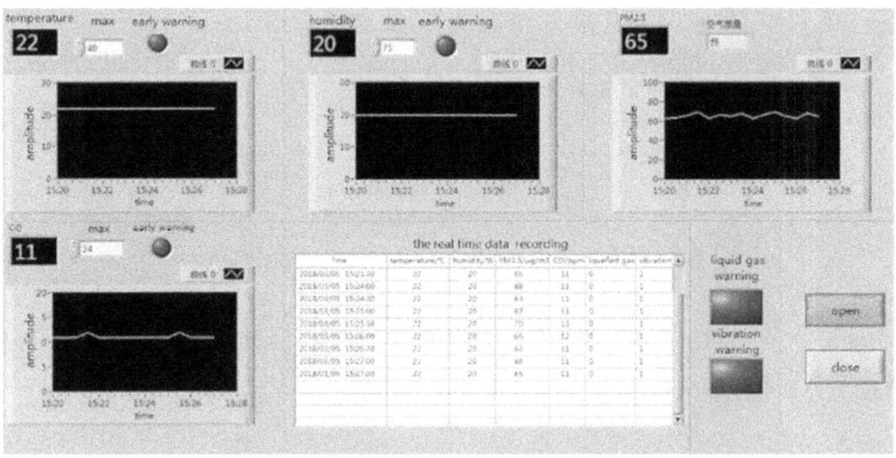

Fig. 5. User monitoring platform.

The changes of temperature and humidity are shown in Fig. 6, the temperature in the room has three large changes in one day, which are 8 am – 12 am, 14 pm – 18 pm, and 21 pm – 23 pm. This is basically consistent with the time of the switch of the air conditioner. Due to the dryness of the indoor environment, the humidity decreased with the rise of temperature. The changes of PM2.5 concentration and carbon monoxide concentration in the room are shown in Fig. 7 and Fig. 8, respectively. It can be seen that the changes of PM2.5 concentration and carbon monoxide concentration in one day are stable. The concentration of PM2.5 fluctuated around 75 μg/m^3, and the concentration of carbon monoxide was around 12 ppm.

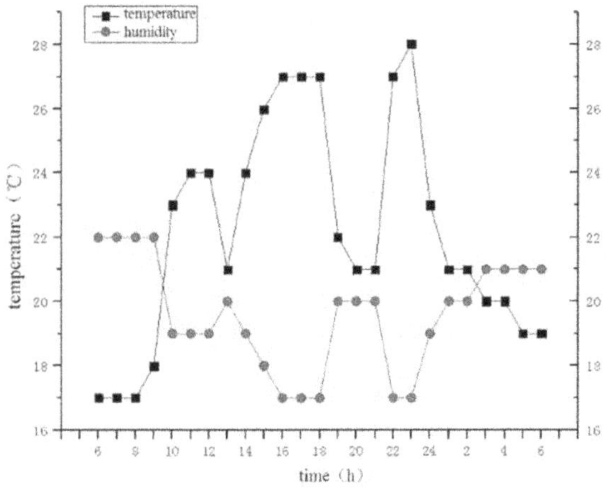

Fig. 6. Variation of temperature and humidity within 24 hours.

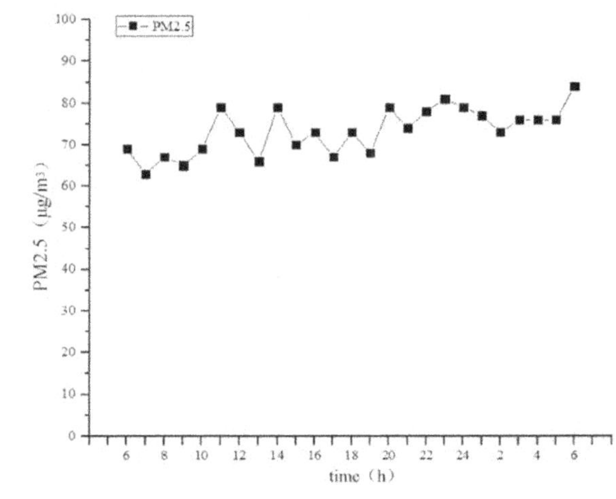

Fig. 7. Variation of PM2.5 within 24 hours.

The measurement of liquefied gas and vibration used a mode of fixed-point warning. And the signals of the liquefied gas sensor and the vibration sensor are transmitted as switching signals. When the liquefied gas is detected, the liquefied gas sensor outputs the number "1", and when there is no liquefied gas, the number "0" is output. The setting of the vibration sensor is opposite to liquefied gas sensor. When there is vibration, the number "0" is output, and when there is no vibration, the number "1" is output. As can be seen from Fig. 9, the liquefied gas sensor outputs the number "0", and the vibration sensor outputs the number "1". This indicates that the liquefied gas and vibration signals were not detected, which was consistent with the actual situation.

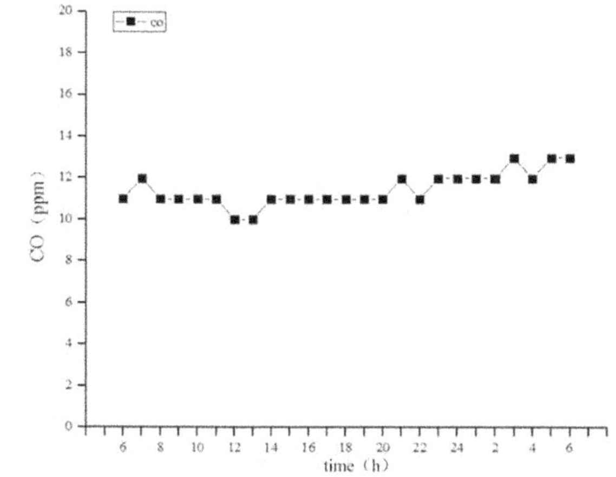

Fig. 8. Variation of carbon monoxide within 24 hours.

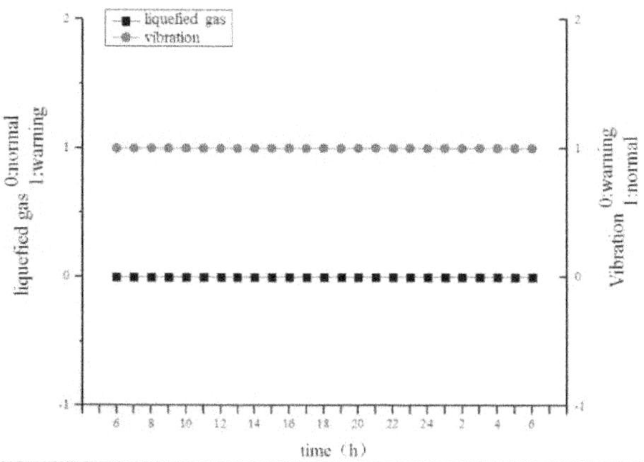

Fig. 9. Variation of liquefied gas and vibration within 24 hours.

4. Conclusions

This paper designs and builds an indoor environmental monitoring system based on ZigBee and wireless sensor network. The system has a simple structure, stable operation, low cost and high reliability. In this paper, a new type of node module based on CC2530 and its package box are designed and fabricated. The double-deck structure separated the node module and sensors to realize the dustproof and waterproof protection of the node module, which is convenient for operation and maintenance. The terminal node module is connected with many sensors, which can greatly reduce the number of terminal node modules. Based on LabVIEW software, the system's online monitoring platform was developed, which has realized the real-time display, storage and over-limit warning of environmental data.

Acknowledgements

This work is supported by 973 Program (No. 2015CB057402), National Natural Science Foundation of China (No.91748207, No.51608437, No. 51720105016), China Postdoctoral Science Foundation (no. 2018T111045 and no. 2017M613114), the Shaanxi Postdoctoral Science Foundation (no. 2017BSHEDZZ69), Shaanxi Natural Science Foundation (no. 2018JQ5156), Natural Science Foundation of Jiangsu Province (no. BK20180238), Key industrial innovation projects in Suzhou (no. SYG201830), the Research Project of the State Key Laboratory of Mechanical System and Vibration (MSV201813), 111 Program (No.B12016). We also appreciate the support from the International Joint Laboratory for Micro/Nano Manufacturing and Measurement Technologies.

References

[1]. L. A. Laranjeira, G. N. Rodrigues, Border effect analysis for reliability assurance and continuous connectivity of wireless sensor networks in the presence of sensor failures, *IEEE Transactions on Wireless Communications*, Vol. 13, Issue 8, 2014, pp. 4232-4246.

[2]. N. K. Chaitanya, G. A. Kumar, P. A. Kumari, ZigBee based wireless sensing platform for monitoring agriculture environment, *International Journal of Computer Applications*, Vol. 83, 2013, pp. 9-12.

[3]. G. Sahitya, N. Balaji, C. D. Naidu, Wireless sensor network for smart agriculture, in *Proceedings of the IEEE International Conference on Applied and Theoretical Computing and Communication Technology (iCATccT'17)*, 2017, pp. 488-493.

[4]. B. Zhang, Z. Zhang, Y. University, Design of a smart agricultural system based on internet of things, *Journal of Yangtze University*, 2016.

[5]. M. Zhang, J. Fang, Y. Han, Design on remote monitoring and control system for greenhouse group based on ZigBee and internet, *Transactions of the Chinese Society of Agricultural Engineering*, Vol. 29, Issue 25, 2013, pp. 171-176.

[6]. S. D. Zhang, J. H. Jiang. Design and Implementation of Watering Control System of Flowers Based on ZigBee, *Hubei Agricultural Sciences*, 2014.

[7]. D. Jiang, L. Yu, F. Wang, X. Xie, Y. Yu, Design of the smart home system based on the optimal routing algorithm and ZigBee network, *PloS ONE*, Vol. 12, Issue 11, 2017,e0188026.

[8]. A. Betzler, C. Gomez, I. Demirkol, J. Paradells, A holistic approach to ZigBee performance enhancement for home automation networks, *Sensors*, Vol. 14, Issue 8, 2014, pp. 14932-14970.

[9]. Q. Gong, G. Li, Y. Pang, Design and implementation of smart home system based on ZigBee technology, *Microcomputer Information*, Vol. 8, Issue 6, 2012, pp. 143-156.

[10]. B. Zhu, Y. Tang, Y. Tan, K. Xia, Design of ZigBee-based data acquiring and transmission system in industrial control network, *Control & Instruments in Chemical Industry*, 2010.

[11]. J. Huang, H. R. Liu, Design of intelligent pH analyzer based on ZigBee, *Control & Instruments in Chemical Industry*, Vol. 6, 2011, 8.

[12]. L. Chen, S. Yang, Y. Xi, Based on ZigBee wireless sensor network the monitoring system design for chemical production process toxic and harmful gas, in *Proceedings of the International Conference on Computer, Mechatronics, Control and Electronic Engineering (ICCMCEE'10)*, 2010, pp. 425-428.

[13]. J. J. Zhao, Y. Bai, R. S. Wang, Application of Zigbee technology in chemical water ultrafiltration system, *Advanced Materials Research*, Vol. 459, 2012, pp. 343-346.

[14]. Y. F. Ni, J. Liu, Based on ZigBee technology remote health care system design and research, *Advanced Materials Research*, Vol. 433-440, 2012, pp. 3439-3444.

[15]. K. A. Khaliq, A. Qayyum, J. Pannek, An emergency alert system for elderly/special people using VANET and WBAN, in *Proceedings of the IEEE International Conference on Emerging Technologies (ICET'17)*, Islamabad, 2017, pp. 1-6.

[16]. M. Clarke, J. D. Folter, V. Verma, H. Gokalp, Interoperable end-to-end remote patient monitoring platform based on IEEE 11073 PHD and ZigBee health care profile, *IEEE Transactions on Biomedical Engineering*, Vol. 65, 2018, pp. 1014-1025.

[17]. T. Yu, Z. Dai, Y. Li, Z. Li, L. Tang, The research on mine safety auxiliary system of wireless sensor networks, in *Proceedings of the International Conference on Instrumentation*, 2013, pp. 313-317.

[18]. P. Blume, LabVIEW Style Book, *Pearson Schweiz Ag*, 2007.

[19]. X. Zhou, Electrocardiogram system performance evaluation based on the LabVIEW virtual instruments development platform. In *Proceedings of the International Conference in Communications, Signal Processing, and Systems (CSPS'17)*, Singapore, 2017, pp. 1901-1908.

[20]. H. E. Ling-Li, D. O. Imaging, Design of virtual ECG monitor system based on LabVIEW, *Chinese Medical Equipment Journal*, 2017.

(093)

Peculiarities of Testing of Measurement Instrument Software in Ukraine

O. Velychko [1], **O. Hrabovskyi** [2] **and T. Gordiyenko** [2]
[1] State Enterprise 'All-Ukrainian State Scientific and Production Centre for Standardization, Metrology, Certification and Protection of Consumer', (SE 'Ukrmetrteststandard'), 4 Metrologichna str., Kyiv, Ukraine
[2] Odesa State Academy of Technical Regulation and Quality, 15 Kovalska str., Odesa, Ukraine
Tel.: + 38 050 464 00 24, + 38 097 177 57 43, fax: + 38 044 526 03 35
E-mail: velychko@hotmail.com

Summary: The developed special SW testing methodology of measuring instruments software (MI SW) with taking into account all essential requirements of OIML document and WELMEC guidelines was presented. This methodology contains the methodology for testing of MI SW on the following generalized indicators. The developed methodology for the testing of six varieties of MI SW with built-in computer was used. The obtained results showed the possibility of applying the Rasch model for the analysis of the scale for evaluating the SW.

Keywords: Software, Measuring instruments, Testing, Conformity assessment, Technical regulations.

1. Introduction

Specialized software (SW) for measuring instruments (MIs) is subject to testing for conformity assessment of MI. The testing of MI SW is devoted to the documents of the International Organization of Legal Metrology (OIML) [1] and the guidelines of the European Organization for Cooperation in Legal Metrology (WELMEC) [2, 3].

Since 2005 in Ukraine, national metrology institutes have been carry out of work on the testing of MI SW [4, 5]. In 2013, the national standard DSTU 7363 [6] has been developed, which sets the general technical requirements for MI SW. This standard only partly meets the requirements of international and regional documents and guidelines. On the basis of DSTU 7363 developed special methodical recommendations for the testing of SW MI.

In [7], a comparative analysis of the general requirements in the documents and guidelines of the international and regional organizations of legal metrology OIML and WELMEC concerning the testing of MI SW was conducted. In this work a universal algorithm for testing of MI SW is proposed.

In [8], an analysis of the normative framework for testing of MI SW at the national level was conducted to determine its suitability for carrying out conformity assessment. It has been established that DSTU 7363 [6] contains only general SW protection requirements and does not define the SW testing methodology.

National metrology institutes and conformity assessment bodies are interested in the existence of effective testing methodology for MI SW, risk assessment and application-related threats. Actual issues are the development of national SW testing procedures for MI SW taking into account all essential requirements of OIML document [1] and WELMEC guidelines [2, 3]. At the same time, it is important to conduct an analysis of the practical testing of national SW testing procedures for MI SW with purpose of its suitability for use.

2. The Main Content of the Research

On the basis of the conducted analysis, a special SW testing methodology of MI SW was developed taking into account all essential requirements of OIML document [1] and WELMEC guidelines [2, 3]. This methodology contains the methodology for testing of MI SW on the following generalized indicators:

- K_P is indicator of software characteristics with built-in computer;
- K_U – is indicator of software characteristics with a universal computer;
- K_L is indicator of checking storage devices;
- K_T is indicator of verification of data transmission devices;
- K_S is indicator of checking of reading;
- K_D is indicator of checking of SW sharing levels;
- K_I – a special indicator of SW testing for a specific MI.

For each of these generalized indicators, the relevant partial indicators are determined and the expressions are set for the numerical value of the indicator.

The numerical value for each generalized indicator is determined by the expression:

$$K_k = \sum_{i=1}^{M} K_i w_i^K / M,$$

where M is the total number of estimated partial indicators for a certain generalized indicator; K_i is partial indicators for a certain generalized indicator; w_i^K weighting coefficient of the corresponding partial indicator.

At the beginning of the evaluation, it is necessary to determine to which basic configuration this SW is related to: with a built-in computer P or with a universal computer U. In the further, it is necessary to use the full set of requirements that relate to the corresponding basic configuration.

The developed methodology for the testing of six varieties of MI SW with built-in computer *P* (meters of electric and thermal energy, gas-dispensing columns, gas chromatograph) was used. The testing results in the evaluation of compliance of these SW by methodology are shown in Table 1 and Fig. 1.

Table 1. Results of testing SW for generalized indicators.

	K_P	K_L	K_T	K_S	K_D	K_I	K_{tot}
SW1	9	8	5	1	1	7	31
SW5	9	8	4	1	1	7	30
SW2	9	7	3	1	1	7	28
SW6	9	7	4	1	1	6	28
SW3	9	6	4	1	1	6	27
SW4	9	8	5	1	1	3	27

The Rasch model suggests that the probability of approving any category of responses to an object depends entirely on the subject's (SW) ability and complexity of the object [9].

The received primary data on these experts were processed using the software WINSTEPS 4.4.1, which implements the Rasch model. The results of the transformation of the input primary data by the subjects – SW in the Rasch measurement are shown in Table 2.

The line AV is average value, the line SD is standard deviation. The columns of Infit and Outfit statistics contain parameters that characterize the matching of the data of the Rasch model: MNSQ is the value that characterizes the level of randomness of the results or the discrepancy between the measurement model data; ZSTD is standardized MNSQ values, that is, the probability of mean-square-statistics, expressed as *z*-statistics (mean-square deviation).

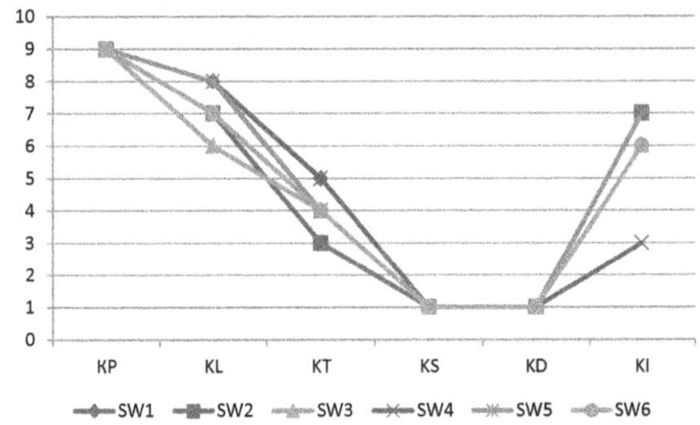

Fig. 1. Results of testing SW for generalized indicators.

The most expected values for MNSQ are near 1.0. The MNSQ values from –2.0 to +2.0 are acceptable. The obtained MNSQ values for the subjects for Infit statistics range from 0.04 to 3.38, and for Outfit statistics from 0.06 to 3.08. This indicates that all these values are acceptable for the analysis by the Rasch model. Only for the SW4, the values of the Infit and Outfit statistics are respectively 3.38 and 3.08, which indicates the presence of "noise" in the input data.

Table 2. Results of data conversion concerning subjects.

Subject	Measurement	Measurement error	Infit statistics		Outfit statistics	
			MNSQ	ZSTD	MNSQ	ZSTD
SW1	1.46	0.68	0.09	-1.72	0.15	-1.35
SW5	1.02	0.66	0.36	-0.73	0.38	-0.70
SW2	0.18	0.65	0.91	0.18	0.77	-0.01
SW6	0.18	0.65	0.04	-1.96	0.06	-1.81
SW3	-0.24	0.64	0.72	-0.11	0.92	0.21
SW4	-0.24	0.64	3.38	2.12	3.08	1.83
AV	0.39	0.65	0.92	-0.40	0.90	-0.30
SD	0.64	0.01	1.15	1.40	1.02	1.20

3. Conclusions

The methodology that are suitable for evaluating the SW were considered in detail. The obtained results showed the possibility of applying the Rasch model for the analysis of the scale for evaluating the SW. A comparative analysis of the results obtained with the use of simple scores method and Rasch model showed convergence, suitability and correlation of the obtained values for SW.

References

[1]. OIML, D 31:2008, General Requirements for Software Controlled Measuring Instruments, *OIML*, Paris, 2008.

[2]. WELMEC 7.2, Software Guide (Measuring Instruments Directive 2004/22/EC), http://www.welmec.org/fileadmin/user_files/publications/WG_07/Guide_7.2_2015_Software.pdf

[3]. WELMEC 2.3, Guide for Examining Software (Non-automatic Weighing Instruments), http://www.welmec.org/fileadmin/user_files/publications/2-3.pdf

[4]. O. N. Velichko, Basic tests, stages, and features in monitoring measuring instrument software, *Measurement Techniques*, Vol. 52, Issue 6, 2009, pp. 566-571.

[5]. O. N. Velichko, Normative base for certification of measurement provision software, *Measurement Techniques*, Vol. 50, Issue 4, 2007, pp. 364-371.

[6]. DSTU 7363:2013, Prohramne zabezpechennia zasobiv vymiriuvalnoi tekhniky, Zahalni tekhnichni vymohy, *Ukraine*, 2013.

[7]. O. Velychko, T. Gordiyenko, O. Hrabovskyi, Testing of measurement instrument software on the national

level, *Eastern-European Journal of Enterprise Technologies*, Vol. 2, Issue 9 (92), 2018, pp. 13-20.

[8]. O. Velychko, V. Gaman, T. Gordiyenko, O. Hrabovskyi, Testing of measurement instrument software with the purpose of conformity assessment, *Eastern-European Journal of Enterprise Technologies*, Vol. 1, Issue 9 (97), 2019, pp. 19-26.

[9]. O. Velychko, T. Gordiyenko, A comparative analysis of the assessment results of the competence of technical experts by methods of Analytic Hierarchy Process and with using the Rasch model, *Eastern-European Journal of Enterprise Technologies*, Vol. 3, Issue 3 (93), 2018, pp. 14-21.

(094)

VarroaCounter – Towards Automating the Varroa Screening for Alleviated Bee Hive Treatment

A. König [1]

[1] TU Kaiserslautern, Institute of Integrated Sensor Systems, Erwin Schrödinger Strasse 12, 67663 Kaiserslautern, Germany

Tel.: + 496312053696, fax: + 496312053889

E-mail: koenig@eit.uni-kl.de

Summary: The importance of insects, and honey bees in particular, for our ecosystem is a well-accepted fact. Currently, environmental problems from pesticides to parasites endanger the well being or even the existence of honey bee colonies and insects in general. This imposes an increasing load on skills and activity of hive keepers. Sensors, instrumentation, and machine learning offer solutions on the one hand to instrument bee hives, and on the other hand to provide efficient assistance systems for hive keepers. The paper presents, improved automation of the tedious and time consuming screening for the varroa infestation of hives, which currently is done manually, and integrate it with hive instrumentation and data analysis activities. This problem is closely related to the frequently visited problem of insect counting for populations estimates, pest control etc. A rich amount of experience and methods can be adopted from that field, though, there are substantial differences in the scenario and the focus of the work presented here is on achieving an economically feasible approach. Thus, a two step procedure, a first attention step for detecting candidate regions and a final classification step of these candidate regions, is applied. It could be shown to perform for sparsely cluttered hive slider units, but requires more effort, e.g., employing color texture classification, to deal with strongly cloaked slider units. The work bears the potential to be extended to continuous in-hive varroa infestation monitoring.

Keywords: Bee health monitoring, Automated varroa screening, Machine vision, Machine learning.

1. Introduction

Though some insect specifies are known to be a pest and a threat to agriculture, the systemic relevance of insects in general, and honey bees in particular, is a well known fact and found its place in public awareness. Screening procedures, and related capture or tracking methods on insects [12], are used in various forms, e.g., to reckon insect populations and give timely information for pest population surges. Predominant manual evaluation is quite tedious and time consuming and often contradicts the need for a swift response, e.g., the just in time deployment of a chemical agent in proportion to pest occurrence.

Insect counting can base on visual count, sweep nets application or various forms of traps, e.g., pheromone, light, sticky layer, liquids, or bait deployment [3, 8, 12]. Numerous of these procedures are destructive in the sense, that counted insects are killed. Further, collateral damage occurs, i.e., in pest control counting applications, also harmless and useful insects are trapped and killed, which is both bad under environmental considerations as well as making the manual or automated recognition and evaluation procedures more hard. Naturally, the field of Machine Vision and Machine Intelligence offer a plethora of similar problems to insect counting and also numerous proven methods immediately or after modification applicable for automated screening solution [1, 2, 4-7, 9, 11].

In numerous activities, that can already deal with amazing number of different species and sizes of insects, e.g., [5], assumptions on background and density and spacing of insects to count are made [5, 9]. Often, real data confronts with strongly varying density of insects, i.e., closely packed or even stapled on top of each other, which gives rise to hard occlusion problems. Further, natural variations in phenotypes, e.g., size and patterning, changes in appearance from live to pass away and decaying entity, orientation as well as dissected limbs etc. make automated recognition not too easy task.

The increasing issues in tending of bees has kindled research and development on instrumenting bee hives and supporting hive keepers by IT infra structure [13]. These activities can be bifurcated in sensors, measurement and instrumentation integrated in modified or "digital" hives and discrete assistance systems development for hive keepers [13], applicable to all hives of an apiary. One contribution of the second activity is pursued in this paper, trying to improve the automation of the tedious screening or counting procedure, which has been scarcely visited in the past [18].

As is well known, the honey bee is threatened by *varroa destructor* parasite and hives need regular screenings or varroa counting to estimate the degree infestation by the mite and initiate treatment procedures according to the screening outcome.

The screening involves the analysis of the hive debris and to detect and count the number of mites on a slider unit as illustrated in Fig. 1. This task also predominantly is delivered by manual work of the hive keepers several times in the year for every hive, which adds to the anyway increasing workload due to rising issues from pesticides to parasites and illnesses.

Fig. 1. Hive with debris on sliders for varroa screening.

In this work, in the IndusBee 4.0 research activity, inspired by related work in the fields of automated industrial quality control and Industry 4.0, a machine vision and learning approach is pursued to improve the automation of the varroa screening process. The goal is to achieve a simple, as possible affordable yet reliable hive keepers assistance system from standard components for varroa screening in bee hives under cost constraints with a sufficient screening coverage for event-driven, right, and effective, treatment decisions and also for improving the house keeping by archiving the images of hive debris together with the evaluation or counting results and providing them for higher level data analysis, also fusing with other sensory information from hive instrumentation, to an Apiary AI [13]. The long term objective is adding hive-integrated continuously repeated varroa counting to advanced hive instrumentation. In this paper, the first step and results of the emerging hive keeper's assistance system will be presented and future steps discussed.

2. General Varroa Screening Procedure

The typical varroa screening process bases on the analysis of the hive debris collected on sliders as given in Fig. 2. They differ in time inserted in the hive and the amount of debris mixing with the mites to be detected. The increasing degree of debris and target varroa objects on the slider unit with insertion time leads to increasing occlusion problems up to literally burying the mites, so that attractive simple methods might fail and more computationally burdensome techniques or even sensory, e.g., [11], registration techniques, e.g., fluorescence photography, and scene improvements have to be considered. As manual inspection also becomes harder, usually a time in hive of 2-3 days is recommended as feasible inspection practice, e.g., [16]. However, this will be hard to meet for the aspired continued automated screening procedure without introducing cleaning steps for the slider units.

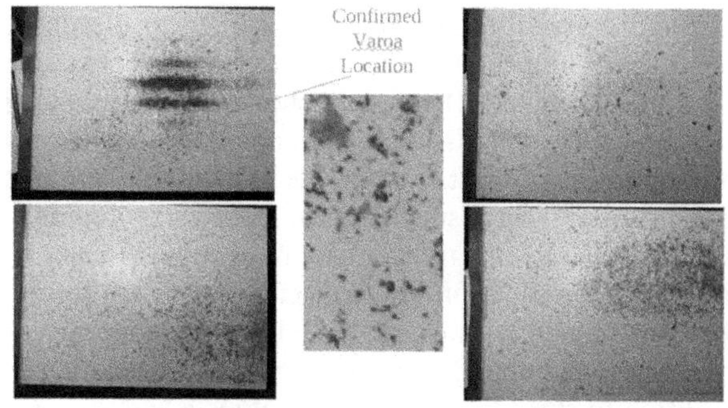

Fig. 2. Hive debris on sliders for varroa screening illustrating possible occlusion problems aggravating with insertion time.

A further significant challenge for an economic system version working with standard camera, e.g., available in contemporary mobile phones, is given by the fact, that the inspection area of the slider is very large compared to the size of a mite. In established manual analysis, hive keepers uses multi-scale analysis by a magnifying glass and hand-eye-coordination literally excavating mites from the debris in an iterative process. Additional complexity will be added for the ambitious goal to do the automated varroa detection and counting repeatedly in hive, as the constrained space will make invariant data registration very hard.

One existing varroa counting system could be found in the web [18] that seems to use a similar approach as the first stage in Section 3. According to some of the comments, the approach works well for rather clean sliders and degrades with increasing cloaking and occlusion. Updates seem not be available.

3. VarroaCounter Assistance System

Our current simple first-cut **VarroaCounter** system concept and implementation bases on standard blob analysis, e.g. [14], which has been employed in numerous machine vision application in the last decades, is illustrated in Fig. 3.

For the first approach, it is assumed for the sake of simplicity, that the slider image is grabbed by a basic commercial still image camera, e.g., the hive keeper's mobile phone camera, under suitable normalization conditions. Perpendicular orientation, identical scale, focus settings, as well as uniform and constant scene illumination are prerequisites for robust recognition. As in related work [2, 5], the obtained color image (see Fig. 3 top) is processed by an segmentation step, identifying possible object, or varroa occurrences, using python and scikit [14] for our simple and computationally effective mite candidate detection algorithm. Instead of computing gray value image for the segmentation step from the color registration [2, 5], here, preliminary analysis showed the red channel itself to be best suited here for the varroa mite candidate segmentation. Under the assumption of the common white slider background and sufficient sparsity, a fixed thresholding with $15 < p_{ij} < 90$ was employed returning candidate regions. A simple morphological filtering step was applied to eliminate singular region detections and smooth remaining significant regions. Then established blob analysis [14] is run, which threads all found segmentation regions and returns a blob list and number of candidate regions for further assessment (see Fig. 3 center). Enthusiastically, the resulting number could already be interpreted as the aspired varroa counting result figure. But unfortunately, this will only be true under severe and somewhat unrealistic constraints, e.g., assuming in addition to sparsity, that no other objects, e.g., insects or insect parts would be present. Thus, in the second stage of the proposed VarroaCounter, the blob regions from the list will be subject to a confirming or rejecting classification by richer features, e.g., blob properties, pixel statistics etc., computed from the initial color image registration (see Fig. 3 bottom), potentially adding structural and/or textural information to the recognition process.

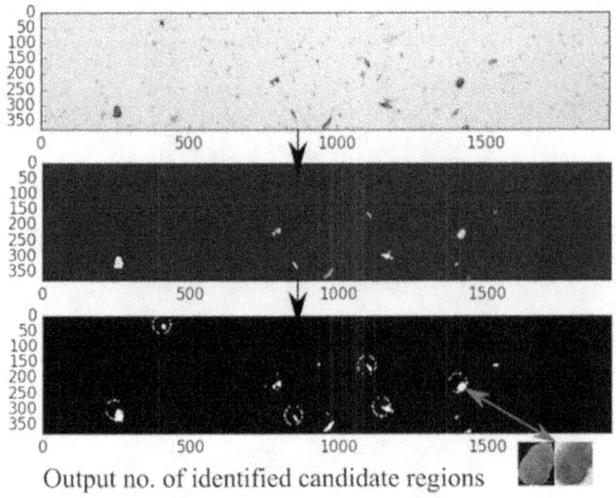

Fig. 3. Main steps of first VarroaCounter system for a subregion of a hive slider unit.

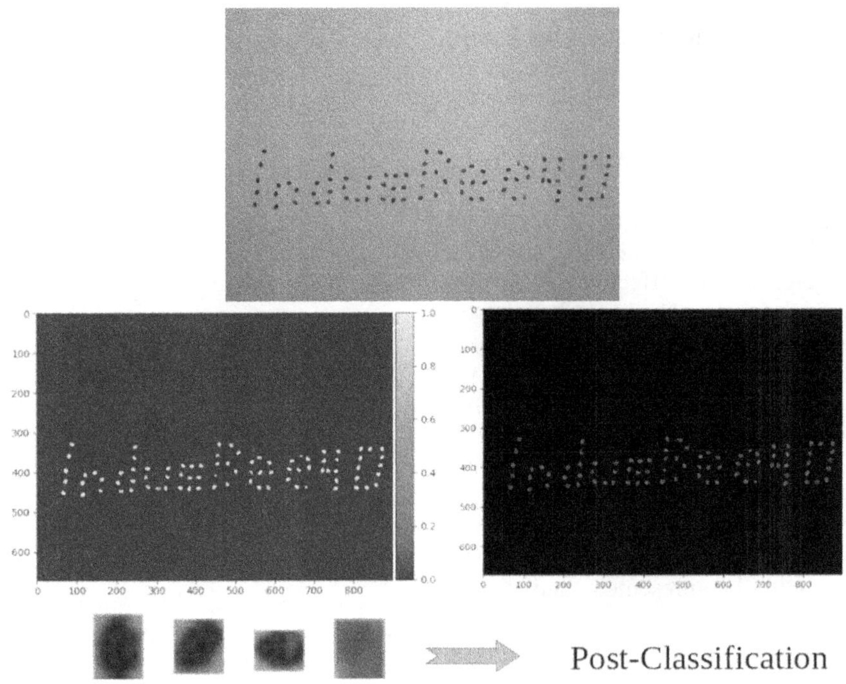

Fig. 4. Clean slider for stage one test and data acquisition for second stage classifier design.

4. Experiments and Results

For the validation and improvement of the first-cut varroa counting assistance system, a collection of sliders have been accumulated from real apiary routine checks. Fig. 2 shows a few example pictures, taken by a simple still image camera. Various insertion times have led to violations of sparsity assumption and resulting occlusion problems (see Fig. 2, center). The slider samples are stored to be able to repeat recordings with different camera, scene, and illumination settings or for research application of available commercial hyper-spectral camera [17] similar to [11]. Both to check stage one of VarroaCounter and to obtain unique and ground truth data for the post classification step, shown in Fig. 4, a clean slider with only sparsely located 84 mites on it was prepared.

The stage one worked well for the easy example and returned both the correct number and list of locations of the 84 mites. For the first instance of the second, post-classification stage, a kNN classifier [15] is employed and trained with mite images given in Fig. 4 and additionally selected non-mite, background patterns. In the first place, histograms of blob regions where computed for post-classification. The corresponding, promising feature space for simple blob region histograms is shown in Fig. 5. Summarizing, the blob analysis, based on color thresholding, acceptably works with sliders sparsely populated by debris and mites.

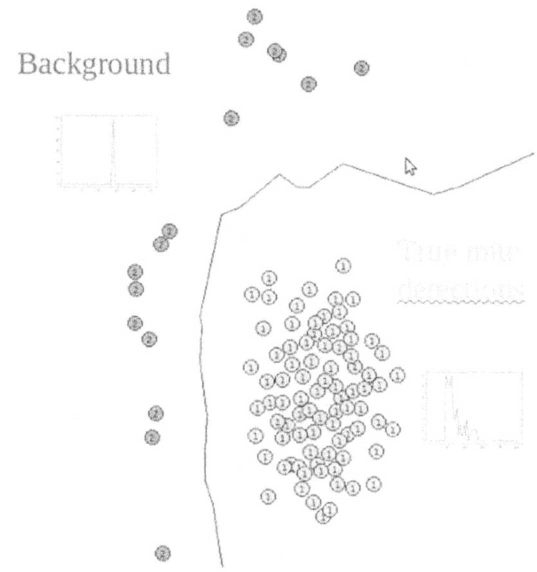

Fig. 5. Feature space of the post-classifier stage.

On densely packed sliders, too large blobs arise, which are harder to rule out and require more information. Structural and color textural features [10] will be regarded next in the work to increase the counting accuracy still with affordable standard equipment.

4. Conclusions

A simple and low-cost approach for improving automated varroa counting [18], pursued as a building blocks in IndusBee 4.0 [13], was presented as first step for bee hive keepers assistance. The suggested two-stage procedure demonstrated by a first prototype works well under some constraints. In future work, in particular for the idea of visual hive-integrated continuous varroa infestation sampling, both stages have to see extension, e.g., by adding structural and color textural features [10] to overcome occlusion issues.

References

[1]. Y. Zhong, J. Gao, Q. Lei and Y. Zhou, A vision-based counting and recognition system for flying insects in intelligent agriculture, *Sensors*, Vol. 18, 2018, 1489.

[2]. C. Zhu, J. Wang, H. Liu, H. Mi, Insect identification and counting in stored grain: image processing approach and application embedded in smartphones, *Hindawi Mobile Information Systems*, Vol. 2018, 2018, 5491706.

[3]. B. Brader, R. C. Lee, R. Plarre, W. Burkholder, G. Barrie Kitto, C. Kao, L. Polston, E. Dorneanu, I. Szabo, B. Mead, B. Rouse, D. Sullins, R. Denning, A comparison of screening methods for insect contamination in wheat, *Journal of Stored Products Research*, Vol. 38, 2002, pp. 75-86

[4]. Zuse Institute, Similarity Search and Insect Classification, https://www.zib.de/features/similarity-search-and-insect-classification

[5]. K. Mele, Insect soup challenge: Segmentation, counting, and simple classification, *in Proceedings of the IEEE International Conference on Computer Vision Workshops* (*NSW'13*), Sydney, 2013, pp. 168-171.

[6]. Y. Ovadia, Y. Halpern, D. Krishnan, J. Livni, D. Newburger, R. Poplin, Ti. Zha, D. Sculley, Learning to count mosquitoes for the sterile insect technique, in *Proceedings of the 23rd SIGKDD Conference on Knowledge Discovery and Data Mining (KDD'17)*, Halifax, NS, Canada, August 13-17, 2017,

[7]. G. Sileshi, A method for estimating insect abundance and patch occupancy with potential applications in large-scale monitoring programmes, *African Entomology*, Vol. 15, Issue 1, 2007, pp. 89-101.

[8]. G. Zehnder, Overview of monitoring and identification techniques for insect pests, *Organic Agriculture*, March 12, 2014, 20140312.

[9]. SpecialVideo, Insects Counting for Integrated Pest Control, https://www.specialvideo.it/en

[10]. C. Palm, Color texture classification by integrative Co-occurrence matrices, *Pattern Recognition*, Vol. 37, Issue 5, May 2004, pp. 965-976.

[11]. J. Fennel, C. Veys, J. Dingle, J. Nwezeobi, S. van Brunschot, J. Colvin, B. Grieve, A method for real-time classification of insect vectors of mosaic and brown streak disease in cassava plants for future implementation within a low-cost, handheld, in-field multispectral imaging sensor, *Plant Methods*, Vol. 14, 2018, 82.

[12]. CDFA, Insect Trapping Guide, https://www.cdfa.ca.gov/plant/PDEP/Insect_Trapping_Guide/docs/Insect_Trapping_Guide_web.pdf

[13]. A. König, Extending bee hive health state monitoring by integrated acoustical sensing and machine learning, in *Proceedings of the 5th International Conference on Sensors Engineering and Electronics Instrumentation Advances (SEIA'19)*, 2019, pp. 168-173.

[14]. Blob Detection, https://scikit-image.org

[15]. kNN-Classification, https://scikit-learn.org

[16]. Bayerische Landesanstalt für Weinbau und Gartenbau, Gemülldiagnose, https://www.lwg.bayern.de/mam/cms06/bienen/dateien/gem%C3 %BClldiagnose_fzbienen2012.pdf

[17]. Photon Focus, http://www.photonfocus.com/de/

[18]. Varroazählen leicht gemacht (Software), http://www.imker-kiel.de/news/molfsee-mit-horst/67varroacount.html

(096)

AlN Nanopowder for Biological and Optical Oxygen Gas Sensors

B. Berzina, L. Trinkler and V. Korsaks
Institute of Solid State Physics, University of Latvia, 8 Kengaraga Str., LV-1063, Riga, Latvia
Tel.: +26157869, fax: + 67132778
E-mail: baiber@latnet.lv

Summary: Spectral characterization of AlN nanopowders (AlN NP) untreated and doped with rare earth elements (Tb, Eu) was done. The photoluminescence and excitation spectra were studied within a temperature region from 10 up to 300 K. The results obtained allowed determination of luminescent defects, revealing of luminescence mechanisms as well as recommendation of AlN NP for possible application for development of optical oxygen gas sensors and biosensors. It was assumed that in AlN NP predominant native luminescent defects are the so-called F-centers. It was found that the F-center luminescence is sensitive to oxygen gas surrounding the material. Oxygen sensing properties of AlN NP were studied allowing ranking the material between those prospective for development of optical oxygen gas sensors. Luminescence of AlN NP doped with Tb or Eu was also studied. The results obtained allow offering the materials for use as biological sensors in biology and medicine.

Keywords: AlN, Nanopowder, AlN-Tb, Photoluminescence, Excitation spectra, Oxygen gas optical sensor, Biosensor.

1. Introduction

AlN is a semiconductor with wurtzit structure and the wide band gap exceeding 6 eV. Different defects - native and dopants of various elements are characteristic for AlN forming their energy levels within the band gap thus ensuring optical properties of the material. Some of the well-known native defects are Al and N vacancies (v_{Al} and v_N, respectively) and oxygen impurity substituting for N (O_N) [1-3]. Research of defect structure and properties allows proposal of the material for different applications.

The report contains our results obtained on defect spectral characterization in AlN nanopowder (NP) untreated and doped with some rare earth elements (Tb and Eu), resulting in revealing of light-induced processes in the material including determination of a structure of the luminescence centers and luminescence mechanisms, allowing to offer AlN NP as the material for application in biomedicine as luminescent marker and optical oxygen gas sensors.

2. Material and Experimental

AlN nanopowder with average grain size 60 nm was used. As synthesized AlN NP and the same doped with Tb or Eu pressed in thin pellets were used.

Photoluminescence (PL) and its excitation (PLE) spectra were examined within a wide temperature range from 10 K up to room temperature (RT) using conventional set-up. Luminescence kinetics was also studied.

3. Results

3.1. Spectral Characterization of AlN NP

It was found that for AlN NP at RT the PL spectrum consists of a wide and complex band situated within the blue spectral region (the blue luminescence – BL) (Fig. 1a, curve 1), which can be excited with ultraviolet (UV) light from a wide spectral range containing several excitation bands shown as I and II on Fig. 1b.

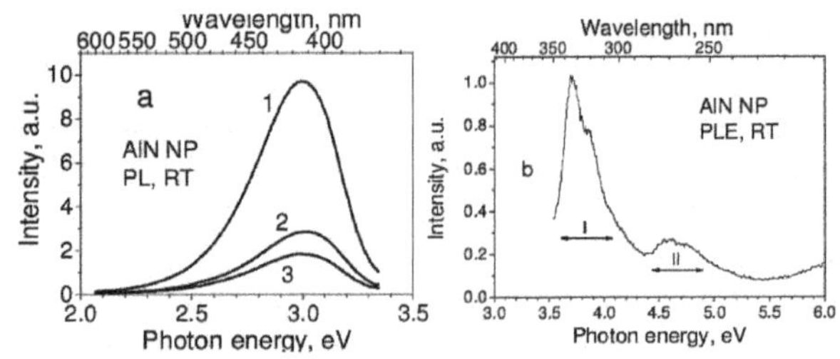

Fig. 1. AlN NP, RT. a – PL spectra at 315 nm excitation, when sample is in: 1 - vacuum (~10^{-5} mbar), nitrogen gas or argon gas; 2 – synthetic air; 3- pure oxygen gas. b – PLE spectrum of BL.

It is also seen (Fig. 1a curves 1 and 3), that the BL is sensitive to surrounding oxygen gas, which is quenching the luminescence intensity. This phenomenon could be related to the surface defects. Nevertheless, the BL is not completely quenched even then, when the sample is surrounded by pure oxygen gas thus allowing assumption that in this case the BL is caused by the luminescent defects from the bulk material.

Analysis of the shapes and structure of the PL and PLE spectra for AlN NP obtained at different temperatures together with their comparison with appropriate data for AlN bulk materials [1, 3 and Refs. therein] allows determination of luminescence centers and revealing of luminescence mechanisms existing in AlN NP. It can be proposed that in AlN NP the main type of luminescent defects can be related to nitrogen vacancy type defects or the so-called F-centers. Variety of the F-center types (F, F^+, F^{2+}) can explain the complex structure of the PL spectra. The main mechanisms originating the BL can be related to an intra-center and recombination ones.

3.2. Oxygen Gas Sensing Properties of AlN NP

It was found that the BL of AlN NP is sensitive to oxygen gas surrounding the material, which can be explained as oxygen adsorption on material surface and its interaction with the F-center type defects (Fig. 1a) situated on the material surface. Oxygen sensing properties of AlN NP were examined. Several of them are demonstrated on Fig. 2.

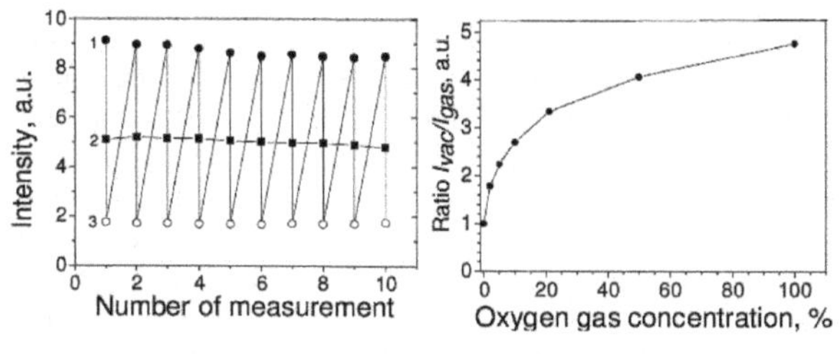

Fig. 2. AlN NP; RT. On the left side - repeatability of the response shown as luminescence intensity, measured when sample is in: vacuum 10^{-5} mbar (1, full circles); in oxygen gas (3, hollow circles) and relation of 1/3 (2, full squares). On the right side - sensitivity of the BL to oxygen gas concentrations in oxygen-nitrogen gas mixtures.

From the left side of the Fig. 2 it is seen, that AlN NP possesses a good repeatability of the results demonstrating oxygen gas sensing measured as luminescence intensity when material is repeatedly surrounded with vacuum (1) and oxygen gas (3).

On the right side of the Fig. 2 the sensitivity of AlN NP to different concentrations of oxygen gas is demonstrated via ratio of the blue luminescence intensities when material is surrounded by vacuum or oxygen containing gas mixture. It is seen that the material presents a good sensitivity to oxygen gas at different concentrations.

In conclusion, the observed spectral properties of AlN NP allow its ranking as the materials prospective for development of oxygen gas optical sensors.

3.3. AlN NP for Biological Sensors

Luminescent nanomaterials are prospective for use as luminescent markers for tracing the processes occurring in biomaterials. Doped AlN NP is prospective candidate for that application possessing: *i)* low toxicity; *ii)* intensive and controllable luminescence within a convenient spectral region; *iii)* low cost of raw materials and *iv)* simple synthesis methods.

The PL and PLE spectra of AlN NP doped with Tb are shown on Fig. 3a and b measured at RT.

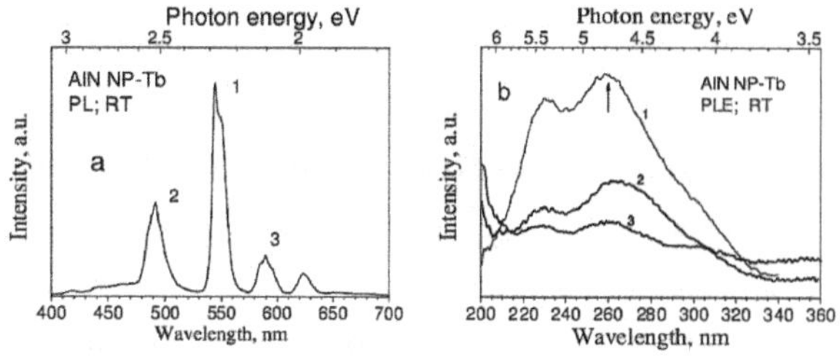

Fig. 3. AlN NP:Tb; RT. PL (a) and PLE (b). The PLE spectra obtained for different luminescence bands are demonstrated on section b using the same numeration.

The green color luminescence at 550 nm is characteristic for AlN NP:Tb, which can be excited with laser or LED light around 260-270 nm (Fig. 3a, and b). Besides, for luminescence of Tb ions the so-called luminescence up-conversion is known allowing use of light with longer wavelengths from the red/IR spectral region for Tb excitation, that is favorable with regard to the transparency of biological materials.

The similar optical properties were found for AlN NP:Eu. It allows to offer both AlN NP:Tb and AlN NP:Eu as prospective materials for use as biosensors.

4. Conclusions

AlN NP is the prospective material emitting blue luminescence predominantly caused by nitrogen vacancy type defects (F-centers). AlN NP is prospective for developing of optical oxygen gas sensors. It was also found that doped AlN NP: Tb and AlN NP:Eu are applicable for biosensors.

Acknowledgements

The present investigations are sponsored from the Project HORIZON 2020-MSCA-RISE, “assymcurv”, Nr. 690853.

References

[1]. T. Koppe, H. Hofsass, U. Vetter, Overview of band-edge and defect related luminescence in aluminium nitride, *Journal of Luminescence,* Vol. 178, 2016, pp. 267-281.

[2]. B. Berzina, L. Trinkler, D. Jakimovica, V. Korsaks, J. Grabis, I. Steins, E. Palcevskis, S. Bellucci, L.-C. Chen, S. Chattopadhyay, K. Chen, Spectral characterization of bulk and nanostructured aluminum nitride, *Journal of Nanophotonics*, Vol. 3, 2009, 031950.

[3]. J. B. Varley, A. Janotti, C. G. Van de Walle, Defects in AlN as candidates for solid-state qubits, *Physical Review B*, Vol. 93, 2016, 161201.

(097)

Polyaniline Nanocomposites Based Sensor for Simultaneous and Selective Measurement of Ammonium Nitrate Aerosol and Ammonia Gas

M. L. Boukhenane [1], **N. Redon** [1], **J.-L. Wojkiewicz** [1] **and P. Coddeville** [1]
[1] IMT Lille Douai, Univ. Lille, Atmospheric Sciences and Environmental Engineering Laboratory, 59000 Lille, France
Tel.: + 33 32771 2296
E-mail: mohamed-lamine.boukhenane@imt-lille-douai.fr

Summary: In the presence of high concentrations of ammonia and nitric acid gas, the formation of ammonium nitrate particles (NH_4NO_3) becomes possible at low temperature. As a result, high concentrations of ammonium nitrate particles are usually observed during the spring and winter period. Due to its semi-volatile nature, the measurement of ammonium nitrate with classical methods based on filter sampling (sampling time $\geq$ 24 h) introduce severe artifacts. Thus, the main objective of this study is to develop a new low cost sensor for simultaneous and selective measurement of ammonium nitrate particles and its precursor ammonia gas in ambient air. The sensor combines two ammonia sensitive surfaces based on polyaniline nano-composites materials. The mass concentration of ammonium nitrate is determined by measuring the concentration of ammonia released by heating one of the sensitive surfaces. Sensors showed a response to ammonia gas at concentrations less than 20 ppb with sensitivity around 0.3 $\%.ppb^{-1}$. Whereas, the limit of detection of sensors to ammonium nitrate particles is around 270 $\mu g/m^3$ with sensitivity of 0.0014 $\%.\mu g^{-1}.m^3$. The ongoing work is focused on the characterization of sensors at lower concentrations of NH_4NO_3 particles.

Keywords: Air quality sensor, Ammonium nitrate aerosol, Ammonia, Conducting polymer, Polyaniline.

1. Introduction

Ammonia gas is one of the main nitrogen compounds found in the troposphere after N_2 and N_2O [1, 2]. Over 80 % of ammonia emissions come from agricultural activities (livestock, fertilizers, soils…) [3]. Although it is not considered as a regulated pollutant in the ambient air in Europe, ammonia emissions are responsible of many adverse effects on the environment including soil acidification and aquatic eutrophication [4]. In addition, ammonia gas can contribute to particulate matter formation by reacting with acidic species (H_2SO_4, HNO_3, HCl) [5].

Atmospheric particulate matter, particularly PM_{10} and $PM_{2.5}$ have a significant impact on human health by causing many respiratory diseases [6]. Spatial and temporal variability of $PM_{2.5}$ composition and mass concentration was studied by many authors [7-10]. Usually, high concentrations of particles are observed during the spring and winter period [11]. In addition, several species can enter into the composition of $PM_{2.5}$. Among them, sulfate, nitrate and ammonium are the predominant water-soluble ionic species of $PM_{2.5}$. They represent approximately 20-40 % of $PM_{2.5}$ [12, 13].

The formation of ammonium nitrate particles NH_4NO_3 becomes possible in areas characterized by high ammonia and nitric acid concentrations at low temperature after full neutralization of acidic sulfate. The process is described by the reversible reaction below at temperature < 170 °C [14]:

$$NH_3\,(g) + HNO_3\,(g) \leftrightarrow NH_4NO_3\,(s, aq)$$

The reaction equilibrium and ammonium nitrate state depends on the temperature and relative humidity. Due to its semi-volatile nature, ammonium nitrate can evaporate at normal temperatures [15]. Furthermore, nitric acid is mainly formed by oxidation of NO_2 by OH radicals (NO_2+OH) during the daytime or by hydrolysis of dinitrogen pentoxide (N_2O_5) during the nighttime [16].

According to Tao and al. [17], the neutralization of nitrate by ammonia to obtain ammonium nitrate particles becomes evident at $[NH_4^+]/[SO_4^{2-}] > 2$. This molar ratio of ammonium to sulfate can be obtained during the spring and winter period.

The continuous monitoring of aerosol composition and their mass concentration with different instruments such us ACSM, AMS and TEOM is relatively expensive, bulky, and require highly qualified personnel. Moreover, most particle sensors rely on an optical sensing principle and don't give any information on the chemical nature of the particles. The mass concentration is estimated from count data, by using an ideal particle (size). Thus, the development of specific sensors for both particles and gas measurement could be complicated.

As known, conductive polymer are widely used to develop gas sensors because they are less expensive, they have tunable electrical properties and they can be synthesized easily [18]. In the present work, we propose a new low cost sensor based on polyaniline nanocomposites for simultaneous and selective measurement of ammonium nitrate particles and its precursor ammonia gas in ambient air.

2. Materials & Methods

2.1. Operating Principle

Sensor consists of two ammonia sensitive surfaces based on polyaniline nano-composites. One of the surfaces is heated with thermoregulation system in a controlled volume at temperature close to 30 °C in order to initiate the evaporation of ammonium nitrate aerosol into ammonia and nitric acid. With the heated surface, the total concentration of ammonia including ammonia released after the thermal decomposition will be determined. Sensing mechanism of ammonia gas with polyaniline is described in the literature and given in Fig. 1 [19]. While, the unheated surface measures only the concentration of ammonia already present in the air. The difference between the two concentrations determined with both surfaces will be correlated to ammonium nitrate concentration.

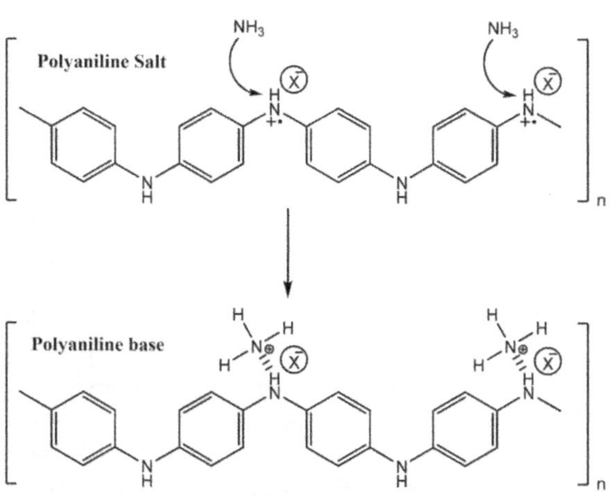

Fig. 1. Sensing mechanism of ammonia gas with polyaniline [19].

2.2. Sensor Films Preparation

In this work, we have synthesized three polyaniline (PANI) nanocomposites: two of PANI/Polyurethane (PU), with two ratios $PANI_{50wt\%}$/PU (PU1) and $PANI_{75wt\%}$/PU (PU2) and one based on $PAni_{10wt\%}$/poly(butyl acrylate) (PBuA) ''Core-Shell''. The polymer matrix has been added in order to improve the mechanical properties of sensitive surfaces. Polyaniline nanofibers (emeraldine base) were protonated by camphorsulfonic acid (CSA) with doping rate of 50 % and then mixed with polyurethane. Dichloroacetic acid (DCAA) was used as solvent for both components and secondary dopant of PANI. For core shell nanocomposite, PANI was synthesized by in situ polymerization of anilinium chloride salt (AniCl) as monomer in the presence of the core latex (PBuA). The synthesis method of these materials is well described in the previous work of [20].

The solutions were drop-coated onto interdigitated gold electrodes patterned on ceramic substrate (Synkera) and then dried in a vacuum oven for 7 days at 100 °C. Three or four sensors of each composite were tested under controlled conditions in order to study the reproducibility.

2.3. Particle Generation and Decomposition

A bench of generation and decomposition of ammonium nitrate particles under controlled conditions has been especially developed for this project in order to determine the sensor response in the presence of ammonium nitrate particles. Particles generation is ensured by nebulization of ammonium nitrate solution with AGK 2000 (PALAS). Particles are then dried with Nafion membrane "Perma Pure" (length: 60 cm). The relative humidity measured after particle drying is around 45 %. The mass concentration of ammonium nitrate particles is measured with TEOM series 1400a operating at 30 °C, while particle size distribution is done with FIDAS 200 particle counter. Particle diameter is comprised between 0.1 µm and 0.4 µm.

The thermal decomposition of particles is carried out using a climatic chamber heated at 30 °C-50 °C when the sensitive surfaces are placed in the exposure chamber (0.5L). Particle decomposition will be ensured with an autonomous thermoregulation system when the detection principle is validated.

3. Results and Discussion

3.1. Ammonia Detection

Firstly, sensors were characterized in the presence of ammonia under controlled conditions in the ppb range in order to be able to detect the low concentrations of ammonia and ammonium nitrate in the ambient air. Sensors response to ammonia gas is determined as a relative variation of the resistance (R) of the sensor exposed to ammonia compared to the initial value (R_0) under air zero [$(R-R_0)/R_0$].

Fig. 2 shows the relative response of $PANI_{75wt\%}$/PU sensors to ammonia at concentration range less than 60 ppb (RT/ 50 % RH) during 30 minutes of exposure to ammonia followed by exposure under air zero during 30 minutes. The curve given by solid line represents ammonia concentration measured in the exposure chamber with ammonia analyzer "PICARRO" model G2103.

The sensors showed response to ammonia at concentrations below 60 ppb. In addition, sensors response is in agreement with the real change in the concentration of ammonia measured with ammonia analyzer.

Calibration curve of $PANI_{50wt\%}$/PU and $PANI_{75wt\%}$/PU sensors obtained by plotting their relative response as function of ammonia concentration measured by the analyzer is given in Fig. 3. Sensors showed a linear response to ammonia in the ppb range with very low limit of detection (<20 ppb). Moreover, better sensitivity to ammonia (slope of the calibration curve) is obtained with $PANI_{50wt\%}/PU$ sensors (0.33 %.ppb^{-1}) compared to 0.2 %.ppb^{-1} in the case of $PANI_{75wt\%}/PU$ sensors.

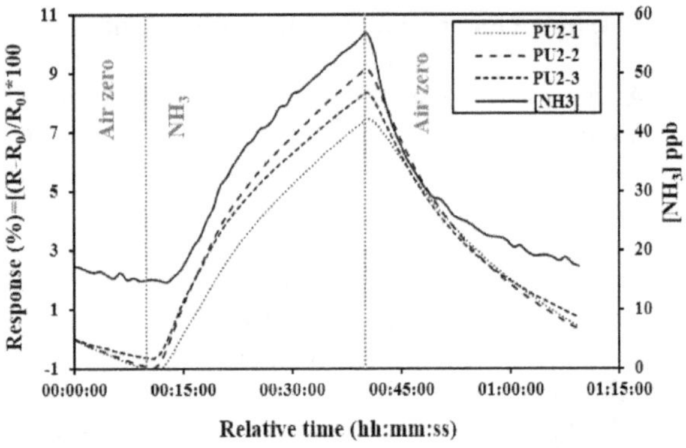

Fig. 2. Response of $PANI_{75wt\%}/PU$ sensors to ammonia at concentrations less than 60 ppb (RT/50 % RH).

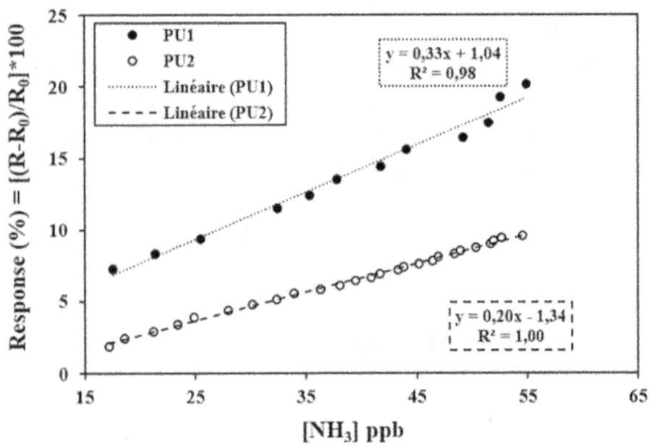

Fig. 3. Calibration curve of $PANI_{50wt\%}/PU$ (PU1) and $PANI_{75wt\%}/PU$ (PU2) sensors (RT/50 % RH).

The better sensitivity of $PANI_{50wt\%}/PU$ sensors can be explained by their porous and structured surface as demonstrated with the SEM images given in Fig. 4.

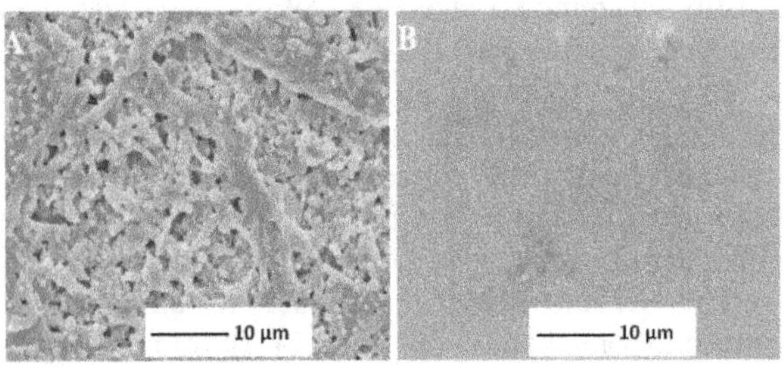

Fig. 4. SEM image of $PANI_{50wt\%}/PU$(A) and $PANI_{75wt\%}/PU$ (B) materials.

Sensors repeatability and reproducibility were also studied. Fig. 5 gives the average sensitivity obtained during four exposures for 3 sensors of each material.

Sensors sensitivity can be considered stable during four exposures in view of the low ammonia concentrations. Similar sensitivities were also obtained by the same sensors. In order to evaluate the sensors reproducibility we calculate the variability of their average sensitivity given by: (standard deviation/average)*100. We determine a variability of 24 % for $PANI_{50wt\%}/PU$ sensors and 8 % for $PANI_{75wt\%}/PU$ sensors. When the variability is less than 25 %, we can consider that the sensors have à good reproducibility. Calibration curve of PANI/PBuA sensors plotted between 200 ppb and 500 ppb of ammonia is given in Fig. 6.

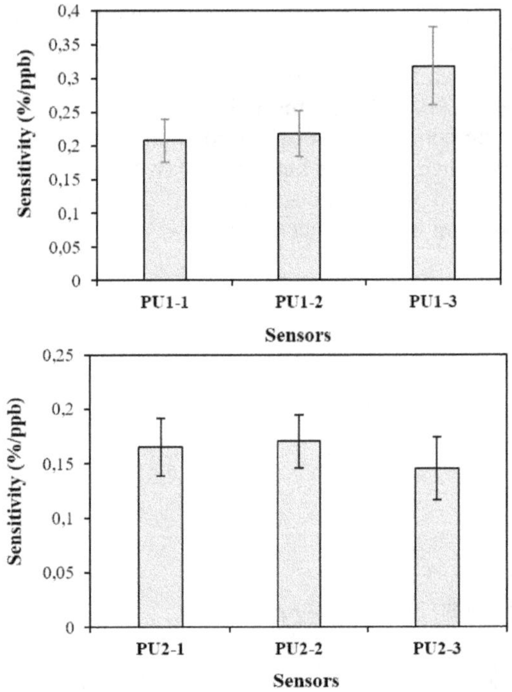

Fig. 5. Average sensitivity of $PANI_{50wt\%}/PU$ (PU1) and $PANI_{75wt\%}/PU$ (PU2) sensors during four exposures (RT/50 % RH).

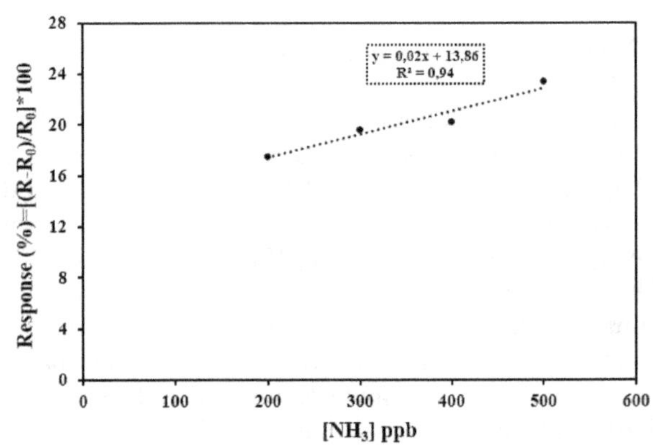

Fig. 6. Calibration curve and sensitivity of PANI/PBuA sensor (RT/50 % HR).

A linear response was also obtained with core shell materials. These sensors have showed a low sensitivity (0.02 %.ppb^{-1}) comparing with PU1 and PU2 sensors. However, this sensitivity can be considered important for ammonia sensors. The detection limit of these materials is around 60 ppb.

3.2. Ammonium Nitrate Detection

The sensors were then tested in the presence of ammonium nitrate particles. Initially, we have worked at high concentrations of particles in order to demonstrate the feasibility of this method. PANI/PBuA sensors are tested at 6.5 mg/m^3 of NH_4NO_3 particles during 90 minutes. First results obtained on four sensors are given in Fig. 7.

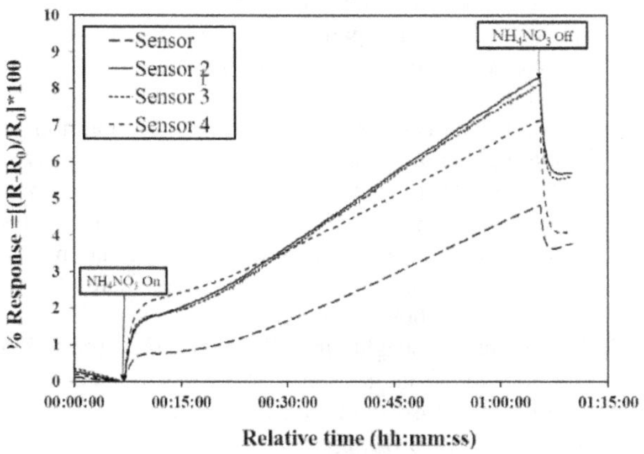

Fig. 7. Response of four PANI/PBuA sensors to NH_4NO_3 particles at 6.5 mg/m^3 (28 °C).

The four sensors react immediately to the presence of NH_4NO_3. The sensitivity between 30 and 60 min is around 6.8±1.2 % per hour and demonstrate the ability to such sensitive surface to detect ammonium nitrate particles.

Sensors based PAni/PU were tested at lower concentrations of ammonium nitrate particles between 270 µg/m^3 and 1100 µg/m^3 at 50 °C. Calibration curve of $PANI_{50wt\%}$/PU sensors obtained by plotting their relative response as function of ammonium nitrate particles concentration is given in Fig. 8. Sensor showed a response of 2.2 % at 1100 µg/m^3 after 15 minutes of exposure. Whereas, sensor response at 270 µg/m^3 is around 1 %. These results could be improved by reducing the relative humidity when particles are generated or by optimizing the decomposition temperature because high temperatures could accelerate the aging of sensitive surfaces. The kinetics of particles decomposition depends also on their diameter. Sufficient decomposition time is required, to achieve this, the particles were trapped in the exposure chamber during the heating period.

The concentration of ammonia released can be deduced from sensor sensitivity determined at 50 °C when sensor is only exposed to ammonia and which is around 0.06 %.ppb^{-1}. For the three mass concentrations of particles: 270 µg/m^3, 580 µg/m^3 and 1100 µg/m^3, the ammonia concentration is estimated at: 16.66 ppb, 30.33 ppb, and 36.66 ppb respectively. These concentrations were compared with the theoretical values of ammonia in equilibrium with ammonium nitrate particles at 50 °C and 45 % RH given by the dissociation constant K_p determined by [21]. The theoretical concentration of ammonia corresponding to 270 µg/m^3 is 81 ppb, and for the both mass concentrations 580 µg/m^3 and 1100 µg/m^3 is 100 ppb. A difference between the theoretical and experimental decomposition rate is obtained. This can be explained by a low particle decomposition under experimental conditions or by the presence of nitric acid that can create a competitive effect with ammonia and reduce the sensor response to ammonia released.

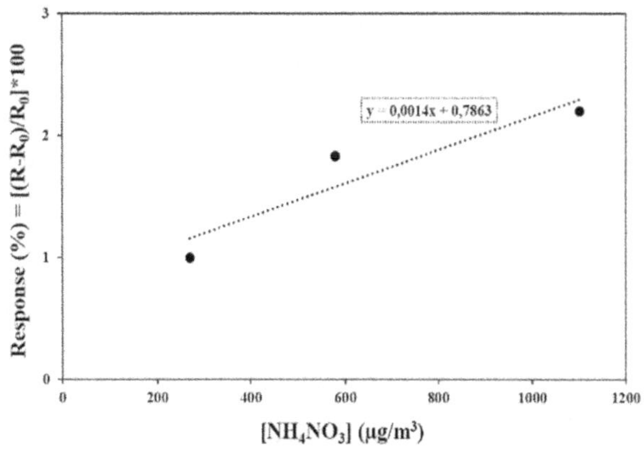

Fig. 8. Calibration curve of $PANI_{50wt\%}$/PU sensors in the presence of NH_4NO_3 particles (50 °C/ particles trapped in the exposure chamber during 15 minutes).

Ammonia concentrations deduced from sensor sensitivity to ammonia at 50 °C should be compared with the ammonia concentrations measured by the ammonia analyzer to verify if sensor sensitivity has been impacted by nitric acid released. Sensors should be characterized also in the presence of nitric acid in order to correct their response to ammonia.

4. Conclusion

Using ammonia sensitive surfaces based on polyaniline we have proposed a low cost sensor for simultaneous and selective measurement of ammonium nitrate particles and its precursor ammonia gas. The sensitive surfaces were synthesized with polyaniline associated with polyurethane or poly(butyl acrylate). The first results obtained in the presence of ammonia showed a good sensitivity of PAni/PU materials at concentrations less than 60 ppb (around 0.33 %.ppb^{-1}) with limit of detection lower than 20 ppb. Moreover, good repeatability and reproducibility were obtained. However, the sensitivity of sensors based on PAni/PBuA is around (0.02 %.ppb^{-1}) and their limit of detection is estimated at 60 ppb. Sensors were after characterized in the presence of ammonium nitrate particles at temperature between 28 °C and 50 °C. PANI/PBuA sensors were able to detect 6.5 mg/m^3 of particles with response of 8 %. While the limit of detection of PANI/PU sensors is around 270 µg/m^3 with sensitivity of 0.0014 %.µg^{-1}.m^3. The ongoing work is focused on the characterization of PAni/PU materials at lower concentrations of NH_4NO_3 particles.

References

[1]. V. P. Aneja, J. P. Chauhan, J. T. Walker, Characterization of atmospheric ammonia emissions from swine waste storage and treatment lagoons, *J. Geophys. Res. Atmospheres*, Vol. 105, Issue D9, May 2000, pp. 11535-11545.

[2]. J. H. Seinfeld, S. N. Pandis, Atmospheric Chemistry and Physics: From Air Pollution to Climate Change, *John Wiley & Sons*, 2016.

[3]. K. W. Van Der Hoek, Estimating ammonia emission factors in Europe: Summary of the work of the UNECE ammonia expert panel, *Atmos. Environ.*, Vol. 32, Issue 3, 1998, pp. 315-316.

[4]. A. Fangmeier, A. Hadwiger-Fangmeier, L. Van der Eerden, H.-J. Jäger, Effects of atmospheric ammonia on vegetation – A review, *Environ. Pollut.*, Vol. 86, Issue 1, Jan. 1994, pp. 43-82.

[5]. X. Tang, X. Zhang, Z. Ci, J. Guo, J. Wang, Speciation of the major inorganic salts in atmospheric aerosols of Beijing, China: Measurements and comparison with model, *Atmos. Environ.*, Vol. 133, May 2016, pp. 123-134.

[6]. T. Li, *et al.*, Fine particulate matter ($PM_{2.5}$): The culprit for chronic lung diseases in China, *Chronic Dis. Transl. Med.*, Vol. 4, Issue 3, Sep. 2018, pp. 176-186.

[7]. X. Querol, *et al.*, Spatial and temporal variations in airborne particulate matter (PM_{10} and $PM_{2.5}$) across Spain 1999-2005, *Atmos. Environ.*, Vol. 42, Issue 17, Jun. 2008, pp. 3964-3979.

[8]. N. Amil, M. T. Latif, M. F. Khan, M. Mohamad, Seasonal variability of $PM_{2.5}$ composition and sources in the Klang Valley urban-industrial environment, *Atmospheric Chem. Phys.*, Vol. 16, Issue 8, Apr. 2016, pp. 5357-5381.

[9]. J. Hu, Y. Wang, Q. Ying, H. Zhang, Spatial and temporal variability of $PM_{2.5}$ and PM_{10} over the North China Plain and the Yangtze River Delta, China, *Atmos. Environ.*, Vol. 95, Oct. 2014, pp. 598-609.

[10]. D. Lv, Y. Chen, T. Zhu, T. Li, F. Shen, X. Li, The pollution characteristics of PM_{10} and $PM_{2.5}$ during summer and winter in Beijing, Suning and Islamabad, *Atmospheric Pollut. Res.*, Vol. 10, Issue 4, July 2019, pp. 1159-1164.

[11]. Y. Wang, *et al.*, In situ measurement of atmospheric particles mass concentration in Anshan, *IOP Conf. Ser. Earth Environ. Sci.*, Vol. 69, Jun. 2017, 012026.

[12]. L. He, *et al.*, Fine particles at a background site in Central France: Chemical compositions, seasonal variations and pollution events, *Sci. Total Environ.*, Vol. 612, Jan. 2018, pp. 1159-1170.

[13]. D. Salameh, *et al.*, $PM_{2.5}$ chemical composition in five European Mediterranean cities: A 1-year study, *Atmospheric Res.*, Vol. 155, Mar. 2015, pp. 102-117.

[14]. A. W. Stelson, S. K. Friedlander, J. H. Seinfeld, A note on the equilibrium relationship between ammonia and nitric acid and particulate ammonium nitrate, *Atmospheric Environ. 1967*, Vol. 13, Issue 3, Jan. 1979, pp. 369-371.

[15]. C. B. Richardson, R. L. Hightower, Evaporation of ammonium nitrate particles, *Atmospheric Environ. 1967*, Vol. 21, Issue 4, Jan. 1987, pp. 971-975.

[16]. R. K. Pathak, W. S. Wu, T. Wang, Summertime $PM_{2.5}$ ionic species in four major cities of China: nitrate formation in an ammonia-deficient atmosphere, *Atmos. Chem. Phys.*, 2009, pp. 1711-1722.

[17]. Y. Tao, *et al.*, Insights into different nitrate formation mechanisms from seasonal variations of secondary inorganic aerosols in Shanghai, *Atmos. Environ.*, Vol. 145, Nov. 2016, pp. 1-9.

[18]. F. M. Kelly, L. Meunier, C. Cochrane, V. Koncar, Polyaniline: Application as solid state electrochromic in a flexible textile display, *Displays*, Vol. 34, Issue 1, Jan. 2013, pp. 1-7.

[19]. S. Mikhaylov, *et al.*, Ammonia/amine electronic gas sensors based on hybrid polyaniline-TiO_2 nanocomposites. The effects of titania and the surface active doping acid, *RSC Adv.*, Vol. 5, Issue 26, Feb. 2015, pp. 20218-20226.

[20]. J. L. Wojkiewicz *et al.*, Nanostructured polyaniline-based composites for ppb range ammonia sensing, *Sens. Actuators B Chem.*, Vol. 160, Issue 1, Dec. 2011, pp. 1394-1403.

[21]. A. W. Stelson and J. H. Seinfeld, Relative humidity and temperature dependence of the ammonium nitrate dissociation constant, *Atmospheric Environ. 1967*, Vol. 16, Issue 5, Jan. 1982, pp. 983-992.

(101)

Detection Level of Honeybee Desease: Varroosis Using a Gas Sensor Array

A. Szczurek [1], **M. Maciejewska** [1], **B. Bąk** [2], **J. Wilk** [2], **J. Wilde** [2], **M. Siuda** [2]
[1] Wroclaw University of Science and Technology, Faculty of Environmental Engineering, Wybrzeże Wyspiańskiego 27, 50-370, Wrocław, Poland
[2] Apiculture Department, Warmia and Mazury University in Olsztyn, Sloneczna 48, 10-957 Olsztyn, Poland
E-mail: monika.maciejewska@pwr.edu.pl

Summary: For humankind, honeybees are primarily important as pollinators. Fulfilling this function shall not be prevented by the uncontrolled occurrence and propagation of insects' deseases. The one particularly dangerous is varroosis, caused by the *Varroa* mites, external bee parasites. Currently, the detection of the desease is based on manual inspection of bee colonies. However, recently it has been shown that an instrumental approach based on gas sensors array measurements and classification is a promising alternative. In this work we investigated the detection level of varroosis by a semiconductor gas sensor array. It was shown that the classification error was smallest when distinguishing between two classes of bee colonis featured by *V. destructor* infestation rate: above two and less or equal two.

Keywords: Gas sensor, Honeybee, Varroosis, *Varroa destructor*, Classification, Measurement.

1. Introduction

Honeybees (*Apis mellifera*), mostly known as honey producers, in fact play major role as pollinators. The EU parliament noted in 2008 (resolution T6-0579/2008) that 79 % of human food depends on honeybee pollination. For food supply protection, honeybee populations need to be maintained in proper condition. However, the problems of honeybees diseases have intensified in recent years, despite increasing efforts to address them. The most serious threat to honeybees is caused by *Varroa destructor* [1]. This mite is a tiny, red-brown, external parasite of honeybees. The disease caused by the mites is called *varroosis*.

V. destructor mites pose an increasing global threat to the apicultural industry and agricultural ecology. For that reason, it is very important to be able to detect mite infection [2]. The traditional approach is based on visual observation and manual annotation. Recently, it has been shown that gas sensor technique may be applied for *varrosis* detection [3, 4]. The proposal was based on the idea that *V. destructor* infestation of bee colony is reflected in the quality of the air inside beehive. The relevant information may be acquired based on gas sensor measurements and the appropriate data analysis, involving classification.

This work is dedicated to the determination of the detection level for varroosis based on gas sensor array measurements. For prevention, it is most desirable to distinguish the null infestation rate from the non-zero one. However, it is thinkable that the changes in the beehive atmosphere, which are recognizable by gas sensors, occur for a different infestation level. The discrimination between colonies infested less and more than such a threshold would be most effective. For practical reasons, the threshold should not be too high.

2. Experimental

The most common honeybee subspecies *A.m. carnica* was considered in the study. It involved 44 bee colonies and their beehives.

The colonies were evaluated in respect of *Varroa destructor* infestation rate. The assessment was done in a traditional manner using a flotation method [5]. The level of infestation with *V. destructor* is defined as the number of mites, found in a sample of bees, divided by the number of bees and multiplied by 100.

The gaseous atmospheres of beehives were examined using prototype gas sensor device. It was equipped with semiconductor gas sensors. The following commercial sensors were used: TGS832, TGS2602, TGS823, TGS826, TGS2603 and TGS2600. The applied semiconductor gas sensors are partially selective (Figaro Engineering Inc.). Such sensors are suitable as the elements of a sensor array, the concept utilised in this study.

Gas sensor measurements were done in a dynamic mode. A single measurement consisted of two phases: 1. gas sensors exposure to the test gas (15 min), and 2. gas sensors regeneration (15 min). The test gas, the air from inside the beehive, was delivered to gas sensor device with Teflon tubing. Ambient air was applied for gas sensors regeneration.

All measurements were done in field conditions in the summer 2018. Multiple test were done for individual beehives, but for technical reasons, their number was diverse. The data set examined in this work included 292 measurements.

3. Classification

The classification task consisted in assigning bee colonies to two groups: 'infested' and 'not-infested', with reference to the threshold value of *V. destructor* infestation rate. If the infestation rate was greater than the threshold, the colony was labelled 'infested'. Otherwise, the colony was labelled 'not-infested'. Several groupings were done based on different values of the threshold.

The classification was based on gas sensor array responses recorded during exposure to beehive air. The feature vector consisted of time point responses of all sensors. For an individual sensor, we used 6 responses, which were distributed over the interval of 30 s with the time step of 5 s. Multiple feature vectors were formed, using the sliding time window technique. Various parts of signal were used as the basis for classification. Differential baseline correction was applied.

The detection level of *varrosis* was defined as the bee colonies grouping threshold for which the classification error was smallest. The following thresholds were examined in this work: 0, 0.5, 1, 1.5, 2, 2.5 and 3.

For classification, *k*-nn algorithm was applied with k = 3. The classifier was evaluated using ten-fold cross-validation. Confusion matrix was utilised for classification performance assessment. In particular, false negative rate (FNR) represented the fraction of data for not infested colonies which were considered infested and false positive rate (FPR) represented the fraction data for infested colonies which were considered not infested. The misclassification rate (MR) was the overall fraction of data which was assigned to the wrong class.

4. Results

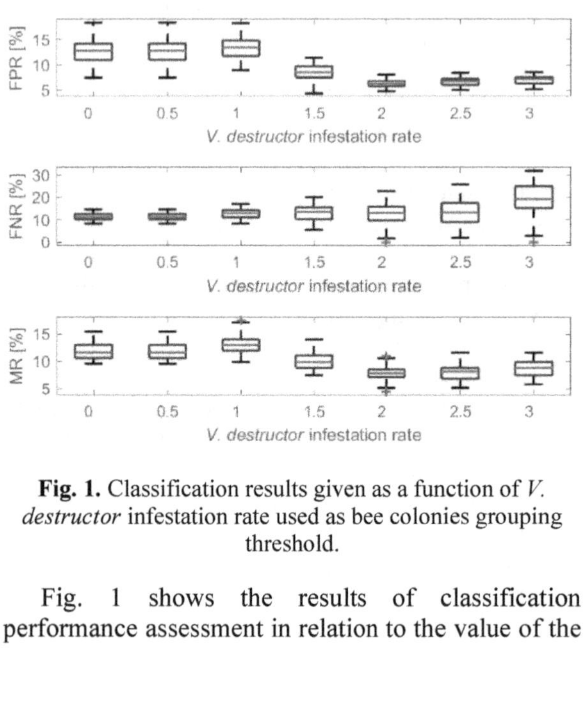

Fig. 1. Classification results given as a function of *V. destructor* infestation rate used as bee colonies grouping threshold.

Fig. 1 shows the results of classification performance assessment in relation to the value of the bee colonies grouping threshold. The box plots were used for aggregating the results obtained for feature vectors associated with vaious stages of gas sensor exposure to beehive air.

As shown, the FPR generally decreased with the increasing threshold of *V. destructor* infestation rate. However, the smallest value of FPR (6.4 %) was noted for the infestation rate equal 2. The FNR was slightly increasing with the value of the grouping criterion. The overall MR showed minimum (7.9 %) at *V. destructor* infestation rate equal 2.

5. Conclusions

Based on the study, it could be proposed that the detection level of *varroosis,* with gas sensor array is the *V. destructor* infestation rate equal 2.

Acknowledgements

This work was supported by the National Centre for Research and Development under the grant number BIOSTRATEG3/343779/10/NCBR/2017 "Developing innovative, intelligent tools to monitoring the occurrence of malignant foulbrood and elevated levels of infestation with *Varroa destructor* in honeybee colonies".

References

[1]. O. Boecking, E. Genersch, Varroosis – the ongoing crisis in bee keeping, *Journal of Consumer Protection and Food Safety*, Vol. 3, 2008, pp. 221-228.

[2]. Varroa Mite – Sampling and Monitoring Infestation Levels, http://www.omafra.gov.on.ca/english/food/inspection/bees/varroa-sampling.htm

[3]. A. Szczurek, M. Maciejewska, B. Bąk, J. Wilk, J. Wilde, M. Siuda, Detection of honeybee disease: Varrosis using a semiconductor gas sensor array, in *Proceedings of the 8th International Conference on Sensor Networks (SENSORNETS'19)*, Prague, Czech Republic, 26-27 February, 2019, pp. 58-66.

[4]. A. Szczurek, M. Maciejewska, B. Bąk, J. Wilde, M. Siuda, Semiconductor gas sensor as a detector of Varroa destructor infestation of honey bee colonies – Statistical evaluation, *Computers and Electronics in Agriculture*, Vol. 162, 2019, pp. 405-411.

[5]. I. Fries, A. Aarhus, H. Hansen, S. Korpela, Comparisons of diagnostic methods for detection of Varroa jacobsoni in honeybee (Apis mellifera) colonies at low infestation levels, *Experimental and Applied Acarology*, Vol. 10, 1991, pp. 279-287.

(102)

Sapphire Optical Fiber Sensor for Ultra High Temperature Measurement

Na Zhao [1], Qijing Lin [1,2,3,4], Zhuangde Jiang [1], Kun Yao [1], Bian Tian [1], Zhongkai Zhang [1]
[1] State Key Laboratory for Manufacturing Systems Engineering, Xi‘an Jiaotong University, Xi’an 710049, China
[2] Collaborative Innovation Center of High-End Manufacturing Equipment, Xi'an Jiaotong University, Xi'an, 710054, China
[3] Xi’an Jiaotong University Suzhou Institute, Suzhou, 215123, China
[4] State Key Laboratory of Mechanical System and Vibration, Shanghai Jiaotong University, Shanghai, 200240, China
E-mail: xjjingmi@163.com

Summary: A sapphire fiber optic sensor for ultra-high temperature is developed, which owns the advantages of high temperature tolerance and compact. The detector fiber is manufactured by fusing the sapphire fiber for detector head. The developed infrared radiation light is detected and the high temperature optical fiber sensor is placed in the ultra-high temperature furnace. The sensor signal will be transmitted to the display device through a photoelectric conversion module, and data acquisition module. As a result, the temperature response sensitivity is 2.292 $uV^{1/2}/K$ The measurement results show that the repeatability of the sensor is good. The maximum temperature measured by the sensor is 1923 K. The sensor can withstand 10 hours at high temperature and the error is less than 1 %. The development of common optical fiber sensor can stably exist at high-temperature. The sensor owns the advantage of simple, compact, easy to fabricate, what’s more, it can tolerate ultrahigh temperature for a long time.

Keywords: Temperature measurement, Sapphire optical fiber, Optical fiber sensor, High temperature sensing, Radiation measurement.

1. Introduction

In extremely harsh environment, the requirement of temperature measurement can reach over 1000 °C. In various fields such as aerospace and hot air detection [1-3], the stability of engine in high temperature environment is affected by temperature. In the process of industrial production and machine condition monitoring, the temperature information is important, which can provide the date for the temperature sensing. The working environment of large generator blades is ultra-high temperature and strong electromagnetic interference [4-8]. The high temperature sensor has good linearity, high precision, corrosion resistance, wide range of temperature response, anti-interference, safety and so on, which can be applied to severe temperature environments, for example, the severe vibration, high temperature and pressure, strong electromagnetic interference [9-15].

The contact type and non-contact type are the two main types of optical fiber temperature sensor. The contact Fabry-Perotsensor(FPI) [9], Bragg grating sensor(FBG), Michelson sensor [10], Mach-Zehnder interference (MZI) [11]. Due to the advantages of wide temperature response range and high precision, anti-interference, corrosion resistance, good linearity and so on, A sapphire fiber FBG is manufactured based on femtosecond laser, although the sensor manufactured by femtosecond laser can withstand high temperatures of 1500 °C [12] for a long time and 1900 °C instantaneously [13], it is difficult for the fiber sensor to be stably work at high temperatures for long-term use. Because of structure and material limitations, FBG structure will lose their sensitive properties.

In this paper, a sapphire optical fiber sensor is developed, which works stably at 1923 K for 10 hours. The sensor has the advantages of compactness, simplicity, convenience and long-term high temperature resistance.

2. Experiment and Discussions

The principle for thermal radiation is one of the basic methods of temperature measurement. Temperature is related to the radiation energy of the object. Therefore, the accurate measurement of temperature can be achieved by measuring infrared energy.

The material characteristics of sapphire fiber determine that it has many advantages that ordinary fiber does not have, such as high strength, high temperature resistance, multimode transmission, and corrosion resistance. Different from the ordinary optical fiber. Since the melting point of sapphire fiber is 2040 °C, we will also pay attention to sapphire fiber in the high temperature field. If we want to add sapphire fiber structure to the sensing system, the coupling point is very important. Therefore, it is of great significance to study the coupling characteristics of sapphire fiber in order to reduce the loss. In addition, its refractive index is between 1.76 and 1.78, and there are certain difficulties in coupling with ordinary fibers.

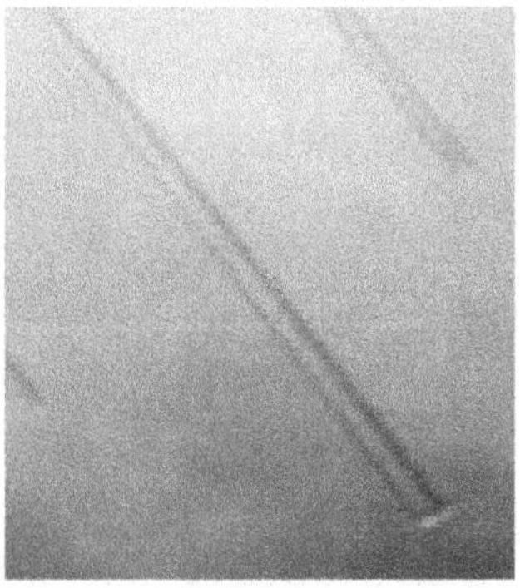

Fig. 1. The sapphire fiber.

Fig. 2. Machine tool.

The sapphire fiber with diameter of 100 um is selected to fabricate the sensor. When light is transmitted in an optical fiber, the optical power gradually decreases with the transmission distance increases. This phenomenon is called the loss of the optical fiber. In order to reduce the loss, the grinding machine is used to pretreat the sapphire fiber end face. A common commercial optical fiber splicing machine (Furukawa S177B) is used to splice the detection optical fiber to a common single mode optical fiber as a transmission optical fiber. As shown in Fig. 4 the thermal radiation intensity can be used to determine the temperature change based on Planck's radiation law.

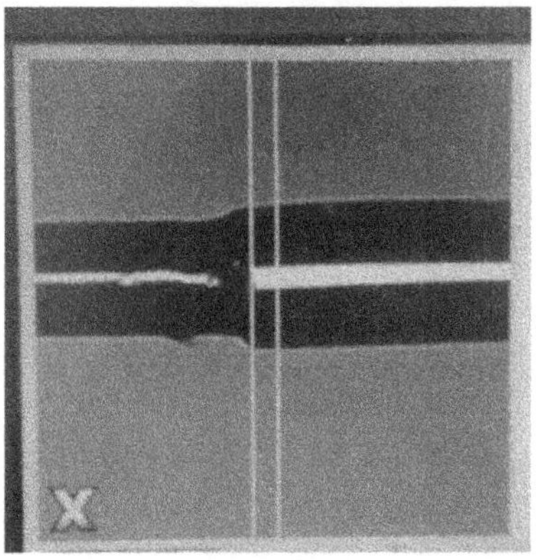

Fig. 3. The melting process between sapphire fiber and ordinary fiber.

Finally, we chose the direct fusion method to weld the sapphire fiber to the common fiber. A high-temperature welding machine with an electrode discharge temperature of 2000 °C or higher was selected. The fabrication process of the sapphire and multimode fiber coupler is shown in the figure. The sapphire fiber radiation is coupled into the multimode fiber on the multi-dimensional precision mobile platform built into the fusion splicer. The thermal imager is used to verify the coupling effect between ordinary fiber and sapphire fiber. Place the coupling point in the center of the field of view of the camera. One end of the fiber is illuminated with a 2 watt laser. The bright line in the middle is the sapphire fiber we tested and the photo taken through the thermal imager. It can be found that there is no bright spot indicating, which means that the light loss is small and the coupling point is well welded.

Fig. 4. The thermogram of the optical fiber point.

The static measurement system has been established to determine the temperature performance of the high temperature sensing point, as shown in Fig. 5. The temperature control precision of high temperature furnace can reach 1 °C, and the highest can be measured to 1750 °C. The system also includes photoelectric conversion module, data acquisition module and display device. The system is used to analyze the response of the high temperature sensor. The optical fiber radiation sensor made by us is put in the furnace, and the temperature is detected by setting different temperatures. First, it is heated to 1823 K, then it is kept at this temperature for 10 hours, and then it is cooled to room temperature. The detected optical signal is acquired and stored by the acquisition device.

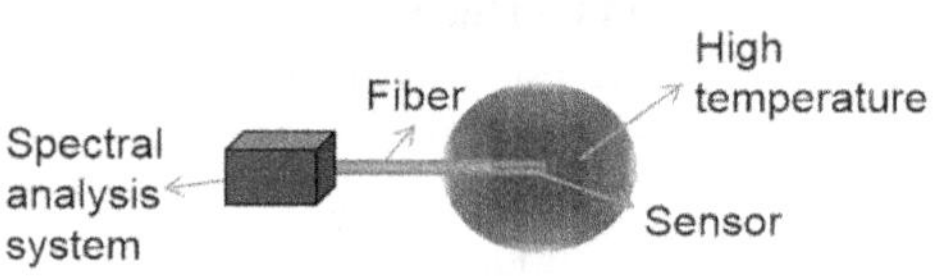

Fig. 5. The infrared radiation sensor of temperature sensing system.

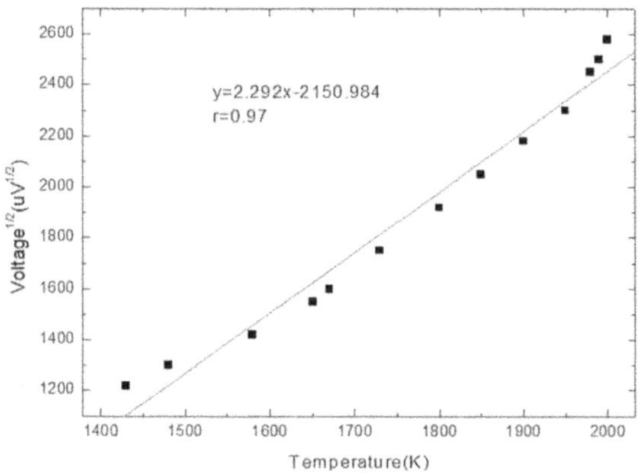

Fig. 6. The temperature response curve of optical fiber temperature sensor.

With the increase of temperature, the spectral intensity of optical fiber high temperature sensor increases. The sensitivity of temperature response is 2.292 $uV^{1/2}/K$. The sensor can withstand 10 hours at 1923 K, and the experimental error is less than 1 %. The experimental results show that the sensor can work stably for a long time at high temperature.

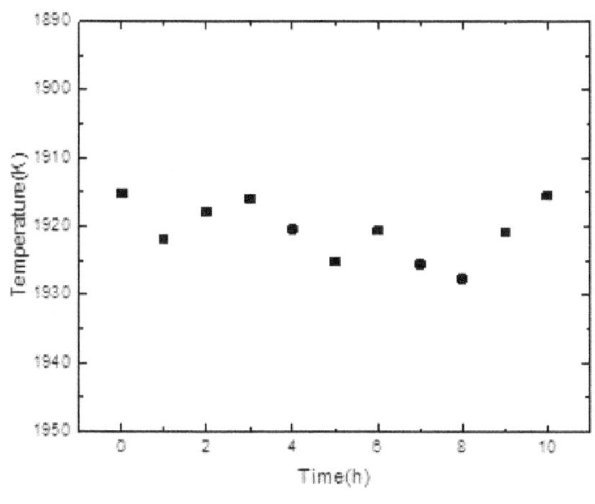

Fig. 7. The temperature response of optical fiber high temperature sensor at 1823 K for 10 h.

4. Conclusions

We fabricated a sapphire optical fiber for the high temperature field measurement. The temperature response sensitivity is 2.292 $uV^{1/2}/K$, which can work in the temperature of 1923 K for more than10 h, and the error is less than 1 %. The sensor owns the advantages of high temperature tolerance and easy to fabricate.

Acknowledgements

This work is supported by National Natural Science Foundation of China (No. 51805421, No.91748207), 973 Program (No. 2015CB057402), China Postdoctoral Science Foundation (no. 2018T111045 and no. 2017M613114), the Shaanxi Postdoctoral Science Foundation (no. 2017BSHEDZZ69), Shaanxi Natural Science Foundation (no. 2018JQ5156), Natural Science Foundation of Jiangsu Province (no. BK20180238), Key industrial innovation projects in Suzhou (no. SYG201830), the Research Project of the State Key Laboratory of Mechanical System and Vibration (MSV201813),111 Program (No.B12016). We also appreciate the support from the International Joint Laboratory for Micro/Nano Manufacturing and Measurement Technologies.

References

[1]. Y. Liu, W. Zhang, H. Fu, et al., Temperature dependency of cladding-etched fiber Bragg grating surrounded with liquid, *Optik*, Vol. 132, 2017, pp. 401-406.

[2]. I. Rozenstrauha, G. Sosins, L. Krage, et al., Elaboration of new ceramic composites containing glass fibre production wastes, *Bol. Soc. Esp. Ceram*, 2013, 52.

[3]. L. Stoica, V. Solomko, T. Baumheinrich, et al., Design of a frequency signal conditioning unit applied to rotating systems in high temperature aero engine control, in *Proceedings of the IEEE International Symposium on Circuits and Systems (ISCAS'15)*, 2015, pp. 1154-1157.

[4]. J. D. Albertson, T. Harvey, G. Foderaro, et al., A mobile sensing approach for regional surveillance of fugitive methane emissions in oil and gas production, *Environmental Science & Technology*, Vol. 50, Issue 5, 2016, pp. 2487-2497.

[5]. J. R. Galvão, A. B. Di Renzo, P. E. Schaphauser, et al., Optical fiber Bragg grating instrumentation applied to horse gait detection, *IEEE Sensors Journal*, Vol. 18, Issue 14, 2018, pp. 5778-5785.

[6]. M. Zipf, J. Manara, T. Stark, et al., Infrared-optical characterization of emitting and absorbing gases at high temperatures and high pressures, *High Temperatures – High Pressures*, Vol. 47, Issue 1, 2018.

[9]. S. C. Warren-Smith, L. V. Nguyen, C. Lang, et al., Temperature sensing up to 1300 °C using suspended-core microstructured optical fibers, *Optics Express*, Vol. 24, Issue 4, 2016, pp. 3714-3719.

[10]. N. Zhao, Q. Lin, Z. Jiang, et al., High temperature high sensitivity multipoint sensing system based on three cascade mach-zehnder interferometers, *Sensors*, Vol. 18, Issue 8, 2018, 2688.

[11]. Q. Rong, Y. Hao, R. Zhou, et al., UW imaging of seismic-physical-models in air using fiber-optic Fabry-Perot interferometer, *Sensors*, Vol. 17, Issue 2, 2017, 397.

[12]. N. Zhao, Q. Lin, W. Jing, et al., High temperature high sensitivity Mach-Zehnder interferometer based on waist-enlarged fiber bitapers, *Sensors & Actuators A Physical*, Vol. 267, 2017, pp. 491-495.

[13]. H. Cao, X. Shu, Miniature all-fiber high temperature sensor based on michelson interferometer formed with a novel core-mismatching fiber joint, *IEEE Sensors Journal,* Vol. 17, Issue 11, 2017, pp. 3341-3345.

[14]. T. Elsmann, T. Habisreuther, A. Graf, et al., Inscription of first-order sapphire Bragg gratings using 400 nm femtosecond laser radiation, *Optics Express*, Vol. 21, Issue 4, 2013, pp. 4591-4597.

[15]. T. Habisreuther, T. Elsmann, Z. Pan, et al., Sapphire fiber Bragg gratings for high temperature and dynamic temperature diagnostics, *Applied Thermal Engineering*, Vol. 91, 2015, pp. 860-865.

[16]. M. Yan, J. Tylczak, Y. Yu, et al., Multi-component optical sensing of high temperature gas streams using functional oxide integrated silica based optical fiber sensors, *Sensors and Actuators B: Chemical*, Vol. 255, 2018, pp. 357-365.

[17]. R. Ruiz-Lombera, I. Laarossi, L. Rodríguez-Cobo, et al., Distributed high-temperature optical fiber sensor based on a brillouin optical time domain analyzer and multimode gold-coated fiber, *IEEE Sensors Journal*, Vol. 17 , Issue 8, 2017, pp. 2393-2397.

[18]. S. C. Warren-Smith, L. V. Nguyen, C. Lang, et al., Temperature sensing up to 1300 C using suspended-core microstructured optical fibers, *Optics Express*, Vol. 24, Issue 4, 2016, pp. 3714-3719.

[19]. V. R. Mamidi, S. Kamineni, R. L. N. S. Prassad, et al., Transducer-based fiber Bragg grating high-temperature sensor with enhanced range and stability, *Optical Engineering*, Vol. 56, Issue 9, 2017, 090502.

[20]. G. H. Zhou, S. W. Wang, X. X. Huang, et al., Unidirectional carbon fiber and SiC particulate Co-reinforced fused silica composite, *Ceramics international*, Vol. 33, Issue 7, 2007, pp. 1395-1398.

[21]. A. Y. Kulesh, M. A. Eronyan, I. K. Meshkovskii, et al., Influence of mechanically stimulated hydrolysis on static fatigue and strength of silica optical fiber, Strength, *Fracture and Complexity*, Vol. 10, Issue 1, 2017, pp. 39-47.

5th International Conference on Sensors Engineering and Electronics Instrumentation Advances (SEIA' 2019), 25-27 September 2019, Canary Islands (Tenerife), Spain

(105)

Ultrasonic Temperature Measurement

M. Reisinger [1], M. Schwarz [1] and B. Zagar [1]
[1] Johannes Kepler University Linz, Institute for Measurement Technology, Altenberger Straße 69, 4040 Linz, Austria
Tel.: + 43 732 2468 5920
E-mail: michael.reisinger@ieee.org

Summary: The paper covers temperature measurement of water averaged over a certain propagation path via the ultrasonic time-of-flight principle. It uses the fact that the speed of sound is a well-studied physical property and is highly temperature-sensitive, thus a very high resolution can be achieved. In addition, since the measurement takes place over a geometrically well-known path, the propagation path, leading to a spatial dependency that offers to also employ tomographic methods in a study. Initial calibration reveals the very high precision attainable by placing two opposing transducers in a cylindrical tank and determining the mean and standard deviation for stationary temperature conditions. It can be shown that transit time measurements with a standard deviation of less than half a nanosecond can be achieved, therefore the temperature can be measured with a resolution of a few thousandth of a Kelvin.

Keywords: Ultrasonic sensing, Time-of-flight measurement, Speed of sound.

1. Introduction

Measuring the speed of sound in water enables very high precision measurements of the temperature along the propagation path of ultrasonic (US) pulses. Fig. 1 shows the schematic setup for such a measurement. The transducers are placed at the positions $\vec{r_1}$ and $\vec{r_2}$ at a distance l to each other.

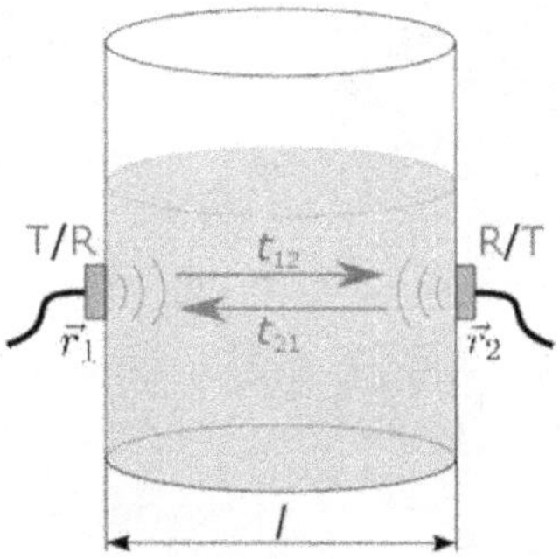

Fig. 1. Measuring scheme.

The propagation times t_{12} of pulses going left to right and t_{21} of pulses propagating right to left can be expressed by the equations (1) and (2).

$$t_{12} = \underbrace{\int_{\vec{r_1}}^{\vec{r_2}} \frac{1}{c(\vartheta(\vec{r}))} dr}_{t_0} + \underbrace{\int_{\vec{r_1}}^{\vec{r_2}} \frac{1}{\vec{v}_f(\vec{r})} d\vec{r}}_{\Delta t_f}, \tag{1}$$

$$t_{21} = \underbrace{\int_{\vec{r_2}}^{\vec{r_1}} \frac{1}{c(\vartheta(\vec{r}))} dr}_{t_0} + \underbrace{\int_{\vec{r_2}}^{\vec{r_1}} \frac{1}{\vec{v}_f(\vec{r})} d\vec{r}}_{-\Delta t_f} \tag{2}$$

The speed of sound c is dependent on the temperature ϑ which in turn is dependent on the position $\vec{r}$. The variable t_0 describes the propagation time attributed to the speed of sound and Δt_f is the propagation delay caused by a possible flow of the medium characterized by the position-dependent flow vector $\vec{v_f}$. Assuming that the flow is constant while carrying out the measurements in both directions, the flow term can be compensated for by subtracting out. This assumption is valid, because the timespan between the measurements is only 200 μs. The average speed of sound along the propagation path can then be calculated according to

$$\bar{c} = \frac{|\vec{r_2} - \vec{r_1}|}{t_0} = \frac{l}{t_0} = \frac{2 \cdot l}{t_{12} + t_{21}} \tag{3}$$

2. Speed of Sound in Water

The speed of sound in water depends on different influencing factors like temperature, pressure and salt concentration. For pure water at atmospheric pressure the dependency of the speed of sound on the temperature can be described by the polynomial [1]

$$c(\vartheta) = c_0 + \sum_{i=1}^{5} a_i \cdot \vartheta^i, \tag{4}$$

where ϑ is the mean temperature along the path in degrees Celsius. The speed of sound c_0 at 0 °C and the polynomial coefficients a_i; $i = 1, \ldots 5$, given by Bilaniuk and Wong [1], are listed in Table 1.

As shown in Fig. 2, Equation (4) exhibits a maximum at around 74 °C, thus the temperature measurement range needs to be restricted either strictly below or above 74 °C.

The measurement accuracy of the temperature depends on both the uncertainty of the propagation time measurement and the sensitivity of the propagation time on temperature changes, besides all other parameters that are assumed constant here. Equation (5) shows the relation between the propagation time, the propagation path length and the speed of sound as a function of the temperature. Equation (6) describes the sensitivity of the propagation time on the temperature which is proportional to the propagation path length l.

$$t_0(\vartheta, l) = \frac{l}{c(\vartheta)}, \tag{5}$$

$$\frac{\partial t_0(\vartheta, l)}{\partial \vartheta} = l \cdot \frac{\mathrm{d}\left(\frac{1}{c(\vartheta)}\right)}{\mathrm{d}\vartheta} \tag{6}$$

Table 1. Polynomial coefficients.

Coefficient	Value	Unit
c_0	$1.40238744 \cdot 10^3$	$\frac{\text{m}}{\text{s}}$
a_1	5.03836171	$\frac{\text{m}}{\text{s }°\text{C}}$
a_2	$-5.81172916 \cdot 10^{-2}$	$\frac{\text{m}}{\text{s }(°\text{C})^2}$
a_3	$3.34638117 \cdot 10^{-4}$	$\frac{\text{m}}{\text{s }(°\text{C})^3}$
a_4	$-1.48259672 \cdot 10^{-6}$	$\frac{\text{m}}{\text{s }(°\text{C})^4}$
a_5	$3.16585020 \cdot 10^{-9}$	$\frac{\text{m}}{\text{s }(°\text{C})^5}$

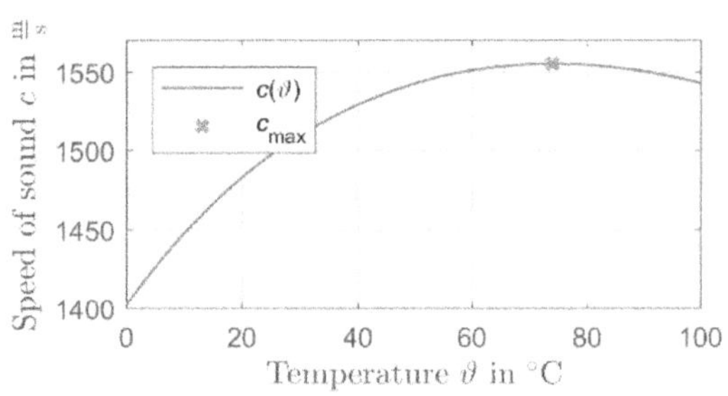

Fig. 2. Speed of sound as a function of the temperature in pure water at atmospheric pressure.

For $l = 150$ mm, as used in the experimental assembly in Fig. 4, the sensitivity is plotted in Fig. 3. For an ambient temperature of $\vartheta = 20$ °C the sensitivity is about $-200\,\frac{\text{ns}}{°\text{C}}$.

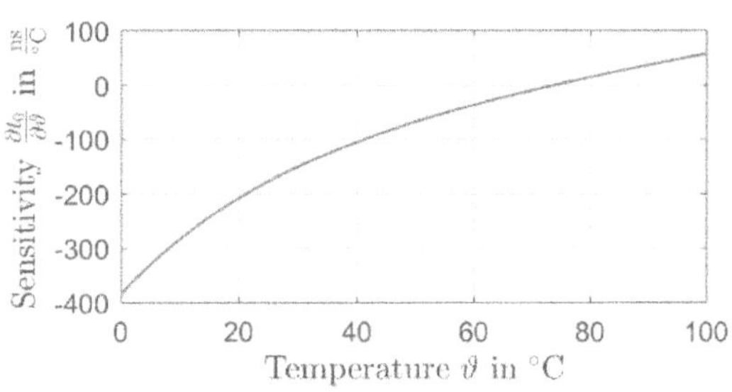

Fig. 3. Propagation time sensitivity in dependence on the temperature for $l = 150$ mm.

3. Measurement Setup

As can be seen in Fig. 4 the measurement setup consists of multiple opposing transducer pairs mounted in the wall of a PVC tubing. The transducers are arranged one above the other at a vertical interval of 25 mm. Below the bottom transducer pair a thin rod with a spirally coiled heating wire is mounted. The distance between the heating coil and the bottom transducer pair is also 25 mm. The transducer pairs are connected to the multiplexer board visible in Fig. 5. The multiplexer board contains two 16-channel multiplexer chips, one for the left row of transducers and one for the right one. The signal input pins of the multiplexers are connected to the analog front end of an EVM430-FR6047 evaluation board from *Texas Instruments* [2] that contains all the necessary analog and digital processing hardware for pulse excitation, data acquisition and data processing. The stimulating signal for ultrasonic pulse excitation generated also by the evaluation board can be switched to a specific transducer pair connected to the multiplexer board. The address of the channel, the signal gets connected to is preset by 4 GPIO-pins of the evaluation board. This enables a time-displaced measurement with up to 16 transducer pairs.

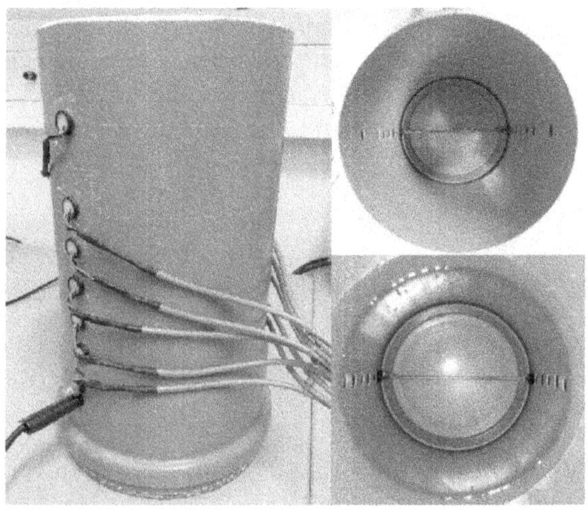

Fig. 4. Measurement setup front view (left) and top view without (top right) and with water (bottom right).

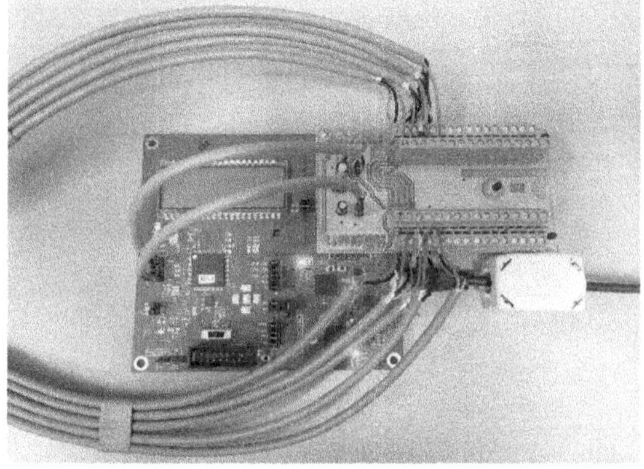

Fig. 5. Hardware for data processing.

4. Time-of-flight Estimation

To determine the speed of sound and consequently the temperature a method for precisely estimating the propagation time of ultrasonic pulses must be found.

4.1. Sensed Signals

The excitation and received signals of the ultrasound transducers are both shown in Fig. 6. Although the US transducers would ideally be excited by a sinusoidal signal burst at exactly the resonance frequency of the transducer in our implementation a burst of rectangularly shaped signals is used. According to the frequency response of the transducers a sinusoidal wave is formed anyway. The signal, the transmitting transducer is excited with, consists of 20 stimulating pulses followed by 2 stop pulses which are in phase-opposition to effectively shorten the ringing of the US signal. The stop pulses ensure a fast settling of the receiving transducers oscillation. The received signal has an amplitude of about 40 mV and is disturbed by noise and interferences as can be seen in Fig. 6 in red. To minimize the estimation uncertainty a matched filter approach is employed for the TOF-estimation [3].

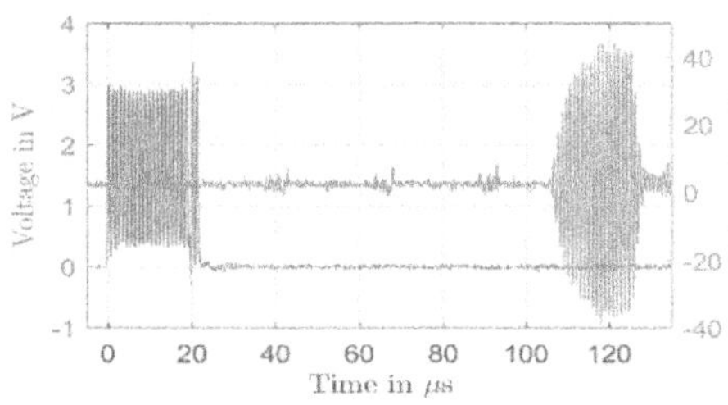

Fig. 6. Transmitted (blue) and received (red) signal.

4.2. Estimation Method

Fig. 7 shows the used theory for TOF-estimation. The time when the signals′ envelope reaches a specified threshold value compared to the maximum value of the envelope is detected. This method enables very low uncertainty.

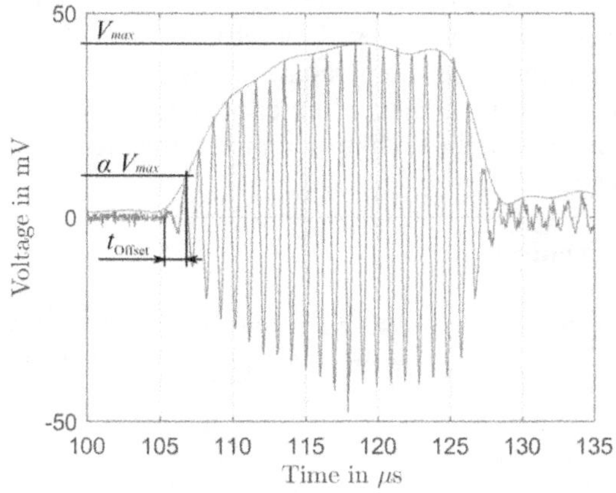

Fig. 7. Estimation theory for absolute time-of-flight detection.

5. Measurement Accuracy

As shown before the achievable accuracy of the temperature measurement depends on the accuracy of the time-of-flight detection. To determine the reached precision, a series of 1000 measurements at constant temperature has been executed. The detected TOF-values are plotted in Fig. 8. The measurements have been executed with a sampling rate of 50 samples per second what means the measuring process for recording 1000 samples takes 20 seconds. Because the temperature can change minimally within this timespan, there can also be a drift of the measured TOF. To obtain accurate, time-invariant distribution statistics of the sample, this value drift has to be corrected for. To estimate the drift, the value drift has been calculated by a moving average filter. The red curve in Fig. 8 shows the drift-corrected data.

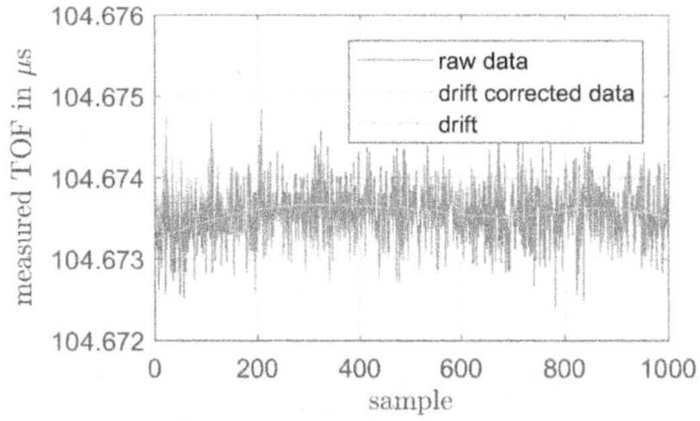

Fig. 8. Raw values (blue), corrected values (red) and value drift (yellow).

The histogram in Fig. 9 was generated using the corrected data of the 1000 values sample. In addition to this a Gaussian probability function curve has been matched to the data set based on the mean value and the standard deviation of the data set. From inspection of the graph in Fig. 9 it is noticeable, that the histogram fits this curve quite good. This deduces, that the measurement is disturbed by white Gaussian noise only. It is visible that the TOF-measurement has a standard deviation of 355 ps. According to Fig. 3 this results in a standard deviation of the temperature measurement of less than 2 thousandths of a Kelvin at ambient temperature.

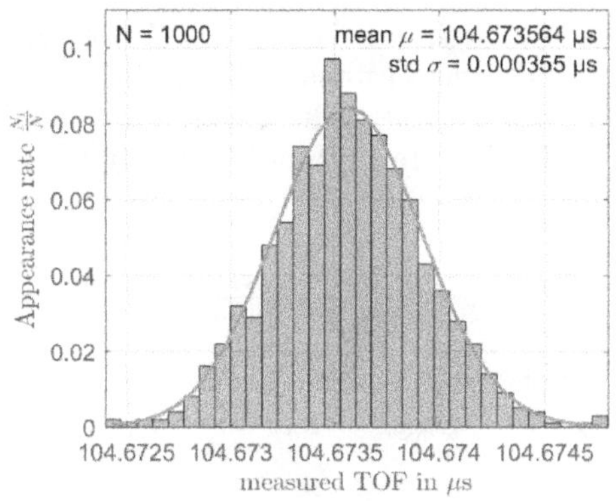

Fig. 9. Histogram of the measurement series with fitted Gaussian probability density function.

6. Calibration

Due to the fact that there are constant delay times caused by the detection method as shown in Fig. 7 and other detrimental factors like the cable length, the dimension of the transducers and the inexact transducer placement, a calibration of the system is essential. The system is quite sensitive to changes in propagation path length, so a very precise distance adjustment would be necessary. To avoid all these needs, the system is calibrated first at two different constant temperatures ϑ_1 and ϑ_2. The measured time $t_{\mathrm{m}}(\vartheta_i)$ at the temperature ϑ_i is the sum of the propagation time $t_{\mathrm{p}}(\vartheta_i)$ at the specified temperature and the constant time delay t_{Offset} as shown in Equation (7).

$$t_{\mathrm{m}}(\vartheta_i) = \underbrace{\frac{l(\vartheta_i)}{c(\vartheta_i)}}_{t_{\mathrm{p}}(\vartheta_i)} + t_{\mathrm{Offset}} \qquad (7)$$

The temperature dependency of the propagation path length l is motivated by the thermal expansion of the pipe. It is only in the range of a few micrometers per degree Celsius, but due to the high sensitivity of the time-of-flight measurement on changes of the propagation path length it is also relevant. The propagation path length at ϑ_2 can be expressed by the propagation path length $l(\vartheta_1)$ and the thermal expansion coefficient of PVC defined with $\alpha \cong 100 \cdot 10^{-6} \frac{1}{°C}$ as in equation (8). For simplicity, the corresponding temperature of the values l, t_{m} and c is given by their index from now on.

$$l_2 = l_1 \cdot [1 + \alpha(\vartheta_2 - \vartheta_1)] \qquad (8)$$

By rearranging equation (7) for the unknown parameters l_1 and t_{Offset} the equations (9) and (10) are obtained.

$$l_1 = (t_{\mathrm{m1}} - t_{\mathrm{Offset}})c_1, \qquad (9)$$

$$t_{Offset} = t_{\mathrm{m2}} - \frac{l_2}{c_2} \qquad (10)$$

By substitution of l_2 in equation (10) by equation (8) and subsequent substitution of t_{Offset} in equation (9) the real propagation path length at ϑ_1 results in equation (11). The real offset-time can be determined by equation (12).

$$l_1 = \frac{(t_{\mathrm{m1}} - t_{\mathrm{m2}}) \cdot c_1 \cdot c_2}{c_2 - c_1[1 + \alpha(\vartheta_2 - \vartheta_1)]}, \qquad (11)$$

$$t_{\mathrm{Offset}} = t_{m1} - \frac{l_1}{c_1} \qquad (12)$$

With the calibrated values, the correct value for the propagation time t_p and concequently the correct speed of sound c can be determined. To recalculate the temperature, an iterative algorithm is needed, because the temperature dependency of the speed of sound in water is a polynom of 5th order.

For a very accurate absolute temperature measurement of course a very exact reference temperature is necessary. Nevertheless, the relative temperature measurement is very exactly. Therefore, the system performs very well in measuring dynamic processes.

7. Measurement Results

To test the systems performance, a current pulse was applied to the heating filament by a power supply. This heats the water in a very small local area around the rod. Consequently the warm water is soaring up. The temperature can be detected at the multiple parallel sound propagation paths of the transducer pairs mounted above the rod. Fig. 10 shows the temperature-time profile in different heights z. While soaring up, the heat diffuses into the volume. This is the reason, why the maximum of the measured temperature is decreasing exponentially with increasing height z.

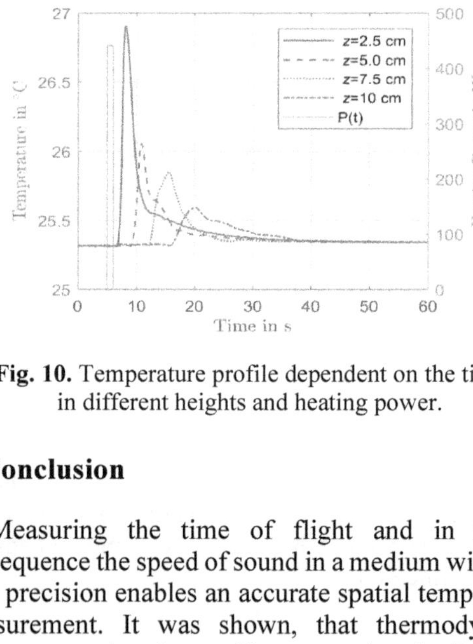

Fig. 10. Temperature profile dependent on the time in different heights and heating power.

8. Conclusion

Measuring the time of flight and in further consequence the speed of sound in a medium with very high precision enables an accurate spatial temperature measurement. It was shown, that thermodynamic processes can be observed by measuring the TOF and consequently the temperature at several propagation paths. With a special configuration the method enables further possibilities like a tomographic temperature detection.

References

[1]. Underwater Acoustics – Technical Guides – Speed of Sound in Pure Water, National Physical Laboratory, 2000.

[2]. Texas Instruments, EVM430-FR6047 Datasheet, http://www.ti.com/tool/EVM430-FR6047

[3]. Texas Instruments, Matched-Filter Ultrasonic Sensing: Theory and Implementation, http://www.ti.com/lit/wp/slaa814/slaa814.pdf

(106)

raw-LoRa vs. LoRaWAN for Urban Monitoring: Two Wireless Sensor Network Architectures in Comparison

S. Tondini [1], **S. Tritini** [1], **S. Croce** [2], **S. Seppi** [3] **and R. Monsorno** [1]
[1] Eurac Research, Center for Sensing Solutions, 12 Volta St., 39100 Bozen, Italy
[2] Eurac Research, Institute for Renewable Energy, 13/A Volta St., 39100 Bozen, Italy
[3] NOI Techpark Südtirol-Alto Adige, Tech-Transfer Digital, 12 Volta St., 39100 Bozen, Italy
Tel.: + 39 0471 055266, fax: + 39 0471 055099
E-mail: stefano.tondini@eurac.edu

Summary: This study aims at comparing the performances of two wireless sensor network (WSN) architectures which share the same end nodes, but differ from the gateway and the communication protocol, namely raw-LoRa and LoRaWAN (868 MHz). Both the architectures rely on open-source components for time-series storage and network management. Moreover, the key physical building blocks have been implemented through open-source compliant hardware elements. All these aspects have made it possible to achieve two flexible and low-cost platforms for environmental monitoring applications, among others. Data collected in the context of two European projects allow us to assess the reliability and the drawbacks of the two WSNs, when applied to an urban monitoring scenario.

Keywords: Low-power wide-area network, Open-source, Time-series management, Environmental monitoring, Urban.

1. Introduction

The use of low-power wide-area network (LPWAN) for environmental monitoring is taking hold because of their high energy efficiency and the long coverage range achievable [1]. Despite the limited data throughput, what makes this technology particularly attractive is the low implementation cost. However, communication robustness is still critical.

Although gap filling methods are becoming more and more satisfactory to cope with data lacks within time-series, communication interference for WSNs remains a crucial point to be addressed (especially in urban areas). In this work, we go through this by taking into account two purpose-built WSNs, where the first one implements only the physical layer raw-LoRa as a communication protocol, while the second one encompasses the LoRaWAN application layer, as well.

2. Urban Monitoring Pilots

2.1. raw-LoRa Network

Within the SINFONIA project [2], a 20 sensor nodes WSN has been set up to evaluate the effects of novel integrated refurbishment interventions at district scale and to calibrate related environmental analysis models. A crucial aspect in this pilot is that the data acquisition must occur at the same time for all nodes. For this reason, we decided to exploit the raw-LoRa protocol, which makes it easier to set a router as master and the end nodes as slaves, so that the master spreads its own time to the slaves making them synchronized before each measurement. The hybrid-star topology of the WSN is shown in Fig. 1, where a gateway (LoRa/wifi) serves as access point to the web. All the hardware building blocks of the WSN are based on Pycom LoPy4 boards (with Semtech SX1276 chipset).

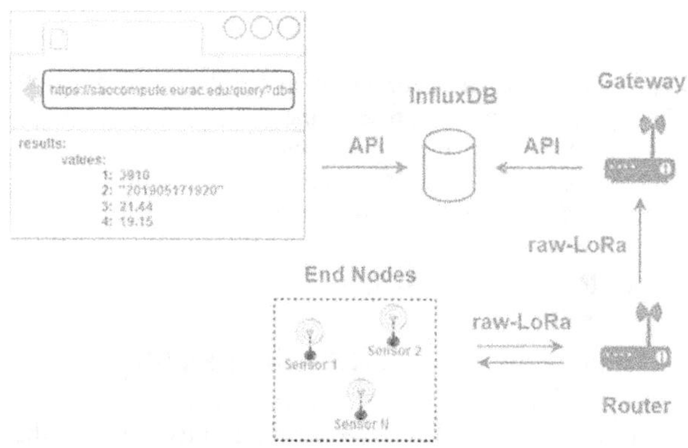

Fig. 1. raw-LoRa wireless sensor network set up for the SINFONIA project. The architecture encompasses 20 end nodes, a router, a gateway, and an InfluxDB database, whose application programming interface (API) can be exploited to retrieve the data acquired by the sensors.

The LoRa packet management performed in the SINFONIA WSN is sketched in Fig. 2. The functioning has been obtained by programming each component of the network, namely the sensor nodes, the router and the gateway, in a stand-alone way. The operating routine starts with the router spreading its own real time clock (RTC) signal for 40 seconds. In the meantime, all the other system components, set to receiving mode, wait for a LoRa packet containing the router's RTC. Once the sensor nodes and the gateway are synced, the whole system goes into wait mode (machine idle) till a preset time. Then all the nodes acquire contemporary a measurement. That done, the nodes forward the data (temperature, relative humidity, battery level) to the router in a staggered way, i.e. one by one every 2 seconds, and then go into deep sleep mode. After 60 seconds from the measurement time (violet vertical bar in the figure), the router wraps all the data received from the sensor nodes in a single LoRa packet and send it toward the gateway, which is still set to receiving mode. The gateway, in turn, posts the sensor data into the InfluxDB database through a Wi-Fi connection via an HTTP request and goes into deep sleep, as well. The only component still on now is the router, which spreads the RTC signal every 1 minute in case of some nodes went out of synchronization during operation. After 9 minutes from the measurement time, the nodes and the gateway wake up, ready for a new synchronization, and the routine starts again.

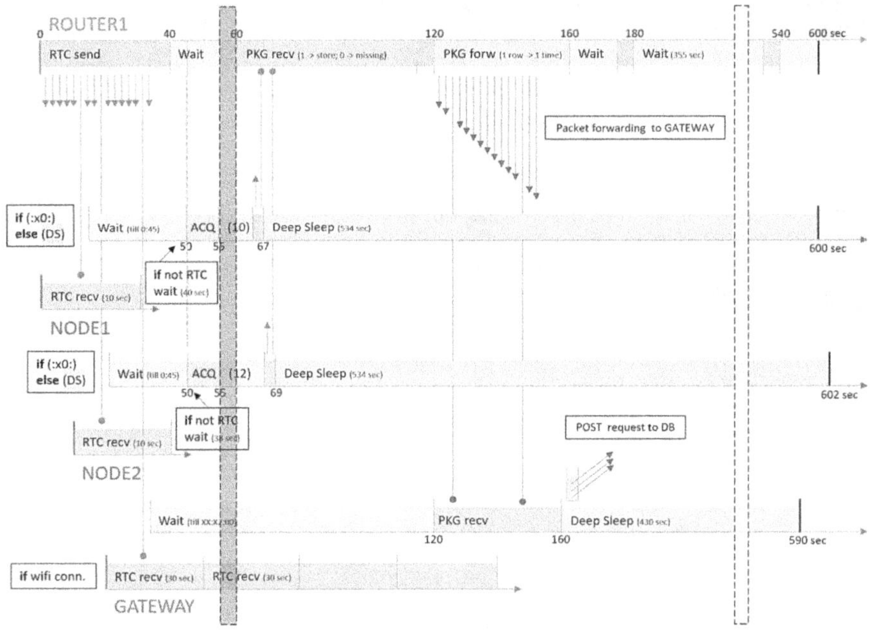

Fig. 2. Sketch of the working plan for the SINFONIA WSN, which repeats itself from left to right in an infinite loop. The layered structure represents the main hardware components of the system, which are the router, the sensor nodes (only two are reported in the figure), and the gateway. The color code use aims at highlighting the different modes in which the components can operate: green for transmitting, red for receiving, yellow for measuring, light orange for waiting, cyan for packet forwarding, and grey for deep sleep mode.

2.2. LoRaWAN network

In the context of the **BEACON Südtirol - Alto Adige** project [3], another architecture has been put in place to grant LoRaWAN connectivity inside the technology park of Bolzano (NOI Techpark). The aim of the pilot was to prove whether an IoT testing network can be set up by using only open source components. LoRaWAN is particularly suited for this, as it can manage many different end nodes at once. Therefore, we exploited a plain LoRaServer installation for the network management, integrated with an InfluxDB instance (see Fig. 3). We also set up a web portal through which both custom and commercial sensor nodes can be added to the network without taking any subscription out. Several Raspberry Pi based gateways (with RHF0M301 LoRa shield) are distributed in the NOI Techpark areal, as access points.

The LoRaWAN gateways have been set up to have direct access to the network of the NOI Techpark (ScientificNetworkSouthTyrol [4]) through a secured virtual private network (VPN) connection at boot.

In this way, the gateways, which are assigned to a stack of static internet protocol (IP) addresses, can reach the LoRaServer from any wireless local area network (WLAN). This strategy allows for a flexible deployment of the LoRaWAN access points, which can also be extended outside the NOI Techpark areal, at will.

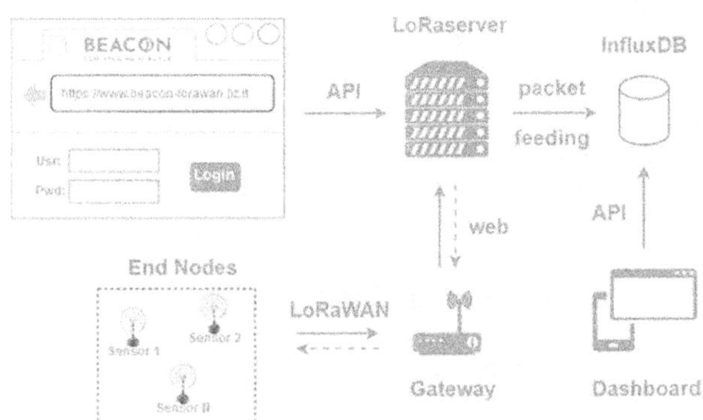

Fig. 3. Schematics of LoRaWAN network deployed at the NOI Techpark within the BEACON BZ project. The topology consists of several gateways that ensure indoor and outdoor connectivity in the technology park area. The network management is achieved through LoRaServer and a dedicated web portal allows for the registration of new end nodes. Data transmitted over the network are available into an InfluxDB database (integrated with LoRaServer).

Fig. 4 shows the registration page (top panel) and the administration page (bottom panel) of the web portal created to make the LoRaWAN network's users autonomous in managing their own end nodes. After registration, the user is redirected to the administration dashboard that encompasses five tabs: *Your devices, Add new application, Add new device, Get data, Settings*. At the first log-in, the devices list (first tab) is empty and the user needs to go through the creation of a new application (second tab) and the registration of new device (third tab). After doing so, a new line will appear in the devices list, which will carry the LoRaWAN device parameters in the format: *Device address, Application session key, Network session key*. If these three values match the ones of a working sensor node, the user will then be able to retrieve the transmitted data through the database API (fourth tab). In the end, the settings page (fifth tab) allows for updating/deleting the user profile.

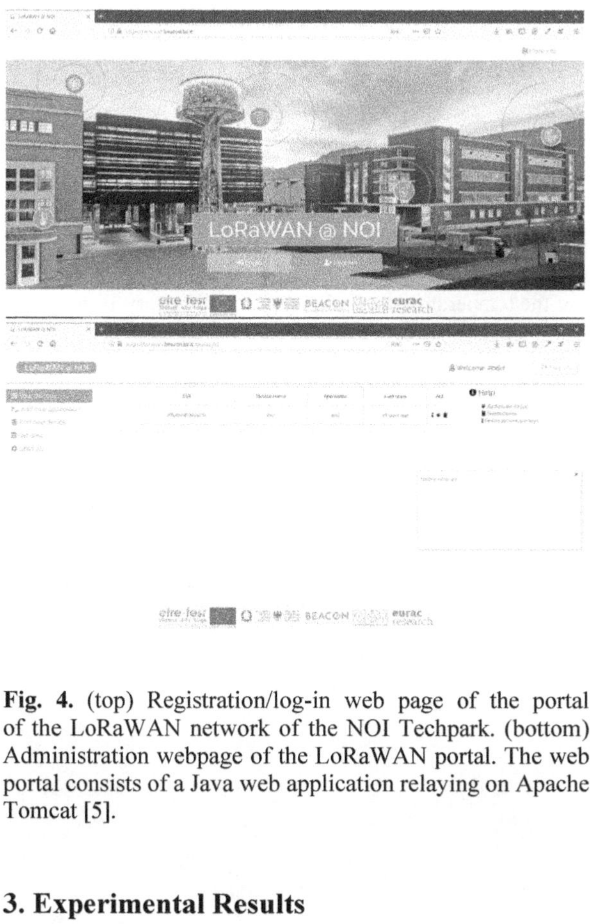

Fig. 4. (top) Registration/log-in web page of the portal of the LoRaWAN network of the NOI Techpark. (bottom) Administration webpage of the LoRaWAN portal. The web portal consists of a Java web application relaying on Apache Tomcat [5].

3. Experimental Results

To assess the performances of the two networks, one sensor node similar to the Sinfonia's ones has been added to the NOI Techpark network. This way, two WSNs are available to us, whose output can be analyzed to compare the reliability of the raw-LoRa and LoRaWAN protocols.

The sensor node schematics is detailed in Fig. 5 (right). It encompasses a Pycom LoPy4 with expansion board version 3.0 as core unit. A DF robot solar LiPo Charger, coupled with a 3.7 V – 5200 mAh battery and a 0.5 W monocrystalline Seeed solar panel, manages the power supply. The sensing element is a Honeywell Humidicon HIH6121 enclosed into a plastic radiation shield and connected to the microcontroller via the I^2C bus. The sensor node carries a microSD flash memory for local storage of acquired data. The system is completed by a Laird 868 MHz 2 dBi antenna and a Bopla IP68 case.

Fig. 6 shows the time-series acquired through the two networks for a period of 52 hours (with 10 min bins). The different temperature and relative humidity trends between the upper and lower plot are since the raw-LoRa node measures the outdoor temperature, while the LoRaWAN node reads out the values from an embedded sensor (within the node housing). The time slots into which lack of communication has occurred for more than 20 min (i.e. for more than two subsequent acquisitions) are represented with the red and grey bars on the two plots. It is worth to note that this issue happens for both the WSNs in like manner, meaning that this is mainly related to interference at the LoRa frequencies. Packet loss ratio figures for both the WSN are reported in Table 1.

Table 1. Packet loss ratio comparison.

	raw-LoRa	LoRaWAN
Packet loss ratio	< 13 %	< 11 %

The value reported in Table 1 for LoRaWAN is consistent with the existing literature [5], as the node is less than 500 m far from the gateway.

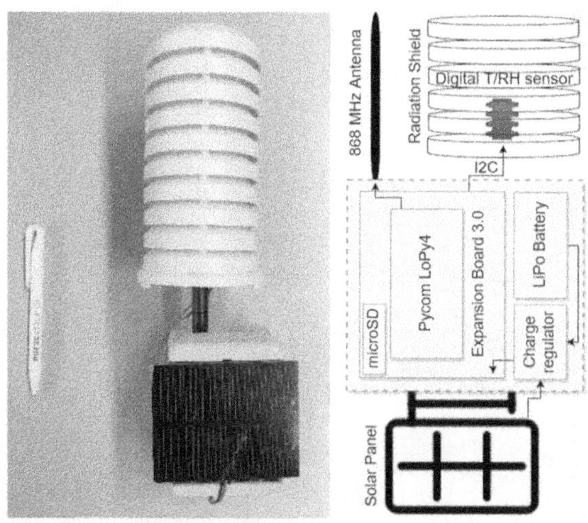

Fig. 5. Picture of one SINFONIA's node (left). Building blocks scheme (right) of the sensor node on the left.

In Fig. 7, a prolonged acquisition period of the two monitoring systems is compared. The 49 days long time-series shown in the charts report the temperature in red solid line, the relative humidity in green solid line and the battery level (mV) in blue solid line. Despite the overall behavior is similar, the data from the two sensor nodes differ in range due to the different placement of the sensing element, which in the raw-LoRa node (SINFONIA's case) is within a radiation shield, while in the LoRaWAN node is inside the IP68 case. This aspect brings also to smoother trends of the temperature and the relative humidity in the second chart, because the sensor suffers from an enhanced sensing inertia [7] within the sealing case.

Concerning the battery level curves, we can ascribe the oscillating tendency in the lower chart to a worse health status of the battery, as the two nodes perform similar operations and the power consumption of the two communication protocols is the same.

The visualization tool used to create the dashboard of Fig. 7 is the open platform Grafana [8], which comes with a native data source plugin for InfluxDB database.

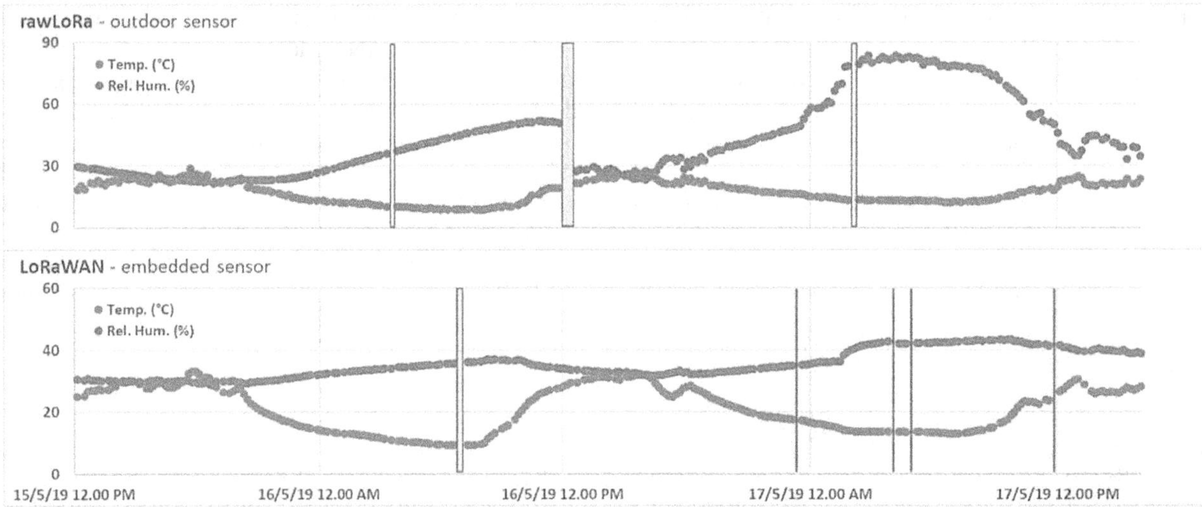

Fig. 6. 52 hours long time-series acquired by one sensor node of the raw-LoRa network (top) and by one sensor node of the LoRaWAN network (bottom). Both the sensor nodes measure the temperature and relative humidity every 10 minutes. The grey/red vertical bars represent the time slots in which prolonged lack of connectivity (>20 min) has occurred.

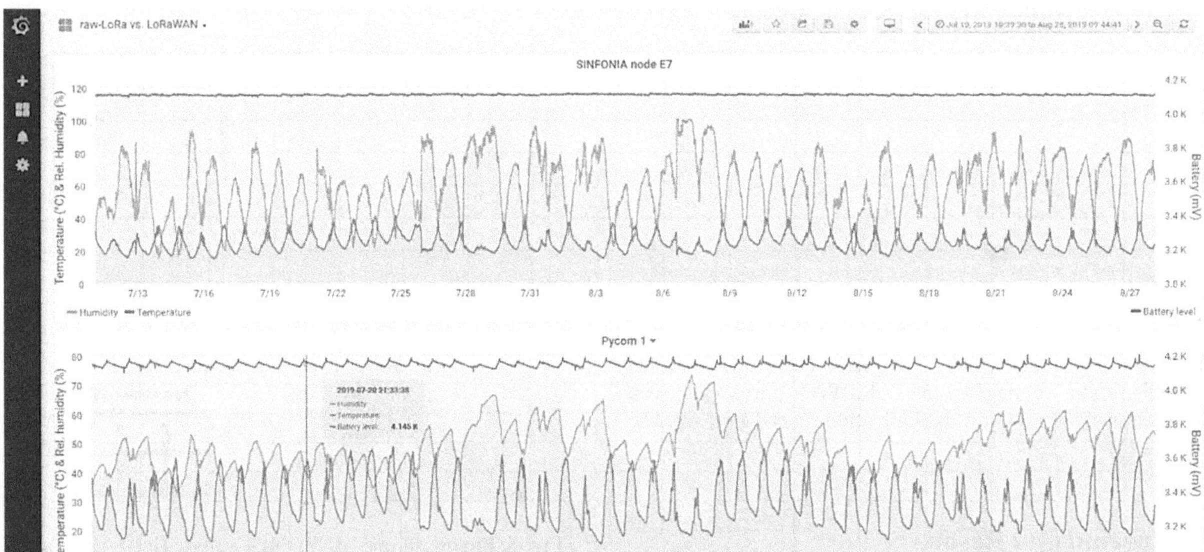

Fig. 7. Grafana dashboard carrying time-series of 49 days long. The upper panel shoes the temperature (red curve), relative humidity (green curve) and battery level (blue curve) acquired by one SINFONIA's node. The lower panel represents the values of the sensor node added to the LoRaWAN network of the NOI Techpark. The difference between the two charts comes from the fact that, despite the hardware is the same, the SINFONIA node measures outdoor parameters while the LoRaWAN node is sensing temperature and relative humidity inside the IP68 case.

4. Conclusions

With this comparative analysis, we show that the raw-LoRa and the LoRaWAN protocols share similar reliability performances. However, the first one happens to be more appropriate for use cases with very specific requirements, e.g. synchronous acquisitions or pear-to-pear communication among end nodes. Instead, the second one is best suited to integrate many different data sources with enhanced information management and sharing capabilities.

A common feature to both the approaches described in this work is that they entirely rely on open-source building blocks. This proves that reliable acquisition systems based on low cost and vendor un-linked components are now within reach.

Acknowledgements

The research leading to these results has received funding by the European Union's Seventh Framework Program FP7/2007–2013 Grant No 609019 and from the European Regional Development Fund (ERDF) Programme 2014-2020, under Project number FESR2023 (Beacon Sudtirol-Alto Adige) and Project number FESR1094 (Data Platform and Sensing Technologies for Environmental Sensing Lab)

References

[1]. N. Harris, J. S. Curry, Development and range testing of a LoRaWAN system in an urban environment, *International Journal of Electronics and Communication Engineering*, Vol. 12, Issue 1, 2018, pp. 43-51.

[2]. SINFONIA Project, www.sinfonia-smartcities.eu

[3]. BEACON BZ Project, https://beacon.bz.it/

[4]. ScientificNetworkSouthTyrol, https://www.sciencesouthtyrol.net/

[5]. Apache Tomcat, http://tomcat.apache.org/

[6]. J. Petäjäjärvi, et al., Performance of a low-power wide-area network based on LoRa technology: Doppler robustness, scalability, and coverage, *International Journal of Distributed Sensor Networks*, Vol. 13, Issue 3, 2017, pp. 1-16.

[7]. Y. Xue, A. P. Cracknell, Advanced thermal inertia modelling, *International Journal of Remote Sensing*, Vol. 16, Issue 3, 1995, pp. 431-446.

[8]. Grafana, https://grafana.com/

(107)

Pulse Shape Dependence on Applied Voltage of Geiger-Mueller Detector

B. Almutairi[1,3,4], **S. B. Alam**[1], **T. Akyurek**[2], **C. Goodwin**[3], **A. Olson**[3], **and S. Usman** [1]

[1] Department of Mining and Nuclear Engineering, Missouri University of Science & Technology, Rolla, MO, 65401, USA

[2] Department of Physics, Faculty of Art and Science, Marmara University, 34722, Kadikoy, Istanbul, Turkey

[3] Rhode Island Nuclear Science Center 16 Reactor Rd, Narragansett, RI 02882, USA

Tel.: + 15599367845

[4] Kuwait Institute for Scientific Research, P.O. Box 24885, Safat 13109, Kuwait

E-mail: bjaxw7@mst.edu

Summary: Geiger-Mueller (GM) detector is one of the oldest radiation detectors that is still being used extensively in the radiation measurement field. The detector was introduced in 1928 by scientists: Geiger and Mueller. It is assumed that all pulses generated by the GM counter are of the same size and shape. There is a general belief that the generated pulse shape characteristics do not carry any useful information. In this study, pulse shape characteristics (Pulse Width, Tail of Pulse, and Amplitude) were observed under various applied voltages under different radiation sources. It has been shown that the amplitude of the generated pulses by the GM detector increases with increasing the applied voltage for Co^{60} and Cs^{137} sources, unlike the background radiation.

Keywords: Deadtime correction, Geiger Muller counter, Radiation measurement, Pulse shape.

1. Introduction

The nature of detector deadtime has baffled researchers since the discovery of the deadtime phenomenon in the early 1940s [1]. There are several proposed deadtime models since that time. The first ideal mathematical deadtime models were proposed by Feller [2] and Evans [3]. These ideal models are known as the paralyzable and non-paralyzable models. The investigation of the deadtime behavior has seen a resurgence in the scientific community since the early 2000s. Other researchers proposed several models since then, i.e., Lee and Gardner [4] and Patil and Usman [5]. In this study, we observe pulse shape characteristics change under various applied voltages for different radiation sources. This is performed in order to develop a correlation between pulse shape characteristics and deadtime phenomenon in future studies.

1.1. Detector Deadtime Models

According to the paralyzable model, the first radiation event triggers a resolving time (deadtime, τ) in which the second radiation event is not detected unless that resolving time is passed. If a second radiation event takes place during the resolving time, the event is lost, and it resets the deadtime. Hence, the radiation measurement system experiences continued paralysis until at least an interval (τ) lapses without the occurrence of a radiation event.

The relationship between the observed and true count rates in the ideal paralyzable model is described by:

$$m = n \cdot e^{-n \cdot \tau}, \tag{1}$$

where m is the measured count rate, n is the true count rate and τ is the detector deadtime.

For the non-paralyzing model, if a second radiation event takes place during deadtime, it is lost; however, it does not extend the deadtime duration. Thus, the detector is dead only for a fixed time τ after a radiation event takes place in the detector. According to the non-paralyzable model, the relationship between the observed count rate and the measured count is given by:

$$m = \frac{n}{1 + n \cdot \tau}, \tag{2}$$

where m, n, and τ are defined similarly to the paralyzable model.

In an effort to develop a more accurate deadtime model, Lee and Gardner [4] proposed a new hybrid model by combining the two traditional ideal models. The hybrid model is given by:

$$m = \frac{n \cdot e^{-n \cdot \tau_P}}{1 + n \cdot \tau_N}, \tag{3}$$

where τ_P is the paralyzing deadtime and τ_N is non-paralyzing deadtime. Nonetheless, Lee and Gardner did not provide any physical explanation behind their hybrid model. Furthermore, Patil and Usman [5] proposed a similar hybrid model; however, it was based on a fixed paralysis factor. The following equation describes the hybrid model:

$$m = \frac{n \cdot e^{-n \cdot \tau \cdot f}}{1 + n \cdot \tau \cdot (1 - f)} \tag{4}$$

where f is the paralysis factor. Another hybrid model was proposed by Hou and Gardner [6]. They modified Lee and Gardner's [4] hybrid model by using a polynomial expansion. It is worth mentioning that none of these models provide any phenomenological explanation of the deadtime behavior. Akyurek et al. [7] investigated deadtime dependence on operating voltages. Their new findings offered a new insight into the deadtime phenomenon. Up to this day, no detector deadtime model exists that is based on valid phenomenological reasoning. Furthermore, Akyurek et al. [7] observed that pulse shape characteristics do change under different applied voltages and further investigation into the observation is required. There is not enough data and literature on pulse shape characteristics of the GM counter under various applied voltages. In this study, we investigate specific pulse shape characteristics under different applied voltages for different radiation sources. Observations from this study will serve as preliminary results for the correlation between pulse shape characteristics and deadtime in order to develop a more accurate deadtime model.

1.2. Typical Voltage Charge vs. Collection Curve

Pulses generated in detector filled gases such as Geiger Mueller (GM) detector are recognized to be voltage-dependent. Fig. 1 below shows various regions of operation. As the applied voltage increases, the region of operation changes from the recombination region to ion Chamber to Proportional counting and finally to the GM counting region. It can be seen that the GM counter operates at higher voltages. In order to understand the deadtime phenomenon, it is essential to understand the behavior of charge creation and charge collection.

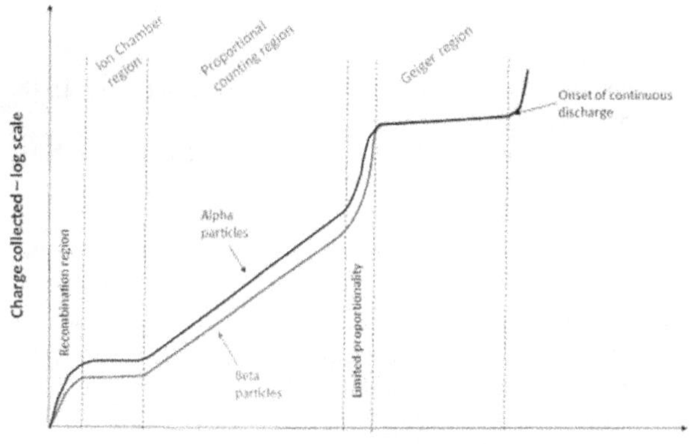

Fig. 1. Gas-filled detector regions of operations.

At lower operating voltages, the first region is the recombination region where recombination of charges prevails. The applied voltage is not sufficient to keep the charges separated. As the voltages increases, it reaches the ion chamber region where all initial charges are collected without any charge multiplication. Increasing the voltage further takes the detector into the Proportional Counter region. In this region, charge multiplication takes place. As the voltage increases further, it leads into the limited proportionality region followed by GM counting region. In the GM region, pulses produced are ideally of the same sizes and shapes. These pulses are independent of the energy and type of the radiation triggering the event. In this region, it is assumed that all pulses generated by the GM counter are of the same size and shape.

2. Methods and Materials

The basic experimental set-up consisted of the following: 1) Radiation sources: two different types of radiation sources were used, Cobalt-60 (Co^{60}) and Cesium-137 (Cs^{137}). The sources were both produced in Feb. 2013 by Spectrum Techniques, LLC. For the purpose of comparison, the pulses generated by background radiation were also recorded separately. 2) GM detector: model 133-2 by Ludlum measurements, Inc. 3) Charge-sensitive preamplifier: model 142A by Ortec Ametek Inc. 4) A high-voltage power supply: model 556 by Ortec Ametek Inc. 5) A Tektronix oscilloscope: model DPO3032 was used to capture and record the generated pulses.

The starting voltage where pulse shape characteristics (Pulse Width, Tail of Pulse, and Amplitude) were recorded was 400 V. The applied voltage was increased incrementally by 50 V until 800V. For every selected voltage, pulse shape characteristics were recorded over 15 sets. The final pulse shape characteristics were then averaged. The data were saved on a USB stick for further analysis on a PC. It is worth mentioning that the recommended operating voltage by the manufacturer is 550 V. Also, it is advised not to operate the detector on voltages above 1000 V because it will damage the detector.

3. Results

For the background radiation, the pulse width and tail of the pulse decreased as the applied voltage was increased, as shown in Fig. 2. On the other hand, the amplitude (height) of the pulse did not exhibit much change when the voltage increased. The maximum amplitude value of the recorded pulses of the background radiation was 12.29 V while the lowest was 11.76 V. The variations in the amplitude values for background radiation is insignificant. This might be due to the low number of radiation events detected by the GM counter.

For the Co^{60} Source, the pulse width and tail decreased as the applied voltage increased. Nonetheless, the amplitude monotonically increased as the applied voltage increased, as shown in Fig. 3. The lowest amplitude value was 13.38 V, while the highest was 17.84 V.

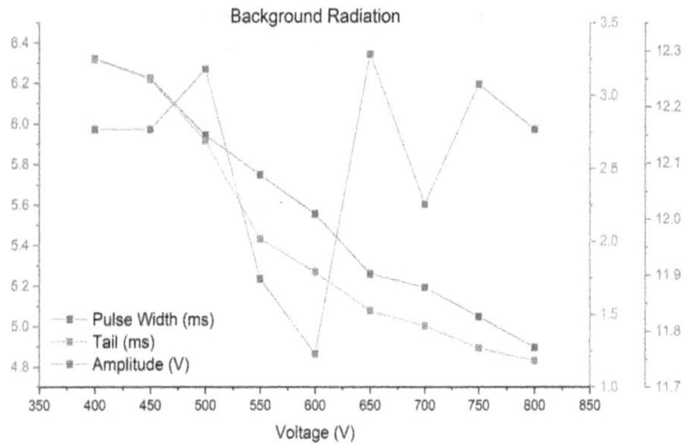

Fig. 2. Change of Pulse Width, Tail of Pulse, and Amplitude under different applied voltages for the background radiation.

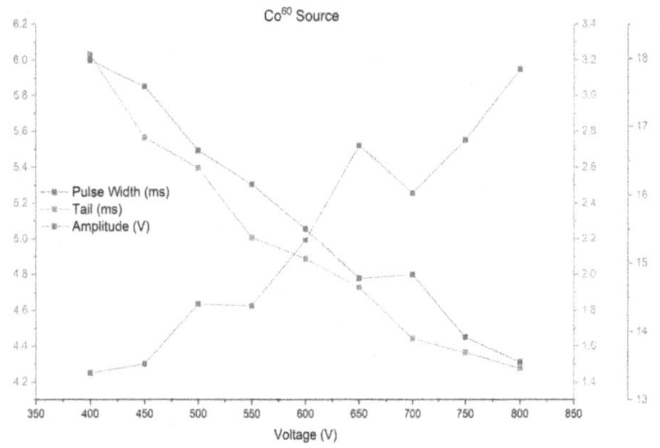

Fig. 3. Pulse width, Tail, and Amplitude values due to Co^{60} source under different applied voltages.

For the Cs^{137} source, pulse shape characteristics showed similar behavior due to the Co^{60} source, as shown in Fig. 4. The lowest amplitude value was 13.09 V, while the highest was 18.08 V. Unlike the background radiation, the increase in the amplitude values due to the Co^{60} and Cs^{137} sources might be due to the higher number of radiation events taking place in the GM counter.

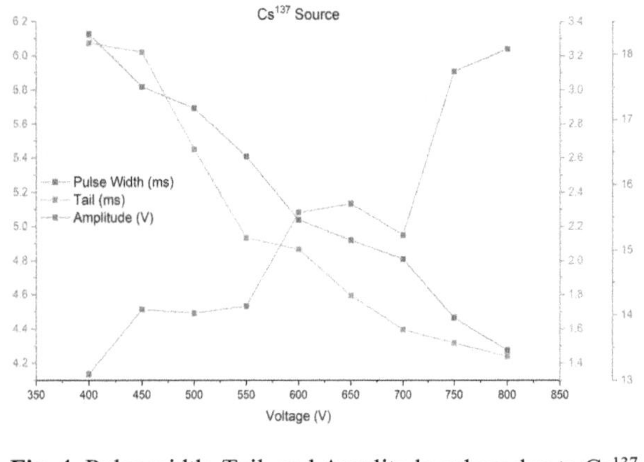

Fig. 4. Pulse width, Tail, and Amplitude values due to Cs^{137} source under different applied voltages.

4. Conclusions

The results are revealing for the inside working of the GM detector. Additional investigation of pulse shape characteristics with careful data analysis would help the radiation measurement community in gaining new insight into the internal operation of the GM detector. The data generated in this experiment show that the general belief that pulse shape properties do not reveal any useful information from the GM detector should be questioned and reevaluated. Further investigation of the pulse shape properties will aid the radiation measurement community on developing a more accurate deadtime model for GM counters.

Nomenclature

τ = deadtime;
m = measured count rate;
n = true count rate;
τ_p = paralyzing deadtime
τ_N = non-paralyzing deadtime;
f = paralysis factor probability based.

Acknowledgments

We thank the Kuwait Foundation for the Advancment of Sciences (KFAS) for funding the participation in the International Conference on Sensors and Electronic Instrumentation Advances.

References

[1]. M. Yousaf, T. Akyurek, S. Usman, A comparison of traditional and hybrid radiation detector dead-time models and detector behavior, *Progress in Nuclear Energy*, Vol. 83, 2015, pp. 177-185.

[2]. W. Feller, On probability problems in the theory of counters, in Studies and Essays, *Interscience,* 1948, pp. 105-115.

[3]. R. D. Evans. The Atomic Nucleus, *McGraw-Hill,* New York, 1955.

[4]. S.H. Lee, R. P. Gardner. A new G-M counter dead time model, *Applied Radiation and Isotopes,* Vol. 83, 2000, pp. 731-737.

[5]. A. Patil, S. Usman, Measurement and application of paralysis factor for improved detector dead-time characterization, *Nuclear Technology,* Vol. 165, Issue 2, 2009, pp. 249-269.

[6]. G. Hou, R. P. Gardner. A new G-M counter hybrid dead-time correction model, *Radiation Physics and Chemistry,* Vol. 116, 2015, pp. 125-129.

[7]. T. Akyurek, M. Yousaf, X. Liu, S. Usman. GM counter deadtime dependence on applied voltage, operating temperature and fatigue, *Radiation Measurements,* Vol. 73, 2015, pp. 26-35.

(108)

Observation of CO Detection Using Aluminum-doped ZnO Nanorods on Microcantilever

R. Nuryadi [1], L. Aprilia [2], M. Hosoda [2], A. Udhiarto [3], D. Hartanto [3], M.A. Barique [2], Y. Neo [2], and H. Mimura [2]

[1] Center for Materials Technology, Agency for the Assessment and Application of Technology, Puspiptek Building #224, South Tangerang, 15314, Banten, Indonesia

[2] Shizuoka University, 3-5-1 Johoku, Naka-ku, Hamamatsu, 432-8011, Shizuoka, Japan

[3] Universitas Indonesia, Depok, Jawa Barat, Indonesia

Tel./fax: +62-21-75791324

E-mail: ratno.nuryadi@bppt.go.id

Summary: A functionalization of aluminum (Al)-doped zinc oxide nanorods (ZNRs) on a microcantilever (MC) was investigated to realize high sensitivity carbon-monoxide (CO) detection. The sensing property was examined by measuring a change in resonant frequency of the vibrating MC at various gases concentrations. Al doping in the ZNRs showed a significant increase in the sensitivity, where the sensitivity on the order of femtograms for CO-gas was obtained. It is also found that an interaction between CO and water vapor on ZnO plays a key role on the sensing mechanism. These findings may contribute to future developments of highly sensitive toxic-CO-gas detectors with a fast response and room temperature operations without a device heating.

Keywords: Microcantilever, ZnO nanorods, Al doping, Resonant frequency change, CO detection.

1. Introduction

Microcantilever (MC) has been extensively studied for molecular gas sensor applications because of their high sensitivity, rapid response and operation at room temperature [1]. To detect carbon monoxide (CO) gas, several materials such as poly(ethylene oxide)/Ni acetate and Fe(III)porphyrin have been used as the sensitive layer on the MC surface. In addition to these polymer-based materials, a functionalization by metal oxides on the MC for CO detection is also very promising from the viewpoint of stability and endurance. So far, zinc oxide nanostructures have been used as sensitive layer on electrical-resistance-change type sensors for sensing NO_2, ethanol, CO, H_2, CH_4, etc. [1, 2], however, only a few studies have been conducted for MC-type sensors, such as ZnO nanorods (ZNRs) for detecting water vapor [3] and volatile organic compounds (VOCs) [4].

Many studies have been undertaken to improve the gas sensor sensitivity such as by adding dopant atoms (Cd, Fe, Ti, Sn, and Al) on the ZnO interface, which affects the gas adsorption and increases the sensing response. Aluminum was doped on ZnO because of the smaller atomic size of Al^{3+} ion compared to that of Zn^{2+} (r Al^{3+} = 0.054 nm and r Zn^{2+} = 0.074 nm) resulting in a decrease in HOMO-LUMO energy gap and an increase in the gas adsorption. The effects of Al atoms in ZnO nanoclusters on CO adsorption have been investigated based on density functional theory. Al doping in ZnO nanoclusters strongly attracted CO molecules, supporting the experimental result of resistance-based sensor where exposure of CO leads to a decrease in the resistance of Al-doped ZnO nanostructures. However, to our knowledge, there is almost no report on the effect of Al dopant atoms in ZNRs on CO adsorption for an MC-type system; and consequently, the gas adsorption mechanism for this system is also not well understood.

In this work, we prepared Al-doped ZNRs (AZNRs) on the piezoresistive MCs using a RF magnetron sputtering and investigated CO adsorption at a room temperature. The sensor sensitivity was observed by measuring the resonant frequency shift of the vibrating AZNR-based microcantilever (AZNMC) under various gas flow rates.

2. Experimental Procedures

ZNRs were grown directly on the MC surface by using hydrothermal method. First, the seed layer (30-nm thick) was deposited on Si substrate by the RF sputtering method, and ZNRs were then grown at 60 °C for 2 hours by dipping the substrate into a mixed solution of 0.04 M zinc nitrate tetrahydrate and hexamethylenetetramine (HMTA) in deionized water. Next, the grown ZnO nanorods was dried at 90 °C for 30 minutes in an oven to evaporate the solvents. Then Al was doped on the ZNR-coated MC (ZNMC) by the RF sputtering.

Commercial piezoresistive MCs (NPX1CTP003, Seiko Instrument Inc., Japan) with a typical piezoresistor resistance of 630 Ω were used for the study [3]. The AZNMC was placed on a piezoelectric inside a stainless-steel chamber and all the measurements were performed at room temperature of 23-25 °C. The vibration of the cantilever modifies the resistance which is detected by a Wheatstone bridge circuit. The resonant frequencies of AZNMC during the measurements were recorded by a system based on direct digital synthesizer (DDS) with 0.1-Hz resolution. Response to the CO gas flow was detected by measuring the shift in the resonant frequency. The relative humidities of inside the chamber was monitored by a hygrometer.

3. Results and Discussions

The surface morphology of ZNRs layer monitored by FE-SEM is shown in Fig. 1. The ZNRs layer on MC surface formed a nanorod structure where the average diameter of the high-density nanorods was approximately 100 nm. After Al doping, distribution of Al was observed by an electron probe micro analyzer (not shown). Although Al was unevenly localized in the whole surface, most Al was found spread out on the ZNRs. High concentration of Al arises where the ZNRs exist, which indicated that Al was selectively bonded with the nanorods.

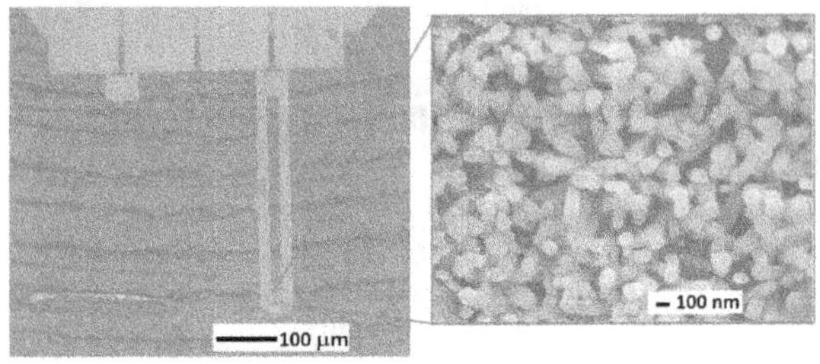

Fig. 1. FE-Scanning electron micrograph of the microcantilever (left) and magnified image of ZNRs on MC surface (right). A vertical nanorod structure with diameter of around 100 nm was observed.

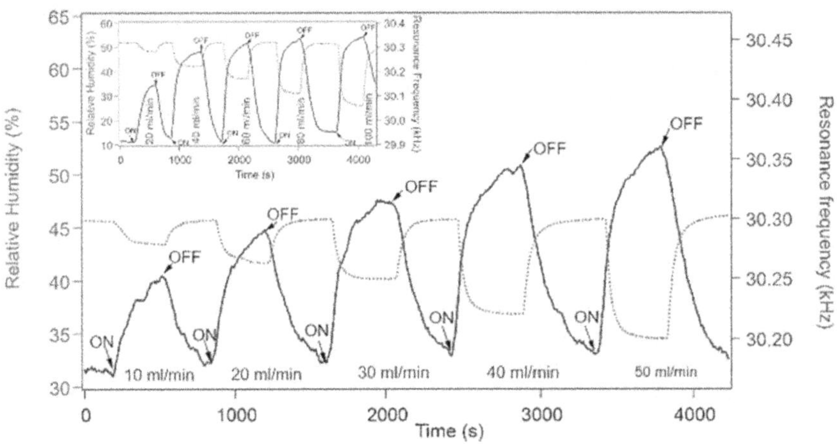

Fig. 2. Changes in the resonance frequency of AZNMC (blue solid line) and RH (red dotted line) due to CO with a 10-50 ml/min flow rate. Lines with both ended arrows indicate CO exposure. Repeated experiment in inset figure shows same tendency of resonant frequency change for CO exposure with various flow rates. The AZNMC used here was sputtered by Al for 105 s.

The resonant frequency shift of AZNMC and the change of a RH due to a CO flow rate of 10-50 ml/min is shown in Fig. 2. At initial condition, the resonant frequency of AZNMC and RH were 30.172 kHz and 45.7 %, respectively. Upon inserting CO, the RH abruptly decreased indicating the desorption of water vapor and then adsorption of CO on AZNR surface. Since the weight molecules of the inserted CO is lighter than the amount of the released water vapors, the resonant frequency of AZNMC increases.

The RH suddenly increased once the gas was switched off (without air flashing) and the adsorbed CO was replaced by water vapor resulting in the decrease of the resonant frequency. It can be also seen that the resonant frequency increased as the gas flow rate (and, hence, the density) increased. It is noted that a saturation of the frequency was reached after time of 350 s to 400 s, depending on the gas flow rate. Fast saturation was obtained for higher flow rate. Such observations agreed with our previous experimental results [3], which indicated that the molecular interface between ZnO and H_2O plays a key role on CO absorption. In addition, inset figure in Fig. 2 shows a different experiment data for CO exposure with various flow rates, indicating that repeated measurements have been performed. From the resonant frequency shift in Fig. 2 (for example, $\Delta f \sim 300$ Hz), the mass change Δm is estimated to be 2.3 pg, and the sensor sensitivity (dm/df) is obtained to be 7.4 fg/Hz.

4. Conclusions

A high-sensitivity CO-sensor was realized using the MC coated by Al-doped ZnO nanorods structures at room temperature. A significant improvement in the gas sensitivity on the order of femtograms was observed. Such sensitivity was enhanced by sputtered Al bonding in the ZnO crystal structure. These results might lead to the development of novel highly sensitive CO gas detectors in the future.

Acknowledgements

This paper was partly funded by an incentive research grant from Ministry of Research, Technology and Higher Education of the Republic of Indonesia which the authors gratefully acknowledge. The authors thank Prof. Y. Hatanaka and Prof. J. Moon for helps of sputtering processes of ZnO and Al.

References

[1]. M. Li, H. X. Tang, M. L. Roukes, Ultra-sensitive NEMS-based cantilevers for sensing, scanned probe and very high-frequency applications, *Nature Nanotechnology*, Vol. 2, Issue 2, 2007, pp. 114-120.

[2]. A. Kooser, R. L. Gunter, W. D. Delinger, T. L. Porter, M. P. Eastman, Gas sensing using embedded piezoresistive microcantilever sensors, *Sensors Actuators B*, Vol. 99, Issues 2-3, 2004, pp. 474-479.

[3]. L. Aprilia, *et al.*, CO Gas-induced resonance frequency shift of ZnO-functionalized microcantilever in humid air, *Journal of Nanomaterials*, Vol. 2017, 2017, 4824607.

[4] N. Kilinc, *et al.*, Fabrication of 1D ZnO nanostructures on MEMS cantilever for VOC sensor application, *Sensors Actuators B*, Vol. 202, 2014, pp. 357-364.

(109)

First Step Towards a Magnetic Polymer-based Ethanol Gas Sensor

B. López-Walle [1], **J. Bravo-Guerra** [1], **S. Gómez-Flores** [1] **and E. Reyes-Melo** [1]
[1] Universidad Autónoma de Nuevo León, Programa Doctoral en Ingeniería de Materiales-FIME, Av. Universidad s/n, Ciudad Universitaria, 66450, San Nicolás de los Garza, Nuevo Léon, Mexico
Tel.: + 528113404000
E-mail: beatriz.lopezwl@uanl.edu.mx

Summary: This paper presents the synthesis, morphological and optical characterizations, and the application as ethanol gas sensor of two magnetic polymer films constituted by a polyvinyl butyral (PVB) polymer matrix and iron oxide (Fe_2O_3) nanoparticles synthetized *in situ* into the matrix. Two nominal contents of iron (Fe) were tested (11 % and 14 %). The analyzed properties are function of this nominal concentration of Fe, as well as the sensing performance. To evaluate this last, an opto-chemical sensor has been developed, being the PVB/Fe_2O_3 sensing material with 14 % of Fe nominal content which obtains the best results. Future works will deal with the improvement of sensing capacities of the magnetic polymer films in order to design a new generation of magnetic polymer-based gas sensors.

Keywords: Polyvinyl butyral, Iron oxide nanoparticles, Opto-chemical sensor, Ethanol gas sensor, Superparamagnetism, Transmittance, Refractive index.

1. Introduction

Magnetic polymers are composed of iron oxide nanoparticles embedded in a polymeric matrix. Their physical and chemical properties can generally be tuned by the relative concentration of the nanoparticles and its spatial dispersion into the polymeric matrix. This paper presents a magnetic polymer film (0.01 mm in thickness) composed of iron oxide nanoparticles (Fe_2O_3) synthetized *in situ* in a polyvinyl butyral (PVB) matrix. Two nominal contents of iron (Fe) were tested (11 % and 14 %).

PVB films and Fe_2O_3 (nanostructures or films), independently, have been largely used for sensing different gases and vapors [1-4]. The objective here is to validate the utility of the proposed films, which combine the sensing properties of the pure components, for ethanol gas sensing applications at room temperature (298.15 K). To operate at room temperature is a crucial factor for this class of sensors [4, 5]. So, this work would be the first step towards a new kind of magnetic polymer-based gas sensors.

The next section describes the synthesis, morphological and optical characterizations of the PVB/Fe_2O_3 sensing material. The nominal content of iron is modified to evaluate their properties in function of the nanoparticles content. Then, we introduce the opto-chemical sensor designed especially to profit the observed properties of the proposed films. Finally, conclusions are depicted.

2. Sensing Material

Sensing material is a magnetic polymer film consisting of a polyvinyl butyral (PVB) matrix with iron oxide (Fe_2O_3) nanoparticles. Two nominal concentrations of Fe (11 % and 14 %) are evaluated.

2.1. Synthesis

The synthesis process, based on [6], is performed in two stages. In the first one, precursor hybrid materials (PVB/Fe^{2+}) were obtained considering two different nominal concentrations of iron (11 % and 14 %). In the second one, a thermochemical treatment was applied to the precursor materials to synthetize *in situ* iron oxide nanoparticles into the polymeric matrix. Thickness of the resulting PVB/Fe_2O_3 films was about 0.01 mm.

2.2. Morphological Characterization

Size and morphology of iron oxide nanoparticles into the magnetic polymer films were examined by using high resolution transmission electron microscopy (HRTEM, FEI Titan 80-300). The samples were prepared by dispersing the powdered film in water using an ultrasonic bath and placing an aliquot of the dispersion onto a cooper TEM grid.

HRTEM images of Fe_2O_3 nanoparticles into PVB/Fe_2O_3 films with 11 % and 14 % Fe nominal concentration are shown in Figs. 1(a) and 1(b), respectively. These results confirm the synthesis *in situ* of the nanoparticles into the polymeric matrix. Although nanoparticles are in contact with each other, they can still be identified as single particles because overwhelming number of them are single crystalline and they are oriented in such a way that lattice planes can be resolved. The degree of homogeneous dispersion of nanoparticles decrease with increasing Fe concentration. The amorphous character of the PVB matrix and the crystallographic planes of the Fe_2O_3 nanoparticles can be clearly distinguished in Fig. 1(b). Fe_2O_3 nanoparticles have almost a sphere-like shape with diameters under 10 nm, as reported in previous works [6, 7]. This size is characteristic of superparamagnetic behavior [8].

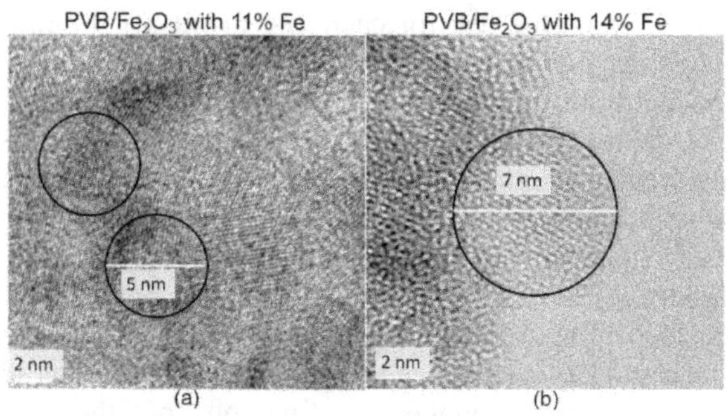

Fig. 1. HRTEM images (scale bar: 2 nm) of PVB/Fe_2O_3 films with Fe nominal concentrations of: (a) 11 %, (b) 14 %.

2.3. Optical Characterization

The quantitative measurements of optical constants may help to tailor the properties of the magnetic polymer films for their use in optical and optoelectronic components and devices. The transmittance spectra and the refractive index of these films were recorded in the wavelength range 200-1000 nm using a Perkin Elmer Lambda 35 UV-Vis spectrophotometer.

Transmittance spectra of PVB/Fe_2O_3 films with a nominal concentration of 11 % and 14 % of Fe are given in Fig. 2. It is evident that both films absorb in the UV region (200-400 nm), then rise rapidly when wavelength increases, and transmit in the near IR region (700-1000 nm). These results combine the absorbance/transmission properties of their constituent materials, PVB and Fe_2O_3. PVB films transmit from 300 nm until IR region, whereas Fe_2O_3 nanoparticles absorb under 400 nm and transmit from 700 nm [9-11]. This last also explains that the PVB/Fe_2O_3 with a Fe nominal concentration of 14 % shifts lightly to the right (higher wavelengths) and its transmittance decreases.

Values of the refractive index were calculated using the theory of reflectivity of light and the methodology proposed by [12]. Refractive index *versus* wavelength of both PVB/Fe_2O_3 films (11 % and 14 % of nominal concentration of Fe) is plotted in Fig. 3. At lower wavelengths (200-400 nm) the refractive index diminishes, followed by a rather increment that seems to stabilize after 600 nm. This is almost the typical response of PVB films [9]. Nevertheless, iron oxide nanoparticles have an impact in the near IR region [11, 13]. Because of this, the refractive index of the PVB/Fe_2O_3 film with 14 % of nominal concentration of Fe is bigger than this with 11 % of Fe. The reduction of the transmittance and the increment of the refractive index with increasing Fe nominal concentration could be explained by the denser distribution of the iron oxide nanoparticles in the polymeric matrix, which causes a bigger dispersion of the electromagnetic radiation [12, 13].

According to these results, the transmittance and refractive index of the magnetic polymer films could be controlled by Fe content. Additionally, properties such as near infrared transmission and a high refractive index (bigger than 1.5) make PVB/Fe_2O_3 films suitable for optical gas sensing applications [5].

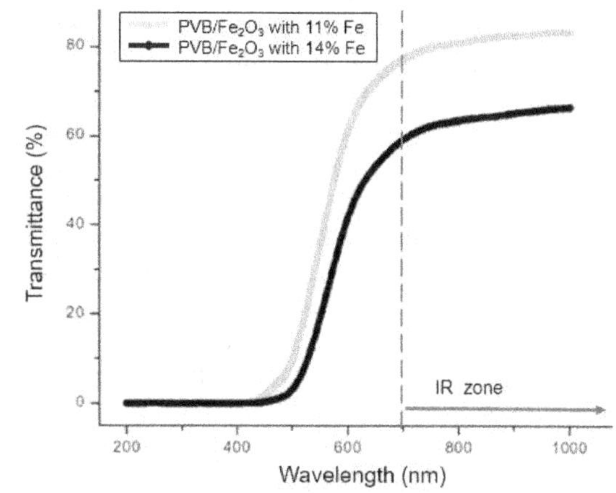

Fig. 2. Transmittance spectra of PVB/Fe_2O_3 films varying the Fe nominal concentration (11 % and 14 %).

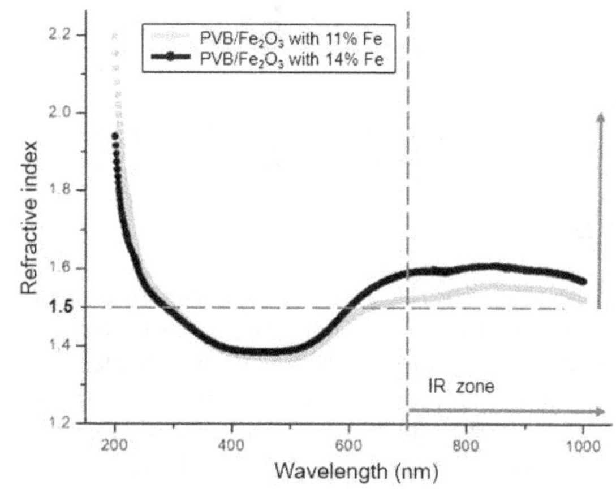

Fig. 3. Refractive index *versus* wavelength for PVB/Fe_2O_3 films varying the Fe nominal concentration (11 % and 14 %).

3. Gas Sensor

Both PVB/Fe_2O_3 magnetic polymers films (11 % and 14 % of Fe nominal content) were tested as the active material of an ethanol gas sensor working at room temperature (298.15 K). A pure PVB film is also tested as reference.

3.1. Principle

The principle of this sensor is a combination of those of chemical and optical gas sensors.

An optical sensor basically consists of an emitter and a detector. The first one is charged of the emission of an electromagnetic radiation whose wavelength varies in the presence of the analyte. This variation is perceived by the detector which transforms it in an electrical signal [5, 14, 15]. In a chemical sensor, the presence of the analyte in the environment is assessed by the chemical interaction of the sensing material and this one, producing an electrical response which can be measured [16].

The proposed opto-chemical sensor is a combination of both kinds of sensors to improve their characteristics. A schematic representation is depicted in Fig. 4. In a closed chamber with the presence of the ethanol gas (A), an infrared radiation is sent by the emitter (B). The outgoing radiation passes through the PVB/Fe_2O_3 sensing material (C) and this last chemically interacts with the ethanol modifying its transmittance in the near IR region. Actually, an oxidation occurs between iron oxide nanoparticles and the ethanol gas yielding a significant decrease on transmission [17, 18]. The resulting response is measured by the transducer (D) which converts optical signal in electrical signal.

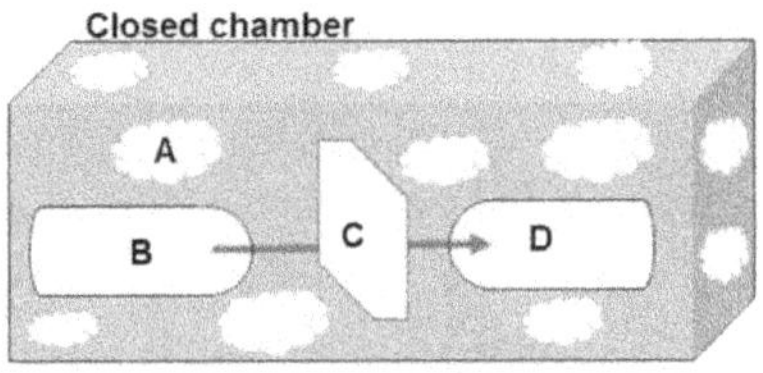

Fig. 4. Representation of the proposed opto-chemical gas sensor. A: analyte, B: emitter, C: sensing material, D: transducer.

3.2. Description

The setup of the polymer-based ethanol gas sensor is shown in Fig. 5. An IR emitting LED and an IR photodetector, both working at a wavelength of 940 nm, constitute the optical elements. Sensing material, PVB/Fe_2O_3 film ($15 \times 10 \times 0.01$ mm^3), is placed between them. Liquid ethanol is placed on two aluminum dishes (6 ml for each one); it is released to the environment by natural convection at room temperature (298.15 K). All these components are enclosed in an acrylic box. Voltage is supplied by a BK PRECISION 1192 power source and the electrical response (voltage) is measured by a Fluke 116 multimeter.

Both PVB/Fe_2O_3 films varying the nominal concentration of Fe (11 % and 14 %) were tested. Results are shown in next section.

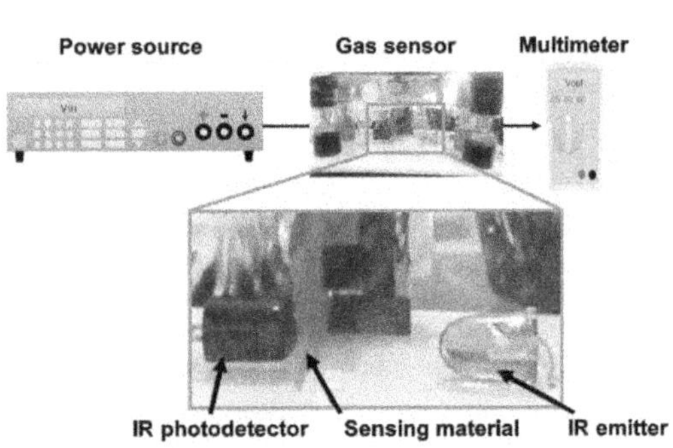

Fig. 5. Experimental setup of the polymer-based ethanol gas sensor. Inset: zoomed view of the gas sensor elements.

3.2. Results

Three tests were carried out at room temperature (298.15 K). One test for each one of the PVB/Fe_2O_3 magnetic polymers with a different iron nominal concentration (11 % and 14 %) and one more for a pure PVB film. This last is used as reference. All the tests required 15 minutes for stabilization. After this period, voltage and gas concentration have been measured every six minutes. The relation between the electrical response (voltage) and the ethanol concentration is plotted in Fig. 6. Straight lines were obtained by least squares. They emphasize the proportional variation of the measured voltage in function of the gas concentration. PVB/Fe_2O_3 film with 14 % of Fe nominal concentration shows higher values of response than the others. This could be presumed as a better sensitivity. Assuming this, PVB film is less sensitive. Consequently, sensing capacity depends on iron oxide nanoparticles concentration.

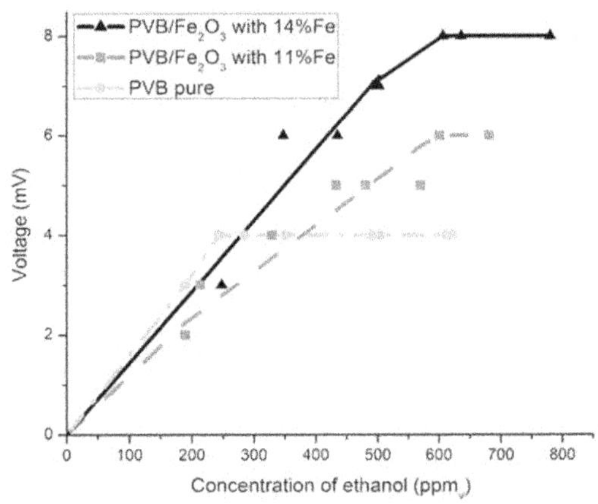

Fig. 6. Measured voltage *versus* ethanol concentration for pure PVB film and both PVB/Fe_2O_3 films with different Fe nominal concentrations (11 % and 14 %), operating at room temperature (298.15 K).

Saturation occurs when almost all the sensing material, and particularly the nanoparticles, have interacted with ethanol [18, 19]. For this reason, the first sensing material to reach saturation is pure PVB, at 243 ppm$_v$ of ethanol concentration. Then, PVB/Fe_2O_3 with 11 % of Fe nominal concentration satures at 601 ppm$_v$, and PVB/Fe_2O_3 with 14 % of Fe nominal concentration attains saturation at 606 ppm$_v$. Even if the difference between the saturation points of PVB/Fe_2O_3 films are almost the same, they depend on iron oxide nanoparticles concentration.

4. Conclusions

PVB/Fe_2O_3 polymer magnetic films were successfully obtained by synthetizing *in situ* Fe_2O_3 nanoparticles in a polymeric matrix of PVB with an iron nominal concentration of 11 % and 14 %. Morphological characterization revealed a superparamagnetic behavior of the spherical-like nanoparticles since their diameter is lower than 10 nm.

Optical analysis exhibits the potential of the synthetized PVB/Fe_2O_3 films to be applied as optical gas sensors because both films transmit in the near infrared region and have a refractive index bigger than 1.5 in this region. Morphological and optical properties depend on the content and spatial dispersion of iron nanoparticles.

Considering the mentioned results, we evaluate the ethanol gas sensing performance of the PVB/Fe_2O_3 films with a proper opto-chemical gas sensor adapted to their characteristics and able to work at room temperature. The response (voltage) and sensitivity of the sensor also depends on the content of iron, showing better results the magnetic polymer with 14 % of Fe nominal concentration.

Results demonstrate that PVB/Fe_2O_3 films are potential candidates to be used as ethanol gas sensors with concentrations under 700 ppm at 278.15 K. During next works we will improve the opto-chemical sensor to test other alcohols in different concentrations and temperatures.

Acknowledgements

Research supported by the Consejo Nacional de Ciencia y Tecnología (CONACYT, Mexico) via project no. 284810 and scholarship no. 444820, and the PAICYT-UANL project no. IT1027-19. Authors thanks Drs. Alejantro Torres-Castro and Rocio Alfaro-Cruz for their help during morphological and optical characterizations.

References

[1]. T. Kinkeldei, C. Zysset, N. Münzenrieder, L. Petti, G. Tröster, In tube integrated electronic nose system on a flexible polymer substrate, *Sensors (Basel)*, Vol. 12, Issue 10, 2012, pp. 13681-13693.

[2]. A. Kaushik, R. Kumar, S. K. Arya, M. Nair, B. D. Malhotra, S. Bhansali, Organic-inorganic hybrid nanocomposite-based gas sensors for environmental monitoring, *Chemical Reviews*, Vol. 115, Issue 11, 2015, pp. 4571-4606.

[3]. S. Zolghadr, K. Khojier, S. Kimiagar, Study of sensitivity and selectivity of α-Fe_2O_3 thin films for different toxic gases and alcohols, *Materials Science in Semiconductor Processing*, Vol. 54, 2016, pp. 6-13.

[4]. Y. Li, H. Zhao, H. Ban, M. Yang, Composites of Fe_2O_3 nanosheets with polyaniline: preparation, gas sensing properties and sensing mechanism, *Sensors and Actuators B: Chemical*, Vol. 245, 2017, pp. 34-43.

[5]. X. Liu, S. Cheng, H. Liu, S. Hu, D. Zhang, H. Ning, A survey on gas sensing technology, *Sensors (Basel)*, Vol. 12, Issue 7, 2012, pp. 9635-9665.

[6]. M. E. Reyes-Melo, J. G. Puente-Córdova, B. López-Walle, A. Torres-Castro, A. F. García-Loera, Synthesis and characterization of a polymeric magnetic hybrid material composed of iron oxide nanoparticles and polyvinyl butyral, *Ingeniería Investigación y Tecnología*, Vol. 19, Issue 1, 2018, pp. 113-123.

[7]. B. López-Walle, J. Romo-Rico, J. Puente-Córdova, M. Reyes-Melo, Synthesis and mechanical characterization of magnetic hybrid materials with PVB as polymeric matrix for micro-actuation applications, *MRS Proceedings*, Vol. 1708, 2014, Mrss14-1708-vv04-37.

[8]. A. S. Teja, P-Y. Koh, Synthesis, properties, and applications of magnetic iron oxide nanoparticles, *Progress in crystal growth and characterization of materials*, Vol. 55, 2009, pp. 22-45.

[9]. M. Mohammadian-Kohol, M. Asgari, H. Shakur, A detailed investigation of the gamma-ray radiation effects on the optical properties of polyvinyl butyral film, *Optik*, Vol. 127, 2016, pp. 7459-7468.

[10]. Z. N. Kayani, E. S. Khan, F. Saleemi, S. Riaz, S. Naseem, Growth and characterization of iron oxide nanocrystalline thin films via sol-gel dip coating method, *IEEE Transactions on Magnetics*, Vol. 50, Issue 8, 2014, 2200304.

[11]. M. Al-Kuhaili, M. Saleem, S. M. A. Durrani, Optical properties of iron oxide (α-Fe_2O_3) thin films deposited by the reactive evaporation of iron, *Journal of Alloys and Compounds*, Vol. 521, 2012, pp. 178-182.

[12]. S. Mahendia, A. K. Tomar, P. K. Goyal, S. Kumar, Tuning of refractive index of poly(vinyl alcohol): effect of embedding Cu and Ag nanoparticles, *Journal of Applied Physics*, Vol. 113, Issue 7, 2013, 07103.

[13]. G. Singh, D. Jalandhara, K. Yadav, Effect of grain size on optical properties of iron oxide nanoparticles, *AIP Conference Proceedings*, Vol. 1728, 2016, 020409.

[14]. A. Lobnik, M. Turel, S. K. Urek, Optical chemical sensors: design and applications, in Advances in Chemical Sensors (W. Wang, Ed.), *InTech*, 2012, pp. 3-28.

[15]. Z. Yunusa, M. N. Hamidon, A. Kaiser, Z. Awang, Gas sensor: review, *Sensors & Transducers*, Vol. 168, Issue 4, April 2014, pp. 61-75.

[16]. M. C. Petty, Molecular Electronics: From Principles to Practice, *John Wiley & Sons*, 2007.

[17]. J. Than, J. Chen, K. Liu, X. Huang, Synthesis of porous α-Fe_2O_3 microrods via in situ decomposition of FeC_2O_4 precursor for ultra-fast responding and recovering ethanol gas sensor, *Sensors and Actuators B: Chemical*, Vol. 230, 2016, pp. 46-53.

[18]. G. Eranna, Metal Oxide Nanostructures as Gas Sensing Devices, *CRC Press*, 2012.

[19]. A. Dey, Semiconductor metal oxide gas sensor: Review, *Materials Science and Engineering: B*, Vol. 229, 2018, pp. 206-217.

(110)

A New Amperometric Enzyme-based Biosensor for Specific Determination of 17β-estradiol

K. Spychalska [1], **J. Cabaj** [1]

[1] Wrocław University of Science and Technology, Faculty of Chemistry, Wybrzeże Wyspiańskiego 27, 50-370 Wrocław, Poland

Tel.: + 48 71 320 24 39,

E-mail: kamila.drzozga@pwr.edu.pl

Summary: A new sensing system for electrochemical 17β-estradiol detection was designed. The miniature biosensor was constructed by immobilizing horseradish peroxidase on an electroactive layer coated platinum electrode. The electroactive film was built of conducting polymer – poly(4,7-bis(5-(3,4-ethylenedioxythiophene)thiophen-2-yl)benzothiadiazole). The detection system is based on the idea, that 17β-estradiol (E2) and pyrocatechol (H_2Q) are co-substrates for horseradish peroxidase (HRP) enzyme. HRP, which does not react with E2, in the presence of H_2O_2, catalyses the oxidation of H_2Q to *o*-benzoquinone. With the optimized conditions, the analytical performance demonstrated a convenient degree of sensitivity, selectivity in a broad linear range - 0.1 – 200 µM with limit of detection found as 105 nM. Such constructed system was successfully applied for hormone determination in presence of potentially interfering compounds (ascorbic acid, estriol, estrone, uric acid and cholesterol).

Keywords: Biosensor, Conducting polymers, Electrochemistry, Estradiol, Horseradish peroxidase.

1. Introduction

The development of civilization deepens the requirements regarding diagnostics, but also the issues of environmental protection. One of the biggest threats related to environmental protection is the presence of endocrine disrupting compounds (EDCs) in water reservoirs. EDCs are factors that interfere with the production, release, transport, metabolism, action or elimination of natural hormones in the body. They are substances that disturb the hormonal balance of organisms, therefore should be treated as dangerous species. In addition to the negative impact on human hormone metabolism, EDCs also have a negative impact on aquatic organisms. These components accumulate in bottom sediments and are absorbed by aquatic organisms. The most common result is the feminization of male individuals and reproductive disorders [1].

Biosensors belong to the group of chemical sensors, which in addition to the transducer element have also a biologically active element that is responsible for the receptor-analyte recognition. Each biosensor or chemical sensor consists of two basic elements: a receptor element, which is responsible for detection of the analyte being tested and transducer element, which transforms the received signal from the receptor element into a measurable value.

2. Methodology

In developed electrochemical biosensor for detection of 17β-estradiol, the semiconductive polymer has been used as a matrix for immobilization of enzyme – horseradish peroxidase. The mechanism is based on the fact, that 17β-estradiol (E2) and pyrocatechol (H_2Q) are co-substrates for horseradish peroxidase (HRP). HRP, which does not react with E2, in the presence of H_2O_2, catalyses the oxidation of H_2Q to *o*-benzoquinone. The electropolymerization process was carried out by chronoamperometry method, in this way a thin layer of polymer film was obtained on the surface of the electrode. Then, by physical adsorption and covalent cross-linking, the electrode was modified with the enzyme. Such modified electrode was introduced into a measurement system, which consisted a measuring cell and a system of three electrodes. The researches were carried out using cyclic voltammetry (CV) and differential – pulse voltammetry (DPV). The changes in the current are inversely proportional to the concentration of E2 in tested samples.

3. Results

Electrochemical character of 17β-estradiol was investigated employing CV and DPV method in oxygen – saturated conditions. Fig. 1 presents the superimposed cyclic voltamperogramms for a bare platinum electrode (Pt), a poly(4,7-bis(5-(3,4-ethylenedioxythiophene)thiophen-2-yl)benzothiadiazole) (Pol) modified Pt electrode and the Pt/Pol/HRP complex, recorded in the presence of a 10 µM solution of 17β-estradiol. In the Pt/Pol/HRP system, characteristic oxidation and reduction peaks are present indicating the catalytic reaction taking place, which confirmed the validity of the enzyme application in the proposed system.

The calibration curve obtained under optimal conditions showed a linear response in a given concentration range, i.e. 0.1 - 200 μM. The calibration curve was constructed as the dependence of peak net current obtained at various concentrations of 17*β*-estradiol on the concentration of 17*β*-estradiol the sample analyzed (Fig. 2). The detection limit (LOD), calculated as the concentration of 17*β*-estradiol, was found as 105 nM.

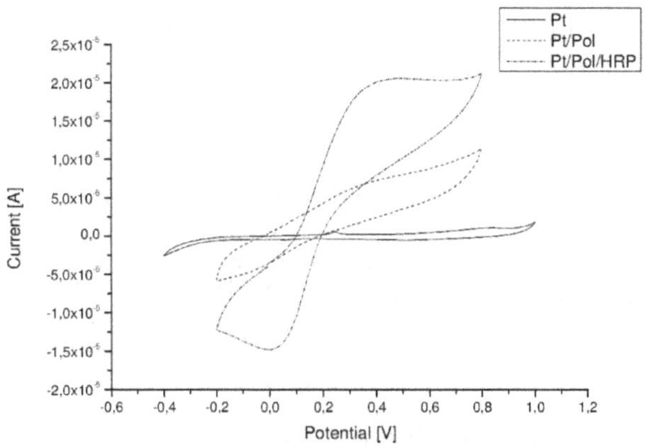

Fig. 1. Representative CV-scans of the bare Pt electrode, Pt electrode modified with polymer and Pt/Pol/HRP system in the presence of 17*β*-estradiol (10 μM) under applied potential range (potential range -0.4 – 1.0 V, scan rate 100 mV/s, *via* Ag/AgCl).

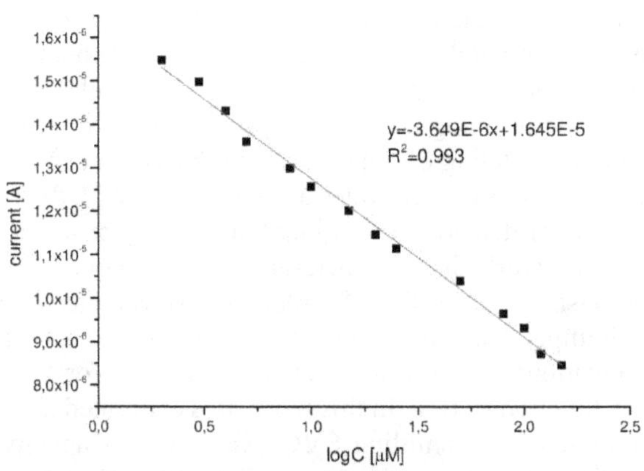

Fig. 2. The linear relationship between ΔI and the logarithm of E2 concentration (linear range from 0.1 to 200 μM, LOD = 105 nM).

The selectivity and specificity of a biosensor are very important parameters in the context of correct design bio-devices. In this study, five different substances (ascorbic acid (AA), estriol (E3), estrone (E1), uric acid (UA) and cholesterol (CH)) have been investigated as a potential interference compounds during the detection of 17*β*-estradiol. Presented results (Fig. 3) confirm negligible impact on the selectivity of fabricated setup and prove that checking interferences does not interrupt the prominence of proposed 17*β*-estradiol test.

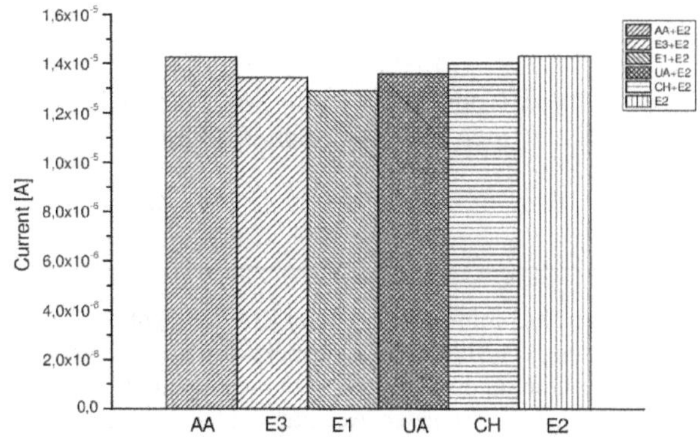

Fig. 3. Effect of interfering substances (50 μM) on the determination of E2 level.

4. Conclusions

The study showed a method for producing the electrochemical horseradish peroxidase biosensor for the detection of 17β-estradiol. The platinum electrode was modified with a previously synthesized conducting polymer, which was the appropriate matrix for immobilization of horseradish peroxidase. The enzyme was attached to the substrate and retained high catalytic activity. Electrochemical measurements have shown that the designed system works with high accuracy and reliability. The detection limit for 17*β*-estradiol was set to 105 nM. No interference was found with AA, E3, E1, UA, CH. The main advantages of the presented sensor are: sensitivity, precision, linearity and simplicity of the structure.

Acknowledgements

The authors gratefully acknowledge for the financial support to Wroclaw University of Science and Technology (0401/0145/18).

References

[1]. W. Nawrot, K. Drzozga, S. Baluta, J. Cabaj, K. Malecha, A fluorescent biosensors for detection vital body fluids' agents, *Sensors,* Vol. 18, Issue 8, 2018, 2357.

(111)

Lung Volumes and Ventilation Assessment Using Multifrequency Bioimpedance Meter

Vladimir Leonov [1] and Huynsoo Ha [2]
[1] Imec, Connected Health Solutions Belgium Department, Kapeldreef 75, 3001 Leuven, Belgium
[2] Imec The Netherlands, High Tech Campus 31, 5656 AE Eindhoven, The Netherlands
Tel.: +32 16 288 367
E-mail: vladimir.leonov@imec.com

Summary: The bioimpedance meter ROSE recently developed at Imec was used in this work for assessing several parameters of human lung. The bioimpedance method was evaluated on slow and forced vital lung capacity (SVC and FVC, respectively), its tidal volume, and maximum voluntary ventilation (MVV). The impedance variation caused by breathing was calibrated using a reference spirometer. Accuracy of the bioimpedance meter coincides with that of off-the-shelf spirometer, which could make lung function testing and its constant monitoring, e.g., during rehabilitation or sleep, much easier and non-obtrusive to the patient. The device also eliminates risk of transmission of illnesses from one patient to another, which still could happen through a spirometer despite its cleaning and disinfection, especially in case of at-home devices with improper sterilization.

Keywords: Lung function testing, Bioimpedance, Multifrequency, Spirometry, Vital capacity, Maximum voluntary ventilation.

1. Introduction

Spirometers are widely used in medicine [1] for assessing respiratory and heart health, as well as neuromuscular deceases and general health. They provide lung function testing and disease diagnosis. For details, see [2]. There is a minor potential risk of pulmonary cross-infection between test subjects through spirometers [2-5], although no large-scale studies have been performed "because of the lack of enthusiasm for this type of study" [4]. Transfer of saliva and other biological matter to spirometer occurs during expiration. Some of droplets and aerosolized particles are deposited inside the device into which the subject is conducting the forced expiration. Protective hygiene measures whilst performing spirometry testing include using a new disposable mouthpiece for each subject, a delay of at least 5 minutes between subjects to allow settling of previously aerosolized particles in the measuring device. Daily visual inspection of spirometer is recommended. If there is *visible* contamination on the device, it must be cleaned and disinfected. To reduce the survival rate of micro-organisms, disinfection is conducted, e.g., weekly by immersing the turbine and air ducts of the spirometer in a disinfecting solution. (The optimal frequency of sterilization has not yet been established.) One must stay alert: these guidelines still leave a space for a very low-probability transfer of illness to a healthy person through a spirometer, because complete disinfection of all its parts after every patient is considered unacceptably time-consuming routine, while barrier filters are not used because of cost constraints, except immunocopromised patient cases [4]. There are commercially available portable electronic spirometers [6] for at-home lung function monitoring, which also represent potential cross-contamination risk to their users because of improper disinfection.

Bioimpedance meters track tiny variations of thorax impedance caused by such body activity as heart beating and breath. Therefore, there are ongoing efforts for developing a method called electrical impedance tomography for visualization of finite element modeling of lung resistivity based on currents between multiple electrodes around the chest [7, 8] and thereby for detection of regional pulmonary problems.

This work has a different target, namely, to demonstrate feasibility of replacing a spirometer with a bioimpedance meter in ambulatory and wearable applications. Such a device was developed for basic lung function testing. In this work, it is calibrated using a manoeuvre resembling SVC (Slow Vital Capacity) manoeuvre with a spirometer. Then, the other basic lung tests, i.e., FVC (Forced Vital Capacity) and MVV (Maximum Voluntary Ventilation) were studied.

2. Bioimpedance Meter ROSE and Spirometry

Previous studies of lung impedance at Imec were conducted using wearable systems based on custom ASICs, at 10-200 kHz [9], and on a multiparameter signal acquisition S oC MUSEIC [10] used at 20 kHz. In this work, a new multifrequency system of Imec is used for accurate lung impedance monitoring. It was designed for 1 – 1024 kHz range and has five bioimpedance readout channels and programmable multiplexer that allows configurable connection to 16 electrodes [11]. In this work, a single channel of this system is used within 2 – 128 kHz. The ROSE board, Fig. 1, features an instrumentation amplifier, with pre-demodulation to intermediate frequencies in front of it. This essentially reduces power consumption, which is important for wearable applications. The current generator supplies square-wave current, where $1/f$ noise manifests itself and usually dominates in the bioimpedance recordings. To improve system noise characteristics, the dynamic element matching was proposed and applied in [11], which enabled $1/f$ noise reduction and outstanding performance characteristics of the bioimpedance meter ROSE. Table 1 shows selectable gain for combination of the measured body resistance and maximum applied current for reaching the lowest system noise. E.g., the system noise was 0.4 mΩ/Hz$^{1/2}$ at 100 μA and a body resistance of 2 Ω, and 1.6 mΩ/Hz$^{1/2}$ at 40 μA and a body resistance of 100 Ω.

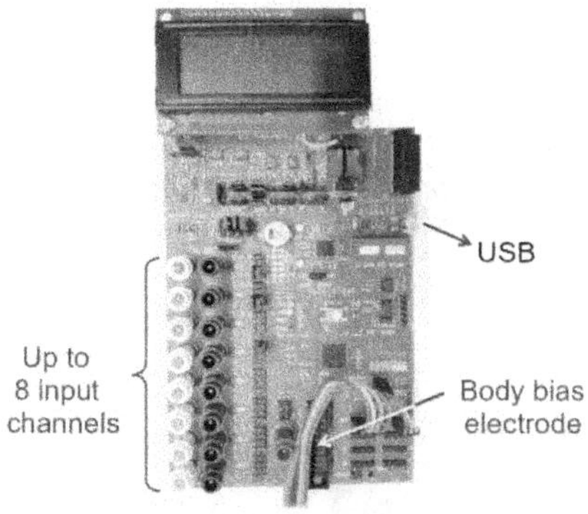

Fig. 1. The bioimpedance meter ROSE with an advanced ASIC S oC fabricated in 0.18 μm TSMC technology.

Accuracy of the device was measured on RC-model of human body. At a gain of 70 and 40 μA current typically used in this work, the measurement accuracy (±3SD) was ±80 mΩ at 2 kHz and ±20 mΩ at 128 kHz.

Input impedance of more than 500 MΩ results in immunity of ROSE to parasitic influence of varying electrode impedance, both current and voltage electrodes, on measurement results typically observed in commercially available impedance meters. A common-mode rejection ratio of 87 dB enables clear recordings. More details can be found in [11].

Table 1. Maximum current and gain selection depending on the measured human body resistance R.

I (μA)	5	10	20	40	100
R (Ω)	**Maximum gain**				
2	220	220	220	220	220
20	220	220	220	220	70
40	220	220	220	70	25
100	220	220	70	70	–
200	220	70	70	–	–
500	70	25	–	–	–
1000	25	–	–	–	–

The measurements were conducted together with the reference pc-based spirometer microQuark (COSMED). The subject tests were conducted according to directions on every standard manoeuvre provided by the spirometer's software, while four electrodes on the body were used for simultaneous impedance recording using ROSE (for four-probe measurements with separate current and voltage electrodes to exclude highly variable skin resistance from the measurement results).

Meditrace 200 electrodes were used for making electrical contact with the skin. Several electrode configurations were used in this study, Fig. 2.

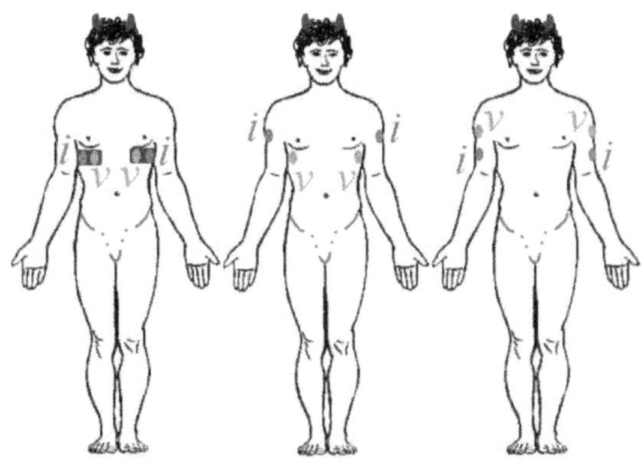

Fig. 2. Location of electrodes suitable for the tidal wave and SVC measurement: left shows using coupled electrodes, at fixed distance of 5 cm between the centers of *i*- and *v*-electrodes, and about 25 cm between potential electrodes; central drawing illustrates current electrodes on arms, at about 10 cm from the shoulders; right picture shows lung impedance measurement with all electrodes on arms, at about 10 cm between adjacent *i*- and *v*-electrodes.

The electrode location on arms, although showing less noise, was not finally adopted because at highest volumes in FVC and MVV tests, the resistance dependence on volume was weak and approaching saturation. It is clear that while approaching the maximum volume, the lung mainly expands downwards. Therefore, if *v*-electrodes are located too high, like in Fig. 2, right, they are not sensitive enough to lung expansion near maximum of inhaled volume.

3. Device Calibration and Measurement of Vital Lung Capacity

Calibration of the bioimpedance system ROSE was performed using a non-standard SVC manoeuvre in spirometry. The manoeuvre was performed starting holding breath for a couple of seconds with muscular relaxation on chest, i.e., while simply stop breathing at the end of exhalation, on tidal wave minimum. Then, breathing restarted at the same pace as the subject was breathing before, but with every inhalation, the amount of inhaled air increased. The exhaling was also conducted with gradual increase of depth to reach both minimum and maximum of air volume in the lung. Fig. 3 shows example of such calibration manoeuvre recorded by the spirometer. Plotting resistance change dR versus wave amplitude measured by spirometer gave calibration of ROSE shown in Fig. 4. As a rule of thumb, the lung resistance variation is 1 Ω per 1 liter of inhaled or exhaled air.

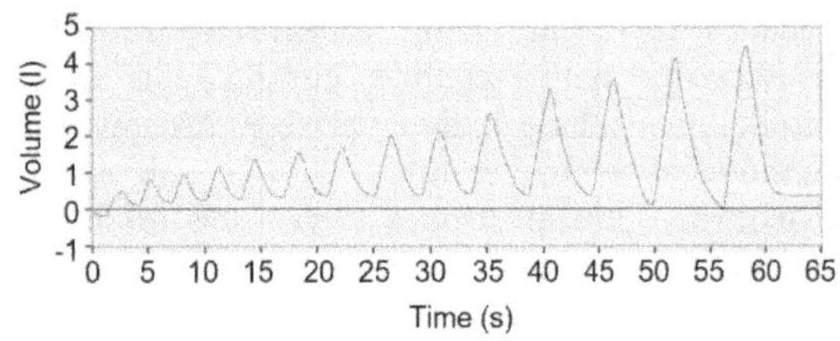

Fig. 3. Calibration manoeuvre recorded by a spirometer. Respiration was performed with gradual increase of inhaled air volume, ending with SVC-like vigorous effort (between the minimum at about 56 s and maximum at about 58 s).

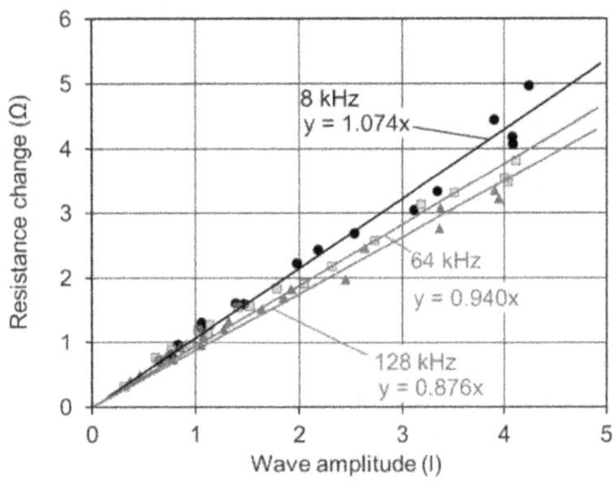

Fig. 4. Calibration of lung resistance measured by ROSE at a gain of 70 and a current of 40 μA versus amplitude measured using a spirometer at SVC-like calibration manoeuvre with gradual increase of inhaled air volume as shown in Fig. 3. Location of electrodes is shown in Fig. 5.

4. Measurement of Maximum Voluntary Ventilation

Unlike the tidal volume and SVC, which can accurately be measured with any reasonable electrode locations on chest, on arms or forearms, MVV required minimization of cable swinging during rapid and energetic chest muscle contractions and the chest shaking. (This problem should not present in a wearable patch-like device, where will be no long free hanging cables.) The cable noise appeared only at MVV of more than 110 l/min, i.e., in the area of interest. Therefore, the electrode configuration shown in Fig. 5 was finally adopted in our tests for the most accurate SVC and MVV. This electrode configuration was found the best in terms of cable swinging minimization, although it adds more body resistance to the net lung resistance. Therefore, the maximum both current and gain slightly decrease, Table 1, in respect to configurations shown in Fig. 2, which causes slightly worse system noise characteristics.

A typical MVV manoeuvre is shown in Fig. 6 for both devices, the spirometer and ROSE. Small deviations of the wave recorded by ROSE from the wave measured using the spirometer are explained by minor modulation of impedance by heart beats seen before and after MVV manoeuvre. The heart beats can be filtered out in the future for increased accuracy. However, they could also be used for simultaneous monitoring of heart rate with the same electrodes.

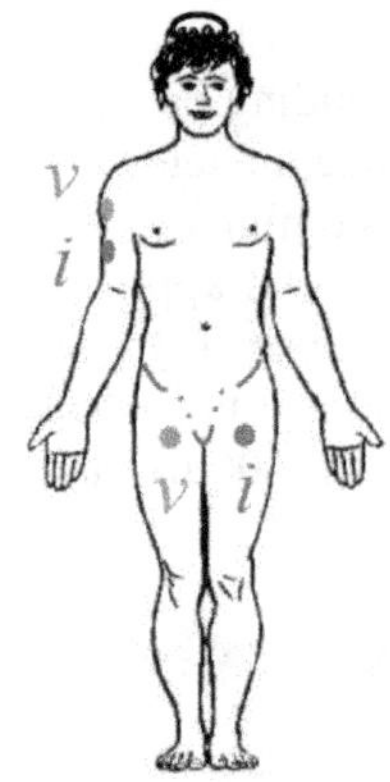

Fig. 5. Electrode configuration chosen for FVC and MVV manoeuvre for motion artifact minimization.

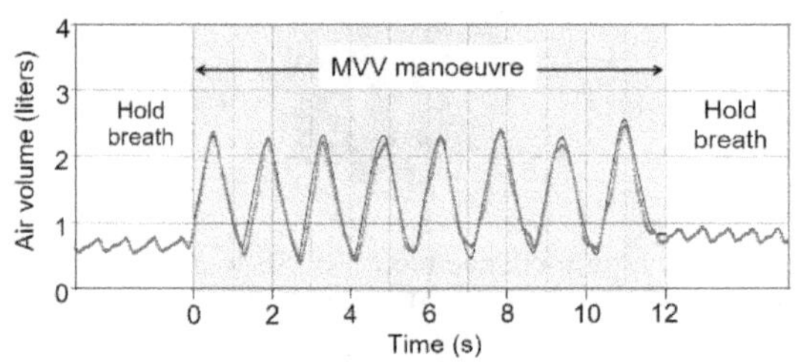

Fig. 6. MVV manoeuvre measured by ROSE (red points, 64 samples per second) and the spirometer (black wave). The aligned resistance is measured at 8 kHz. The breath is hold before and after the manoeuvre for clarity of this figure to the reader and visibility of heart beats.

The MVV manoeuvre was performed with different efforts to observe dependence of resistance on ventilation measured using a spirometer. Fig. 7 shows a sum of resistance change in all peaks during MVV manoeuvre (see Fig. 6), one experimental point per manoeuvre. Using regression equation shown in Fig. 7, the MVV measurements have been performed. Fig. 8 corresponds to vigorous effort and shows that accuracy of ROSE is the same as that of the reference spirometer, so one could call the ROSE sensor a bioimpedance spirometer.

5. Discussion

The measurement in this work have been performed on sitting people, because a reference spirometer was always used. However, as it is clear from the principle of bioimpedance method, there is no need to specify the posture of the subject. Furthermore, it can be used on a walking person, e.g., on an elderly person, or during rehabilitation period. Appropriate algorithms for artifact reduction are usually used in such case in our devices.

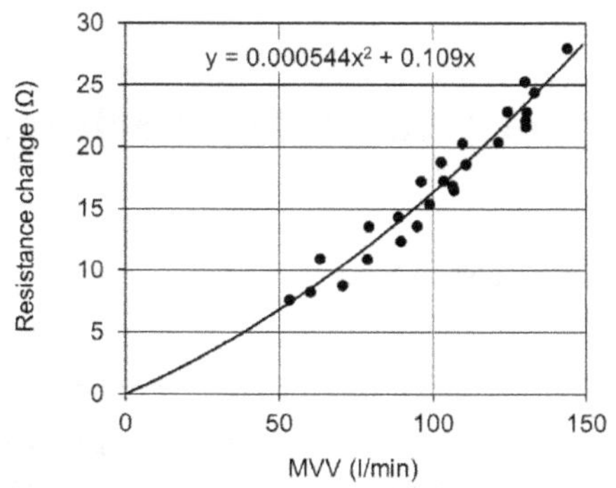

Fig. 7. Sum of resistance variations at 8 kHz in all breaths during MVV manoeuvre with different efforts, from medium to maximum, plotted against ventilation volume measured using the reference spirometer.

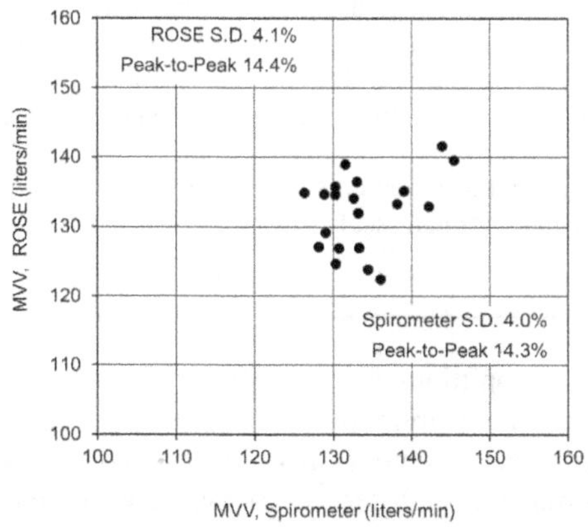

Fig. 8. MVV measured by ROSE and the spirometer on 5 days. On each day, the resistance was measured at 2, 8, 32, and 128 kHz, i.e., four points per day, 20 points in total.

6. Conclusion

The new bioimpedance device ROSE developed for lung function testing enabled accurate measurement of basic pulmonary tests, SVC, FVC, tidal wave and MVV. The other typical lung characteristics such as PEF (Peak Expiratory Flow), FEV_1 (volume exhaled during the first second), etc., can easily be extracted from the results in the same way as it is done in a spirometer. The bioimpedance calibration performed on one person will not suit to another person because of different body impedance.

Therefore, initial calibration with a spirometer for every patient is needed. After that, the device can for a long time be used safely compared to spirometers that require periodic inspection, proper cleaning and disinfection. The device similar to the one used in this work has recently been miniaturized to the size of a patch or a pendant [12].

Acknowledgements

The authors are thankful to N. Van Helleputte and B. Grundlehner (Imec) for discussions and support of the project.

References

[1]. G. J. Gibson, Spirometry: Then and now, *Breathe*, Vol. 1, Issue 3, 2005, pp. 207-216.

[2]. M. R. Miller, J. Hankinson, V. Brusasco, et al., Standardisation of spirometry, *European Respiratory Journal*, Vol. 26, Issue 2, 2005, pp. 319-338.

[3]. M. R. Miller, R. Crapo, J. Hankinson, et al., General considerations for lung function testing, *European Respiratory Journal*, Vol. 26, 2005, pp. 153-161.

[4]. A. H. Kendrick, D. P. Johns, J. P. Leeming, Infection control of lung function equipment: A practical approach, *Respiratory Medicine*, Vol. 97, 2003, pp. 1163-1179.

[5]. S. A. Rasam, K. K. Apte, S. S. Salvi, Infection control in the pulmonary function test laboratory, *Lung India*, Vol. 32, Issue 4, 2015, pp. 359-366.

[6]. D. M. Carpenter, R. Jurdi, C. A. Roberts, et al., A review of portable electronic spirometers: Implications for asthma self-management, *Current Allergy and Asthma Reports*, Vol. 18, Issue 10, 2018, 53.

[7]. I. Frerichs, M. B. P. Amato, A. H. van Kaam, et al., Chest electrical impedance tomography examination, data analysis, terminology, clinical use and recommendations: consensus statement of the Translational EIT development study group (Review), *Thorax*, Vol. 72, 2017, pp. 83-93.

[8]. B. Lobo, C. Hermosa, A. Abella, F. Gordo, Electrical impedance tomography (Review), *Annals of Translational Medicine*, Vol. 6, Issue 2, 2018, 26.

[9]. S. Lee, C. Agell, S. Polito, et al., Demonstration paper: A low-power and convenient bio-impedance monitor, and its application to respiration monitoring, in *Proceedings of the 4th Conference on Wireless Health*, Baltimore, MD, 1-3 November 2013.

[10]. H. De Cannière, C. J. P. Smeets, P. Vandervoort, et al., A wearable bioimpedance device for respiratory monitoring. in *Proceedings of the Biomedica Summit*, Genk, Belgium, 2-3 June 2015.

[11]. H. Ha, M. Konijnenburg, B. Lukita, et al., A bio-impedance readout IC with frequency sweeping from 1k-to-1MHz for electrical impedance tomography, in *Proceedings of the Symposium on VLCI Circuits*, Kyoto, Japan, 5-8 June 2017, pp.174-175.

[12]. S. Lee, B. Grundlehner, R. Garcia van der Westen, et al., Nightingale V2: Low-power compact-sized multi-sensor platform for wearable health monitoring, in *Proceedings of the 41st IEEE Conference of the Engineering in Medicine and Biology Society (EMBC'19)*, Berlin, Germany, 23-27 July 2019, pp. 1290-1293.

(114)

Cognitive Measurements: a New Paradigm of Measurement Science

Valery B. Tarassov
Bauman Moscow State Technical University, 5 2nd Baumanskaya st., Moscow, 105005, Russia
E-mail: vbulbov@yahoo.com

Summary: A new paradigm of measurement science – Cognitive Measurements – is discussed. An approach of Russian metrological school is presented; it considers measurements as cognitive processes together with judgments, reasoning and computations. Basic principles of organizing cognitive measurements are proposed. The problem of creating «cognitive» and «understanding» sensors is formulated; such sensors should be able to automatically make transitions from data to information and from information to knowledge and meta-knowledge. The definition of cognitive sensor is suggested by giving its three main capacities: measurement of quantitative values of object parameters, their pragmatic evaluation and visualization of obtained data. The logical-algebraic method of interpreting measurement results is proposed, both logical pragmatics and pragmatic logics together with product lattices are envisaged. Basic logical scales and logical worlds to represent measurement pragmatics are constructed. Colored Hasse diagrams for three-valued and four-valued logical pragmatics, based of color metaphor for many-valued logics, are introduced. Some interpretations and logical-graphic pragmatics for minimal-valued cognitive sensors are developed. Logical analysis of measurement results obtained by sensor networks is performed too.

Keywords: Intelligent measurements, Cognition, understanding, Cognitive measurements, Cognitive sensor, Cognitive measurement-information system, Granule, Information granulation, Basic logical scale, Logical world, Multi-valued pragmatics, Color metaphor in measurement.

1. Introduction

In the late XX – early XXI century, there is a significant impact of new computing concepts on the methodology of measurement science (see also [1]). In 1994 L.A. Zadeh published a seminal paper on *Soft Computing* [2], and three years after by analogy S. V. Prokopchina and A. N. Averkin introduced the idea of *Soft Measurements* [3]. The concepts of *Information Granulation* by L. Zadeh [4] and *Granular Computing* by T. Lin [5] initiated the arrival of *Granular Measurements* age [6, 7]. At last, the development of *Cognitive Computers* [8], *Cognitive Informatics* [9] and Cognitive Computing (CC) [10], especially CC based on IBM Watson [11], led to the formation of *Cognitive Measurements* [12-14] using *Cognitive Sensors* [15]. Recent advent of both cloud computing and fog computing allows us to predict the occurrence of cloud measurements and fog measurements.

The objective of the paper consists in presenting a new trend of measurement science called Cognitive Measurement with Cognitive Sensors and Cognitive Measurement-Information Systems. It is organized in the following way. In Section 2 two important sources of innovations concerning measurement science are discussed. On the one hand, we deal with the transition to a new standard for Measurement Uncertainty in accordance to the Guide to the Expression of Uncertainty in Measurement (GUM). On the other hand, the well-known NBICS convergence concept supposes the fusion of nano-, bio-, info-, cogno-, sociotechnologies, where the cogno- component remains a bottleneck. So a family of new cognitive micro-sciences is being generated, including cognitive semiotics, cognitive linguistics, cognitive graphics, etc. In our opinion, cognitive metrology focused on cognitive measurements, performed with using cognitive measurement-information systems, should take its place in such series. So basic cognitive aspects of measurements are highlighted, and the problem of constructing «understanding» sensors and sensor networks is stated. Four main principles of cognitive measurement are formulated.

Section 3 begins with the definition of cognitive sensor together with the presentation of both logical pragmatics and pragmatic logics approaches enhanced by cognitive graphics to interpret quantitative measurements results. Both basic logical worlds and their visualization by colored Hasse diagrams are introduced. They explain the idea of automatic transition to adequate coarse-grained pragmatics while interpreting measurement results. Basic types of interpreting sensors are specified by applying appropriate logical pragmatics – Vasiliev's sensor, Kleene's sensor, Belnap's sensor, and so on.

In Conclusion the formation of cognitive sensor networks is related to complex, multidimensional logical worlds. It is pointed out that product lattices and multi-lattices (in particular, bilattices) can be used in order to process and interpret multi-sensor information.

2. Why a New Paradigm in Measurement Science?

Since the early 1990s, there has been rapidly growing a steady trend of extending and rethinking the foundations of metrology [16-19], that means an inclusion into this field in addition to traditional probabilistic methods the techniques of interval analysis, fuzzy set theory, linguistic and fuzzy variables [6, 20], development of intelligent measurement technologies and tools [16, 21].

One of the main catalysts of this trend was the new ISO standard [22], adopted in 1990's, where the usual term «Measurement Error» was replaced by a broader and multifaceted concept «Measurement Uncertainty» (see [23)]. Two types of measurement uncertainty were specified: a traditional stochastic type *A* uncertainty and a non-stochastic type *B* uncertainty (more exactly, a family of non-stochastic variants of uncertainty). So a need for a generalized vision of different uncertainties appeared, and, consequently, the development of new approaches to both measurement organization and formal representation of measurement results became crucial

In this context the variants of considering measurement as a generic cognitive technique [24] and a basic knowledge acquisition tool for next generation intelligent systems are of special concern [14].

2.1. Measurements as Cognitive Process

The representation of measurement as a cognitive process has a long tradition in Russian metrology. To justify this assertion, it is worth reminding the concept of measurement in «Technical Encyclopedia» by P. A. Florensky: «Measurement is a basic cognitive process in science and technology, by means of which an unknown value is quantitatively compared with another one, considered as homogeneous and well-known» [25, p.777-778].

In [26] measurement is defined as experimentation intended to obtain truth judgments about the object under study. It means that the phase of interpreting and analyzing measurement results plays a leading part in measurement process; it is worth using logical approaches and evaluations for this purpose.

Cognitive aspects of measurement are tightly connected with such operations as comparison, matching, classification, comprehension, explanation, interpretation. It is worth stressing that cognitive structures of measurements enable the chain of transitions «Data – Information – Knowledge – Meta-Knowledge» (Fig. 1) [27]. In other words, generalized measurement structures suppose the unity of such procedures as collecting data from primary measurements, discovering useful information in these data, forming beliefs and performing reasoning with respecting norms in intelligent systems.

The implementation of these transitions into artificial cognitive systems requires the modeling of multi-leveled understanding mechanisms (Fig. 1): a) understanding the links between objects and events; b) understanding behavior patterns and examples; c) understanding situation as a whole. Here purposeful actions suppose understanding both links and norms, and the whole activity performance needs situation understanding.

Understanding measurement results is inseparably linked with its evaluation from some positions, on the basis of some pattern, principle, standard, norm, etc.

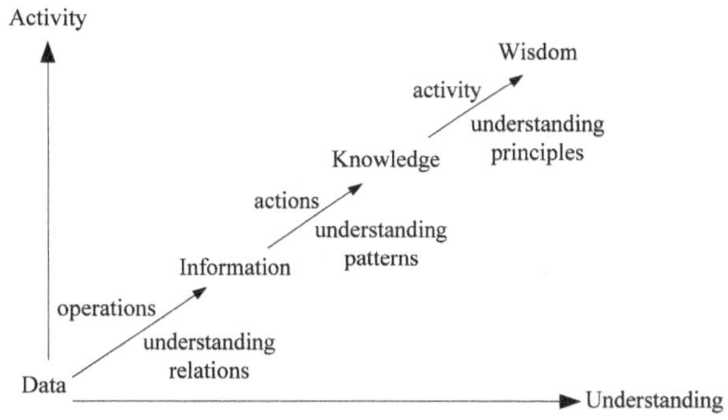

Fig. 1. Representation of Transitions «Data –Information – Knowledge – Meta-Knowledge» in «Understanding – Activities» co-ordinates.

According to M. Bakhtin, understanding without evaluation is impossible, because we cannot divide them: they are simultaneous and constitute a holistic act [28]. Two basic operations making understanding possible are: 1) search for a norm (evaluation standard) and its representation; 2) justification of its application in a given situation.

2.2. Main Principles of Cognitive Measurements

The term «Cognitive Measurement» was proposed in 2010 by S. V. Prokopchina [12] as a kind of intelligent measurements, where a metrologically certified knowledge was acquired by using Bayesian technologies.

In this paper the concept of cognitive measurements means the construction and use of cognitive sensors as special devices for hierarchical granulation of measurement data. It is based on the following principles: 1) principle of open nature of measurement as cognitive process; 2) principle of the unity of measurements, evaluations and reasoning; 3) principle of synthesizing truth theories while interpreting measurement results; 4) principle of measurement information granulation.

An open nature of measurements means their dependence on selected purpose, environment, instrumentation, and finally, their inclusion into some larger processes such as monitoring or decision-making. So measurement is not an isolated process. For instance, the integrated system of monitoring complex object (the bridge) is based on measurements of both object and its environment parameters. Moreover, it is also responsible for object state diagnostics, prognosis of future state changes, making recommendations.

For example, let us measure the wind speed on the bridge by anemometer. We obtain the result as 20-21 m/s, and also receive the information on expected deterioration of weather and increase in wind speed. The problem is not an exact measurement of wind speed itself, but a joint evaluation of received data, specification of a trend and making conclusion, e.g. «traffic on the bridge is limited».

The principle of the unity of measurements, evaluations and reasoning (Fig. 2) is tightly related to the principle of open measurement.

Let us measure two quantities – the wind speed by anemometer and the bridge structure deformation by strain gauge. The measurement results can be interpreted by fuzzy linguistic labels «The wind speed is rather large» and «The magnitude of the bridge deformation is close to the maximum permissible».

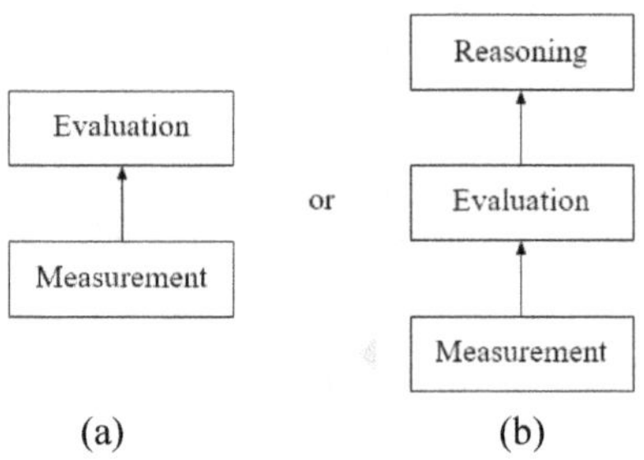

Fig. 2. Two Ways of Constructing Cognitive Sensors: a) Principle of the unity of measurements and evaluations; b) Principle of the unity of measurements, evaluations and reasoning.

Thus, we can build a fuzzy production rule: «IF the wind speed is rather large AND the magnitude of bridge deformation is close to the maximum permissible, THEN impose restrictions 3500 T on the mass of trains moving on the bridge».

In accordance with the principle of the unity of measurements, evaluations and reasoning, any measurement is viewed as a cognitive process associated with the formation of granular information structure represented by two leveled hierarchy [15]. On the lower level ordinary measurements are performed by using families of sensors; their indications give fine-grained information. On the higher level these indications are mapped into a pragmatic scale of aggregated evaluations, providing with coarse-grained, non-numerical information (obtained values of measured parameter are «normal», «almost normal», «not normal», «far from the norm» and so on). Typical examples of building both rather simple homogeneous and complex heterogeneous granules are depicted in Fig. 3.

After constructing these basic logical-linguistic values, their further analysis can be performed by using axiological («good-bad») or deontic («permitted»-«forbidden») scales.

Let us take the principle of synthesizing two truth theories as a basis for logical-pragmatic interpretation of measurement results. We deal with: 1) Tarski's correspondence truth theory; 2) Peirce's pragmatic truth theory (see [29]).

To make clear the principle of cognitive sensor operation as «understanding» sensor let us define two worlds (Fig. 4): a Measurement World W and a Norm World M. To enable a sensor-based understanding, it is necessary to show the links between two suggested interpretations of truth: 1)' Truth as a correspondence of logical values to measured results (*measured truth*); 2)' Truth as a useful norm (and falsity as a useful anti-norm) that allows take into consideration some designated or anti-designated logical values.

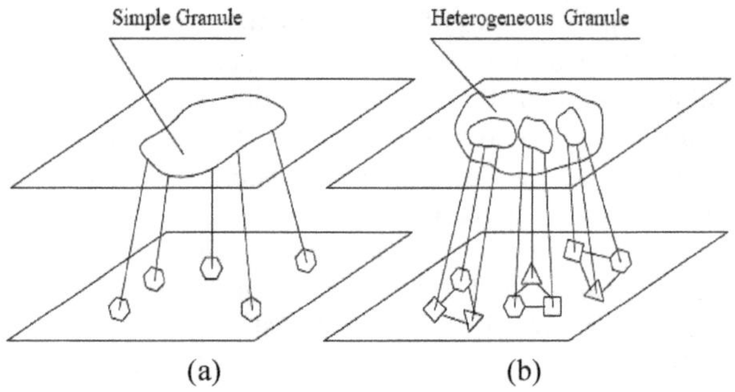

Fig. 3. Simple homogeneous and complex heterogeneous granules: a) Homogeneous data granulation issued from the same types of sensors; b) Heterogeneous data granulation from different types of sensors.

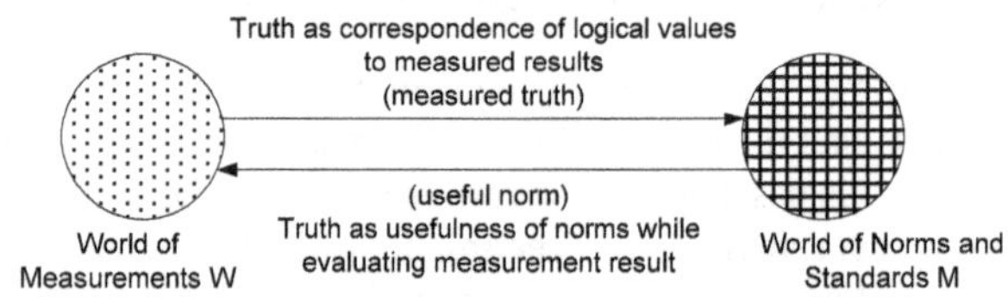

Fig. 4. Illustration of two dual truth concepts for cognitive sensors.

3. What is «Cognitive Sensor» and How to Construct It?

Let us define Cognitive Sensor as the sensor that measures quantitative values of object parameters, gives their pragmatic interpretation and visually represents obtained data. Models of understanding measurement results are tightly connected with measurement pragmatics, and, specifically, logical models of understanding – with logical pragmatics.

3.1. Logical Pragmatics and Pragmatic Logics for Cognitive Sensor

Among novel concepts and approaches in measurement science approaches such terms as semiotics, specifically cognitive semiotics, and pragmatics are worth mentioning [30-32].

In semiotics and information theory semantics reflects the link between a message and its author (sender), whereas pragmatics expresses the link between a message and its recipient, specifying the usefulness of this message for his activities. To differ from semantics, pragmatics always focuses on message user and message context. Therefore, the development of logical models for understanding requires a promotion of pragmatic ideas in logic.

Below we shall use the terms «pragmatic logic» and «logical pragmatics» (in particular, the notion of granular logical pragmatics). By *pragmatic logic* we mean any logical system based on value (utility). Typical examples of pragmatic logics are inductive logics, evaluation logics, preference logics, deontic logics, illocutive logics and so on.

The term «pragmatic logic» is not new. It was coined by K. Ajdukiewich in the mid-1970's. His pragmatic logic is mainly reduced to logical foundations of teaching, whereas A. A. Iwin [33], B. N. Pyatnitsyn, et al. proposed a wider vision of pragmatic logics as logical systems driven by values. These logical systems are based on axiological, utilitarian, deontic, epistemic definitions of basic logical concepts.

A granular logical pragmatics is related to pragmatic truth interpretations, for instance, Finn's «factual truth» [34], our «concerted truth» [35] and finally «measured truth» [36]. A well-known way of logical granulation consists in following Dunn's strategy with defining generalized truth values on the set of subsets 2^V or, more generally, Zadeh's approach by describing truth values as terms of linguistic variable TRUE, i.e. fuzzy variables.

We shall visually represent the pragmatics of measurement results in various logical worlds. Informally logical world can be viewed as a set of logical (truth, modal) values. Let us take as a basis of logical world a primitive bipolar (opposition) scale – the three-valued lattice $\{1, 0.5, 0\}$ or the isomorphic lattice $\{+1, 0, -1\}$. Different types of logical scales can be specified depending on the opposition strength between the poles of basic three-valued scale and the status of its neutral (mean) value.

In the case of strong opposition between poles a negative part of logical scale is a mirror reflection of positive oart, and these two areas are seen as mutually exclusive.

Let us take standard denotations for truth and falsity as logical poles: 1 = T, 0 = F. For a «gray» Pospelov's logical scale [37] the neutral value is a contradiction (both truth and falsity), where for any $p \in P$, $F(p) = 1 - T(p)$, but for a «black-white» scale the neutral point is ignorance (neither truth nor falsity).

A weakening of «truth–falsity» opposition opens the opportunity of independent coexistence of the pairs $(T(p), F(p))$. Furthermore, new links between elements of positive and negative areas can appear (for example, the links between strong positive and weak negative, weak positive and strong negative modal values).

A color metaphor is quite suitable both to represent values of logical lattices in Hasse diagrams and understand the results of measurements. The traffic lights pragmatics is natural for basic bipolar scale: T – «green color», F – «red color», B $=T \wedge F$ – contradiction as transitory value, i.e. «yellow color». These colors express the pragmatics of three-valued paraconsistent Vasiliev's logic. When 0.5 = N (total uncertainty or ignorance), this value can be colored by dark blue. Such a «green-blue-red» palette can give the pragmatics of Kleene's minimal paracomplete logic K_3.

An intensive black color of neutral value can be attributed to Bochvar's three-valued nonsense logic B_3 with its property of «nonsense infection» (by analogy with «black holes» in astronomy). The light-green color of neutral value in Heiting's minimal intuitionistic logic H_3 expresses the idea of a «half truth», and the pink color of neutral value viewed as a «half falsity» reflects the pragmatics of its dual – Brower's logic Br_3. Similarly, the dark-green and light-green colors can be used to illustrate respectively strong positive modalities (necessary, obligatory) and weak positive modalities (possible, permitted). The dark-red color can represent strong negative modalities (impossible, forbidden) and pink color – weak negative modalities (contingent, non-obligatory).

3.2. Logical-Graphical Pragmatics of Measurement Results for Cognitive Sensors

The structures of logical worlds for basic cognitive sensors are given in Fig. 5. A natural pragmatics for cognitive sensor begins with the concept of Vasiliev's sensor [6]. Here the obtained measurement results are classified into three pragmatic values (the traffic lights graphics – Fig. 5a): 1) T – «measured truth» («norm» – sensor indications are located in a «green zone»); 2) F – «measured falsity» («out of norm», «failure» – sensor indications are located in a «red zone»); 3) B – «measured ambiguity» («boundary situation» or «pre-failure» – sensor indications are located in a «yellow zone»).

By analogy, Kleene's sensor is specified by replacing the yellow pragmatic value with the dark-blue color: 4) N – total uncertainty or ignorance (sensor's resources are exhausted and there are no indications, or the sensor is sleeping).

By considering all four cases with $V = \{T,B,N,F\}$, we obtain the so-called Belnap's sensor (Fig. 5c).

It is worth stressing that generally we have not strict boundaries between green and yellow, yellow and red, red and dark-blue, dark-blue and green zones. It justifies the use of Pedrycz's shadow sets [38] to model such a situation.

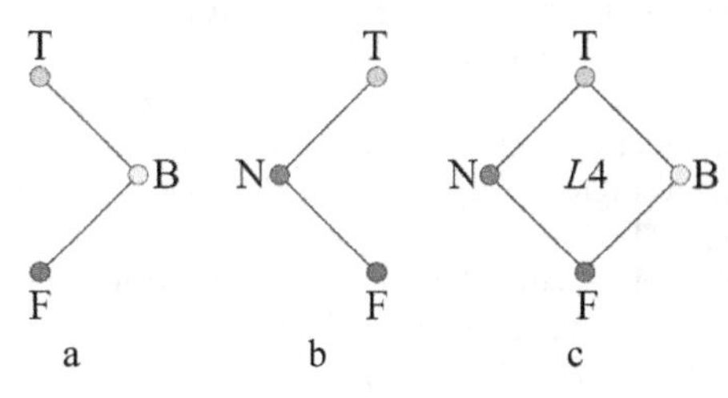

Fig. 5. Structures of colored logical worlds for Vasiliev's sensor (a), Kleene's sensor (b) and Belnap's sensor (c).

4. Conclusions

Further prospects of the suggested approach to building cognitive sensors are related to the development of cognitive sensor networks, including homogenous and heterogeneous sensors. In particular, logical pragmatics of a network with n Vasiliev's sensors is given by 3^n, with n Belnap's sensors – 4^n, where n is an integer, $n > 1$. A simple non-trivial network with two Belnap's sensors has the set of pragmatic values $4^2 = 16$, a network with three Belnap's sensors – $4^3 = 64$, and so on.

To construct logics and logical pragmatics of sensor networks we can use product logics in the form of product lattices and construct multi-lattices [39] of logical values. So a multi-dimensional logical world can be specified by n-dimensional logical multi-lattice (or simply logical n-lattice) $ML = \langle V, \leq_1, \ldots, \leq_n \rangle$, where V is a non-empty set of logical values, for instance, $V = V_1 \times \ldots \times V_n$, and $\leq_1, \ldots, \leq_n$ – order relations defined on V, such that $(V, \leq_1), \ldots, (V, \leq_n)$ form various lattices. As a special case, we have a bilattice of logical values $BL = \langle V, \leq_v, \leq_i, \neg_G \rangle$, where $V = V_1 \times V_2$ is a Cartesian product of two sets of pragmatic values, $\leq_v$ is a truth order, $\leq_i$ is an information order, and $\neg_G$ is a heterogeneous Ginsburg negation [40] (it can be also viewed as a semi-negation, semi-affirmation operation). Bilattices were extensively used to model non-monotonic reasoning and default reasoning (see [40]). No doubt, they represent a perspective tool to formalize cognitive measurements.

Acknowledgements

The research work is supported by Russian Foundation for Basic Research, grant No 17-07-01374, and the State Order (Goszadaniye) No 2.7918.2017

References

[1]. A. C. W. Finkelstein, L. Finkelstein, The convergence of measurement science and computer science: A scientific conversation, *Measurement*, Vol. 46, Issue 8, 2013, pp. 2946-2949.

[2]. L. A. Zadeh. Fuzzy logic, neural networks and soft computing, *Communications of the ACM*, Vol. 37, Issue 3, 1994, pp. 77-84.

[3]. A. Averkin, S. Prokopchina, Soft computing and measurements, *Intelligent Systems* (*MSU*), Vol. 2, Issues 1-4, 1997, pp. 93-114 (in Russian).

[4]. L. A. Zadeh, Toward a theory of fuzzy information granulation and its centrality in human reasoning and fuzzy logic, *Fuzzy Sets and Systems*, Vol. 90, 1997, pp. 111-127.

[5]. T. Y. Lin, Granular computing on binary relations I: Data mining and neighborhood system, in Rough *Sets in Knowledge Discovery* (A. Skowron, L. Polkowski, Eds.), *Physica-Verlag*, Heidelberg, 1998, pp. 107-140.

[6]. L. Reznik, Measurement theory and uncertainty in measurements: Application of interval analysis and fuzzy set methods, in Handbook of Granular Computing (W. Pedrycz, A. Skowron, V. Kreinovich, Eds.), *John Wiley and Sons Ltd,* 2008, pp. 517-532.

[7]. V. B. Tarassov, Granular measurement structures in ambient intelligence: Vasiliev's and Belnap's sensors and their communication models, *Informatsionno-Izmeritel'nye i Upravlyayuschiye Systemy* (*Information-Measurement and Control Systems*), Issue 5, 2013, pp. 65-74 (in Russian).

[8]. R. C. Shank, P. J. Childers, The Cognitive Computer: on Language, Learning and Artificial Intelligence, *Addison Wesley,* 1984.

[9]. Y. Wang, The theoretical framework of cognitive informatics, *International Journal of Cognitive Informatics and Natural Intelligence*, Vol. 1, Issue 1, 2007, pp. 1-27.

[10]. Y. Wang, On cognitive computing, *International Journal of Software Science and Computational Intelligence*, Vol. 1, Issue 3, 2009, pp. 1-15.

[11]. *D. Ferrucci, E. Brown, J. Chu-Carroll, et al.,* Building Watson: An overview of the DeepQA project, *AI Magazine*, Vol. 31, Issue 3, 2010, pp. 59-79.

[12]. S. V. Prokopchina, Cognitive measurements on the basis of bayesian intelligent technologies, in *Proceedings of the XIIIth International Conference on Soft Computing and Measurements (SCM'10)*, St. Petersburg, June 23-25, 2010, pp. 28-34 (in Russian).

[13]. V. B. Tarassov, Granular measurement structures in ambient intelligence: Vasiliev's and Belnap's sensors and their communication models, *Informatsionno-Izmeritel'nye i Upravlyayuschiye Systemy (Information-Measurement and Control Systems)*, 2013, Issue 5, pp. 65-74 (in Russian).

[14]. V. Tarassov, M. Svyatkina, B. Karabekov, Cognitive measurements for new generation knowledge acquisition systems, in *Proceedings of the 9th Conference on Application of Information and Communication Technologies (AICT'15)*, Rostov-on-Don, Russia, October 14-16, 2015, pp. 122-127.

[15]. M. N. Svyatkina, V. B. Tarassov, Logical-algebraic methods of constructing cognitive sensors, in *Proceedings of the VIth International Scientific and Technical Conference on Open Semantic Technologies for Intelligent Systems (OSTIS'16),* Minsk, BSUIR, February 18-20, 2016, pp. 331-348 (in Russian).

[16]. L. Finkelstein. Intelligent and knowledge-based instrumentation, an examination of basic concepts, *Measurement,* Vol. 14, 1994, pp. 23-30.

[17]. V. G. Knorring, G. N. Solopchenko, The theory of measurements as an independent field of knowledge: historical prerequisites, the entry into XXIst Century, *Izmeritelnaya Technika (Measurement Technology)*, Issue 5, 2003, pp. 12-16 (in Russian).

[18]. L. Mari, Epistemology of measurement, *Measurement*, Vol. 34, Issue 1, 2003, pp. 17-30.

[19]. G. Solopchenko, V. Kreinovich, L. Reznik, Development of mathematical structure of the modern measurement science, in Measurement Science: A Discussion (K. Kariya, L. Finkelstein, Eds.), *IOS Press*, Tokyo, 2000, pp. 23-36.

[20]. G. Mauris, L. Foulloy, A fuzzy symbolic approach to formalize sensory measurements: An application to a comfort sensor, *IEEE Transactions on Instrumentation and Measurement, Vol. 51, Issue 4, 2002, pp. 712-715.*

[21]. L. Reznik, What is Intelligent Measurement? Computing with Words in Information/Intelligent Systems, *Physica Verlag*, Heidelberg, 1999, pp. 78-89.

[22]. Guide to the Expression of Uncertainty in Measurement (GUM), *International Organization of Standardization*, Geneva, 1993.

[23]. L. Mari, G. Zingales, Uncertainty in measurement science, in Measurement Science: A Discussion (K. Kariya, L. Finkelstein, Eds.), *IOS Press*, Tokyo, 2000, pp.135-147.

[24]. V. G. Knorring, Gnoseotechnique – A technique of cognition, *Izmereniya, Control, Automatizatsiya (Measurement, Monitoring, Automatization)*, Issue 1-2, 1992, pp. 3-12 (in Russian).

[25]. P. A. Florensky, Measurement, Technical Encyclopedia, Vol. 8, *Soviet Encyclopedia*, Moscow 1929, pp. 777-781 (in Russian).

[26]. V. Y. Rozenberg, Introduction to the Accuracy Theory of Measuring Systems, *Sov. Radio*, Moscow, 1975 (in Russian).

[27]. R. Ackoff, From data to wisdom, *Journal of Applied Systems Analysis*. Vol. 16, 1989, pp. 3-9.

[28]. M. M. Bakhtin, Toward a Methodology of Humanitarian Sciences, Aesthetics of Verbal Creativity, *Iskusstvo*, Moscow, 1979, pp.361-373, 409-412 (in Russian).

[29]. Ch. S. Peirce, Issues of Pragmatism. Vol. II. Logical Foundations of the Theory of Signs, *Dover Publications*, Mineola, NY, 1958.

[30]. L. Finkelstein, Foundational problems of measurement science, in Measurement Science: A Discussion, (K. Kariya, L. Finkelstein, Eds.), *IOS Press*, Tokyo, 2000, pp. 13-22.

[31]. L. Mari, Principles of Semiotics as Related to Measurement, Lectures on Measurement Sciences, *Università Carlo Cattaneo*, 2005.

[32]. L. Mari, P. Carbone, D. Petri, Measurement fundamentals: A pragmatic view, *IEEE Transactions on Measurement and Instrumentation*, Vol. 61, Issue 8, 2012, pp. 2107-2115.

[33]. A. A. Iwin, Logic of Evaluations and Norms, *Prospect Publishers*, Moscow, 2016 (in Russian).

[34]. V. K. Finn. Standard and non-Standard Logics of Argumentation. Part I, Vol. 13, *Logical Investigations*, Moscow, Nauka, 2006. pp. 157-189 (in Russian).

[35]. V. B. Tarassov, Dia-logics and dia-semantics: A bilattice-based approach, in *Proceedings of the 5th World Congress and School on Universal Logics (UNILOG'15)*, Istanbul, Turkey, June 20-30, 2015, pp. 291-295.

[36]. V. B. Tarassov, M. N. Svyatkina, Logical pragmatics in cognitive measurements, in *Proceedings of the 6th Russian Scientific and Practical Conference on Fuzzy Systems and Soft Computing (SPIIRAS'14)*, Vol. 1, June 27-29, 2014, St. Petersburg, pp. 155-168 (in Russian).

[37]. D. A. Pospelov, The grey and/or the black-and-white, *Applied Ergonomics, Special Issue on Reflexive Processes*, Vol. 4, 1994, pp. 25-29.

[38]. W. Pedrycz, Shadow sets: Representing and processing fuzzy sets, *IEEE Transactions on Systems, Man and Cybernetics (Part B),* Vol. 28, 1998, pp. 103-109.

[39]. Y. Shramko, H. Wansing, Truth and Falsehood: An Inquiry into Generalized Logical Values, *Springer-Verlag*, Berlin, 2012.

[40]. M. Ginsberg. Multi-valued logics: A uniform approach to reasoning in AI, *Computer Intelligence*, Vol. 4, 1988, pp. 256-316.

(116)

Magnetic Polymer Minirobots for Aquatic Operation: Design and First Characterizations

L. Miranda-Vázquez [1], **B. López-Walle** [1], **S. Zapata-Reyes** [1], **J. Cisneros-Hinojosa** [1] **and E. Reyes-Melo** [1]

[1] Universidad Autonoma de Nuevo Leon, Facultad de Ingenieria Mecanica y Electrica, Av. Universidad s/n, Ciudad Universitaria, 66450 San Nicolas de los Garza, Nuevo Leon, Mexico

Tel.: + 528113404000

E-mail: liliana.mirandavz@uanl.edu.mx

Summary: This paper presents two magnetic square-shaped minirobots able to operate in aqueous medium without contact of the driving signal. These minirobots (3×3 mm^2) are fabricated with two magnetic polymer films (thickness: ~0.04 mm) constituted by iron oxide nanoparticles (Fe_2O_3) in a polymeric matrix of either polyvinyl butyral (PVB) or ethylcellulose (EC). The Fe_2O_3 nanoparticles are synthetized *in situ* with a nominal concentration lower than 10 % of Fe. The magnetic properties were evaluated showing a superparamagnetic behavior of the nanoparticles into both films. So, an experimental setup was developed to drive the minirobots without contact in an aqueous medium. Their linear displacement and pushing force were evaluated. These minirobots have a precision of 96 % when they follow a specific path. Their maximum pushing force is 106 mN and 182 mN for EC/Fe_2O_3 and PVB/Fe_2O_3 minirobots, respectively. These capacities depends on minirobots mass. Further characterizations will be done in future works.

Keywords: Polyvinyl butyral, Ethylcellulose, Iron oxide nanoparticles, Superparamagnetism, Minirobots.

---

1. Introduction

Compounds with excellent magnetic response features can be obtained by combining magnetic materials with polymeric materials. Magnetic polymer films react in the presence of an external magnetic field which allows their control remotely from a relatively long distance [1, 2]. Control without contact is an important characteristic in a variety of actuating and sensing applications [3-6]. Mini and microrobots with magnetic actuation can be used in tiny and complex environments where safety is critical. For these reasons, they are frequently operated in fluidic media providing the additional advantage of reducing adhesion and friction between surfaces, making transportation and manipulation tasks easier at these scales. As magnetic interactions scales favorably to micro world [7], potential applications point to microfluidic, biomanipulation and microassembly systems [8-13]. In this sense, micromanipulation and microtransportation tools with multiple degrees-of-freedom motions, high dexterity and precision, and large range of displacement and force are needed [11, 13]. During previous works, the nonlinear dynamics of a magnetic polymer cantilever actuator have been experimentally characterized in dry conditions (air) [14]. This work presents two magnetic polymer films and their first characterizations as aquatic minirobots. These films are composed by either polyvinyl butyral (PVB) or ethylcellulose (EC), used as matrices, and iron oxide nanoparticles (Fe_2O_3) synthetized *in situ* with a nominal concetration lower than 10 % of Fe. We evaluate the ability of the minirobots to follow linear displacements defined by the driving magnetic field in aqueous medium, and their maximal pushing force.

Next section describes the magnetic polymer minirobots and the experimental setup that leads the movement of the driving magnetic field. Then, magnetic, displacement and force characaterizations are analysed. Finally, some conclusions are outlined. Future works will deal with the complete characterization of the minirobots, testing also others geometries.

2. Experimental

2.1. Magnetic Polymer Minirobots

The synthesis of the PVB/Fe_2O_3 and EC/Fe_2O_3 magnetic polymer films is based on [15] but, for the first time, each one of these films has a nominal concentration of Fe lower than 10 %. This synthesis method has two stages: in the first stage, hybrid precursor films are obtained; in the second stage, these hybrid precursor films are treated to synthesize the iron oxide nanoparticles *in situ* into the polymeric matrixes which results in the magnetic polymer films. Both stages are detailed below.

For the first stage, the polymeric matrix (PVB or EC) and the precursor salt $FeCl_2$ are dissolved, independently, in tetrahydrofuran (THF). Quantities and suppliers for each magnetic polymer film are described in Tables 1 and 2. PVB-THF, EC-THF and $FeCl_2$-THF solutions are stirred 1 hour on a hot plate at a temperature of 45 °C, 40 °C and 45 °C/40 °C depending on the polymer matrix used, respectively. Then, each polymeric solution (PVB-THF and EC-THF) is mixed with the precursor salt solution by 3 hours at the corresponding polymer temperature. The PVB-Fe(II) and EC-Fe(II) hybrid precursor films are obtained by tape casting with the mixture of each polymer solution at room temperature for 24 h to eliminate the solvent by a natural convection process.

Table 1. Quantities and suppliers of PVB/Fe_2O_3 film.

Material	Supplier	Quantity
PVB	Sigma Aldrich	2.06 g
THF	Fisher Scientific	21.00 ml
$FeCl_2 \cdot 4H_2O$	Fermont	1.14 g
THF	Fisher Scientific	11.00 ml

Table 2. Quantities and suppliers of EC/Fe_2O_3 film.

Material	Supplier	Quantity
EC	Sigma Aldrich	2.06 g
THF	Fisher Scientific	25.00 ml
$FeCl_2 \cdot 4H_2O$	Fermont	1.14 g
THF	Fisher Scientific	14.00 ml

The second stage involves a thermo-chemical treatment to sinthesize *in situ* the Fe_2O_3 nanoparticles into the polymeric matrixes. For this, PVB-Fe(II) and EC-Fe(II) precursor films are immersed, during 20 min and at 70 °C and 50 °C, respectively, in 40 ml of a 3.3 M NaOH aqueous solution. Right after, 40 ml of a H_2O_2 solution is added during 10 min at the same temperatures. Finally, the films are washed with distilled water and dried by natural convection, obtaining the PVB/Fe_2O_3 and EC/Fe_2O_3 magnetic polymer films.

Two square-shaped minirobots were fabricated with each one of the magnetic polymer films, PVB/Fe_2O_3 and EC/Fe_2O_3. Respective dimensions are described in Table 3. Both minirobots were designed to operate in an aqueous medium.

Table 3. Aquatic minirobots dimensions.

Film	Length (mm)	Width (mm)	Thickness (mm)
PVB/Fe_2O_3	3	3	0.043
EC/Fe_2O_3	3	3	0.036

The following section describes the experimental setup which lets the displacement of the external magnetic field to drive the aquatic minirobots.

2.2. Experimental Setup

A proper experimental setup was developed for the contact-free operation of the aquatic minirobots (Fig. 1). The aim of this setup is to lead the displacement of the driving magnetic field and to record the movements of the minirobots in aqueous medium. For this, a cartesian robot was fabricated. It carries a permanent magnet which generates the magnetic field that drives the minirobots through several paths indicated by a program developed for this purpose. Main components of this experimental setup are described in the next paragraphs.

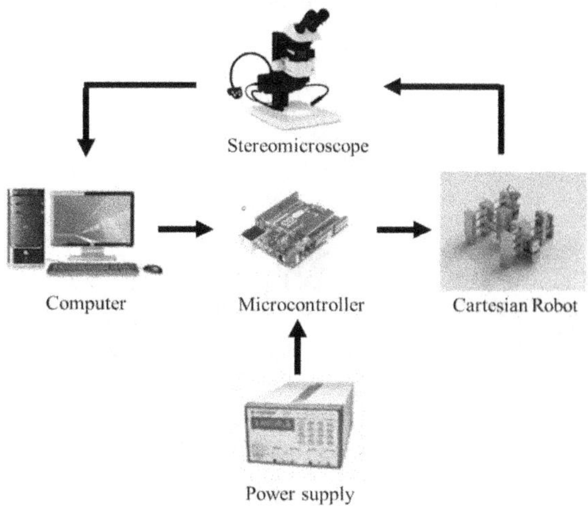

Fig. 1. Scheme of information flow of the experimental setup.

The hardware of the experimental setup is shown in Fig. 2. The cartesian robot, Fig. 2(a), is able to trace planar movements in x and y axes and their diagonals. Displacement tolerances are ±0.0038 mm, ±0.006 mm and ±0.81 mm, respectively. A gripper mounted on the cartesian robot, Fig. 2(b), holds a magnet which produces the minirobots' driving magnetic field. A M80 stereomicroscope with an IC80 HD camera (both Leica), Fig. 2(c), let record the minirobots movements. The minirobots are deposited on the free surface of a water container filled with 8 ml of water and where the working magnetic field is about 1 kOe. The water container ($7.6 \times 2.4 \times 5.2$ cm^3) and its support, Fig. 2(d), are placed between the magnet and the stereomicroscope.

All the supports (for the cartesian robot and the water container) are polylactic acid (PLA) pieces printed in a MakerBot replicator desktop 3D Printer MP05825. Cartesian robot supports are complemented with bearings, spindles, guides and couplings to allow movement. This last is achieved with four stepper motors (NEMA 17HS4401) controlled with a Pololu A4988 stepper motor driver and an Arduino Uno microcontroller board, Fig. 2(e). They are energized by a BK Precision 1770 power supply at 12 V (1.1 A).

A computer with LabView and Leica LAS software are used, respectively, to control the cartesian robot movement and to analyse the minirobots displacement images. User's interface, designed also in LabVIEW, lets control the cartesian robot and therefore the PVB/Fe_2O_3 and EC/Fe_2O_3 minirobots in two different modes: manual or automatic. In the first one, movents are directed by the user in real time. In the second one, the direction and the distance of the displacement (in mm) are first specified by the user, and then the robot execute them. Besides, an emergency stop to shut down the robot in case of any risk of collision is considered.

Next section presents the results concerning the magnetical analysis of the PVB/Fe_2O_3 and EC/Fe_2O_3 films and the displacement and force characterizations of their corresponding minirobots.

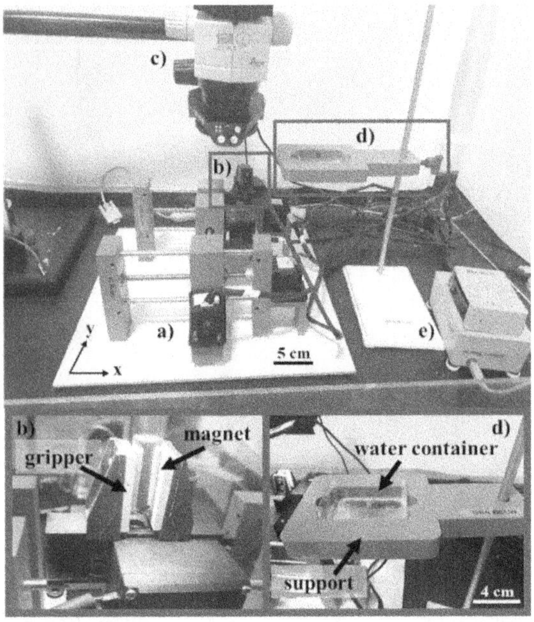

Fig. 2. Experimental setup: a) Cartesian robot; b) Gripper and magnet; c) Stereomicroscope; d) Support with water container; e) Stepper motor controls.

3. Results and Discussion

3.1. Magnetic Characterization

Magnetic properties were evaluated by a Vibrating Sample Magnetometer (MP SQUID VSM, Quantum design). Resulting curves in Fig. 3 expose a characteristic superparamagnetic behavior of the Fe_2O_3 nanoparticles into the polymer matrixes; so on, their diameter is lower than 20 nm [15]. Maximal measured magnetizations for embedded nanoparticles into PVB/Fe_2O_3 and EC/Fe_2O_3 films are ~0.82 emu/g and ~0.48 emu/g, respectively. Lower magnetization on EC/Fe_2O_3 could be due to the synthesis process which causes a smaller amount of nanoparticles in EC film. A similar behavior has been reported on [16]. Inset in Fig. 3 shows the magnetization around the working magnetic field. It is evident that magnetization in EC/Fe_2O_3 is about 10 times lower than that in PVB/Fe_2O_3. A deep analysis of morphological and magnetic properties must be done in future works.

3.2. Displacement Characterization

Linear displacements of PVB/Fe_2O_3 and EC/Fe_2O_3 minirobots in x, y and diagonals directions were tested. Two goal distances (5 mm and 10 mm) were considered for each direction. In all cases, movements were controlled in the LabView user's interface developed for the cartesian robot using the automatic mode. The displacement of the miniriobots was recorded in order to determine their precision by image analysis.

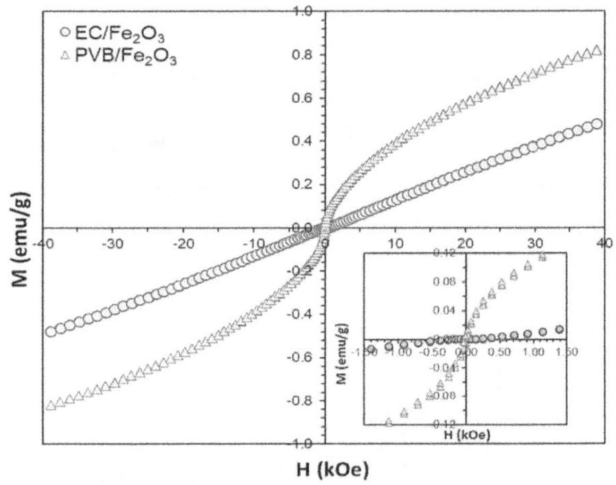

Fig. 3. Magnetization curves at room temperature (300 K).

Both minirobots were able to follow contactless the driving magnetic field and describe each one of the paths etablished by the magnet on the cartesian robot. Fig. 4 displays a sequence of the EC/Fe_2O_3 minirobot displacement on x direction along a distance of 10 mm. The minirobots show a precision of 96 % with respect to the position of the magnet, considered as the target.

A delay response of the minirobots was also observed. PVB/Fe_2O_3 and EC/Fe_2O_3 minirobots take 78 ms and 137 ms, respectively, to move after the magnet begins to displace. These times are related to the magnetization of the films and the viscoelasticity of the polymer matrix, PVB or EC.

Results concerning force characterization are described in the following section.

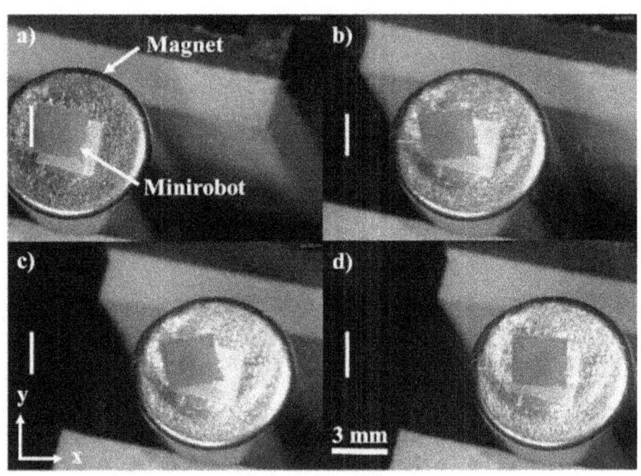

Fig. 4. Displacement in x direction of the EC/Fe_2O_3 minirobot. Distance from the start position (vertical white line): a) 0 mm; b) 4.11 mm; c) 7.40 mm; d) 9.55 mm.

3.3. Force Characterization

Force characterization was done by evaluating the ability of the minirobots to push objects with different mass along 5 mm and 10 mm of distance. Movements were also controlled in automatic mode via the user's interface of the cartesian robot. Three spheres, which mass is reported in Table 4, were considered.

PVB/Fe_2O_3 minirobot pushed S1 and S2, but it could not overcome the hydrodynamic forces to push S3. EC/Fe_2O_3 minirobot only could push S1. Pushing force depends directly on the minirobots mass, and PVB/Fe_2O_3 minirobot has about 30 % more mass than EC/Fe_2O_3 minirobot. The maximum pushing force are 106 mN and 182 mN for EC/Fe_2O_3 and PVB/Fe_2O_3 minirobots, respectively. Fig. 5 shows EC/Fe_2O_3 minirobot pushing mass S1 (10.8 mg).

Table 4. Mass of spheres for the force characterization.

Spheres	Mass (mg)
S1	10.8
S2	18.6
S3	31.2

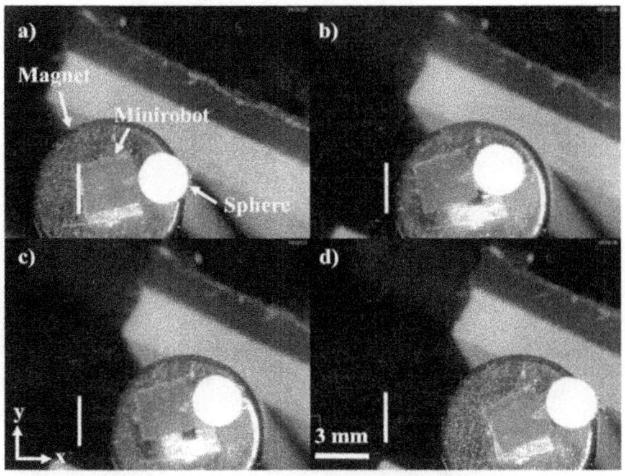

Fig. 5. Displacement in *x* direction of the EC/Fe_2O_3 minirobot pushing mass S1 (10.8 mg). Distance from the start position (vertical white line): a) 0 mm; b) 1.3 mm; c) 3 mm; d) 4.94 mm.

4. Conclusions

Magnetic films of PVB/Fe_2O_3 and EC/Fe_2O_3 with a nominal concentration of Fe lower than 10 % were synthetized. Magnetic characterization curves shown a superparamagnetic behavior for both films, indicating that Fe_2O_3 nanoparticles have dimensions lower than 20 nm. An experimental setup able to displace the driving magnetic field in *x* axis, *y* axis and their diagonals was developed. The magnetic polymer minirobots can hold an imposed trajectory during 5 and 10 mm with a precision of 96 % when an external magnetic field of 1 kOe is applied. EC/Fe_2O_3 and PVB/Fe_2O_3 minirobots push objects with a maximal force of 106 mN and 182 mN, respectively. So, the challenge to pushing objects in aqueous medium using contact-free magnetic polymer minirobots was addressed. Further characterizations of the magnetic polymers properties and its relation to the minirobots behavior, as well as more deeper displacement and force evaluations, testing also others geometries, must be done.

Acknowledgements

Research partially supported by the Consejo Nacional de Ciencia y Tecnología (CONACYT, Mexico) through scholarship no. 486968 and project project no. 284810, and by the PAICYT-UANL project no. IT1027-19. Authors would like to thank the research group "Tecnología e Innovacion Mecatrónica" for its assistance during 3D printing.

References

[1]. J. Thévenot, H. Oliveira, O. Sandre, S. Lecommandoux, Magnetic responsive polymer composite materials, *Chemical Society Reviews*, Vol. 42, Issue 17, 2013, pp. 7099-7116.

[2]. Z. Nagy, B. J. Nelson, Lagrangian Modeling of the magnetization and the magnetic torque on assembled soft-magnetic MEMS devices for fast computation and analysis, *Transactions on Robotics*, Vol. 28, Issue. 4, 2012, pp. 787-797.

[3]. W. Hilber, Stimulus-active polymer actuators for next-generation microfluidic devices, *Applied Physics A*, Vol. 122, Issue 751, 2016, pp. 1-39.

[4]. M. R. Aguilar, J. San Román, Smart Polymers and Their Applications, *Elsevier*, 2019.

[5]. S. Shrivastava, N. Jadon, R. Jain, Next-generation polymer nanocomposite-based electrochemical sensors and biosensors: A review, *Trends in Analytical Chemistry*, Vol. 82, 2016, pp. 55-67.

[6]. S. Cichosz, A. Masek, M. Zaborski, Polymer-based sensors: A review, *Polymer Testing*, Vol. 67, 2018, pp. 342-348.

[7]. J. J. Abbott, Z. Nagy, F. Beyeler, B. J. Nelson, Robotics in the small, Part I: Microbotics, *IEEE Robotics & Automation Magazine,* Vol. 14, Issue 2, 2007, pp. 92-103.

[8]. T. Xu, J. Yu, X. Yan, H. Choi, L. Zhang, Magnetic actuation based motion control for microrobots: An overview, *Micromachines*, Vol. 6, Issue 9, 2015, pp. 1346-1364.

[9]. C. Bi, M. Guix, B. V. Johnson, W. Jing, D. J. Cappelleri, Design of microscale magnetic tumbling robots for locomotion in multiple environments and complex terrains, *Micromachines*, Vol. 9, Issue 68, 2018, pp. 1-17.

[10]. H. Li, Fabrication of magnetic two-dimensional and three-dimensional microstructures for microfluidics and microrobotics applications, PhD Thesis, *University of Kentucky*, United Kingdom, 2014.

[11]. G. Hwang, I. A. Ivan, J. Agnus, H. Salmon, S. Alvo, N. Chaillet, S. Régnier, A. M. Haghiri-Gosnet, Mobile microrobotic manipulator in microfluidics, *Sensors and Actuators A: Physical*, Vol. 215, 2014, pp. 56-64.

[12]. H. Kim, J. Ali, U. K. Cheang, J. Jeong, J. Seok Kim, M. Jun Kim. Micro manipulation using magnetic microrobots, *Journal of Bionic Engineering*, Vol. 13, Issue 4, 2016, pp. 515-524.

[13].D. Wong, I. B. Liu, E. B. Steager, K. J. Stebe, V. Kumar, Directed micro assembly of passive particles at fluid interfaces using magnetic robots, in *Proceedings of the*

International Conference on Manipulation, Automation and Robotics at Small Scales (MARSS'16), Paris, France, 18-22 July 2016, pp. 1-6.

[14]. M. Llamas-Hernández, B. López-Walle, M. Rakotondrabe, E. Reyes-Melo, Dynamic behavior of magnetic hybrid films of polyvinyl butyral/iron oxide nanoparticles (PVB/Fe2O3) for their control as microactuators, *Physica B: Condensed Matter*, Vol. 549, 2018, pp. 113-117.

[15]. J. G. Puente Córdova, Síntesis y caracterización del comportamiento eléctrico de materiales híbridos magnéticos, PhD Thesis, *Universidad Autónoma de Nuevo León*, 2018.

[16]. A. S. Teja, P. Y. Koh, Synthesis, properties, and applications of magnetic iron oxide nanoparticles, *Progress in Crystal Growth and Characterization of Materials,* Vol. 55, 2009, pp. 22-45.

[17]. M. Kryszewski, J. K. Jeszka, Nanostructured conducting polymer composites – superparamagnetic particles in conducting polymers, *Synthetic Metals*, Vol. 94, Issue 1, 1998, pp. 99-104.

(118)

Suspended Carbon Nanotubes Gas Sensor

S. Arunachalam, F. Nabki and R. Izquierdo
Ecole de technologie Superieure, Electrical Engineering Department, 1100, rue Notre-Dame Ouest, Montreal,Canada
Tel.: + 15143968751
E-mail: ricardo.izquierdo@etsmtl.ca

Summary: A MEMS based suspended Carbon Nanotube (CNT) sensor was fabricated by a low temperature surface micromachining process. The sensor was tested in various gaseous environments to determine the sensing characteristics of the ionization sensor. The resultant sensor demonstrated excellent sensing capabilities.

Keywords: Ionization, Carbon nanotubes, Sensor, Suspended.

1. Introduction

Carbon Nanotube based gas sensors have been the subject of much research in the past few decades owing to their superb electrical and mechanical properties. A myriad of sensing principles and device designs have been proposed to develop gas sensors based on CNTs with high performance metrics. Ionization gas sensors operate based on detecting gases based on their ionization properties. The main advantage of ionization gas sensors is their independence on absorption characteristics of the sensing material or chemical interaction between species. Thus, they have faster response and recovery times and are sensitive to inert gases. CNT based ionization sensors work on the principle of the formation of a corona on the tips of the individual CNTs due to very high electric fields near the tips that helps in discharge at low voltages. On the contrary, traditional ionization sensors based on metal electrodes suffer from high power consumption and high voltage operation. A CNT based ionization sensor was first proposed in 2003 by [1], using vertically aligned Multi-walled CNTs. The sensor showed good sensitivity to various gases. A comparative study showed that the breakdown voltages for CNT based sensor are much lower than a sensor with metal electrodes. Other CNT based ionization sensors have been developed to detect humidity, NO_2, etc. [2-5]. In all the works mentioned above, the CNTs are grown vertically using very high temperature-based vapor deposition techniques. These growth techniques have a risk of yield that can have an impact on device performance. In this work, we present an ionization gas sensor based on suspended single walled CNTs that are arranged horizontally, instead of being vertically aligned like previously mentioned works. The resultant sensor showed similar behavior to vertically aligned CNT ionization sensors.

2. Experimental

The developed fabrication process uses SU8 photoresist as a sacrificial material. The process was originally developed to obtain suspended nanotube beams. The detailed fabrication process has been described elsewhere [6]. In short, SU8 photoresist was partially cross-linked to form anchors for the CNT beam. CNTs were obtained as prepared from Carbon Solutions Inc. and the thin film was prepared by vacuum filtration. The film was transferred onto the substrate and the CNTs were patterned followed by sacrificial layer removal to release the CNT beams. To fabricate the ionization sensor, a 30 nm thick Aluminum layer was deposited, and a metal electrode was patterned beneath the CNT beam which acts as one of the electrodes. The separation between the metal electrode and the CNT beam is of 3.6 um. The SEM Micrograph of the device Fig. 1.

Fig. 1. SEM Micrograph of the device.

3. Results

Measurements were first performed in air at room temperature. To determine the breakdown voltage of air, the voltage was increased from 50 V to 490 V in steps of 40 V. After 490 V, the voltage was increased in steps of 2 V. The breakdown in air was observed at a voltage of 500 V with a discharge current of 0.44 mA. The breakdown was characterized by the formation of a visible "spark" in the device. Once the breakdown voltage in air was established, the chamber was evacuated, and the voltage measurement was performed again. Under vacuum, there was no observed breakdown at any voltage. This is expected since there are no gaseous molecules in the chamber for ionization to occur. Following measurements in vacuum, Nitrogen was introduced in the sensor and the breakdown was observed at 646 V with a discharge current of 0.64 mA. Then, Nitrogen was flushed out of the chamber and Argon was introduced. In Argon, breakdown was observed at 378 V with a current of 0.32 mA. Further, in Helium, the breakdown was observed at 233 V with a discharge current of 0.19 mA. The measurements were performed multiple times to verify the breakdown voltages and to determine the repeatability of the sensor. A plot of the breakdown voltages of the gases as a function of the discharge current is shown in Fig. 2 and Table 1 gives an overview of the breakdown voltages and resultant discharge currents for different gases.

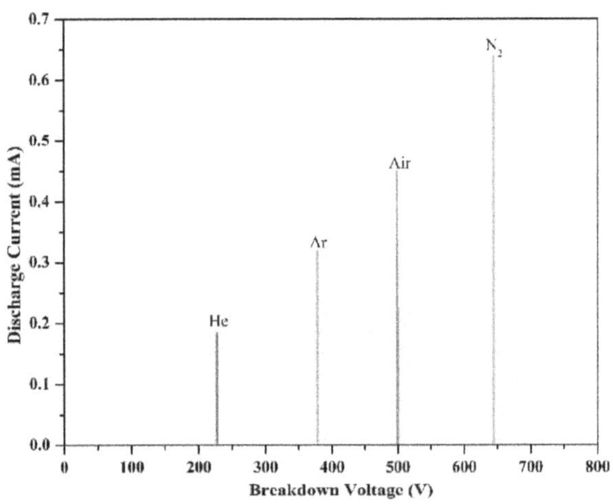

Fig. 2. I-V plot showing distinct breakdown voltages of various gases.

Table 1. Breakdown voltages of different gases as tested.

Gas	Breakdown Voltage (V)	Discharge Current (mA)
N_2	646	0.64
Air	500	0.44
Ar	378	0.32
He	233	0.19

The breakdown process in CNT ionization sensors have been discussed in prior works. Individual CNTs that constitute the CNT beam create very high nonlinear electric fields near the tips. As a result, the breakdown process is accelerated, and a corona of highly ionized gas is formed. The one-dimensional structure of CNTs aid in converging the electric field and induce discharge at a lower voltage as compared to traditional metallic ionization sensors.

4. Conclusions

In this work, we present an ionization sensor based on suspended single walled Carbon Nanotubes fabricated by a room temperature MEMS based surface micromachining process. In comparison to previous works where the CNTs are vertically grown using high temperature processes, this work uses horizontally aligned CNTs deposited as a film. The resultant CNT beam is used as one of the electrodes for the ionization sensor. The sensor showed repeatable and stable sensing characteristics for several gases, comparable to prior works of the same nature. The device design also eliminates the use any high temperature processes for CNT growth and thus can be used in conjunction with other MEMS and CMOS technologies.

References

[1]. A. Modi, N. Koratkar, E. Lass, B. Wei, P. M. Ajayan, Miniaturized gas ionization sensors using carbon nanotubes, *Nature*, Vol. 424, Issue 6945, 2003, pp. 171-174.

[2]. W. Jiahao, L. Hai, W. Yanyan, X. Dong, Z. Yafei, A MEMS-based ionization gas sensor using carbon nanotubes and dielectric barrier, in *Proceedings of the 3rd IEEE International Conference on Nano/Micro Engineered and Molecular Systems (NEMS'08)*, 2008, pp. 824-827.

[3]. H. Song, K. Li, C. Wang, Selective Detection of NO and NO_2 with CNTs-based ionization sensor array, *Micromachines*, Vol. 9, Issue 7, 2018, 354.

[4]. H. Song, K. Li, Q. Li, A tripolar-electrode ionization gas sensor using a carbon nanotube cathode for NO detection, Journal of Micromechanics and *Microengineering*, Vol. 28, Issue 6, 2018, 065010.

[5]. H. Bohr-Ran, L. Tzu-Ching, Y. Ying-Kan, Gas ionization sensors with CNT/Ni field cathodes, in *Proceedings of the 3rd International Nanoelectronics Conference (INEC'10)*, 2010, pp. 535-536.

[6]. S. Arunachalam, A. Gupta, R. Izquierdo, F. Nabki, Suspended carbon nanotubes for humidity sensing, *Sensors*, Vol. 18, Issue 5, 2018, 1655.

(120)

Continuous Sorting of Submicron Particles in a Pre-analytical Device Based on Acousto-fluidic Microsystem

A. Chaalane [1], D. Guneysu [1], M. Addouche [1], R. Zeggari [2], F. Lardet-Vieudrin [1], C. Elie-Caille [1], W. Boireau [1] and A. Khelif [1]
[1] FEMTO-ST Institute, UBFC, CNRS, ENSMM, UTBM, 15B Avenue des Montboucons, 25030 Besançon, France
[2] FEMTO Engineering, 15B Avenue des Montboucons, 25030 Besançon, France
Tel.: + 33363082432
E-mail: amar.chaalane@femto-st.fr

Summary: In this work, we present a pre-analytical module based on the combination of microfluidic and electroacoustic technologies to intercept and sort submicron biological particles. It is composed by assembling a lithium niobate (LN) substrate for acoustic function and a micromachined glass substrate for microfluidic channel. The interference between the two surface acoustic waves (SAWs) generated by inter digital transducers (IDTs) create a distribution of an acoustic radiation force (ARF). This force affects differently particles depending on their physical proprieties. The device is powered by an electronic circuit with a phase shifter to move the node of the standing surface acoustic wave (SSAW) along the channel width. When the device is powered at resonance frequency, experience has shown a submicron particles alignement at a fixed position along the channel. By shifting the SSAW phase, the particles are driven to one of the three channel outlets.

Keywords: Acoustic particle manipulation, Continuous sorting system, Microfluidic MEMS, Standing surface acoustic wave, Acoustic radiation force, Extracellular vesicles.

1. Introduction

Manipulating micro and nano-biological particles like extracellular vesicles (EVs), without extracting them from their media, presents a big challenge for the biological, clinical and pharmaceutical communities. The goal behind EVs manipulation is to better detect and qualify them for diagnosis purposes. Multiple technics have been used in literature to improve the efficacy of sorting particles devices. In acoustic microfluidic devices, either bulk acoustic waves (BAWs) [1] or surface acoustic waves (SAWs) [2] can be used to generate acoustic standing wave fields to trap particles realizing meso-scale particle patterning. This concept has become possible by integration, in the same chip, of acoustic and microfluidic systems. It allows to control the interaction between the acoustic wave and the submicron particles at such small scale.

In this work, we will present the design, microfabrication and characterization of a particle separation device based on the combination of microfluidic and electroacoustic systems. Our objective is to intercept and sort EVs, which represent highly important mediators in cell-to-cell communication [3, 4]. Tracking those EVs "messengers" represents an attractive way to convey fundamental information between cells.

2. Methodology and Principals

Based on the interference of two identical SAW in the microchannel, a standing surface acoustic wave (SSAW) is accured. It generates an acoustic radiation force (ARF) that affects differently particles in fluid depending on their properties. The heavy and dense particles tend to tighten towards the nodes while the light and compressible particles remain on the antinodes [5, 6]. This allow to align the large particles in the nodes and the small ones in the anti-nodes. By shifting the nodes position, we can drive the particles to the desired channel output.

3. Experimentation

3.1. Design and Fabrication Process

In Fig. 1, we draw a schematic of the design of submicron particles separation micro-device. It is composed of two inter digital transducers (IDTs) deposited on a lithium niobate (LN) substrate. A microfluidic channel with six outlets is fabricated by dry etching of the glass substrate. Both LN and glass substrates are then bonded to ensure the tightness of the fluidic circuit. The two IDTs generate SSAW that create ARF inside the microchannel. The width of the acoustic channel (50 µm) was designed to match the acoustic half-wavelength. Based on the speed of sound in LN and the acoustic wavelength, the frequency of 40 MHz is used to activate the SAW transducer. In these conditions, the acoustic chip has one pressure node located at the center and two pressure antinodes located at both channel walls.

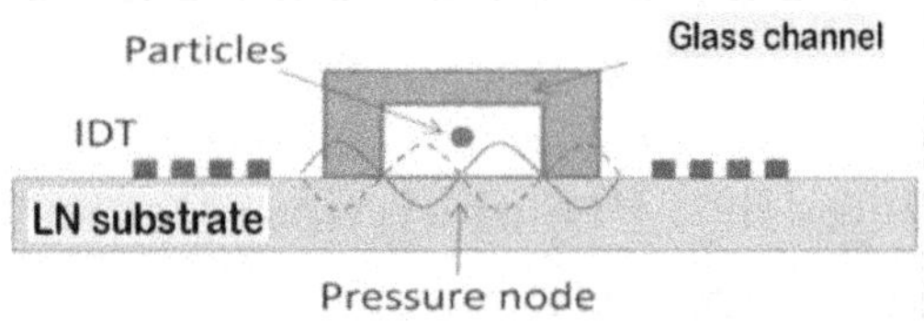

Fig. 1. Schematic of the acousto-fluidic device: it is composed of two IDTs patterned on LN substrate, the microchannel is realised by dry etching of a glass substrate.

3.2. Experimental Setup

In order to characterize the microsystem we have manufactured a mechanical support (Fig. 2) allowing: (i) To host the chip; (ii) Ensure the six fluidic connections to a controlled injection system, and (iii) Electrical connection of the power supply. In Fig. 3 we represent a schematic view of the experimental setup. The device is powered by an AC sinusoidal signal generated by a, Anritsu 68147C synthesized signal generator. A power splitter divides the signal in two ways (0° and 180°) to power the two IDTs. The second one contains a continuously variable attenuator to balance the signal amplitude at the inputs of the IDTs. The first way contains an analog phase shifter to move the signal node and thus the SSAW pressure node along the channel width.

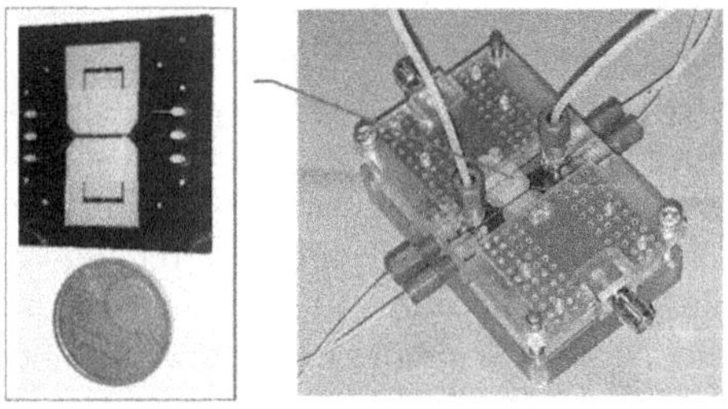

Fig. 2. The realized acousto-fluidic device (left) with electric and microfluidic connections (right).

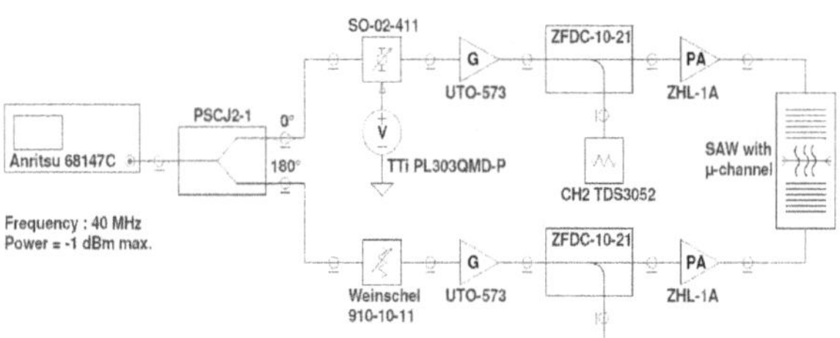

Fig. 3. Electronic circuit power supply.

4. Results and Discussion

We used two sizes of synthetic fluorescent spherical particles of 480 nm and 920 nm in diameter in a circulating buffer. Initially, the 480 nm in diameter particles are randomly dispersed along the channel in both static and dynamic fluidic mode. When the power radiation is switched on, the randomly dispersed particles are aligned along the channel. By shifting the SSAW phase, we realize sorting by driving 480 nm in diameter particles to one of the three channel outlet (Fig. 4).

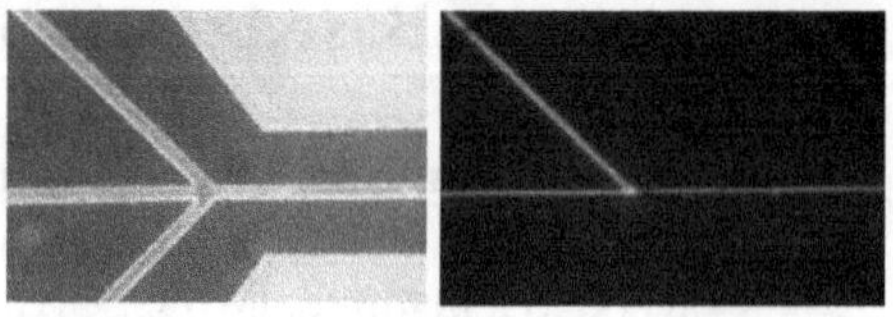

Fig. 4. Some results of the 480 nm in diameter particles sorting using SAW in 50 µm width microfluidic channel: fluorescence microscopy presents the capability of the device to modify the particles trajectory along the channel.

5. Conclusion

The main idea of this project is to be able to manipulate the nanoscale vesicles by acoustic waves. In this work, we experimentally validate the submicron particles alignment and sorting using SSAW microsystem. Experience has shown a submicron particles alignement at a fixed position along the channel. By shifting the SSAW phase, we can drive particles to one of the three channel outlet.

Acknowledgements

We would like to thank the EUR EIPHI program for the project financial support (Contract No. ANR-17-EURE-0002). We also thank MIMENTO staff and CLIPP platform members for facilities.

References

[1]. A. Haake, A. Neild, G. Radziwill, J. Dual, Positioning, displacement, and localization of cells using ultrasonic forces, *Biotechnology and Bioengineering*, Vol. 92, 2005, pp. 8-14.

[2]. D. J. Collins, B. Morahan, J. Garcia-Bustos, C. Doerig, M. Plebanski, A. Neild, Two-dimensional single-cell patterning with one cell per well driven by surface acoustic waves, *Nature Communications*, Vol. 6, 2015, 8686.

[3]. D. Song, D. Yang, C. A. Powell, X. Wang, Cell-cell communication: old mystery and new opportunity, *Cell Biology and Toxicology*, Vol. 35, 2019, pp. 89-93.

[4]. I Huang-Doran, C-Y. Zhang, A. Vidal-Puig, Extracellular vesicles: Novel mediators of cell communication in metabolic disease, *Trends in Endocrinology & Metabolism*, Vol. 28, Issue 1, 2017, pp. 3-18.

[5]. M. Groschl, Ultrasonic separation of suspended particles. I. Fundamentals, *Acustica-Acta Acustica*, Vol. 84, 1998, pp. 432-447.

[6]. L. Johansson, J. Enlund, S. Johansson, I. Katardjiev, V. Yantchev, Surface acoustic wave induced particle manipulation in a PDMS channel – principle concepts for continuous flow applications, *Biomedical Microdevices*, Vol. 14, Issue 2, 2012, pp. 279-289.

(122)

An Ontological Approach in Measurement Science: Towards a System of Measurement Ontologies

V. B. Tarassov, M. N. Koroleva
Bauman Moscow State Technical University, 5b1 2nd Baumanskayast., 105005 Moscow, Russia
E-mail: maria.svyatkina@gmail.com

Summary: A knowledge management and ontological engineering vision of measurement science is developed. A special attention is paid to the analysis of uncertainty types in measurement. In order to represent and engineer measurement knowledge a top-down ontological approach to measurement specification is presented. A hierarchical system of measurement ontologies is proposed, some basic ontologies, including measurements ontology itself, ontology of measured properties, ontology of sensor networks, are constructed by using mind maps. A granular meta-ontology is selected as a unifying framework for dealing with uncertainty in the context of GUM general provisions.

Keywords: Measurement science, Ontological modeling, Measurement ontology, Ontological hierarchy, Measurement uncertainty, Meta-ontology, Sensor networks.

1. Introduction

The new industrial revolution under the name of Industry 4.0 based on Cyber-Physical Systems and Internet of Things [1, 2] supposes the concept formation and implementation of the so-called Ubiquitous Measurements with using Wireless Sensor Networks. Such a network can be viewed as a community of autonomous agents located in different places and maintaining communications to generate a distributed cognition system. Enabling mutual understanding and joint work of these agents requires a system of measurement ontologies.

An ontological consideration of measurement science is not new. A good example of foundational measurement ontology is QUDT (Quantity, Unit, Dimension, Type) ontology [3], where class properties and restrictions are defined to model physical quantities, units of measure and their dimensions in various measurement systems. The goal of the QUDT ontology is to provide a unified model of measurable quantities, units for measuring different kinds of quantities, the numerical values of quantities in different units of measure and the data structures and data types used to store and manipulate these objects in software.

Among other interesting ontological approaches to measurements Kuhn's functional ontology of observation and measurement [4] should be mentioned. To summarize, it is worth stressing that most of existing measurement ontologies are devoted to measurement units.

The creation of JCGM 100:2008 (GUM 1995 with minor corrections) [5] has not considerably changed the general situation with the lack of integrating theoretical and formal framework for measurement science (see, for instance, the introductory article by L.van Biesen in [6]). Nowadays a new round of discussion on the basics of measurement theory [7-9] is required followed by a broad ontological engineering of measurement information systems.

This problem is faced in our paper. Its objective consists in constructing a system of interrelated measurement ontologies. First of all, in Section 2 main reasons for rethinking measurement uncertainty concept and taking into account its non-stochastic types are discussed with basing on the GUM provisions. The classification of type B uncertainties is introduced. Section 3 is devoted to the presentation of ontological approach to measurements. Two basic definitions of ontology are discussed, various ontological classes are considered and two principal approaches to developing ontological systems (top-down and bottom-up) are specified. The concept of granular meta-ontology is presented in details, and a three-leveled system of measurements ontologies is proposed. The mind maps for the ontology of measurements, ontology of measured properties, ontology of sensor network is constructed.

2. Measurement Uncertainties: Reasons and Types

2.1. On Some GUM Provisions

Any measurement is an experiment under uncertainty. Thus it always has an error. Measurement results depend on measurement-information system, measurement techniques, external conditions, human-operator's qualification and so on. An old idea of measurement with obtaining exact result has outlived itself.

Measurement is never exact. In GUM (see, for example, [5]) the usual term «measurement error» has been replaced by a more general concept of «measurement uncertainty». The matter is not only terminological differences, but a quite new presentation of measurement results.

The *uncertainty* as a quantifiable attribute of measurement has various interpretations. On the one hand, the word «uncertainty» means doubt, and thus in its broadest sense «uncertainty of measurement» means doubt about the validity of the result of a measurement [5]. On the other hand, the uncertainty of the result of a measurement reflects the lack of exact knowledge of the value of the measurand [5].

Formally uncertainty is a non-negative parameter associated with the result of measurement that characterizes the dispersion of the values that could reasonably be attributed to the measurand.

In any case, the result of every measurement has two components: a) a measurand value $x_0$and b) its estimated uncertainty u_x. Generally we have $(x_0 \pm u_x)$ *mu*, where *mu* means measurement unit.

To deal with these uncertainty factors, the probabilistic and statistical methodologies are usually applied; they consider such values as a confidence level, an error probability, the significance level. The criticism of the probabilistic approach is mainly reduced to the following remarks: 1) a unified representation of measurement uncertainty by random errors is not well justified; 2) in some circumstances the postulation of errors normal distribution law is not realistic.

The measurement procedures are often based on rather restrictive and approximate suppositions, and the measurement instruments have limited resolution, in particular, for complex and dynamic object structures.

In [5] the difference is made between type A and type B uncertainty. Type A uncertainty evaluation is performed by the statistical analysis of series of observations. Type B uncertainty requires a new evaluation method other than the statistical analysis of series of observations, for instance, fuzzy variable, fuzzy interval, fuzzy number, possibility distribution. Here, type B uncertainty (more exactly, a complex of various types of non-stochastic uncertainty) encompasses such factors as: a) an incomplete or inaccurate definition of the measurand; b) an instrument imperfection due to its finite resolution or discrimination threshold; c) an inadequate knowledge of the effects of environmental conditions on the measurement results or imperfect measurement of environmental conditions; d) a non-representative sampling – the sample measured may not represent the defined measurand; e)an open character and dynamics of measurement procedures, their dependence on measurement goals, constraints and available instrumentation.

2.2. Classification of Type B Uncertainties

Let us take as a basis Borisov's uncertainty classification [10] that was introduced in 1980's. In this classification uncertainty was also divided into two classes – stochastic uncertainty and non-stochastic (linguistic) uncertainty. It is easy to modify it for measurements with using sensor networks by taking into account inaccuracy, incompleteness, ambiguity, contradictoriness, fuzziness in measurement (Fig. 1).

Measurement inaccuracy means a limited sensor resolution capacity due to the nature of measured parameter. In every measurement there exists an unrecoverable error depending on the sensor threshold. For instance, an acoustic range finder determines the distance to objects within the limit of 0.2-80 m with an error 2 %.

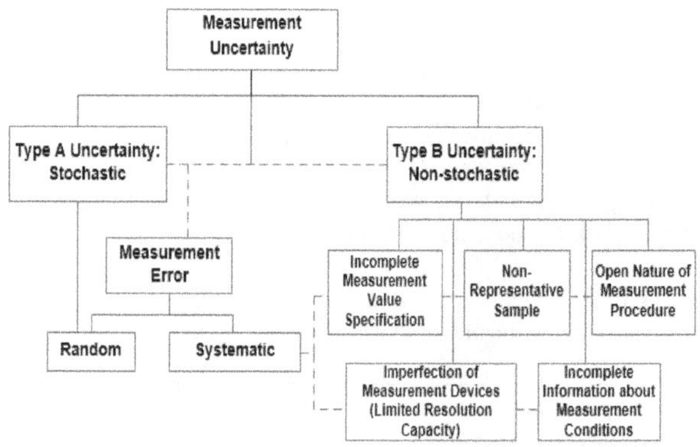

Fig. 1. Classification of type B measurement uncertainties.

Incomplete measurement information supposes that in sensor network a few sensors are inaccessible, and total ignorance means a loss of communication with the sensor or malfunction of measuring element.

Contradictory indications may appear for homogeneous sensors measuring the same parameter. For example, two strain gauges measure structural stress: the first sensor indication is interpreted as «normal», and the second sensor indication is seen «as out of norm».

Measured data ambiguity supposes the use of some (non-probabilistic) distribution. Fuzzy value is attributed to the terms of linguistic variable. For instance, «a measured wind speed on the bridge is almost in the norm».

Thus, the result of measurement is only an approximation of measured value. It cannot be expressed by a singleton and is characterized by a distribution on confidence interval. Moreover, various types of uncertainty are present in all components of measurement information systems and processes.

So the idea of finding a more general formal framework for measurement theory and, in particular, measurement uncertainties, for instance, interval analysis and granular computing, seems to be quite reasonable and timely (see [11, 12]). Our proposal is to construct a system of interrelated measurement ontologies and start by defining a granular meta-ontology of measurement [13].

3. An Ontological Approach to Measurement Theory

Measurement theory encompasses knowledge about measurement types, methods, tools, results, conditions. In order to develop intelligent metrological systems on the basis of knowledge engineering let us consider an ontological approach to measurements.

Ontology (in Computer Science and Artificial Intelligence) is usually seen as a formalized description of some problem area. Two classical approaches in ontological modeling are well-known – relational and logical ones; these approaches rise to the pioneering papers by T. Gruber [14] and N. Guarino [15]. T. Gruber defines ontology as an explicit formal specification of a conceptualization shared by the members of some community.

Below we shall consider the basic keywords of this definition in the scope of modern measurement theory. *Conceptualization* means the synthesis of abstract conceptual model of measurement types, methods, instruments, situations, results and other concepts together with studying main relations between them. The word «*formal*» focuses on the important requirement that any measurement conceptualization should be expressed in a machine readable format to be understood by computer (sensor) system, whereas «*explicit*» supposes that both measurement concepts used and the constraints of their use are explicitly defined. Finally, the term «*shared*» shows that primarily ontology captures consensual knowledge accepted by a group in order to enable mutual understanding and joint work of various agents (a communicative aspect of ontology).

According to N.Guarino, an ontology is a logical theorythat gives an explicit partial account of a conceptualization. It includes some basic terms forming taxonomy, their definitions and attributes, related axioms and inference rules. In other words, ontology is seen as both a formal view to semantics and a thesaurus used by logical theory.

It is often very difficult or even impossible to construct single, comprehensive, coherent and practically useful ontology. To simplify ontology development and reuse, a modular approach is taken and some hierarchies of ontologies are formed [16-18]. On the low level apart from domain ontology, both task ontology and application ontology are constructed, and on the high level upper ontologies [19] are viewed to represent general categories encountered in many problem areas. These ontologiesdescribe such concepts as space, time, resources, events, etc., which are independent of particular problem or domain.

Besides, meta-ontology («ontology of ontologies») is given that provides both an exact mathematical specification of various ontologies and formal analysis of their properties. Specifically, it includes methods and forms of representing, developing and merging different ontologies.

There are two principal approaches to ontology development: top-down approach and bottom-up one. Bottom-up ontology development begins with rather specific classes and continues with their further composition into more general categories. Its implementation in a semi-automatic mode is related to selecting some pre-ontological units (thesauri, Excel tables, ER or UML data models, frame systems, and so on) and translating them into some popular ontological language, for example, OWL. This approach proposed by N. Guarino [16] starts with designing domain ontology followed by task and application ontologies, and finally, leads to upper ontologies.

Inversely, top-down design supposes the use of most general categories such as meta-ontology and upper ontology. Here meta-ontology specifies the properties of lower ontologies, in particular, it is responsible for the choice of ontological primitives, relations and operations.

3.1. A Few Words on Granular Meta-Ontology for Measurements

Sometimes the confusion is observed between the concepts of meta-ontology and upper ontology. In practice,meta-ontology is often reduced to some ontological unit or formal structure, for instance, «entity–relation–actor» [20]). To show the difference, let us take some analogies between meta-ontology on the one hand and meta-model, meta-knowledge and meta-mathematics on the other hand. Here meta-mathematics is viewed in two senses. In a wide sense, it is a family of mathematical theories applied to the studies offormal calculus. Internal characteristics of various models are studied by using mathematical methods. Such a meta-theory of mathematics does not suppose any special constraints on employed methods and tools.

In the narrow Kleene's sense, meta-mathematics is a part of mathematical logic that considers a general structure of mathematical proofs and mathematical theories by using formal methods.

Similarly, meta-model in a complex system specifies both operating principle and basic structure of lower level models. In particular, it determines model construction tools, for instance, formal languages or graphics notations to specify classes, relations and attributes. Specifically, meta-knowledge as meta-model for intelligent system contains knowledge about both declarative and procedural knowledge circulating in the knowledge lifecycle.

It is clear that on the basis of measurement meta-ontology some adequate models, languages and tools for representing the information and knowledge about the measurement upper ontologies (for example, uncertainty, instrumentation ontologies) are selected. In their turn, upper ontologies provide reusable constructs for low level ontologies. Finally, application ontology is obtained by integrating domain ontology and task ontology.

We can take as representative examples the concepts of singular and granular meta-ontologies. In case of singular meta-ontology lower level ontological primitives are singletons, and the ontology itself can be depicted as an ordinary graph (tree).

The concept of *measurement granule* and *measurement information granulation* seems to be more suitable as a formal framework of measurement science. According to L.Zadeh [21], granule isa collection of objects which are drawn together by indistinguishability, similarity, proximity or functionality.Let us outline basic components of granulation theory in measurement science. These are: general granulation principles and criteria in measurement science, classification and interpretation of measurement granules, approaches and methods of granulation in measurements, formal models of granules, mappings between granules, quantitative measures and indices for granules.

The granules differ one from another by theirs sizes, complexity, abstraction level. A basic granulation principle consists in finding tradeoffs between the size of granule and the number of granules. Here such formal criteria as specificity and covering are useful. The concept of measurement information granulation is closely related to data abstraction and derivation of knowledge from information [22].

There are various classifications of granules: crisp and fuzzy granules, one-dimensional and multidimensional, data and knowledge granules, time and space granules, etc.Typical interpretations of granules are: part of the whole, sub-problem of the problem, uncertainty degree, variable constraint.

Among classical formal approaches to construct crisp granules we may cite partitions, neighborhoods, interval analysis, rough set theory, random set theory, Dempster-Shafer theory, etc.The main tools for generating fuzzy granules are fuzzy sets, linguistic variables, as well as fuzzy partitions, fuzzy neighborhoods, fuzzy numbers, fuzzy rough sets, et al.Following L.Zadeh, we shall consider various procedures of fuzzification, randomization, usualization as granulation instances.

The selection of granular meta-ontology for measurement means the use of intervals or distributions as ontological primitives.

3.2. The System of Measurement Ontologies

Below a three-leveled system of measurement ontologies is suggested (Fig. 2). Here we take a granular meta-ontology that induces granular uncertainty ontology as upper ontology.

The spectrum of low level ontologies encompasses measurement domain ontology, ontology of measurement properties, ontology of sensor networks as measurement tools, measurement applications ontology (for example/ measurement for monitoring). No doubt, depending on the task to perform, another types of low level and upper ontologies can be constructed, for example, instruments, sensors, and so on.

Such visualization tools as mind maps [23] are worth employing to represent ontologies. A basic idea of mind map is to automatically transform some text fragments in a graphical form. Such a map possesses the following features: a) it has a form of a bush; b) a studied object is placed in the center of the picture that corresponds to the center of attention; c) primary topics related to the object of study diverge from the center as branches explained with keywords; d) secondary topics are also branching; e) the branches form a connected nodal structure.

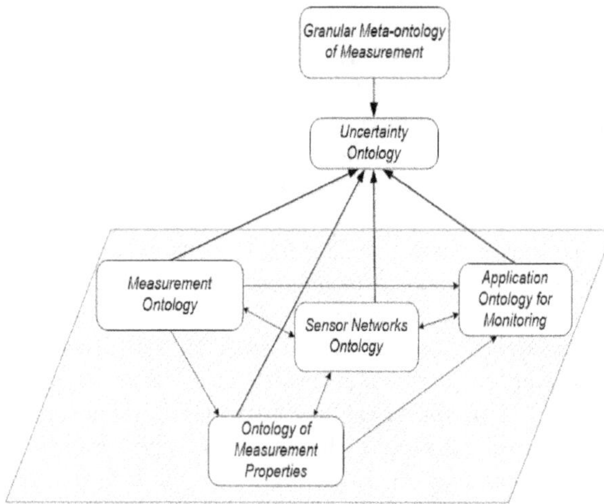

Fig. 2. The System of Measurements Ontologies.

Three low-level ontologies are depicted in Figs. 3, 4 and 5: ontology of measurements, ontology of measured properties (with using Doynikov's classification of measured properties [24]) and ontology of sensor networks.

It is seen from Fig. 5 that the problem area «Sensor Networks» is revealed through such concepts – classes as «Sensors», «Networks», «Environment», «Application Goal».

4. Conclusions

The choice of components for constructing the system of measurement ontologies depends on measurements application area. Specifically, measurement tasks and measurement properties are tightly connected with measurement use. We have considered measurement ontologies in the framework of monitoring problem [25]. Main measurement tasks in monitoring situation are both formation of judgments and support of reasoning. These judgments and reasoning procedures concern diagnostics of current state of monitoring object (bridge), prognosis of their change tendencies, decision-making and recommendations. For example, «if the wind speed measured by the anemometer on the bridge is 21-22 m/s and considerable bridge vibrations are observed, then the traffic on the bridge is prohibited». It is the case of joint, multiple, dynamic, open measurements, included into a wider monitoring system.

In our opinion, the development of ontological systems in measurement theory and the modeling of different sides of measurement uncertainty on the basis of granular meta-ontology are quite necessary steps on the way to advanced measurement science in the Industry 4.0 age.

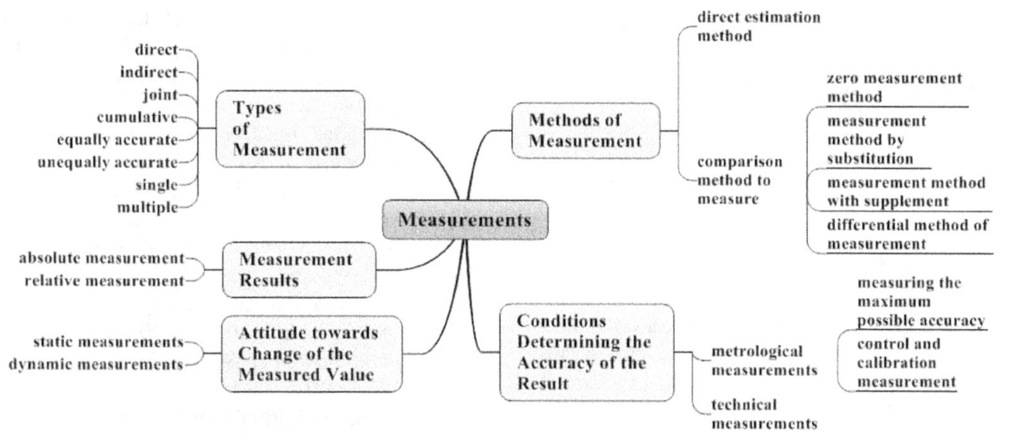

Fig. 3. The Mind Map for Measurements Ontology.

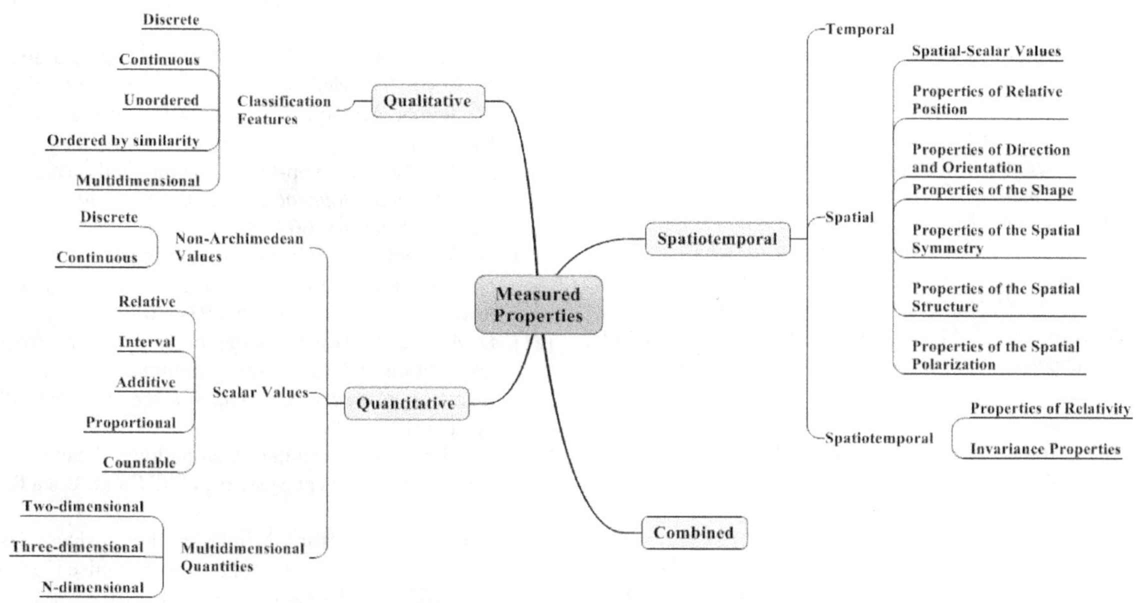

Fig. 4. The Mind Map for the Ontology of Measured Properties.

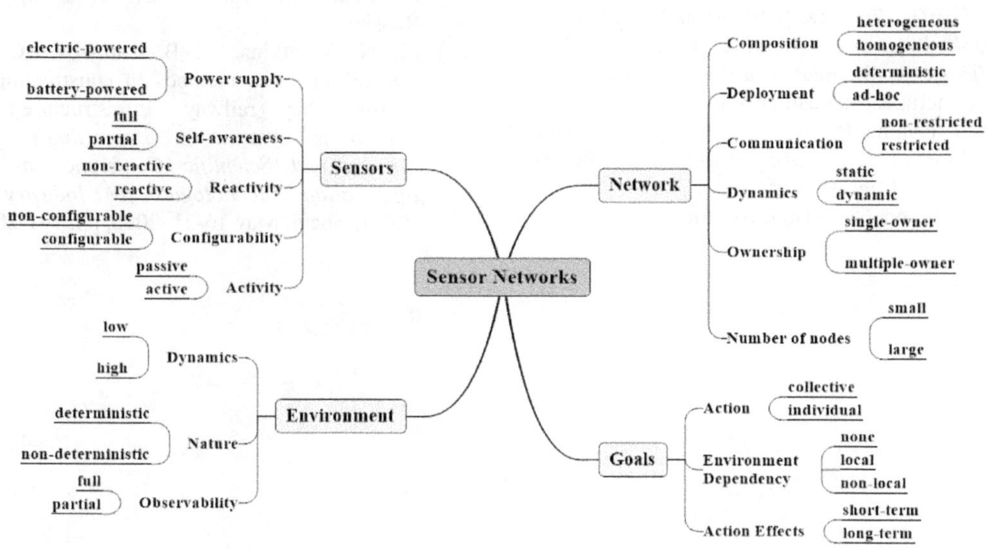

Fig. 5. Visual Representation of Sensor Network Ontology.

Acknowledgements

This work is supported by Russian Foundation for Basic Research, grants No 18-07-01311 and No 19-07-01208.

References

[1]. K. Schwab, The Fourth Industrial Revolution, *World Economic Forum*, Geneva, 2016.

[2]. V. B. Tarassov, Enterprise total agentification as a way to industry 4.0: Forming artificial societies via goal-resource networks, in *Proceedings of the 3rd International Scientific Conference Intelligent Information Technologies for Industry (IITI'18)*, Vol. 1, Sochi, September 17-20, 2018, pp. 26-42.

[3]. R. Hodgson, et al., The NASA QUDT Handbook, *NASA*, 2014.

[4]. W. Kuhn, A Functional Ontology of Observation and Measurement, in *Proceedings of the 3rd Workshop on GeoSemantics (GeoS'09),* Mexico City, December 3-4, 2009, pp. 26-43.

[5]. JCGM 101:2008, Evaluation of Measurement Data – Guide to the Expression of Uncertainty in Measurement (GUM: 1995), Geneva, 2008.

[6]. L. Van Biesen, Advances of Measurement Science, in Measurement Science: A Discussion (K. Kariya, L. Finkelstein, Eds.), *IOS Press*, Tokyo, 2000, pp. 3-9.

[7]. L. Finkelstein, Fundamental concepts of measurement, *Acta IMEKO*, Vol. 3, Issue 1, May 2014, pp. 10-15.

[8]. A. Giordani, L. Mari, Modeling measurement: Error and uncertainty, in Error and Uncertainty in Scientific Practice (M. Boumans, G. Hon, A. C. Petersen, Eds.), *Pickering & Chatto*, London, 2014, pp. 79-86.

[9]. S. V. Prokopchina, Modern theory of measurements: a classification of measurements types, *Soft Measurement and Computing*, Vol. 1, Issue 1, 2017, pp. 4-16.

[10]. A. N. Borisov, A. V. Alexeev, O. A. Krumberg, et al., Decision-Making Models on the Basis of Linguistic Variables, *Zinatne*, Riga, 1982 (in Russian).

[11]. L. Reznik, Measurement Theory and Uncertainty in Measurements: Application of Interval Analysis and Fuzzy Set Methods, in Handbook of Granular Computing, (W. Pedrycz, A. Skowron, V. Kreinovich, Eds.), *John Wiley and Sons Ltd,* 2008, pp. 517-532.

[12]. V. B. Tarassov, Granular measurement structures in ambient intelligence: Vasiliev's and Belnap's sensors and their communication models, *Informatsionno-Izmeritel'nye i Upravlyayuschiye Systemy* (*Information -Measurement and Control Systems*), Issue 5, 2013, pp. 65-74 (in Russian).

[13]. V. B. Tarassov, A. V. Fedotova, B. S. Karabekov, Granular meta-ontology, fuzzy and linguistic ontologies: enabling mutual understanding of cognitive agents, in *Proceedings of International Conference on Control, Automation and Artificial Intelligence (CAAI'15)*, Phuket, Thailand, August 23-24, 2015, pp. 253-261.

[14]. T. R. Gruber, A translation approach to portable ontologies, *Knowledge Acquisition*, 1993, Vol. 5, Issue 2, pp. 199-220.

[15]. N. Guarino, Formal ontology, conceptual analysis and knowledge representation, *International Journal of Human-Computer Studies*, Vol. 43, Issues 5-6, 1995, pp. 625-640.

[16]. N. Guarino, Formal ontology and information systems, in *Proceedings of the 1st International Conference on Formal Ontologies in Information Systems (FOIS'98)*, Trento, Italy, June 6-8, 1998, pp. 3-15.

[17]. A. Gomez-Perez, M. Fernandez-Lopez, O. Corcho, Ontological Engineering, *Springer-Verlag*, Heidelberg, 2005.

[18]. G. S. Plesniewicz, V. B. Tarassov, B. Karabekov, N. T. Min Vu, Methods and Languages for Ontological Modeling, Almaty, *IICT MRS KR Editions*, 2017 (in Russian).

[19]. J. F. Sowa, Top-level ontological categories, *International Journal of Human-Computer Studies*, Vol. 43, 1995, pp. 669-685.

[20]. M. Uschold, M. Gruninger, Ontologies: Principles, methods and applications, *Knowledge Engineering Review*, Vol. 11, Issue 2, pp. 93-136.

[21]. L. A. Zadeh, Toward a theory of fuzzy information granulation and its centrality in human reasoning and fuzzy logic, *Fuzzy Sets and Systems*, Vol. 90, 1997, pp. 111-127.

[22]. W. Pedrycz, Granular Computing: Analysis and Design of Intelligent Systems, *CRC Press*, Boca Raton, 2013.

[23]. T. Buzan, C. Griffiths, Mind Maps for Business. Using the Ultimate Business Tool to Revolutionise How You Work, 2nd Ed., *Pearson*, New York, 2014.

[24]. A. Doynikov, Classification of measured properties, in *Proceedings of the 5th International Conference on Soft Computing and Measurements (SCM'02)*, St. Petersburg, June 25-27, 2002, pp. 46-49 (in Russian).

[25]. M. N. Svyatkina, V. B. Tarassov, A. I. Dolgiy, Logical-algebraic methods in constructing cognitive sensors for railway infrastructure intelligent monitoring system, in *Proceedings of the 1st International Scientific Conference on Intelligent Information Technologies for Industry (IITI'16)*, Vol. 1, Sochi, May 16-21, 2016, pp. 191-206.

(123)

Rapid Strain Differentiation of E. coli-inoculated Urine Using Olfactory-based Smart Sensors

Aminat Adebiyi [1], Nam Than [2], Sarath Swaminathan [1], Mohammed Abdi [1], Amy N. Bowers [1], Andrea Fasoli [1], Alberto Mannari [1], Luisa Bozano [1]
[1] IBM Research, Almaden. 650 Harry Road, San Jose, CA USA 95120
[2] Department of Biomedical Engineering, San Jose State University, San Jose, CA USA
Tel.: + 14089271579
E-mail: aminat.adebiyi@ibm.com

Summary: Urinary tract infections (UTIs) have a large effect on the public healthcare system. Millions are affected each year, and are likely to have the infection recur with increased resistance to antibiotics. This is partly due to the lack of a targeted healthcare regimen that requires pathogen identification. As traditional cultures can take several days, there is a need for a real-time diagnostic tool for identifying uropathogens. In our study, the use of rapid olfactory-based smart sensors for the strain identification was explored by taking measurements of in-vitro assays involving the inoculation of two strains of E. coli (K12 and Uropathogenic E. coli) in urine. Using a Support Vector Machine classifier to train on two previous data collections yielded 95.83 % accuracy when tested on the third subsequent collection. These results demonstrate the promise of smart sensing in differentiating close metabolomic signatures of E. coli and are expected to improve with more extensive data collection, assay development and sensor augmentation.

1. Introduction

With over 3 million cases a year in the United States alone, Urinary Tract Infections (UTIs) are the most prevalent bacterial infection and poses a great risk of morbidity in vulnerable populations, including children, pregnant women and the elderly [1]. In 2011, the estimated economic burden was estimated at \$2.8 billion in the U.S. and the cost is only expected to increase with increasing hospitalizations [2].

Accurate diagnosis of UTIs is difficult at the point of care, as it depends on both the presence of symptoms and a positive urine culture, which takes a minimum of 24 hours or and even up to multiple days for results [3]. A rapid test would lead to the correct empirical course of treatment targeted to the specific bacterial strain – significantly improving treatment outcomes, reducing medical costs and the likelihood of antimicrobial resistance.

Olfactory-based sensors like electronic noses have been widely studied and applied across industries to associate volatile organic compound profiles with the emitting analyte [4]. In healthcare, metabolomic markers can be used for studying disease diagnostics and have motivated the study of volatile organic compounds that correspond to these biomarkers [5].

In this paper, we introduce the Electronic Volatile Analyzer (EVA), a rapid, gas sensing IoT platform under development at IBM Research, Almaden – designed to deliver classification results under two minutes. A main objective of this study was to demonstrate the utility of rapid gas sensing IoT platforms in differentiating K12 and Uropathogenic (UPEC) strains of Escherichia coli (E. coli) inoculated into urine at comparable cellular counts.

2. Methods

2.1. The IBM Electronic Volatile Analyzer™

The IBM Electronic Volatile Analyzer™ (EVA, Fig. 1) is an electronic nose under development at IBM Research Almaden. The design of the platform is modular: each gas sensor is mounted on its own sensor module, a Printed Circuit Board (PCB) of common design carrying an integrated microcontroller and the circuitry needed to operate the sensor. The sensor modules communicate via I2C protocol with a central hub (Beagle Bone Black), a single-board computer that manages the sensor modules and processes the multisensorial output. A set of six commercial Metal Oxide (MOX) gas sensors was used to collect the measurements described in the following sections. The selection of the sensor portfolio was based on the indications of their respective manufacturers regarding the nominal target gases of each sensor. Consequently, the as-selected sensor array was expected to respond to a variety of volatile molecules, including alcohols, hydrocarbons, ammonia, methane, as a well as a wide range of volatile organic compounds (VOC), with a lower limit of detection at parts per million (ppm) level. Each MOX sensor was operated using an individualized multi-step, periodic voltage profile applied to the heating element, which resulted in stepwise modulation of the device temperature (Figs. 1 b(i) & (ii)). Modulation of the temperature of MOX sensing elements is a well-known technique that can be used to enhance and tune the dependence of the sensing element resistance to the surrounding environment (Fig. 2b) [REF added]. The heater voltage profiles applied to the EVA sensors were optimized to maximize the sensitivity of each sensor with respect to a subset of target VOC, as well as to improve response orthogonality. Although the duration and amplitude of the individual waveforms was adjusted independently for each sensor, all waveforms were synchronized to a period of 80 s to simplify the handling and processing of the sensor array outputs. The resistance of each MOX sensing element was monitored at fixed voltage at a rate of 10 Hz. During operation, a constant flow of 150 ± 10 sccm) was established by means of a small pump, drawing air from the environment in a continuous fashion while also collecting the headspace of the sample of interest The vapor-carrying flow was directed towards an enclosed chamber containing the six MOX gas sensors, for the exposure and detection to take place.

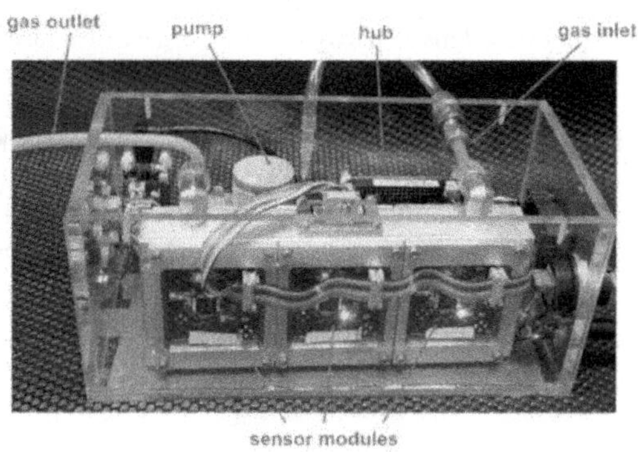

Fig. 1a. Electronic Volatile Analyzer (EVA) Sensor Array prototype with parts labeled.

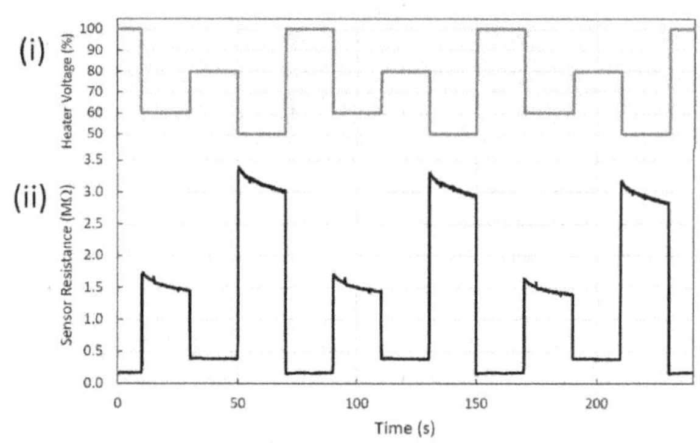

Fig. 1b. Example of a temperature profile modulation for MOX sensors for IBM EVA™: (i) periodic waveform of heater voltage, expressed as a percentage of the maximum operating voltage recommended by the sensor manufacturer; (ii) corresponding variations in MOX sensor resistance under constant environment.

2.2. Assay Development and Sample Preparation

Two liquid UTI assays were developed by culturing Escherichia coli (Migula) Castellani and Chalmers (ATCC® 29425™) and Escherichia coli (Migula) Castellani and Chalmers (ATCC® 700928™) in Normal Human Urine (purchased from UTAK Laboratories). The Normal Human Urine (NU) was centrifuged and filtered with 0.2 μm filters prior to inoculation, to remove sediments and to aid in the homogeneity of samples. Each of the E. coli strains were then separately inoculated into NU at a fixed concentration, and incubated at 37 °C over five days to guarantee a terminal concentration of 10^9 CFU/ml. After samples were plated, they were further filtered to remove cellular bodies of the bacteria, leaving only their emitted metabolomic volatiles behind [6]. Sample volumes of 5 ml were dispensed in 20 ml Wheaton Septa Top vials for measurement. The list of samples measured were filtered Uropathogenic E. Coli (UPEC) in NU, K12 E. coli in NU, Normal Urine, and Lab Air as reference.

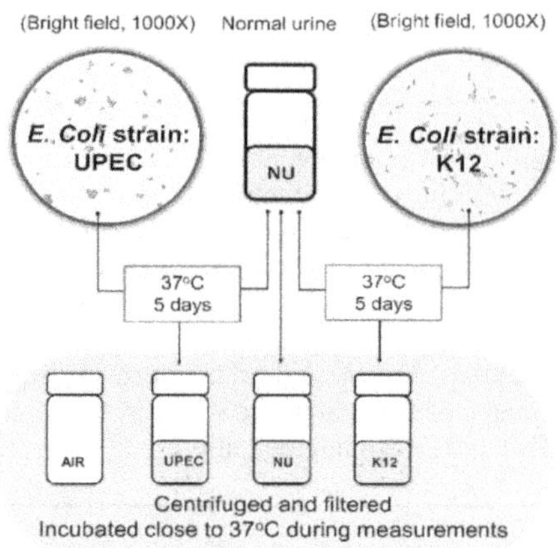

Fig. 2. Diagrammatic representation of assay development and conditions of data collection. Each strain of E. coli (K12 and UPEC) was inoculated separately into Normal Urine (NU) and incubated over five days, where it was filtered to trap volatiles. Each data collection involved four sample vials with Lab Air, NU, K12 and UPEC. Samples were incubated close to 37 °C during measurements.

2.3. Data Collection and Processing

Collection of data for the training and testing datasets were made at different times, with a minimum of eight hours gap in between. The training dataset was comprised of two experiments, and the test dataset was the third sequential acquisition. Headspace measurements of each of the four sample vials were made using the IBM EVA™. The sequence of collection from the sample vials was randomized over the course of an eight hour collection period, with an air flush between vials. Samples were incubated close to 37 °C during measurements, to aid in the evaporation of trapped volatiles. The response of the sensors was saved using Data Acquisition System (DAS) before feature extraction and labeling of the sensorial data were performed. The resultant input consisted of the fused mean features extracted from sensor resistance curves to capture sensitivity and amplitude differences in the samples. Machine learning was performed mainly on the extracted feature set to identify the odor classes from the testing dataset (Testing Batch) by training on the two prior collections of the training set (Training Batches A and B), which is the main analytical model in view of the intended application. Given the limited number of total examples, small classifiers were selected and included

Support Vector Machine (SVM), Logistic regression, and Random Forest. Implementation of these classifiers was from the Scikit-learn 0.19.1 package on Python [8]. The entire dataset was cross validated using three-fold validation with each batch serving as a fold. The parameters for cross validation are shown in Table 1.

Table 1. Cross validation parameters for selected "small" classifiers.

Classifier Type	Tuning Parameters
Logistic Regression	'l1' and 'l2' loss, C values from 0.001 to 1.0, optimization algorithms – 'newton-cg', 'libfgs', 'liblinear', 'sag', 'saga' and multiclass classifications using 'one-vs-rest' and 'multinomial'
Support Vector Machine	'linear' and 'rbf' kernels with C values from 0.001 to 1.0 and 'one-vs-one' and 'one-vs-rest' for multiclass classification
Random Forest	estimators from 5 to 20, maximum depth of tree from 2 to 10 or None, max_features – 'auto', 'sqrt', 'log2' and None, and criterion – 'gine' or 'entropy'

3. Results and Discussion

3.1. Data Visualization

Resultant data from the projected input set by sample label is shown in Fig. 3. The two main principle components of the dataset represented 98.7 % of variation in the data, showing the dataset is separable and easily defined. Training Batch A showed more variation across labels compared to Training Batch B and Testing Batch. Based on the sequence of collection, measurement drift was experienced in Training Batch A but stabilized between Training Batch B and Testing Batch. This phenomenon could also represent the correlation of concentration of volatiles in samples between the batches.

3.2. Machine Learning

Based on the tuning parameters the best parameters for each classifier is outlined in Table 2. Based on the linearity of the data shown in Fig. 3, the crossvalidation parameters for the classification show high separability based on the C values and the number of estimators.

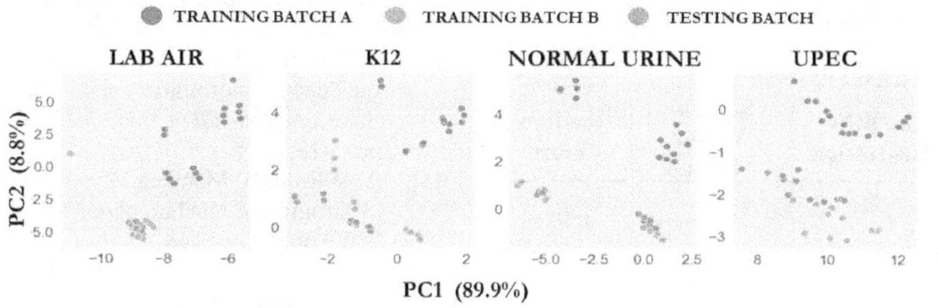

Fig. 3. Principal Component Analysis (PCA) visualization of each sample label (Lab Air, Normal Human Urine with K12 E.coli strain, Normal Human Urine and Normal Human Urine with UPEC E. coli strain) by data collection batch. The two axes represent Principal Components (PCs) 1 and 2 which describes 98.7 % percent of the variation in the data. Training Batches 1(blue) and 2 (orange) were collected in sequence before the terminal collection of the Testing Batch (green). Both Training Batches were used to develop the classifier for which the Testing Batch was tested on. Training Batch B and Testing Batch overlap closely, while Training Batch A is more distant with respect to the other batches.

The results of the performance of the classifiers on the dataset for the three-fold validation with each batch serving as a fold is showing in Table 3. Table 3a shows the accuracy of using the indicated batch as a hold out while training on the remaining batches. Table 3b shows the F1 scores, which represents the harmonic mean of the precision and recall, measuring the accuracy of the classifier on the dataset. In addition to the main sequential model for our application, the results for the remainder of the batches as hold-outs are shown.

For the main sequential model based on Training Batches A and B, SVM performed the best with Testing Batch with 96 % accuracy with a similar F1 score. However, Random Forest and Logistic Regression performed best in classifying Training Batches A and B respectively. On average, Random Forest performed the best on all batches, followed by Logistic Regression and SVM, with the same trend repeated with the F1 scores.

SVM was selected as the best performing model in view of application (Testing Batch), and the confusion matrix is shown in Fig. 4. The classification accuracy was 95.83 % when trained on first two collections (48 examples per class) and tested on the third (12 examples per class). The model had an overall precision of 95.75 % and sensitivity of 96.5 %. There were 100 % classification on all labels apart from NU inoculated with K12 E. coli, which was confused as Normal Urine.

Table 2. Cross-validation parameters for classifier type for the training data.

Classifier Type	Parameters
Logistic Regression	C = 0.5, multi_class = 'multinomial', penalty = 'l2', solver = 'newtoncg'
Support Vector Machine	C = 0.5, decision_function_shape = 'ovo', kernel = 'linear'
Random Forest	criterion = 'gini', max_depth = 2, max_features = 'auto', n_estimators = 7

Table 3. Normalized accuracy and F1 scores of leave-one-batch out folds by classifier type when trained on two other batches and tested on indicated sample set.

(a) Normalized Accuracy Scores.

Batch	Logistic Regression	Support Vector Machine	Random Forest
Training Batch A	0.80	0.55	0.98
Training Batch B	0.93	0.86	1.00
Testing Batch	0.85	0.96	0.75
Average	0.86	0.79	0.91

(b) F1 scores.

Batch	Logistic Regression	Support Vector Machine	Random Forest
Training Batch A	0.76	0.48	0.98
Training Batch B	0.93	0.85	1.00
Testing Batch	0.84	0.96	0.67
Average	0.84	0.76	0.88

4. Conclusions

The results of the pilot experiments presented here highlight the promise of olfactory-based smart sensors in pathogen identification for bacteria of the same genus. Further studies including extensive data collection, assay development and sensor augmentation are warranted to validate this device as a precision antibiotic matching tool.

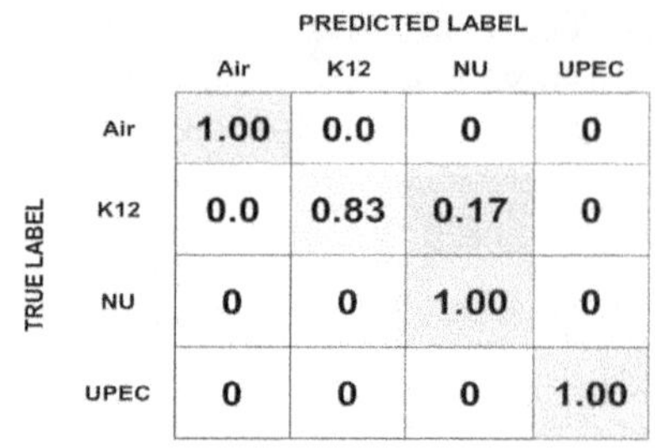

Fig. 4. Normalized confusion matrix outlining prediction results on Testing Batch for each label with highest accuracy classifier (SVM) built on Training Batches A &B. All labels apart from K12 were predicted with 100 % accuracy, which was mislabeled as NU 17 % of the time.

References

[1]. A. Flores-Mireles, et al., Urinary tract infections: epidemiology, mechanisms of infection and treatment options, *Nature Reviews Microbiology*, Vol. 13, Issue 5, 2015, 269.

[2]. J. E. Simmering, et al., The increase in hospitalizations for urinary tract infections and the associated costs in the United States, 1998-2011, in Open Forum Infectious Diseases, Vol. 4, Issue 1, *Oxford University Press*, 2017.

[3]. B. Foxman, Epidemiology of urinary tract infections: Incidence, morbidity, and economic costs, *The American Journal of Medicine*, Vol. 113, Issue 1, 2002, pp. 5-13.

[4]. A. Wilson, B. Manuela, Applications and advances in electronic-nose technologies, *Sensors*, Vol. 9, Issue 7, 2009, pp. 5099-5148.

[5]. A. Amann, D. Smith, Volatile Biomarkers: Non-Invasive Diagnosis in Physiology and Medicine, *Elsevier*, 2013.

[6]. R. B. Caldwell, T. L. Claus, Simple and reliable method to precipitate proteins from bacterial culture supernatant, *Appl. Environ. Microbiol.*, Vol. 70, Issue 1, 2004, pp. 610-612.

[7]. A. Schutze, T. Baur, M. Leidinger, W. Reimringer, R. Jung, T. Conrad, T. Sauerwald, Highly sensitive and selective VOC sensor systems based on semiconductor gas sensors: how to?, *Environments*, Vol. 4, Issue 1, 2017, 20.

[8]. F. Pedregosa, G. Varoquaux, et al., Scikit-learn: Machine learning in Python, *Journal of Machine Learning Research*, Vol. 12, 2011, pp. 2825-2830.

(124)

Communication Anomaly Detection in Cyber-physical Systems

P. Blazek [1], R. Fujdiak [1], M. Hodon [2], I. Zolotova [3], P. Mlynek [1] and J. Misurec [1]
[1] Brno University of Technology, Technicka 12, 61600 Brno, Czech Republic
[2] University of Zilina, Univerzitna 8215/1, 01026 Zilina, Slovakia
[3] Technical University of Kosice, Letna 9, 04200 Kosice, Slovakia
Tel.: +420 54114 6921
E-mail: blazekpetr@feec.vutbr.cz

Summary: Sensors networks are widely representative of using wireless and power line communication. Each network infrastructure contains specific anomalies that may or may not affect communication. In the case of wireless and power line communication, communication is also significantly influenced by various environmental influences. In this paper, it is aimed to present the anomaly detection model for sensors networks that correlates anomalies with transmission parameters of communication media for more accurate detection.

Keywords: Anomaly, Communication, Detection, Traffic modeling, Parameters.

1. Introduction

Digitization and automation are significant parts of Industry 4.0 that is closely related to an area of Operational Technology (OT). In conventional IT networks, legitimate anomalies can only mean a change in the pattern of network infrastructure behavior (increased communication due to legitimate action, infrastructure upgrade). These anomalies are very unlikely in OT networks because lifecycle is significantly higher than in conventional IT networks. According to NIST 800-82 or IEC 62443-4-1, the updates in OT should be systematic and controlled [1]. However, the OT and IT areas are merging into one system. The OT systems were long-time isolated and security measures were not a concern. This means to deal with an "old" issue, which was already set-up several decades ago in IT world. Besides IT, the OT technologies are highly connected to the physical environment, where security issues might have significant impact on safety.

From the technological point of view, we might mention the mobile ad-hoc networks (MANET) and wireless sensor networks (WSN), which walked as well a significant technological step forward [15]. The MANET networks serve mostly as an infrastructure solution. On the other hand, the WSN collect data from various devices and thus control the operation of the system. Sensor networks are characterized by periodic messaging throughout almost the entire transmission. For these type of networks, simple messages with a small data size (temperature, status, voltage, current, etc.) are transmitted [2]. The only non-periodic events may be signaling the exceeding of the interval limit (e.g., critical temperature or pressure), possible influences affecting the communication medium or human influence (attack, device damage). Also, various transmission media are used in OT and specifically in sensor networks. Wireless and PLC media are a significant representative, as they are much easier to apply to already realized systems against metallic and optical cables [3, 4]. However, the disadvantage of these transmission media is highly susceptible to multiple adverse effects (interference, outages) [5]. Another disadvantage of PLC and Wireless networks is significantly lower transmission speeds than in metallic and optical networks [6]. However, most OT networks do not need large data transfers because simple data such as voltage, current, power, etc. are usually transmitted [3, 4]. All this information might be together used for detecting anomalous behavior in network communication, which is a valuable input in the Intrusion Detection/Prevention Systems (IDS/IPS).
The cyber-physical systems recently emerged as a promising direction to provide sufficient solution for human-to-human, human-to-object and even object-to-object communication or interaction in the physical as well as virtual world [16]. Apparently, the CPS adopt and also nurture, the areas of MANET and WSN, thanks to the need for much more sensors and data [17]. This paper focuses on anomaly detection as a vital part of security measures in the CPS systems. The rest of this paper is organized as follow. Section 2 provides brief introduction of current state of the art for CPS and security, followed by Section 3, which already describes our detection model for anomaly detection. Section 4 brings information for anomaly detection based on communication media. Finally, Section 5 summarized our findings.

2. Related Work

There are several papers focused on the CPS/WSN, which might help to understand the environment. For example, Fang-Jing Wu et al. [17] complexly described the environment of CPS, which uses the WSN, and divide it to the several parts: (i) Network formation; (ii) Communication pattern; (iii) Power management, (iv) Network coverage; (v) Node mobility; (vi) Knowledge mining, and (vii) Quality of service. Moreover, the connection of virtual space, physical space and intelligence is explained over the data flow and interaction in a CPS application, together with clear model of CPS with adopted WSN and MANET systems. This information is valuable if considering security in CPS, because this environment is becoming very complex with including many different systems and technologies as shown.

Many current papers [8-13] deal with the detection of anomalies to detect illegitimate communication (attacks) or other anomalies. However, most systems are based on detection models for classical IT networks. These systems are then transformed into protocols used in OT networks, but no longer consider specific parts of OT such as transmission media, which have a significant impact on communication. This can cause false alarms in anomaly detection (false positive or false negative errors) or even miss the actual system intrusion, which overall significantly reduce the efficiency of the detection model. Other papers, today already outdated, were also focused on the MANET/WSN issues [18-24], but again, these papers are mostly focused on routing protocols and other IT issues.

Nowadays, we are having two different approaches to the anomaly detection – parametric and non-parametric methods [25]. The parametric methods include, i.e., Multivariate Distribution Functions (MDFs) [26], density-based spatial clustering of applications with noise (DBSCAN) [27]. The non-parametric methods include, i.e., K-Nearest Neighbor (kNN) [28], Support Vector Machines (SVMs) [29], Artificial Neural Networks (ANNs) [30], Genetic Algorithms (GA) [31], and hybrid systems, which connect more than one method.

In this paper, we focus on the correlation between network communication parameters and transmission media parameters. For functionality in various systems, all parameters have been selected to be independent of any OT protocol. Based on this, the proposed model can be used in almost any network infrastructure. These should be used in IDS/IPS systems to enhance the detection model and detection techniques.

3. Detection Model

Detection of non-standard network traffic is divided into two main methods – signature-based detection and anomaly detection. Both methods detect anomalies occurring in network traffic, but each focuses on a different type of detection. Signatures are more accurate for detecting known anomalies that are described by the pattern. This type is very effective, but only in cases of known anomalies, where the slightest change may make detection impossible. In contrast, anomaly detection responds to any change in network communication, but cannot reliably determine what anomaly it is (attack, change in legitimate traffic, etc.) This article focuses on this detection.

Fig. 1 shows the main parts of the detection model, which is based on network communication parameters and the parameters of the transmission media. The main components of the model are described in the chapters below.

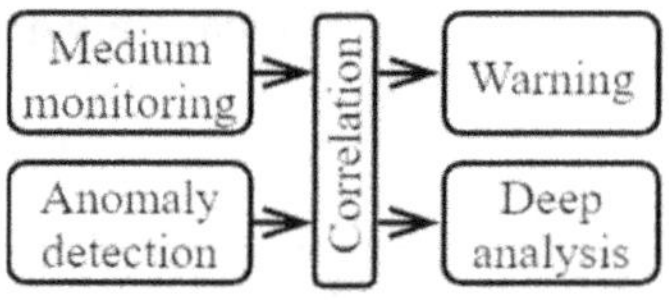

Fig. 1. Detection model block diagram.

3.1. Anomaly Detection

The most crucial part of anomaly detection is input parameters that define the monitored content in communication. Without appropriately defined parameters, the detection model will not be successful. The parameters presented in this work are used to DDoS (Distributed Denial of Service) detection. These attacks are characteristic by the transition of the communication model from relatively non-periodic (legitimate traffic) to almost periodical (DDoS attack). However, in the OT network, an opposite model is used. Network communication consists of almost periodic traffic, and in case of any change, it is an anomaly in communication. A disadvantage of anomaly detection is also the generation of so-called False Positive (FP) alarms, which are difficult separated from True positive (TP) anomalies.

Anomaly detection consists of two parts (Fig. 2) – Training and Real-time. Reference values (p_0, h_0, t_0, n_0, and c_0 - detailed description at individual parameters) are defined during the training session and represent the legitimate network communication of the infrastructure. Thus, each parameter monitors and records specific values from network communication in a defined time interval. This interval is the same for all parameters and is not fixed. The time interval is based on the requirements (network size, communication period, etc.) of the network infrastructure. To determine the optimal reference value, the Training phase is performed repeatedly and the median is calculated from these values to eliminate the extreme values. The measured values from the Real-time part are computed from the monitored network communication, which is compared with the reference values from the test part to determine the deviation from legitimate traffic.

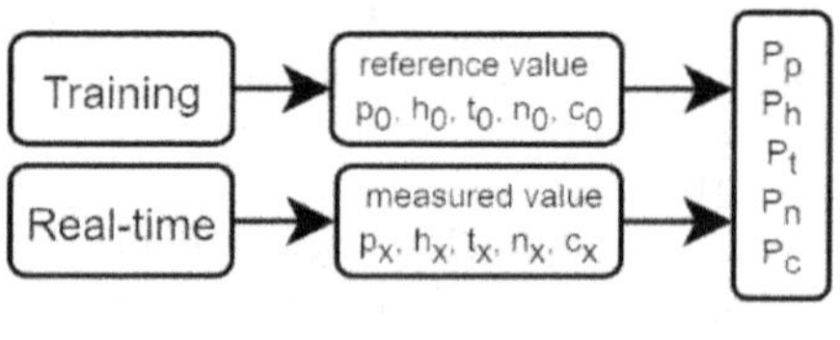

Fig. 2. Anomaly detection block diagram.

A total of five parameters were selected for the Anomaly detection section, which generally describes network communication [7].

The Number of Packets: In sensors networks, in most cases, messages are sent at a periodic time interval that is constant for the entire communication period. Non-periodic messages can be caused only by critical values on the sensors (exceeding the range of typical values) or by human external influences (attack, device damage). However, both options should be classified as TP anomalies because they identify major network events. The value of the parameter is calculated as

$$P_p = \frac{p_x}{p_0}, \tag{1}$$

where p_0 is the reference value that represents the legitimate communication of the system. Ideally, the number of valid traffic packets should be fixed or within a small interval. The p_x value represents the calculated value of the monitored communication. By dividing these two values, the resulting ratio of the number of packets in the monitored interval to the reference value. This value is an input to the correlation section.

The entropy of packet size: The size of transmitted data is unchanged in sensors networks. In most protocols, objects specifying the message type with a given payload size are implemented. Accordingly, there is less variation in packet length in sensor network communication. The computed entropy from the probabilities of packet lengths should be within a small interval. In the case of an anomaly, the value of entropy should increase (higher representation of packets of different lengths) or a decrease (higher representation of packets of the same length). In this case, anomalies may be caused by communication media (e.g., TCP retransmission) or human (attack).

In Formula 2 below, the Shannon entropy expression is used, which is calculated entropy from the size of the recorded packets. The total value of the parameter P_h (formula 3) is calculated by dividing the reference value h_0 and the value h_x, which is computed from the monitored communication. The parameter output is an input to the correlation section.

$$h = \sum_{i=1}^{n} p_i \log_2 p_i, \tag{2}$$

$$P_h = \frac{h_x}{h_0} \tag{3}$$

The interval between two events: In regular periodic communication, the delay between two messages varies at a small interval. Interval change may be caused by a deteriorating quality of the transmission medium. However, there should be no peak values exceeding several times defined the interval. Anomalies can be caused by interruptions on the communication medium or can be influenced by human intervention (attack). Both options should be classified as TP. The output value of the parameter (P_t), which describes the deviation between the defined reference value (t_0) and the obtained value of the monitored network traffic (t_x), is given in formula 4.

$$P_t = \frac{t_x}{t_0} \tag{4}$$

Non-standard packets: As mentioned, in sensors networks, there is little packet variation. The communication should not contain non-standard packets such as TCP datagrams with non-standard flags, packets of a different protocol than used for communication, etc. Non-standard packets can be caused by errors in the communication medium or by human activity (attack). The parameter specifying the number of non-standard packets is calculated as

$$P_n = \frac{n_x}{n_0}, \tag{5}$$

where n_0 is a reference value, which is defined during the training phase and indicates the average number of non-standard packets in legitimate communication. The value of the second parameter n_x is calculated from the monitored network communication.

The number of TCP connections: In the case of sensor network devices, a permanent TCP connection is maintained for the duration of the entire communication or is set to the long interval. New connections are usually established only in case of loss due to, e.g., restart of the device, loss of connection due to communication media or human influence. TCP connections terminated before a defined time interval should be classified as TP in correlation with media parameters. The parameter (P_c) calculates the number of TCP connections as in the case of the previous parameters.

$$P_c = \frac{c_x}{c_0}, \tag{6}$$

where c_0 determines the value that corresponds to the legitimate traffic of the communication infrastructure. The value of the second parameter c_x represents the monitored network communication.

3.2. Medium Monitoring

As mentioned in the previous sections, wireless or PLC transmission media are typical for communication between sensor network devices. These media are well usable in an area where it is not possible to use the most common metallic or optical fiber. However, the main disadvantage is the susceptibility to environmental influences, which causes a reduction in the quality of the transmission medium (delay increase, packet loss, etc.).

The structure of the Medium monitoring (Fig. 3) section is very similar to the Anomaly detection section. The content is the Training section, where the reference values for the parameters described below are scanned. The second part is Real-time, where the parameters for the monitored network communication are calculated. As in the Anomaly detection section, reference values are calculated as the median of the n-samples to determine the optimal reference value.

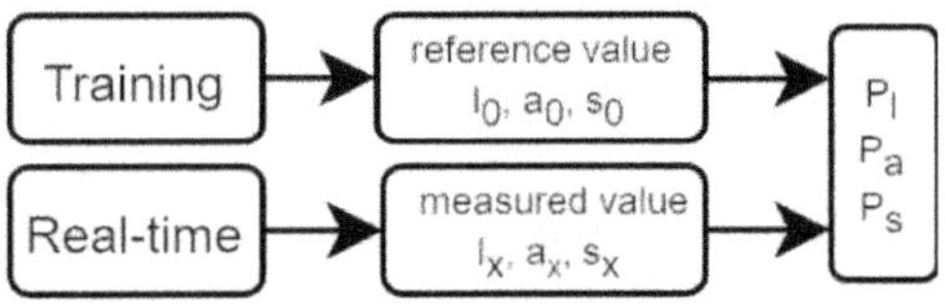

Fig. 3. Medium monitoring block diagram.

During the selection of parameters, the emphasis was placed on the parameters that are most commonly available on Wireless and PLC devices. For this reason, three parameters describing the quality of the communication medium were selected.

Packet loss: affects TCP communication (TCP retransmission) behavior and disturbs the frequency of sent messages.

$$P_l = \frac{l_x}{l_0} \tag{7}$$

Latency: A significant delay can affect the behavior of TCP communication (TCP retransmission) and affect the detection capabilities of the parameters – Number of packets and the interval between two events.

$$P_a = \frac{a_x}{a_0} \tag{8}$$

SNR/RSSI: The SNR (Signal to Noise Ratio) and RSSI (Received Signal Strength Indicator) parameters are a commonly used indicator of the quality of Wireless and PLC communication media. In connection with packet loss and latency can reliably determine whether the quality of the communication medium.

$$P_s = \frac{s_x}{s_0} \tag{10}$$

3.3. Correlation

The Anomaly detection described above achieves positive values in the unchanged network model. In the case of sensor networks, communication should be almost unchanged. This assumption is influenced by the communication medium, which can cause FP alarms at low throughput. For FP alarms reduction, the anomaly detection parameters are correlated with the transmission medium parameters. Furthermore, the correlation distinguishes whether an anomaly has occurred or is legitimate traffic with low-throughput.

Fig. 4 shows a block diagram of the Correlation section. The main part consists of a forward Neural network (NN), where the input is parameters from Anomaly detection and Medium monitoring. Before detection, NN is started in the training mode, when the output options of NN (legitime traffic and medium anomaly) are simulated. Reference values are used to simulate legitimate traffic. For the simulation of the Medium anomaly event, it is necessary to interrupt the communication medium. The remaining cases are evaluated as Unknown anomaly.

During detection, the neural network evaluates whether it is legitimate traffic (the input parameter values approach the reference values) or an anomaly (the input parameters converge to infinity). In the case of an anomaly caused by the transmission medium, all parameters converge to infinity. In the case of unknown anomalies, only parameters from the Anomaly detection section converge to infinity.

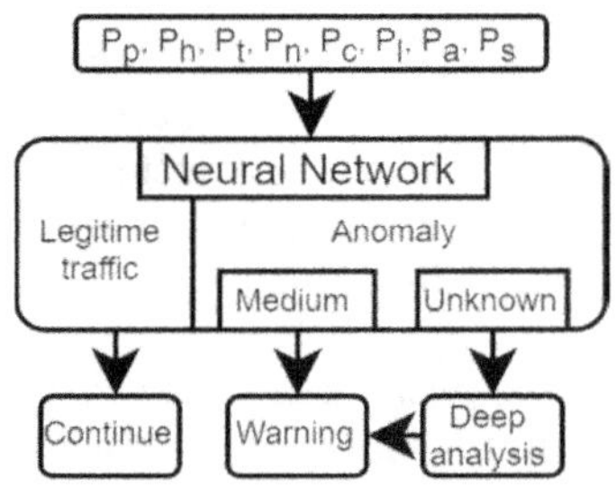

Fig. 4. Correlation block diagram.

3.4. Deep Analysis and Warning

The last part of the detection model is Deep analysis and Warning. The in-depth analysis provides a closer analysis of the payload on the application layer. In-depth analysis should be performed for anomalies that cannot be classified as legitimate communication or anomaly caused by the transmission medium. In-depth analysis is divided into two parts – full-traffic record and human analysis. During each period, all communication is recorded in *.pcap format. In the case of an unknown anomaly, it is possible to determine by reverse analysis whether it is a false alarm or an attack has been detected.

In-depth analysis can also be extended to detect critical values. At present, the detection model identifies non-periodic messages (threshold exceedance, device error, etc.) as anomalies, which is not an error due to the importance and need for logging. In-depth analysis, however, thanks to the signatures could distinguish these states from anomalies and thus better classify into the Warning section.

The second block Warning is used to record a given event (medium anomaly and unknown anomaly) and eventual notification of a critical condition on the device. The Warning is divided into two parts logs and messages. Logs contain records of all events that have been evaluated as anomalies. The messages part allows sending event reports to authorized persons. The logging and messaging methodology is based on the model described in the paper [14].

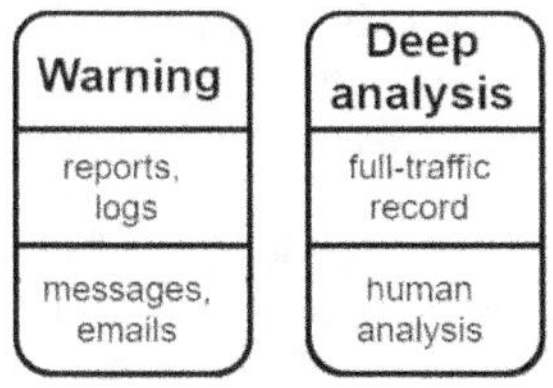

Fig. 5. Warning and Deep analysis block diagram.

4. Influence of Transmission Medium on Network Communication

The communication between the universal monitor Mega44PAN and the concentrator was measured to verify the influence of communication media. For the transmission was chosen Broadband over power lines (BPL), which is a method PLC and allows relatively high-speed data communication over the public electric power distribution wiring. The distance between the devices was approximately 5.4 km, with a total of 19 repeaters. It was a real connection of the metering location (Distribution transformer station – MEga44PAN) of the power company with data transmission to the central node (Concentrator).

Communication on the application layer was performed using the IEC 60870-5-104 protocol where concentrator was sent a request on a set of objects and responses containing the measured data of the Mega44PAN. The total volume of one request and one response was 796B, and the request was periodically sent every minute for five days.

Latency was monitored at intervals of one second to include transmission media parameters. Subsequently, the mean value and the mean deviation were calculated from all values at a range of ten minutes. The resulting values were shown in Fig. 6. One sample on the X-axis corresponds to the calculated mean value and the mean deviation. As can be seen in the graph of approximately 600 samples, there was an increase in latency and even a complete loss of communication (zero values in the graph) with the Mega44PAN unit.

After analyzing the communications, it was confirmed that they communicate periodically with a small number of packet types during legitimate standard communications without interference. The graph in Fig. 7 shows the packet sizes that occurred during transmission. The graph shows more than the pattern of sent packets, as well as the number of packet sizes used (only eight packet sizes are repeated).

The breakdown of the formula of Fig. 7 was observed, in the case of legitimate communication, which was affected by interference on the communication medium, Packet lengths recorded during interference are shown in Fig. 8. The graph shows that another packet size was recorded, including packets recorded during legitimate communication without interference. These packets were caused by attempts from the Concentrator side to communicate with Mega44PAN or incomplete Mega44PAN responses.

Furthermore, during the interference, an increase in TCP SYN, ACK, and RST packets was detected as a result of unconfirmed communication by Mega44PAN and attempted to establish a new connection by the Concentrator. Also, during the interference, packet growth was observed in connection attempts and packet drop when Mega44PAN did not respond to TCP connections, and IEC 60870-5-104 requests.

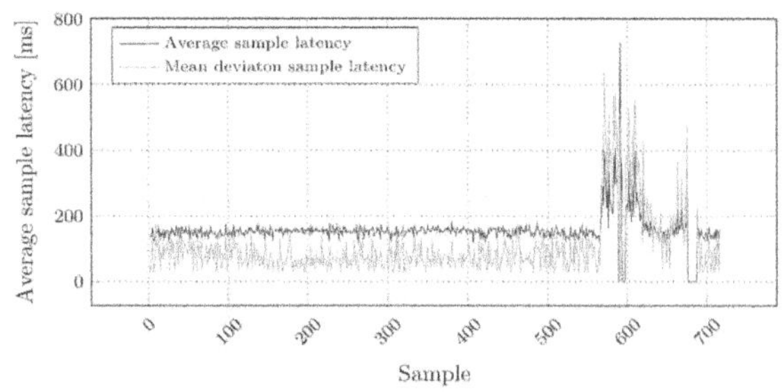

Fig. 6. Mean value and mean deviation of latency.

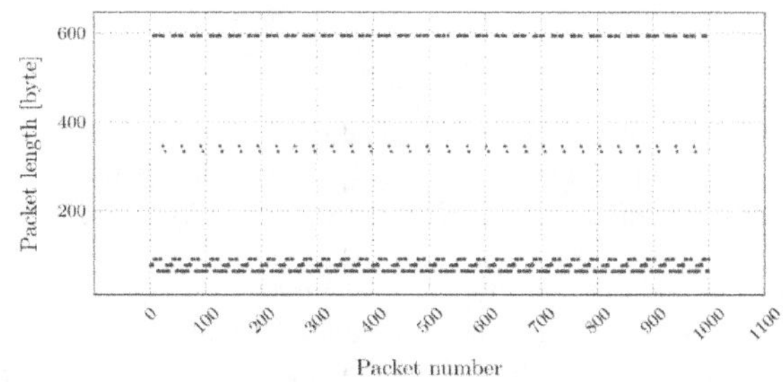

Fig. 7. Legitime IEC 60870-5-104 communication.

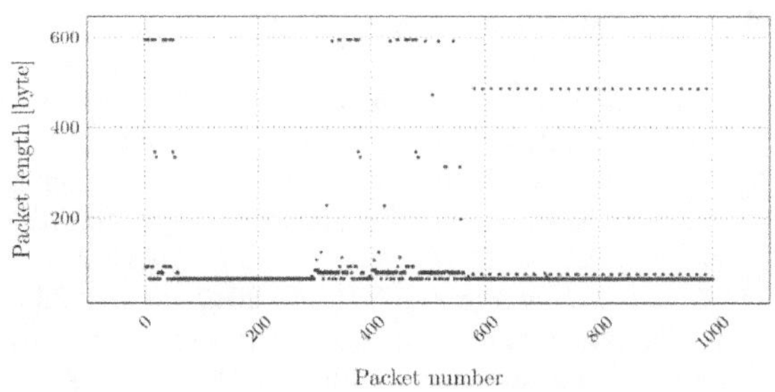

Fig. 8. Interference IEC 60870-5-104 communication.

5. Conclusion

Anomalies are part of every network communication. However, it is essential to distinguish the type of anomaly. This work is comprehensively focused on anomaly detection in OT sensors networks. Therefore, the work presents a model which classifies anomalies based on the detected anomaly and transmission parameters of the communication medium. Further research should focus on extending the Deep Analysis part for signatures record representing the critical values of the system.

References

[1]. Z. Drias, A. Serhrouchni, O. Vogel, Analysis of cyber security for industrial control systems, in *Proceedings of the International Conference on Cyber Security of Smart Cities, Industrial Control System and Communications (SSIC'15)*, 2015. p. 1-8.

[2]. B. Galloway, G. P. Hancke, Introduction to industrial control networks, *IEEE Communications Surveys & Tutorials*, Vol. 15, Issue 2, 2012, pp. 860-880.

[3]. Y. Kabalci, A survey on smart metering and smart grid communication, *Renewable and Sustainable Energy Reviews*, Vol. 57, 2016, pp. 302-318.

[4]. E. U. Ogbodo, D. Dorrell, A. M. Abu-Mahfouz, Cognitive radio based sensor network in smart grid: Architectures, applications and communication technologies, *IEEE Access*, Vol. 5, 2017, pp. 19084-19098.

[5]. P. Mlynek, et al., Simulation of achievable data rates of broadband power line communication for smart metering, *Applied Sciences*, Vol. 9, Issue 8, 2019, 1527.

[6]. M. Kuzlu, M. Pipattanasomporn, S. Rahman, Communication network requirements for major smart grid applications in HAN, NAN and WAN, *Computer Networks*, Vol. 67, pp. 201474-201488.

[7]. C.-J. Hsieh, T.-Y. Chan, Detection DDoS attacks based on neural-network using Apache Spark, in *Proceedings of the International Conference on Applied System Innovation (ICASI'16)*, 2016. p. 1-4.

[8]. S. N. Shirazi, et al., Evaluation of anomaly detection techniques for SCADA communication resilience, in *Proceedings of the Resilience Week (RWS'16)*, 2016, pp. 140-145.

[9]. W. Ren, T. Yardley, K. Nahrstedt, EDMAND: Edge-Based Multi-Level Anomaly Detection for SCADA networks, in *Proceedings of the IEEE International Conference on Communications, Control, and Computing Technologies for Smart Grids (SmartGridComm'18)*, 2018. p. 1-7.

[10]. I. A. Siddavatam, et al., An ensemble learning for anomaly identification in SCADA system, in *Proceedings of the 7th International Conference on Power Systems (ICPS'17)*, 2017, pp. 457-462.

[11]. J. Vávra, M. Hromada, Evaluation of anomaly detection based on classification in relation to SCADA. in *Proceedings of the International Conference on Military Technologies (ICMT'17)*, 2017, pp. 330-334.

[12]. I. Ullah, Q. H. Mahmoud, A hybrid model for anomaly-based intrusion detection in SCADA networks, in *Proceedings of the IEEE International Conference on Big Data (BigData'17)*, 2017, pp. 2160-2167.

[13]. S.-J. Kim, et al., Network anomaly detection for m-connected SCADA networks, in *Proceedings of the Eighth International Conference on Broadband and Wireless Computing, Communication and Applications (BWCCA'13)*, 2013, pp. 351-354.

[14]. Z. Martinasek, et al., Methodology for correlations discovery in security logs, in *Proceedings of the 9th International Congress on Ultra Modern Telecommunications and Control Systems and Workshops (ICUMT'17)*, 2017, pp. 294-298.

[15]. I. F. Akyildiz, et al., A survey on sensor networks, *IEEE Communications Magazine*, Vol. 40, Issue 8, 2002, pp. 102-114.

[16]. A. Vadla, Cyber physical systems by using wireless sensor networks, *International Journal of Science and Research*, Vol. 79, Issue 57, 2016, pp. 1380-1396.

[17]. F.-J. Wu, Y.-F. Kao, Y.-C. Tseng, From wireless sensor networks towards cyber physical systems, *Pervasive and Mobile Computing*, Vol. 7, Issue 4, 2011, pp. 397-413.

[18]. S. Rajasegarar, C. Leckie, M. Palaniswami, Anomaly detection in wireless sensor networks, *IEEE Wireless Communications*, Vol. 15, Issue 4, 2008, pp. 34-40.

[19]. S. Rajasegarar, et al., Distributed anomaly detection in wireless sensor networks, in *Proceedings of the 10th IEEE Singapore International Conference on Communication Systems (ICCS'06)*, 2006. p. 1-5.

[20]. M. Xie, et al., Anomaly detection in wireless sensor networks: A survey, *Journal of Network and Computer Applications*, Vol. 34, Issue 4, 2011, pp. 1302-1325.

[21]. V. Bhuse, A. Gupta, Anomaly intrusion detection in wireless sensor networks, *Journal of High Speed Networks*, Vol. 15, Issue 1, 2006, pp. 33-51.

[22]. S. Rajasegarar, et al., Quarter sphere based distributed anomaly detection in wireless sensor networks, in *Proceedings of* the *IEEE International Conference on Communications (ICC'07)*, 2007, pp. 3864-3869.

[23]. I. Onat, A. Miri, An intrusion detection system for wireless sensor networks, in *Proceedings of the IEEE International Conference on Wireless And Mobile Computing, Networking And Communications (WiMob'05)*, 2005, pp. 253-259.

[24]. C. Eik Loo, et al., Intrusion detection for routing attacks in sensor networks, *International Journal of Distributed Sensor Networks*, Vol. 2, Issue 4, 2006, pp. 313-332.

[25]. D. Ramotsoela, A. Abu-Mahfouz, G. Hancke, A survey of anomaly detection in industrial wireless sensor networks with critical water system infrastructure as a case study, *Sensors*, Vol. 18, Issue 8, 2018, 2491.

[26]. R. Magán-Carrión, J. Camacho, P. García-Teodoro, Multivariate statistical approach for anomaly detection and lost data recovery in wireless sensor networks, *International Journal of Distributed Sensor Networks*, Vol. 11, Issue 6, 2015, 672124.

[27]. H. S. Emadi, S. M. Mazinani, A novel anomaly detection algorithm using DBSCAN and SVM in wireless sensor networks, *Wireless Personal Communications*, Vol. 98, Issue 2, 2018, pp. 2025-2035.

[28]. M. Xie, et al., Scalable hypergrid k-NN-based online anomaly detection in wireless sensor networks, *IEEE Transactions on Parallel and Distributed Systems*, Vol. 24, Issue 8, 2012, pp. 1661-1670.

[29]. S. Rajasegarar, et al., Centered hyperspherical and hyperellipsoidal one-class support vector machines for anomaly detection in sensor networks, *IEEE Transactions on Information Forensics and Security*, Vol. 5, Issue 3, 2010, pp. 518-533.

[30]. S. A. Yusuf, et al., Application of dynamic neural networks with exogenous input to industrial conditional monitoring, in *Proceedings of the International Joint Conference on Neural Networks (IJCNN'13)*, 2013, p. 1-8.

[31]. Z. Bankovic, et al., Improving reputation systems for wireless sensor networks using genetic algorithms, in *Proceedings of the 13th Annual Conference on Genetic and Evolutionary Computation (GECCO'11)*, 2011.

(126)

New Trends in Measurement Science. Bayesian Intelligent Measurements

S. V. Prokopchina
Financial University under the Government of the Russian Federation,
Moscow, Russian Federation
E-mail: svprokopchina@mail.ru

Summary: The development of measurement theory is primarily associated with the creation of new types of measurements, such as intelligent, fuzzy, soft, cognitive, polysystem, entropy and other types of measurements. The distinctive features of these types of measurements is the possibility of using different types of knowledge in the implementation of measurements. In this regard, methods of measurement of these species include methods and models of artificial intelligence, fuzzy set theory, cognitive theory. From a methodological point of view, this diversity of new types of measurements together with classical ones determines the necessity of their systematization and classification for their effective and purposeful application. This issue is devoted to the article. The concept of the Bayesian Intelligent Measurement (BIM), based on the Regularizating Bayesian Approach and methodology of Bayesian Intelligent Technologies (BIT) and Soft Measurement, is considered. Parametrical (for data integration), functional (for models integration), multy-factors (for recognition of the situations) BIM as well as metrological aspects of obtaining decisions are considered. The analytical equations for these types of BIT on concept level are given. The "probability-possibility" integrant technologies, features of BIT for models integration, making decisions under conditions of great a priory uncertainty on the basic of BIT, definition of the situations and main influence factors are discussed. BIT-versions for integration of the perspective information technologies and distributed information recourses is given. The article deals with the methodology and methodology of creating scales with dynamic constraints, focused on applying in the face of uncertainty. Formula dependences for definition of borders of classes of reference points of a scale are resulted. Defined and formalized the terms of the distinctiveness of the classes of reference points of the scale.

Keywords: Measurement Science, Bayesian Approach, intelligent measurement, soft measurement.

1. Introduction

The development of Measurement Science stimulated the expansion of metrological requirements and nomenclature of indicators of data quality, as well as the emergence of new types of measurements, metrological certification and control, in particular, Metrology algorithms, models, objects and measurement conditions.

As a rule, the practice of modern measurement tasks are accompanied by complex experimental conditions associated with the presence of a considerable a priori uncertainty about the properties of objects and of the factors affecting the environment of their functioning, the relationships between them, the inaccuracy and incompleteness of the experimental information, inaccessibility to direct observation of many properties of objects, or influencing factors, what distinguishes the cognitive function of the methodology of their solutions as fundamental. Therefore, the formulation of these problems as measuring causes the strengthening of the role of cognitive function of measurements and demands the results of their decisions in the form of knowledge (analytical expressions for models, as well as conclusions and decisions) on the basis of taking into account the entire amount of a priori information and information received during the measuring experiment, including non-numeric. Thus, the process of understanding the results of measurements in the form of numbers, as well as analytical conclusions and decision-making are included in the contour of the measuring processes. However, *this is permissible only in the presence of metrological support at all stages of decision-making. Only under this condition, the processing of information can be attributed to the measuring.*

We define this type of measurement tasks as a type of *complex measurement tasks*, understanding this complex dimension objects, which is a system of interrelated properties that are actively interacting with the ambient environment, complex and multi-step analytical measuring processes, difficult conditions of measurement for information uncertainty and instability as the objects themselves, and their environment.

The fulfillment of this requirement contributed to the attraction of the theory of artificial intelligence apparatus, optimal solutions, computer science to the creation of new information-measuring technologies for processing heterogeneous flows of experimental data by means of measuring equipment.

2. Classification of Types of Measurement

One of the main points of construction of measuring technologies is to determine the type of information situation in which measurements were made. There are three types of measurement information situations.

This is a situation with completely certain information, and the conditions of measurement (type I); the situation with incomplete certain information and stable measurement conditions (type II), but this situation with considerable uncertainty, and partial uncertainty can be removed in the iterative mode , adapting the model of the object of measurement according to the information received (type II) and situations with significant uncertainty, instability of the measurement conditions , the active influence of the external environment (type III).

Identifying this specificity of methods, there appeared, in addition to the existing direct measurements, which are implemented in the first information situation, the concept of:

- Integrated measurement based on extended concepts of the measurement and thesauri, which are implemented as structures model; (situation (type II);
- Statistical measurements [based on the probability-theoretical approach; situation (type II);
- Dynamic adaptive measurements, emphasizing the time characteristics of the properties of objects of measurement; situation (type II);
- Adaptive measurements with correction of the result in the process of measurements by a given algorithm; situation (type II);
- Interval measurements of the situation (type II) and (type III);
- Processor measures that allocate the class of it tools necessary for their implementation; situation (type II);
- Algorithmic measurements determining the relationship with the computational aspects of solving measurement problems; situation (type II).
- Measure, reflecting their applied focus (industrial measurement, bioisoprene, radio astrophysics, aerobalance, social dimension, economic dimension, etc.) situation (type II), and intellectual dimensions of the situation (type III).

In further works of the author of the given article concepts were offered, theoretical bases, the principles and information technologies and means of the following types of measurements which are realized in situations are developed (type III):

- Bayesian intelligent measurements based on the Bayesian regularization approach;
- Fuzzy measurements in the form of a set of alternative solutions with varying degrees of reliability;
- Soft measurements with flexible logic output measurement solutions;
- The cognitive dimension with the inclusion of the subject of the probe in the measuring technology;
- Polysystem measurements with the object of measurement in the form of complex Autonomous systems. actively interacting with each other and external environment for this population;
- Entropy measurements in which the object of measurement is the amount of information received;
- Retrospective measurement based on the deferral and technologies BIM;
- Prospective measurement based on Bayesian intelligent technologies.

For the above mentioned new types of measurements it is possible to derive the corresponding equations of measurement explaining their methodology, similarities and differences.

The object of the measurement entropy measurement $G_t^{(H)}$ is the amount of information $I_t^{(H)}$, obtained at each new measurement of properties of the observed real object G_t. The difference between the a priori value of the amount of information $I_{t-1}^{(H)}$ and again received $I_t^{(H)}$, allows you to define a change in conditional entropy $\Delta H_t\,^{(G)} \mid D_t\,^{(G)}$ assuming the existence of solutions for the selected dynamic compact measurement solutions $D_t\,^{(G)}$. at a certain point in time t. The amount of information obtained in the experiment is determined by the difference between a priori and a posteriori regularized Bayesian estimators (mles) on a scale of dynamic constraints

Thus, the equation for entropy measurements has the following form:

$$\Delta H_t\,^{(G)} | \, D_t\,^{(G)} = 1 - (I_t^{(H)} - I_{t-1}^{(H)}), \qquad (1)$$

where $I_t^{(H)}$ is the amount of Fisher information for the corresponding a priori and a posteriori regularized Bayesian estimator of the observed Gt object properties, the Calculation of these parameters is not difficult, since RBOs are essentially an estimate of the discrete distribution law of the observed object properties. According to the approximation algorithm estimates distribution laws of the model laws of the system of curves K. Pearson, proposed in, estimates can be approximated and the habit of the amount of information formula for these laws can be used to improve the accuracy of calculations.

In the works of the author on Bayesian statistical measurements and Bayesian mathematical statistics [1-6], theoretical bases, information technologies are developed and applied problems of measurement of numerical characteristics of random variables and processes, correlation measurements, measurements of test hypotheses are solved. measurement of risks and potentials, functional measurements of the distribution laws, measurement of trends, the regression of measurements, entropy measurements in conditions of uncertainty.

For processing of archive data and measurement of events of the past time has developed a methodological framework for retrospective measurement based on the deferral. Historical dimensions allow you to determine estimates of an object's properties at past times or to restore its dynamic model if you have current and archived information of any type about the object and external factors.

For perspective measurements (measurements for future periods and situations) , the basic equation of measurements is constructed taking into account the forecast values associated with the measured other properties of the object and environment $G_{t+1}^{(OE)}$.

Asa type of applying measurement the global measurements have been suggested [5] by the author of this paper.

It is obvious that the above types of measurements are different for objects, methods and application-specific measurement tasks. The systematization and classification of these types of measurements will improve the efficiency of using the full range of types of measurements. Figure 1.the variant of such classification is offered.

As features of the classification are the following:

1. On the method of using knowledge in the measuring process:
   - Classical measurements based on current experimental numerical data;
   - Intelligent measurements using different types of knowledge along with the incoming measurement information in the measurement process.
2. By the method of output of the measuring solution:
   - Probabilistic approach (statistical measurements, Bayesian intelligent measurements);
   - With flexible output logic (soft measurements)
3. According to the method of organization of the measuring process:
   - Without the inclusion of the subject of the probe in the measuring process;
   - With the inclusion of the subject-meter in the contour of the measuring process (cognitive measurement);
4. If possible, the implementation of algorithmic processing:
   - Without the possibility of implementing algorithmic processing (direct measurements);
   - With the possibility of implementing algorithmic processing (indirect measurements, algorithmic measurements).
5. On the complexity of the measurement object:
   - Simple property;
   - Complex system (system measurements, combined, joint measurements);
   - A set of systems (polysystem measurements).
6. On the complexity of the measurement conditions:
   - Stable measurement conditions;
   - Instability conditions (dynamic measurements);
   - Conditions of uncertainty and instability (retrospective, prospective intellectual measurements).
7. The type of application-specific measurement tasks:

Different types of applied measurements (for example, listed above).

This classification, as well as a brief historical review of the development of modern measurement theory, certainly do not claim completeness. Therefore, the author would be grateful to colleagues for discussions and continuation of this topic.

3. The Methodological Aspects of Bayesian Intelligent Measurement

For the purpose of methodological formalization, a generalized equation is obtained for all types of the above types of measurements. So, for Bayesian intelligent technologies, the equation is written as follows [1-3]:

$$\{h_{kt}^{(Q)} | \{MX\}_{kt}^{(Q)}\} | (Y_t^{(O)}\} * Y_t^{(OE)} * G_t^{(OE)}) = \{\text{argexstr}\, C^{(B)}[\varphi_j(f_{it}\{X_{it}\} | (Y_t^{(Q)})\} * Y_t^{(OE)} G_t^{(OE)})]\}, \quad (2)$$

where $\{h_{kt}^{(Q)} | \{MX\}_{kt}^{(Q)}\} | (Y_t^{(O)}\} * Y_t^{(OE)} * G_t^{(OE)})\}$ is the fuzzy solution to the problem under the conditions $Y_t^{(O)}$, (determined by a priori information, requirements, constraints and assumptions) of the functioning of the object; under conditions of environmental factors $Y_t^{(OE)}$; in compact sets of properties of the $G_t^{(O\ )}$ object and the $G_t^{(OE\ )}$ environment; $\{MX\}_{kt}^{(Q)}$ is the complex of metrological characteristics of solutions, containing indicators of accuracy, reliability, reliability, risk of solution, entropy, Fisher information; $C^{(B)}$ is the criterion that defines the logic for solving the problem; in Bayesian smart technologies, this is a Bayesian criterion for the average risk of a decision; soft dimensions is Zadeh's logic; in classical dimensions - this is Lukasevich's logic; φ_j is the Bayesian intelligent technologies that define and implement regularization of solutions; f_{it} is the computing or algorithmic technologies used to process the available primary information; $\{X_{it}\}$ is the set of information flows (data and knowledge) used to obtain decisions.

The equation for fuzzy measurements differs in that in the general case it may have non-Bayesian criteria for deriving solutions of C. But its result, like in equation (1), is presented by a list of alternative solutions with different degrees of confidence, which is calculated for any indicator, for example, Estimates of environmental factors, according to the modified Bayes formula [4]:

$$P(h_{kt}^{F_t} | G_t^{(OE)}) = \frac{P^a(h_{kt\text{-}1}^{(F_t)} | G_{t-1}^{(OE)}) \overset{L}{\circ} P(G_t^{(OE)} | h_{kt}^{(F_t)})}{\sum_{j=1}^{J} P^a(h_{jt\text{-}1}^{(F_t)} | G_{t-1}^{(OE)}) \overset{L}{\circ} P(G_t^{(OE)} | h_{jt}^{(F_t)})}, \quad (3)$$

Cognitive measurements are organized by combining experimental information $\{X_{it}\}$ and available data with the estimates, knowledge and solutions of $\{X_{it}^{(st)}\}$ of the subject-meter, which was proposed and considered in detail in [5]. Based on this, the equation for cognitive measurements is written as:

$$\{h_{kt}^{(Q)} | \{MX\}_{kt}^{(Q)}\} | (Y_t^{(O)}\} * Y_t^{(OE)} * G_t^{(OE)}) = \{\text{argexstr}\, C^{(L)} [\varphi_j(f_{it}\{X_{it}\} * \{X_{it}^{(st)}\} | (Y_t^{(Q)})\} * , \quad (4)$$
$$* Y_t^{(OE)} G_t^{(OE)})]\}$$

In the articles of the author [5, 6] the cognitive Bayesian networks were suggested.

4. About the Difference between Classical and Bayesian Intelligent Measurement

Differences from the measurement equation in the classical scheme of measuring physical quantities are: explicitly introducing the criterion for choosing a measurement solution into the equation, information and technological transformations in accordance with the measurement methodology and methodology, influencing external factors, as well as the conditions formalizing the measurement situation.

Among the main differences from the classical methodology for measuring physical quantities are also the following. Ensuring the development of methods, models, measurement algorithms depending on changing measurement conditions.

Cognitiveness and interpretability is provided directly in the process of forming a measurement solution. In the process of measurement, it is possible to use, process and convolve both numerical and high-quality, linguistic information.

Ensuring the stability of the resulting solutions during integration, identification of system models and their characteristics. Obtaining measuring solutions not only in the form of numerical values of parameters as in classical measurements, but also in the form of models, audit decisions, recommendations, assessments, conclusions, scenarios.

The conditional nature of the solutions obtained, that is, the measurement solutions will be valid only if the experimental conditions are met, namely: within the framework of this compact solutions, the constraints, metrological requirements, and a priori information available.

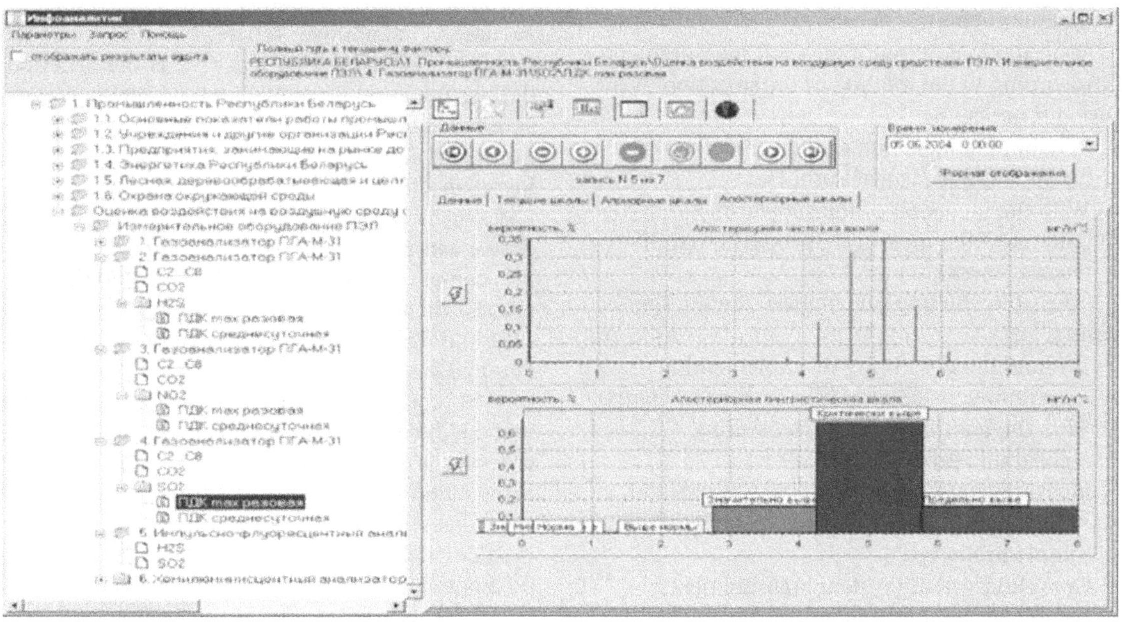

Fig. 1. An example of SDC with a measuring solution for the state of the air in the form of multi-alternative quantitative and interpreted linguistic results.

The measuring process SM and BIM has some more differences from the classical measurement scheme. They are as follows:

1. Measurement is implemented as a decision-making process on conjugate numerical and linguistic scales, called scales with dynamic constraints (SDC). Such scales are multidimensional metric compacts that can change their structure over time. The view of the SDC is shown in Fig. 1.

2. Each value of the measured parameter corresponds to a value of its reliability (probability or possibility). Thus, the scale for measuring the value of a quantity has the form of a two-dimensional scale, on which the values of magnitude are plotted along the abscissa axis, and the ordinates – their ordinates – their probability or possibility (see Fig. 1). Moreover, the measuring solution can be obtained both on a numerical and on a conjugate linguistic scale.

3. Each SM and BIM decision is a multi-alternative series of values with the corresponding certainties.

4. The SDC may include criteria scales, conjugated with the main scales, as illustrated in Fig. 2. Simultaneously with the determination of the measuring solution on the linguistic scale, the solution is checked according to the norm criterion. Several criteria conjugate scales can be used to carry out nogocriteria verification.

5. Models of measurement objects have the form of models with dynamic constraints (MDC), the properties of which can change over time, in accordance with which the SDC scales will change. A detailed description of the MDC models and the SDC scales is given in [1-6].

6. In the process of forming a measuring solution of BIM and SM, the knowledge and expert assessments can be used.

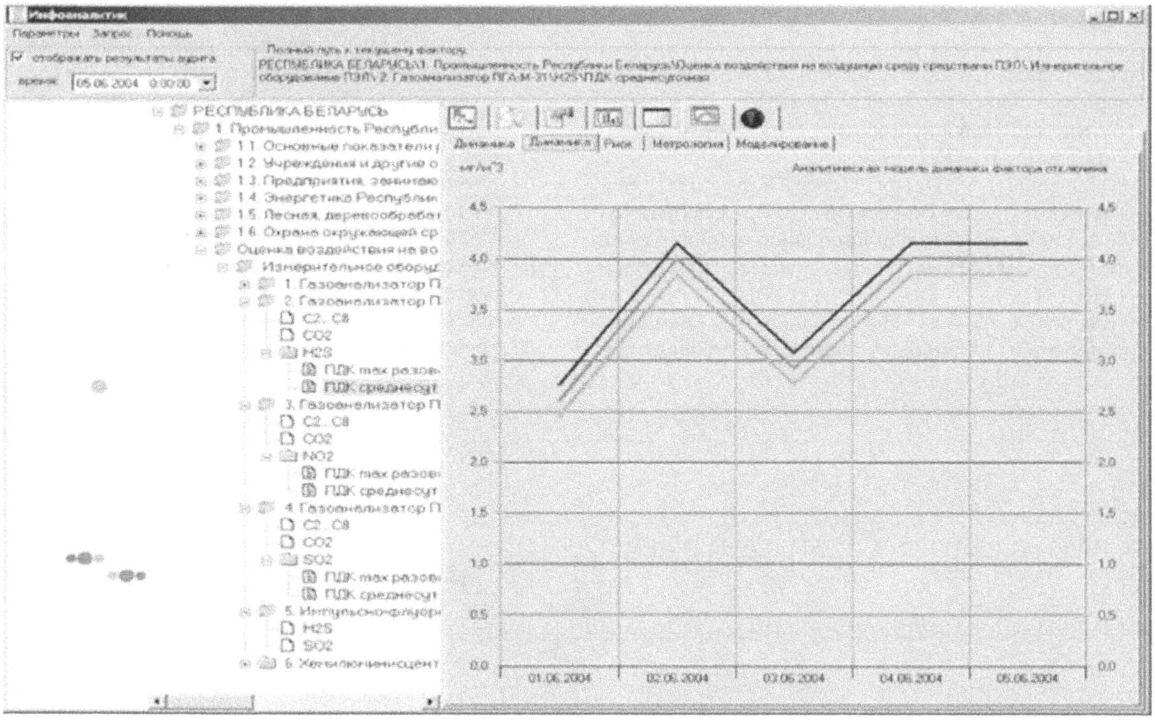

Fig. 2. Measurement of indicator dynamics based on BIM.

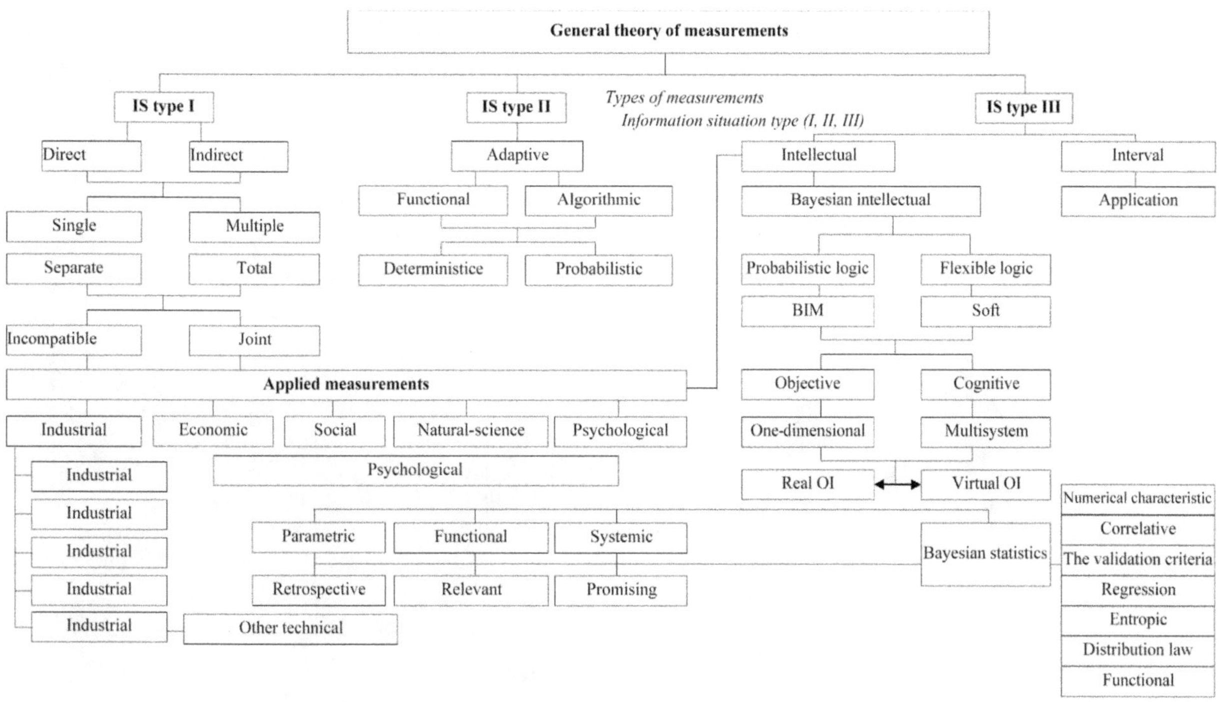

Fig. 3. Classification of modern types of measurements.

This types of dynamical models cfn be considered as the fuzzy regression model and can be useful for solving tasks of econometrics under conditions of uncertainty.

5. Conclusions

Thus, the new classification of the types of measurement has given. The methodology and methodology of Regularizating Bayesian Approach and Bayesian Intelligent Measurement (BIM) are considered. The main principles of BIM and Soft Measurement and creating scales with dynamic constraints, focused on applying in the conditions of uncertainty have discussed. Set of main formulas of BIM methodology are resulted. An example of using this type of scales to measure dynamical and nonphysical items have given.

References

[1]. Prokopchina S. V., Development of the method of approximation of the probability density of a random variable and a random process by typical distributions, PhD Thesis , L., *LETI*, 1976.

[2]. Prokopchina S. V., Doctor Science Thesis , SPb, 1996.

[3]. Nedosekin D. D., Prokopchina S. V., Cherniavsky E. A. Intellectualization of information technologies of measuring processes, *Energoatomizdat*, Leningrad, 1996, 187 p.

[4]. Prokopchina S. V., The creation of emerging information technologies based on regularizing Bayesian approach, in *Proceedings of the IX International Conference 'Soft Calculations and Measurements'*, Saint-Petersburg, LETI, 2007.

[5]. Prokopchina S. V., The organization of the measurement process under uncertainty. Regularizing Bayesian approach, in *Proceedings of the International Conference on Soft Computing and Measurements SCM-98*, Saint-Petersburg, 22-26 June 1998, Vol. 1.

[6]. Prokopchina S. V., Basic principles of solving economic problems in the conditions of information uncertainty, *Scientific Works of the Free Economic Society of Russia*, Vol. 144, 2010.

(127)

Financial Measurement by the Means of Bayesian Regularization Approach

V. B. Dernovoi [1], **S. N. Mishchenko** [2]
[1] Publishing house "Science library"
[1] 123022 Moscow, Zvenigorodskoye Shosse, 5, Building 1
Tel.: +7 (495) 592-29-98 / +7 (915) 087-73-76
E-mail: idnb11@yandex.ru
[2] Financial University under the Government of the Russian Federation
E-mail: SNMishchenko@yandex.ru

Summary: The article considers the use of the analytical system 'Infoanalyst' for the analysis of the composition and structure of the balance sheet liability based on BITS. The audit of the analysis of the balance sheet liabilities composition and structure dynamics for 2007, 2008 and 2009 is presented.

Keywords: Bayesian intelligent technologies, System "Infoanalyst", BIT-model, Factor tree, Integral factor, Audit analysis of the dynamics and structure of liabilities of the balance.

1. Introduction

The analysis of financial condition carried out on the basis of data of the accounting and statistical reporting for 2008-2009 allows to estimate property condition, solvency, financial stability and business activity of LLC Management on preparation of technological liquid for maintenance of reservoir pressure on the basis of a number of generalizing analytical indicators.

For the implementation of financial analysis were applied platform "Infoanalyst". The main evaluation factor is the factor "analysis of financial situation" and depends on the analysis of financial results, control of economic activity and tariff calculation. Since the factor of analysis of financial results can not be estimated immediately, it is an integral factor, which, in turn, depends on the analysis of the dynamics of the composition and structure of the balance sheet liabilities, described in detail in this article.

The factor "Analysis of the balance sheet liabilities composition and structure dynamics" is integral, a BIT model is formed for it.

$$Q_{1.5} = Q_{1.5.1} * Q_{1.5.2} * Q_{1.5.3}, \quad (1)$$

where $Q_{1.5}$ is the analysis of the composition and structure of the balance sheet liabilities, $Q_{1.5.1}$ is the total sources of formation of assets of the organization, $Q_{1.5.2}$ is the borrowed sources,

$$Q_{1.5.1} = Q_{1.5.1.1} * Q_{1.5.1.2}, \quad (2)$$

where $Q_{1.5.1.1}$ is the long-term borrowings, $Q_{1.5.1.2}$ is the short-term borrowings, $Q_{1.5.3}$ is the own capital.

The tree of the integral factor is shown in Fig. 1.

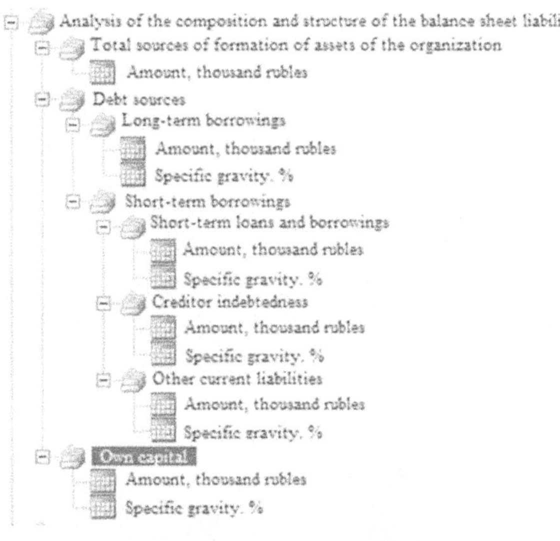

Fig. 1. The tree is the integral factor "analysis of dynamics and structure of liabilities of the balance".

The analysis of the dynamics of the composition and structure of the balance sheet liabilities for 2007-2009 is shown in Figs. 2-4.

Analysis of the composition and structure of the organization's liabilities leads to the conclusion that the main source of formation of the organization's property is the company's own funds. At the end of the analyzed period, the share of equity in the structure of liabilities was 80 %. This is slightly lower compared to the beginning of the analyzed period – 92, 31.

At the end of the analyzed period, the share of equity, the main source of formation of the enterprise's property, in the structure of liabilities decreased.

Table 1. Analysis of sources of formation of assets of the organization.

Indicators	Year 2008		Year 2009		Deviation
	The sum, thousand RUB	Specific weight, %	The sum, thousand RUB	Specific weight, %	The sum, thousand RUB
1. Total sources of formation of assets of the organization	802054	100	1066601	100	264547
Among them:					
1.1. Own capital	740775	92.36	853254	80	112479
1.2. Debt sources	61279	7.64	213347	20,00	152068
1.2.1. Long-term borrowings	0	0.00	0	0,00	0
1.2.2. Short-term borrowings	61279	7.64	213347	20,00	152068
- short-term loans and borrowings	0	0.00	0	0,00	0
- creditor indebtedness	14693	1.83	122082	11.45	107389
- other current liabilities	46586	5.81	91265	8.56	44679

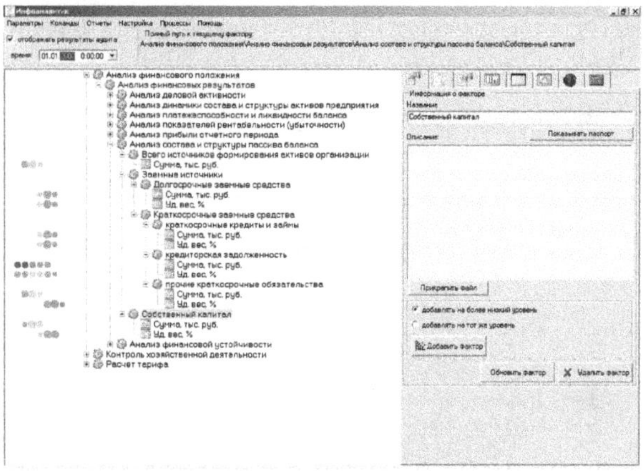

Fig. 2. Audit of analysis of the composition and structure of the balance sheet liabilities for 2007.

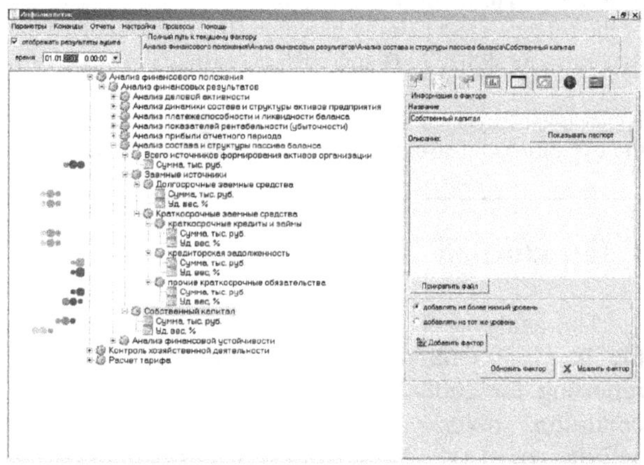

Fig. 4. Audit analysis of the dynamics of composition and structure of liabilities of the balance sheet for 2009.

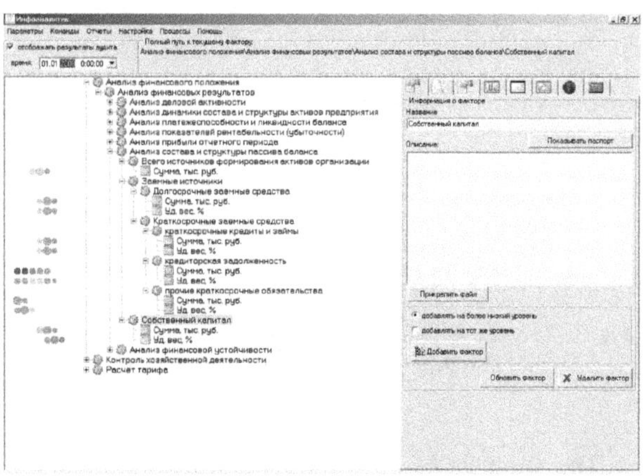

Fig. 3. Audit analysis of the dynamics of composition and structure of liabilities of the balance sheet for 2008.

The largest share in the structure of accounts payable at the end of the analyzed period is occupied by debt to suppliers and contractors – 108385 thousand rubles, debt to other creditors – 9070 thousand rubles, debt to the staff of the organization 3139 thousand rubles.

Graphical analysis of the structure of the sources of formation of assets of LLC "Management for the preparation of process fluid to maintain reservoir pressure" at the end of 2009 is shown in Fig. 4.

Short-term debt sources the company at the end of 2009 short-term payables in the amount of RUB 122082 thousand, debt on farm settlements in the amount of RUB thousand 91265 Growth of accounts payable may have been the result of untimely repayment of receivables.

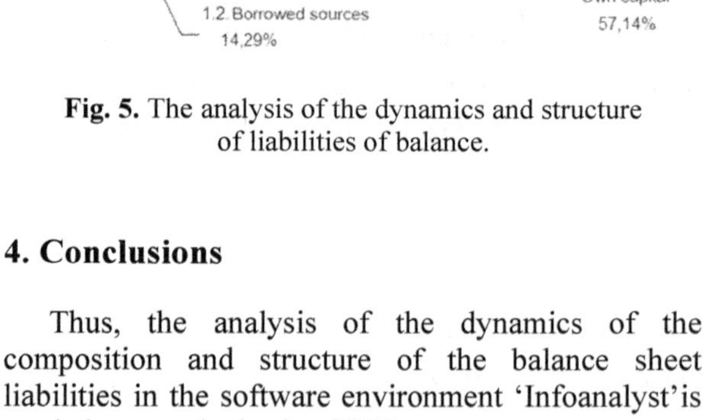

Fig. 5. The analysis of the dynamics and structure of liabilities of balance.

4. Conclusions

Thus, the analysis of the dynamics of the composition and structure of the balance sheet liabilities in the software environment 'Infoanalyst' is carried out on the basis of BIT.

References

[1]. V. V. Kovalev, Financial Analysis, Money Management, Investment Choice, Reporting Analysis, 2nd Ed., *Finance and Statistics*, Moscow, 1997.

[2]. S. V. Prokopchina, A. I. Fedichkin, Application of Bayesian Intelligent Technologies (BITS) for estimation of integral indices, in *Proceedings of the International Conference on Soft Computing and Measurements (SCM'06)*, Vol. 2, St. Petersburg, 27-29 June 2006, pp. 20-22.

[3]. S. V. Prokopchina, A. I. Fedichkin, Forecasting of characteristics of complex objects on the basis of Bayesian Intelligent Technologies, in *Proceedings of the International Conference on Soft Computing and Measurements (SCM'06)*, Vol. 2, St. Petersburg, 27-29 June 2006.

(128)

The Global Measurement. The New Trend in Measurement Science

Ilia V. Iljin [1], Svetlana V. Prokopchina [2]
[1] Moscow State University
[2] Financial University under the Government of the Russian Federation
Moscow, Russian Federation
E-mail: svprokopchina@mail.ru

Summary: A wide range of processing technologies and information uncertainty based on methods of artificial intelligence, soft computing, Bayesian and measurement approaches provide a meaningful choice among them those that will be most effective in the modeling and processing global information. For elaboration of methodological and instrumental bases, as well as the number re-surveyed applied problems in the areas of modeling of global systems, one of the leaders is the methodology of regularizing Bayesian approach.

Keywords: Global systems, Bayesian approach, Bayesian Intelligent Measurement.

1. Introduction

Modern conditions of global systems parameters and processes measure and modeling have a significant degree of uncertainty. Further development of global analytics is possible only with the deepening of its technological specialization and the introduction of new advanced technologies on this basis, which, in turn, brings additional instability and uncertainty associated with innovative processes in the evaluation of the above indicators. Therefore, currently the most urgent tasks for the global measurement and modeling is the development of methods and tools for assessing the indicators of preparation and construction of models and decisions under uncertainty.

The uniqueness and specialization of modern models of global systems, as well as the influence of all the above factors lead to the need to apply and develop special methods and techniques to modeling and making decisions processes.

2. The Principles of Global Measurement and Modeling on the base of Regularization Bayesian Approach

One of the main directions was the development and use of intelligent computing and measurement technologies in the design and manufacture of complex global objects in conditions of uncertainty and risk. Achievements of this direction are considered in the works of domestic and foreign scientists.

One of the leaders in the development of methodological and instrumental bases, as well as in the number of solved applied problems in the fields of social and economical tasks is the methodology of Regulariziting Bayesian Approach (RBA). Classes of Bayesian Intelligent Technologies (BIT), including technologies of Global Measurement (GM), for solving problems of measurement, monitoring, auditing and management of complex global objects and processes under the conditions of considerable uncertainty are based on RBA [1]. Based on RBA system models can be represented as dynamic constraint models (MDC).

According to the active interaction global systems $G^{(o)}$ and impact factors of the environment $G^{(e)}$ under conditions $Y^{(oe)}$ of uncertainty we present global system and processes conceptual model $G^{(moe)}$ as following:

$$G^{(moe)} = G^{(mo)} * G^{(me)} * Y^{(oe)}, \qquad (1)$$

where:
* is the symbol of the convolution (for BIM - Bayesian convolution);
$G^{(mo)}$ is the model of global processes;
$G^{(me)}$ is the model of system of impact factors of the environment under conditions of uncertainty.

Following the methodologies issue of BIM and BIT, considered in [1-3], it possible to develop the concrete algorithms of Global Measurement. As the example of GM, we suggest the processes of GM of globalization.

3. Measurement of the Globalization under Condition of Uncertainty

The measurement of globalization, as a rule, is based on the definition of a set of indicators (indices) that allow you to measure the degree of globalization of different countries and to rank these indices.
Well-known indices are the following:

1. $G(_1^{mr})$ – Maastricht Index (developed by the International Center for Integrated Assessment and Sustainable Development (ICIS). He became one of the first indices to measure globalization.
2. $G(_2^{mr})$ – KOF (Developed by Swiss Economical Institute and Swiss Institute of Technology.
3. $G(_3^{mr})$ – Index A.T. Kearney proposed by Foreign Policy magazine.
4. $G(_4^{mr})$ – Ernst & Young Company Index.

5. $G(_5^{mr})$ – University of Warwick Index.
6. $G(_1^{mr})$ – Index developed by the Trans Europe Network.
7. And others, given, for example, in [1].

As noted in [1], the main difficulties when working with the above indices are their instability, the impossibility of integrating their solutions, the lack of indicators of geographic and vertically oriented specifics, volatility and other shortcomings.

The explanation for this is the informational specificity of indices. They are based on incomplete, inaccurate, fuzzy information, inconsistent according to the methods of collecting information, often objectively and subjectively distorted. All this causes uncertainty in the conditions of their measurements, incompleteness and inadequacy of models of global processes and systems.

Therefore, to measure globalization, it is necessary to use measurement methods oriented to uncertainty conditions, which ensure the fulfillment of three basic conditions for the correctness of Hadamard decisions.

These methods include methods based on the RBP, that is, Bayesian intelligent measurements. In relation to the specifics of the processes, they can be called global measurements (GM), as given in [5].

The conceptual model of globalization measurements can be presented as a convolution of the indices given above. Such a model, implemented according to the BIM methodology in the form of a model with dynamic constraints, will be stable on a multitude of characteristic indicators and in dynamics.

It will provide stable, highly reliable estimates and measurement solutions, identify trends in integrated and private indices with enough reliability, and will also allow to identify risks and potentials of globalization in different countries, the main influencing factors, the reasons for their formation and management recommendations.

A GP model with dynamic constraints in accordance with the above notation for the indices included in it has the form:

$$G(mr) =* i=1,..6\ Gi(mo) * G(me)*Y(oe), \qquad (2)$$

When measuring for the implementation of the measurement process according to MDC, scales with dynamic constraints (SDC) should be built, which will allow, in addition to numerical indicators in the indices, to use qualitative indicators. Such as the depth and breadth of globalization and others.

The most basic issue of integrating various indices is solved by applying the Bayesian convolution of indicators in the indices and the indices themselves in formula (2).

The important issue of determining the weight of indicators in indices when measuring globalization is solved quite simply. Integration of identical repeatable indicators increases their weight automatically when their convolution is implemented in proportion to the frequency of their use in various integrable indices.

The system in the generalized indicators can be presented in the form of a set of the following interconnected systems.

4. Conclusions

New type of applying type of measurement, as a branch of Bayesian Intelligent Measurement, named as global measurement is represented. The concept models of these measurement are given.

All these models are used in determining different parameters and status of globalization objects under conditions of instability and uncertainty.

A wide range of processing technologies and information uncertainty based on methods of Bayesian measurement approaches provide a meaningful choice among them those that will be most effective in monitoring, modeling, making decisions under conditions of uncertainty.

References

[1]. I. Iljin *at all.*, Historical Globalistics, *Publishing Comp. Uchitel*, Moscow, 2017, 415 p.

[2]. Prokopchina S. V., Methods of mathematical statistics and econometrics under uncertainty based on the regularizing Bayesian approach, *Soft Measurements and Calculations*, 7, 2018, 7, pp. 30-51.

[3]. Prokopchina S. V., Soft Measurements: Methodology and Application in the Scientific, Technical, and Socio-Economic Tasks of the Digital Economy, *Soft Measurements and Computations*, 9, 2018, pp. 4-33.

[4]. Prokopchina S. V., A system approach in conditions of uncertainty. From system measurements to system synthesis, *Soft Measurements and Calculations*, 11, 2018, pp. 3-13.

[5]. Prokopchina S. V., Measuring Aspects of the Digitization of Socio-Economic Processes Based on the Regularizing Bayesian Approach, *Economics and Management: Problems, Solutions*, 8, 3, 2019, pp. 41-45.

[6]. Prokopchina S. V., Bayesian Intellectual Technologies as a Methodological Basis for Big Data Processing under Uncertain Approach Conditions, *Economics and Management: problems, solutions*, 11, 3, 2019, pp. 105-109.

(131)

The Methodologies Aspects of Measurement in Energy and Housing Applications

S. Sheremet'ev, A. Frolov
Tver State University, Tver, Russian Federation
E-mail: sheremetev_sa@mosvodokanal.ru, frolov_aleksey.19@mail.ru

Summary: The article deals with the process approach to water management on the basis of Bayesian intelligent technologies. This is an example of the use of ARIS notation in preventive maintenance. The basic processes of the water supply enterprise are allocated.

Keywords: Bayesian intellectual technologies; the process approach; management efficiency increase; the water economic enterprise; reconstruction of pipelines

1. Introduction

The water enterprise, as well as any structure or organization, has the goal of improving the efficiency of the management system. One of the main tasks of the transition to a new management system is the transition to the management of the enterprise on the basis of the process approach.

Process approach today is the most famous and popular management concept. But in the field of water supply management, the application of the process approach has not yet been implemented. There are a number of difficulties in moving to a process approach, as:

- Management for water management enterprise, based on the process approach, is just starting to develop;
- Professional level of education of managers is traditionally focused on planning and execution of management functions, not on processes;
- The most information technologies are also focused on the application of a functional approach to management, rather than process;
- There is a long-established practice of managing executors ' functions rather than the processes in which they participate;
- Senior management's attention is mainly focused on financial indicators, which are a priori considered indicators of success or failure.

But to date, the process approach based on international standards is already being implemented as pilot projects at some enterprises of water management complexes, for example, at Mosvodokanal. And also cascade training of heads of the enterprises to process approach and quality management system is carried out. In this regard, for the effective and early implementation of the process approach, as well as the creation of a quality management system, it is necessary to develop an information technology basis for the production processes of the enterprise. These issues are reflected in this Chapter.

The main processes of the water supply company include the following processes: preventive maintenance, supervision of the condition and maintenance of the network, maintenance and preventive maintenance, etc.

To apply the process approach in preventive maintenance, it is necessary to determine all the processes that arise during its planning, organization and implementation. Their description begins with the definition of external inputs and outputs of preventive maintenance, with the fixation of their external environment. For preventive maintenance, the external environment is other processes of the enterprise, departments involved in this process, labor, financial and administrative resources. External inputs and outputs are information and material flows through which preventive maintenance interacts with its environment.

The purpose of the system of preventive maintenance is to prevent accidents, improve the reliability and sustainability of the urban water supply network, providing customers with services of good quality and in a timely manner while ensuring the safety of their lives and health.

The objectives of the preventive maintenance system are to prevent premature wear and tear of structures, equipment and pipelines, damage and accidents on them; to increase the overhaul life; to reduce the cost of repair and operation of the network by improving the maintenance and care of structures, equipment and pipelines.

The system of preventive maintenance provides:

a) Definition of the list of the constructions and the equipment which is subject to repairs; definition of the type and the maintenance of repair work;

b) Determination of the duration of repair periods, the structure of repair cycles for different types of structures and equipment, taking into account the specifics of their work;

c) Repair planning;

d) Organization of repair works;

e) Provision of technical and estimate documentation;

f) Provision of repair and maintenance works with necessary materials, spare parts;

g) Organization of production base for repair works;

h) Organization of preventive maintenance service;

i) Application of progressive methods of repair using mechanization and methods of restoration of worn parts;

k) Application of rules of operation of constructions and the equipment, observance of requirements of labor protection and safety measures;

l) The organization of control of care of constructions, the equipment;

m) Organization of quality control of repair.

The works carried out on planned preventive maintenance are divided into:

a) Maintenance, including daily rounds and inspections of trails and structures (scheduled and unscheduled) and preventive maintenance (pre-planned, without disassembly of major components of equipment and assemblies);

b) Repair work - current repair;

Bypasses of the city water supply network form the basis of supervision over the condition and safety of the network and include both inspections of pipeline routes, without lowering people into wells, and inspection of wells and chambers with a check of technical condition (acceleration of valves) and the action of fittings and equipment.

The schedule of preventive maintenance is drawn up for a year with a monthly breakdown of the preventive maintenance plan.

Currently for the description of processes within the enterprise processes of the third level are used as different methodologies (BWPin, UML diagrams, IDEF0, IDEF3, DFD, ARIS) and the standard block diagrams, text, tables, process map, matrix, or combinations thereof. Any of these forms of process description gives a clear idea of the sequence of works, requirements for the implementation of stages of work, responsible performers.

You can use the ARIS eEPC (extended event-driven process chain) notation to describe third-level processes, which represents a process as a sequence of procedures, activities performed by different departments or departments in an enterprise. An example of using ARIS eEPC is shown in Fig. 1.

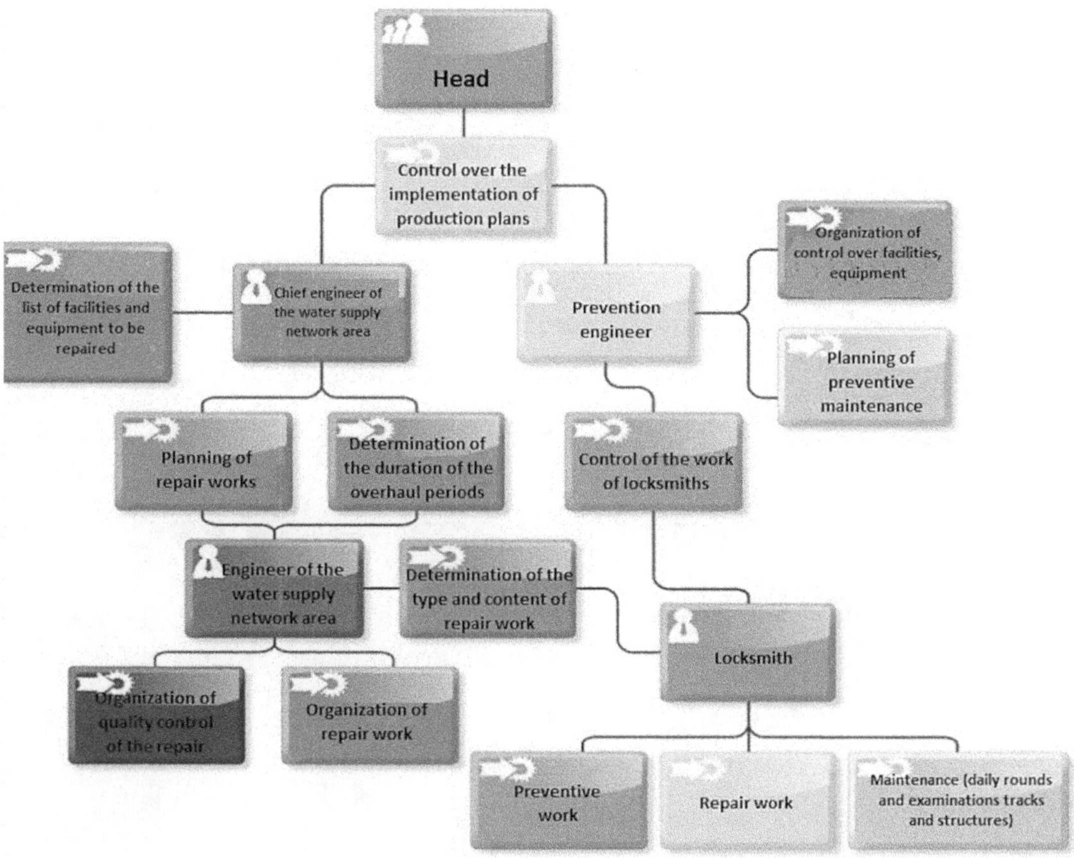

Fig. 1. Example of using ARIS notation in preventive maintenance.

4. Conclusions

In order of execution, ARIS eEPC notation with logic symbols allows you to reflect branching and merging processes. The workflow view diagram is used to create process models in an organization in the workplace. When building process models in ARIS eEPC, the condition must be met that each function must be triggered by an event and must end with an event, and that each function cannot contain more than one arrow, and no more than one arrow describing the completion of the function can come out.

References

[1]. Abramov L. N., Manaenkova E. A., Organization and planning of construction production, management of construction organization, *Stroizdat*, Moscow, 1990.

[2]. Prokopchina S. V., Principles of developing information technologies based on the regulatory Bayesian approach, in *Proceedings of the International. Conference on Soft Computing and Measurements (SCM-2005),* St. Petersburg, June 27-29, 2005, Vol. 1.

[3]. Prokopchina S. V., The Concept of Bayesian intellectualization of measurements in the problems of monitoring of complex objects, *News of Artificial Intelligence,* 3, 1997, pp. 7-56.

(065)

Estimation of Uncertainty in Calibration of Precision LCR-meters on the State Primary Standard of Units of Inductance and Tangent Angles of Losses

Iu. Kuzmenko [1], O. Velychko [1], S. Shevkun [1] and M. Dobroliubova [2]
[1] State Enterprise "All-Ukrainian State Scientific and Production Centre for Standardization, Metrology, Certification and Protection of Consumer", (SE "Ukrmetrteststandard"), 4 Metrologichna str., Kyiv, Ukraine
[2] National Technical University of Ukraine "Igor Sikorsky Kyiv Polytechnic Institute", 37 Peremogy ave., Kyiv, Ukraine
Tel.: + 38 050 387 56 55, + 38 097 596 89 74, fax: + 38 044 526 55 68
E-mail: shevkun@ukrcsm.kiev.ua, m.dobroliubova@ukr.net

Summary: Materials are presented on the estimation of uncertaintyin the calibration of precision LCR meters on the State primary standard of units of inductance and tangent angles of losses. The general approaches to calibration of LCR-meters are presented, the equation (model) of measurements in calibration is presented, as well as the budget of uncertainty. Details of uncertainty budget elements during calibration are disclosed. The presented approaches to estimation of uncertainty allow us to calibrate a wide range of LCR-meters by values such as inductance, capacity and active resistance in a wide frequency range and range of values. Materials can be used by laboratories that carry out calibration, testing and conformity assessment of measuring equipment of electromagnetic values.

Keywords: Uncertainty budget, Standard, Calibration, LCR-meter, Estimation of uncertainty, Parameters of the impedance.

1. Introduction

One of the most important tasks for all branches of science and technology related to the use of measurements of electrical quantities is the measurement of inductance, capacity and active resistance in a wide range of frequencies.

High accuracy and traceability of impedance parameters measurements are essential in such fields as electrical engineering, electronics, machine tools, instrument making, construction, transport, telecommunications, metallurgy, machinery and aircraft industries, the security industry, as well as science research and defense [1, 2].

At present, there are a large number of high-accuracy standard and working measurment equipment of the impedance parameters, such as: precision LCR-meters Keysight (Agilent) E4980 AL of various modifications, IET / QuadTech 7600 Plus, GW INSTEK LCR-6300, Applent Instruments AT810A etc.

Combined devices and multimeters got the most widespread use in the creation, set up and operation of a variety of electronic equipment.

They require calibration over a wide frequency range and a range of values of inductance, electric capacity, active resistance.

The purpose of the research is to create a unified approach to determining the main components of the uncertainty budget and to presentation of the results of calibration of LCR meters of precision and other meters of the impedance parameters [3].

2. The Main Content of the Research

Calibration of precision LCR meters in Ukraine is carried out using highly stable measures of inductance, electrical capacity and active resistance, which are part of the State primary standard of units of inductance and tangent angles of losses.

These measures have metrological traceability to international standards on fundamental physical constants by means of calibration or participation in international comparisons [4, 5].

Let's consider the calibration of the LCR-meter precision when using the measure of inductance as a standard. The equation (model) of measurements has the following form:

$$\Delta L_{LCR} = L_{LCR} - L_S + \Delta L_{S\gamma} + \Delta L_{QLCR} + \Delta L_{ST} + \\ + \Delta L_{TLCR} + \Delta L_l \quad (1)$$

where ΔL_{LCR} is the deviations of the calibrated LCR meter from the actual value of the inductance reproduced by the reference measure; L_{LCR} is average value of the inductance shown on the LCR-meter screen when measuring the inductance of the reference measure; L_S is the value of the inductance of the reference measure from the calibration certificate; $\Delta L_{S\gamma}$ is the correction for the drift of the reference measure of inductance since the last calibration; ΔL_{QLCR} is the correction for quantizing error of the calibrated LCR meter; ΔL_{ST} is the correction for the temperature dependence of the reference measure; ΔL_{TLCR} is the correction for the temperature dependence of the calibrated LCR meter; ΔL_l is the correction for parasitic inductance of connecting cables and contacts.

Based on this measurement model, an example of calculating the uncertainty budget when calibrating a precision LCR meter when measuring the inductance using a reference measure of 100 mH on the frequency of 1 kHz may be as follows:

Table 1. Uncertainty budget.

Quantity X_i	Estimate x_i, *mH*	Standard uncertainty $u(x_i)$, *mH*	Probability distribution	Method of evaluation (A, B)	Sensitivity coefficient c_i	Uncertainty contribution $c_i \cdot u(x_i)$, *mH*
L_{LCR}	100.0144	3.52E-04	normal	A	1	3.52E-04
L_S	100.0082	4.00E-04	normal	B	-1	-4.00E-04
$\Delta L_{S\gamma}$	0	1.10E-05	normal	B	1	1.10E-05
ΔL_{QLCR}	0	5.77E-06	rectang.	B	1	5.77E-06
ΔL_{ST}	0	3.60E-04	normal	B	1	3.60E-04
ΔL_{TLCR}	0	5.00E-04	rectang.	B	1	5.00E-04
ΔL_l	0	1.00E-04	normal	B	1	1.00E-04
ΔL_{LCR}	**0.0062**					
		Combined standard uncertainty				8.21E-04
		Effective degrees of freedom			ν_{eff}	>200, k=2
		Expanded uncertainty ($p \approx 95$ %)				1.64E-03

It must be noted that presented calculations of calibration result correspond to only one measurement point – 100 mH at a frequency of 1 kHz. At the same time, the calibration of the meter, which is the precision LCR meter, must be carried out in the full range of frequencies and values. An exception is made when the range of calibration is narrowed on customer requirement.

Therefore, calibration must be performed at several points (at least three) of each sub-range. The result of the calibration should be presented in the form of a graph showing the dependence of the deviation of the LCR meter outputs from the actual value of the measured value and the associated uncertainty of measurements.

The graph is obtained by approximating the calibration results at the corresponding calibration points. In areas of the range where there can be extremums of graphic, or substantial nonlinearity, the number of points of calibration should be increased.

When calibrating a LCR meter, calculations of the measurement result and the uncertainty budget are performed in a similar manner using values such as electric capacity, active resistance and other impedance parameters.

3. Conclusions

As a result of the conducted research, a method for calibrating LCR meters of precision on the State primary standard of units of inductance and tangent angles of losses was developed. The method fully allows calculations of the calibration and uncertainty budget when calibrating LCR meters, multimeters and combined instruments based on values such as inductance, electrical capacitance and active resistance in the full range of frequencies and values. The method can be used when conducting international comparisons of reference standards of national metrology institutes. Research results will facilitate the organization of metrological support for measurements of impedance parameters in all fields of science and technology in the state.

References

[1]. ISO/IEC 17025:2017, General requirements for the competence of testing and calibration laboratories, *Derjspojivstandart of Ukraine*, 2017.

[2]. O. Velychko, S. Shevkun, A support of metrological traceability of inductance measurements in Ukraine, *Eastern-European Journal of Enterprise Technologies. Information and controlling systems*, 5/9, 89, 2017, pp. 12–18.

[3]. ISO/IEC Guide 98–3:2008, Uncertainty of measurement. Part 3. Guide to the expression of uncertainty in measurement, *ISO/IEC*, 2008.

[4]. O. Velychko, S. Shevkun, Final Report on COOMET supplementary comparison of inductance at 10 mH and 100 mH at 1 kHz (COOMET.EM-S14), *Metrologia*, Vol. 53, Issue 1A, Technical Supplement 01009, 2016 (http://iopscience.iop.org/article/10.1088/0026-1394/53/1A/01009).

[5]. E. Dierikx, et. al., Final report on the supplementary comparison EURAMET.EM-S26: inductance measurements of 100 mH at 1 kHz (EURAMET project 816), *Metrologia*, 49, 01002, 2012.

Authors

Index

C

D

E

F

G

H

I

J

K

L

M

N

O

P

Q

R

S

T

U

www.ingramcontent.com/pod-product-compliance
Lightning Source LLC
Chambersburg PA
CBHW082347230926
51846CB00004B/938
* 9 7 8 8 4 0 9 1 7 4 9 2 8 *